MECHANICS OF FIBER AND TEXTILE REINFORCED CEMENT COMPOSITES

MECHANICS OF FIBER AND TEXTILE REINFORCED CEMENT COMPOSITES

Barzin Mobasher

CRC Press
Taylor & Francis Group
Boca Raton London New York

CRC Press is an imprint of the
Taylor & Francis Group, an **informa** business

CRC Press
Taylor & Francis Group
6000 Broken Sound Parkway NW, Suite 300
Boca Raton, FL 33487-2742

First issued in paperback 2019

ISBN-13: 978-1-4398-0660-9 (hbk)
ISBN-13: 978-0-367-38238-4 (pbk)

Library of Congress Cataloging-in-Publication Data

Mobasher, Barzin.
  Mechanics of fiber and textile reinforced cement composites / author, Barzin Mobasher.
     p. cm.
  Includes bibliographical references and index.
  ISBN 978-1-4398-0660-9 (hardcover : alk. paper)
  1. Fiber cement. 2. Fiber cement--Testing. 3. Fiber-reinforced concrete. I. Title.

TA438.M63 2011
624.1'833--dc22   2011009972

"Let yourself be silently drawn by the stronger pull of what you really love. There are a thousand ways to kneel and kiss the ground. Be a lamp, a lifeboat, a ladder. Help someone's soul heal. Walk out of your house like a shepherd."

Mawlana Jalal-al-Din Rumi,
AD 1207–1273

Contents

Preface

A safe and secure shelter is a human right.

The production and use of cement-based concrete materials have remained virtually unchanged for the past 50 years. Despite the fact that the global usage of these materials exceeds a staggering 6 billion tons per year, two major drawbacks still limit the use of cement-based materials. These include the manual nature of concrete placement, which affects quality control, cost, and the speed of production, and the low tensile strength of the brittle material, which makes it susceptible to cracking. The low tensile strength limits the functioning of these materials in areas where tensile stresses are the dominant forces. Given the challenges our society faces in providing sustainable shelters and civil infrastructure systems with improved seismic resistance, durability, and increased service life, we must demand materials with better performance than what is conventional. The enormous investment in the infrastructure systems must be underwritten with innovations that address improved properties in terms of carbon footprint, life cycle cost, durability, corrosion resistance, strength, ductility, and stiffness.

The main objective of the book is to present mathematical models, experimental results, and computational algorithms, which are the basis for efficient designs with fiber and textile reinforced composite systems. Each chapter includes a detailed introduction and a thorough overview of the existing literature on the topic and sets forth the reasoning behind the experimentation and theory development. The methods for manufacturing, testing, analysis, and design are developed and explained. Decades of attempts to develop guidelines for fiber reinforced concrete using traditional reinforced concrete approaches—which inherently neglect the tensile capacity of concrete—are abandoned in this book, and criteria are developed for the use of the postcracking response as a material and structural property in the design procedures. An adoption of empirical approaches is, thus, at a minimum. Although it is acknowledged that these composites are prone to controlled and stable cracking, formulations are presented that capitalize on their large strain capacity and postpeak strength characteristics. The effect of interface properties on the toughening response of the composite are of paramount importance; hence, tools available within the mechanics of composite materials, finite element modeling, plasticity based, and nonlinear fracture mechanics are used to study the toughening mechanisms.

The need for a multifaceted-multidisciplinary approach is quite pronounced in the case of textile reinforced composites because the complex multidimensional nature of textile materials results in reinforcing efficiency and strain-hardening responses. Although two-dimensional and homogenized one-dimensional approaches are used for a majority of the analysis and design approaches, it is essential to note that the complex nature of textiles makes it necessary to consider the full three-dimensional nature of deformations on the micromechanical aspects of failure, cracking, and crack-bridging mechanisms. The depiction of textile materials as plain one-dimensional reinforcement systems providing a residual stiffness may limit the potential areas of application and interaction of reinforcements in multiple directions, so we hope that the present approach will chart the path for future opportunities.

Chapters 1 and 2 present the case for cement-based composites as sustainable construction materials. An overview of the historical aspects of more than 50 years of conventional fiber reinforced concrete systems, including asbestos cement, hatscheck process, ferrocement, and glass fiber reinforced concrete—using alkali resistant glass, polymeric, natural fibers, and textile materials—are presented.

Chapter 3 presents an introduction to the manufacturing and general properties of various high-volume fraction cement composite systems using processes such as extrusion, injection molding, compression molding, rheology control, rapid setting, spin casting, high shear mixing, and pultrusion. These developments are supported by the basics of an understanding of the mechanics of toughening. An overview of the terminology and classification systems in textile reinforcement using various hierarchical textile forms is addressed in Chapter 4. Chapter 5 is a microstructural evaluation of the deformation fields that take place in a woven textile structure. This chapter documents the modes of deformation and interlock mechanisms. The strength of a single yarn is also addressed using a formulation to compute the Weibull parameters. Chapter 6 is a review of mechanics of composite laminate theory. Various stages of cracking and ply discount method are characterized. Chapter 7 presents an introduction to mechanical testing using concepts in closed-loop testing. An overview of tensile, compression, fracture, and cyclic tests are presented. Additional tests such as round panel, shrinkage, and impact are also shown.

The significance of interfacial modeling is addressed by an overview of the analytical derivation for fiber pullout in Chapter 8. Analytical techniques are compared with the finite element, finite difference, and fracture mechanics approaches. The use of the fiber pullout mechanism as one of the main components that affects the fracture process is addressed in Chapter 9, which discusses the introduction to nonlinear fracture mechanics with an emphasis on cohesive crack models, closing pressure formulations, and the concept of R-Curves.

Chapters 10 and 11 present strain-hardening cement composites using experimental and theoretical models as can be applied to the tensile response of continuous and cross-ply composites. Chapter 12 addresses strain-hardening microfiber and hybrid composites. Major differences between the flexural and tensile responses are addressed. Characterization of the strain-hardening composites must be associated with the nature and mechanism of damage in these systems. Chapter 13 presents a methodology to correlate the distributed damage parameters (such as parallel cracking with stiffness degradation mechanisms) for continuous fiber and textile reinforced systems. These approaches are also applicable to short-dispersed fiber composites that exhibit a strain-hardening response.

To develop modeling for the flexural performance of strain-hardening and strain-softening systems, an approach that correlates the stages of tensile response during flexural behavior is presented in Chapter 14. Using a closed-form solution to examine the moment–curvature relationship, various stages of cracking are simulated. The algorithm is then used to predict the load–deflection response under flexural loading conditions for a variety of fiber composite systems. Using this approach, a back-calculation procedure of material properties from flexural tests is also presented in Chapter 15.

Applications of the finite element method are addressed in Chapter 16 by showing that the moment–curvature relationship that was derived in Chapters 14 and 15 can be implemented within a finite element framework. Chapter 17 presents the procedure for the flexural design of strain-softening materials, addressing serviceability concerns dealing with a minimum postcrack tensile capacity, deflection, shrinkage, and temperature concerns using several case studies. Similar approaches can be used for the design of strain-hardening cement composite systems.

Chapters 18 and 19 address several special cases of cement composite systems that can be integrated in the broader set of design and analysis procedures. These include aerated concrete and natural fiber composites. Chapters 20 and 21 address the performance of textile composites under severe environments, including restrained shrinkage cracking and flexural impact tests. In the area of repair and retrofit, Chapters 22 and 23 address strategies for reverse cyclic loading of walls and beam-column connections. Finally, the dynamic tensile characteristics of textile composites are addressed in Chapter 24.

This work compiles the research activities of my collaborators, colleagues, and students during the past 25 years; its preparation could not have been possible without the support of so many gifted individuals. Starting with my elementary and high school education in Iran, I am humbled to acknowledge the contributions of two outstanding educators, my K-12 principals Mrs. Touran Mirhadi and the late Dr. Mohammad Ali Mojtahedi. They have sown the seed of prosperity through education to thousands of students in my generation. My postgraduate advisors, Professors Menashi

Cohen of Purdue University and S. P. Shah of Northwestern University, taught me the many intricacies of cement-based materials. My former students have contributed the most to this work and include the following: Dr. Cheng Yu Li, Joanne Situ, Andrew Pivacek, Garrett Haupt, Jitendra Pahilajani, Nora Singla, Sachiko Sueki, Dnyanesh Naik, Juan Erni, and Mehdi Bakhshi. A special thank you to Drs. Chote Soranakom, Deju Zhu, Flavio Silva, and Amir Bonakdar who, during the past 6 years, consistently excelled on every set of algorithms, test plans, computations, and analysis procedures we worked on. The laboratory facilities for the mechanical testing of materials at ASU could not have supported my research without the knowledge and professional support of Mr. Peter Goguen who has consistently done a superb job in maintaining them facilities. I am grateful to my collaborations with several international colleagues, including Professor Alva Peled of Ben Gurion University, Israel; Professor Romildo Toledo Filho of Federal University of Rio De Janeiro, Brazil; Professor Viktor Mechtcherine of Technical University Dresden, Germany; Professor Antoine Naaman of the University of Michigan, and Professor Subramaniam Rajan of ASU. Support from various funding agencies, including the National Science Foundation, Federal Aviation Administration, US-Israel Bi-National Science Foundation, Salt River Project, Meccano, Rio-Tinto Resolution Copper Mining, and the Arizona Department of Transportation have also been indispensable to this project. In addition, I am grateful to the support and diligent work of Ms. Flurije Salihu in technical editing of the manuscript. I am also grateful to the staff of CRC Press, especially Mr. Joe Clements and Ms. Amber Donley, for their consistent patience, encouragement, and support in the preparation of this work. I could not have asked for better support.

I reconsidered my research direction in 2004 while conducting a reconnaissance tour of the ancient city of Bam, Iran within a week after a modest earthquake devastated the entire region causing more than 50,000 deaths and leaving an equal number of people injured and homeless. The emotional toll of lives lost and misery due to construction failures was too much to bear. I shifted direction toward the development of ductile construction products, especially geared toward the sustainable residential applications. The drafting of this book began after the passing of my stepdaughter Sheila due to complications of cerebral palsy in 2008 and was, ironicaly, completed during the final days of my mother's battle with Alzheimer's disease in 2010. It was their struggle with incurable medical conditions that motivated me to channel my inability in improving their condition into this contribution. I am grateful for all the lessons I've learned from them. This book is dedicated to my wife, Tina, and our children, Keon Solomon, and Jasmin Lily, for the abundance of love, support, and encouragement they have given me throughout.

Barzin Mobasher
Tempe, Arizona

Author

Barzin Mobasher obtained his BS and MS in civil engineering from the University of Wisconsin–Platteville and Northeastern University in 1983 and 1985, respectively. He received his PhD in civil engineering from Northwestern University in 1990 and was a member of the Technical Staff at USG Corporation during 1990–1991. He joined the Department of Civil and Environmental Engineering at Arizona State University in 1991 as an assistant professor of structural materials. He has been a professor of engineering at ASU since 2004. Dr. Mobasher has led programs involved with the design, analysis, materials testing, and full-scale structural testing of construction and structural materials. His list of publications include more than 150 research papers in leading professional journals and conference proceedings. He has made fundamental contributions to the field of fiber and textile reinforced concrete materials and mechanics of toughening in cement-based systems, modeling the mechanical and durability of materials, and experimental mechanics. He has served as the Chair of the American Concrete Institute, ACI Committee 544 on Fiber Reinforced Concrete, and has been a reviewer for a variety of journals.

1 Cement-Based Composites—A Case for Sustainable Construction

INTRODUCTION

The continuous development of civil infrastructure systems in support of economic productivity is a challenge faced by all nations. Among all building materials, concrete is the most commonly used and there is a staggering demand for its utilization. The exponential growth of infrastructures, especially in developing regions further increases the demand for concrete materials such that the worldwide production and use of concrete will soon surpass the 10 billion tons per year mark. In the United States, concrete production has almost doubled from 220 million cubic yards per year in early 1990s to more than 430 million cubic yards in 2004. Such growth rates, in addition to the aforementioned increased demand for infrastructure development in emerging markets like China and India, strains the production capacity of cement products. To sustain such growth trends, the ever-increasing demand by these developing nations must be balanced against the environmental aspects. From a sustainability point of view, we must therefore address material shortages along with technically feasible production technologies and life cycle maintenance costs of construction projects.

From a completely different perspective, one must appreciate that the need for shelter is an inherent global problem. A safe and secure shelter is simply a human rights issue. According to UN-Habitat estimates, 924 million people worldwide or 31.6% of the global urban population lived in slums in 2001 [1]. In the next 30 years, this figure is projected to double to almost 2 billion, unless substantial policy changes are put in place. Numerous challenges remain to meet this demand through sustainability, intelligent construction practices, novel product manufacturing, and the promotion and use of alternate sources for concrete materials.

From a structural safety perspective, there is quite a bit of work that has yet to be accomplished. For example, approximately 74,000 people died during the earthquake that struck near the Pakistan–India border on October 8, 2005. In addition, approximately 70,000 were injured and more than 3 million people became homeless. Just 2 years earlier, the historic city of Bam in Kerman Province of Iran was devastated by a similar earthquake resulting in as many as 50,000 deaths. An evaluation of earthquake records indicates that approximately 60% of the deaths due to earthquakes in the past 60 years stem from the collapse of unreinforced masonry structures [2]. Although these wall systems are the standard traditional construction choice in many parts of the world, there is a need to find safer alternative structural systems.

One of the reasons for the extensive use of cement-based systems is the design versatility that can be tailored for each application. Various constituent materials and processing techniques can be used to achieve desired performances from fresh state properties to superior mechanical properties and durability. Numerous technical challenges remain for the production and use of blended cements as sustainable cost-saving alternatives in addition to the value-added ingredients such as fibers for concrete production [3]. However, we must appreciate the complexity of the integration of cement chemistry, early age and long-term properties, and specifications when using cement products in construction projects [4].

This book addresses some of the recent trends and future directions in cement and concrete manufacturing, design, and analysis. New research findings enable opportunities for the development of innovative and cost-effective products to address the ever-expanding demands of the public. By understanding, predicting, and controlling microstructural changes and chemical reactions during the life cycle of these products, we can better address premature degradation of infrastructure systems. This would allow us to better use our resources.

CEMENT AND CONCRETE PRODUCTION

The cement industry is a relatively small, but comprises a significant component of the U.S. economy, with annual shipments valued at around $8.6 billion. In the United States, 39 companies operate 118 cement plants in 38 states. The largest company produces just more than 13% of the industry total, and the top five companies collectively produce around 56%. Foreign companies now own approximately 81% of U.S. cement capacity. Worldwide, the United States ranks third in cement production, behind China—the world's leading producer—and India.

Approximately 94 million tons of portland cement were produced in 2006, while the total cement production capacity was approximately 115 million tons. Sales prices increased significantly during the year and reflect a value of cement production, excluding that of Puerto Rico, to be approximately $9.8 billion, an increase of $1.2 billion in comparison with 2002 figures. Most of the cement was used to make concrete, worth at least $54 billion. Inflation-adjusted construction spending was $755 billion in 2005, a 4.2% increase from 2004 levels, and accounted for a 9.2% share of the national economy.

Although cement compositions and fundamental production methods have remained relatively unchanged for a century, there have been enormous leaps in the efficiency of cement production. The average kiln in use today produces more than 60% more cement than an average kiln of 20 years ago: 468,000 metric tons in 2002 compared with 287,000 metric tons in 1982. In 2002, the cement industry employed 19,140 workers—a 29% reduction compared with 1982 levels. By 2005, the cement industry employed 16,859 workers; this drop in employment is the result of industry efforts to increase efficiency by automating production and closing small kilns. Approximately 73% of all cement shipments are sent to ready-mix concrete operators, 14% to concrete product manufacturers, 6% to contractors, and 3% to building material dealers (http://minerals.er.usgs.gov/minerals/pubs/commodity/cement/cemenmcs07.pdf).

These growth trends have encouraged the cement industry to undertake an aggressive capacity expansion. By 2010, more than 20 million metric tons of new capacity will come online, representing more than $5 billion in investments [5]. A significant amount of energy is required to produce cement, and for approximately each ton of cement produced, a ton of carbon dioxide is emitted into the environment.

Traditionally, it has been quite easy to increase the supply side of the cement and concrete manufacturing to meet the demand. Modern cement production mills are highly efficient. Using state-of-the-art industrial and technical manufacturing processes, portland cement and ready-mix plants can be built in a matter of few years to meet the construction demands of a region. Having said that, factors such as environmental aspects of cement production are becoming a source of concern. There are also limitations due to the availability of quality aggregates for concrete production. The increased use of recycled concrete for aggregates and also demand for high strength concrete have added to the interest for the study of aggregates research.

CURRENT TRENDS

As we strive to build taller structures with improved seismic resistance or durable pavement with an indefinite service life, we require materials with better performance than the conventional materials used today. As we consider the enormous investment in the infrastructure system and society's need to sustain it, the need for new and innovative materials for the repair and rehabilitation of civil infrastructures becomes more evident. These improved properties may be defined in terms of economics, durability, corrosion resistance, strength, ductility, and stiffness. This book focuses on the development of new and innovative materials, their manufacturing and analysis techniques, and the opportunities for their economical implementation. Such lightweight, durable, technically feasible, and environmentally friendly cement-based materials may replace nonrenewable resources such as wood-based or metallic-based systems, allowing for a more robust construction methodology. The potential rewards of these materials are enormous.

A review of the current trends and speculations of research and development efforts as they apply to the sustainable design philosophy of construction materials leads to the following conclusions:

- Durability—The current trend is to address the demand for durability based design through proper engineering complexity, design efficiency, and an increased service life through more durable construction.
- Quality—Quality control (QC) parameters affect the cost of a project. There is a need to better understand the QC measures and incentives for the payment based on early age and long-term properties of concrete materials. Life cycle cost modeling combined with statistical QC measures could identify potential savings [6].
- Innovations—It is imperative that new guidelines and alternatives for concrete materials be analyzed so that economical alternatives such as materials with a higher specific strength, ductility, and stiffness are considered during the preliminary design of a project. As the cost of raw materials changes, many new materials become feasible alternatives.
- Cross-disciplinary efforts—The development of cross-disciplinary software tools that integrate various design considerations may guide us toward a more sustainable, intelligent, and economical engineering and construction policy. By combining analysis and design procedures, both the materials and the cross-section sizes can be designed to meet a specified loading condition.
- Education—There is a need to reduce the gap between research and development and become proactive through the use of appropriate analysis, design, and technology transfer tools. New design tools need to be developed to harness the special characteristics of the materials proposed.

The materials presented here address to a large degree many of the parameters that have been defined in the context of high-performance concrete (HPC) materials. The definition and applicability of "high performance" has been evolving, though. It is defined as "concrete which meets special performance and uniformity requirements that cannot always be achieved by using only the conventional materials and mixing, placing and curing practices. The performance requirements may involve enhancements of placement and compaction without segregation, long-term mechanical properties, early-age strength, toughness, volume stability, or service life in severe environments" [7]. In the past 20 years, there has been an enormous amount of research on this subject, and thousands of papers have been published [8]. What is evident, is that the HPC materials are expected to meet stringent performance standards beyond high strength values. Having said that, some truly HPC materials do not meet any of the above definitions.

Chemical attack such as corrosion, sulfate attack, and alkali silica reactions in concrete are among the major durability concerns in civil infrastructure systems [9]. These mechanisms affect the service life and long-term maintenance of concrete. For example, the average annual direct cost of corrosion for highway bridges was estimated to be $8.29 billion in the United States [10]. A majority of these durability issues in concrete structures deal with the ingress of one or several different ions into the material through the pore structure and microcracks. Although there are many test methods developed for characterization of durability, it is often difficult to develop guidelines for correlation between the laboratory results and the service life performance of the structures. Models for a universal predictive tool are not fully developed because of the complexities of the interaction of chemical and mass transport, nonequilibrium cement chemistry, and the inter-action of phases, temperature, and humidity effects, in addition to volumetric changes, cracking, and the changes in properties due to cracking. The interaction of cement chemistry, concrete phys-ics, and mechanics are required for a model development that includes predictions of the expan-sion or shrinkage of cement-based materials exposed to a various ionic or humidity conditions. A major challenge is in incorporating durability test methods into a standard acceptance test protocol [11–13]. What is already established is the fact that the addition of fibers significantly reduces the crack width and that, in turn, will directly affect the diffusivity of various ionic species. The role of fiber reinforced materials in extending the durability service life is still one of the unanswered questions in the field.

Recent developments of rheology modifying admixtures have led to the development of self-consolidating concrete (SCC) that allow for a better utilization of fibers in concrete. Concretes with a high slump flow are prone to segregation and bleeding. Alternatively, additions of reasonable amounts of fibers have been limited to the rheology and clumping or balling of fibers. To minimize segregation, a large amount of fine material, a small nominal maximum size aggregate size, a uni-form grading, or a low water–cementitious material ratio are needed. Alternatively, conventional mixtures with viscosity modifying admixtures are used [14]. SCC, also known as self-compacting concrete, is a highly flowable, nonsegregating concrete that spreads and fills into formwork and encapsulates even the most congested reinforcement, all without any mechanical vibration. Reasons for the popularity of SCC mixtures are attributed to several factors. Cost reductions are, in part, due to the increased speed of construction, improvements in formed surface finish, reduced repair and patching costs, reduced maintenance costs on equipment, and faster form and truck turnaround time. Reduced labor costs are also realized due to the lack of skilled workers who perform the rigo-rous work of quality concrete construction. In addition, to improve aesthetics SCC eliminates the use of vibrators for concrete placement and improves safety during construction [15, 16].

SCC's workability is a function of its rheology. SCC is required to be highly flowable such that the conventional slump test cannot distinguish between different levels of SCC flowability, while it is also viscous enough so that the mortar suspends and carries coarse aggregate, maintaining a homogenous, stable mixture, resistant to segregation, bleeding, excessive air migration, or paste separation [17].

In principle, SCC should allow for the easy movement of concrete around flow obstructions under its own weight without the use of external vibration. The flow of concrete around barriers will depend on a variety of factors, including the concrete or mortar viscosity, the yield stress, and the size distribution and shape of the coarse aggregate [18, 19]. It is therefore intended that the use of short fiber systems developed in this book will be accompanied with proper concrete technology, as discussed earlier.

STRUCTURE OF THIS BOOK

The structure of this book is based on several interrelated sections. Historical and innovative methods for the manufacturing and testing of cement composites are presented to document the composite action of fibers and textiles. The processing techniques include pultrusion, filament winding, cross-ply lamination, extrusion, compression molding, and high energy mixing. Analysis and design with cement-based composite materials are addressed next. A multiscale approach based on micromechanics of failure in composite materials is presented next. By predicting and control-ling microstructural changes, both mechanical and environmental stresses can be modeled during the life cycle of products.

The primary highlight of this work is to demonstrate that by using a higher volume fraction of fiber systems, composites are manufactured that can present themselves as quasi-elastic plas-tic materials. Then, the area of continuous fiber systems and the development of textile reinforced cement composites using angle ply laminates by means of filament winding technique are addressed.

These composites that contain as much as 95% portland cement-based matrix can withstand tensile stresses as much as 50 MPa (more than 10 times the strength of ordinary concrete). The strain capacity of these composites, as high as 2% is more than one hundred times the strain capacity of ordinary concrete. The fracture toughness is as much as two orders of magnitude higher than the conventional concrete materials, making them ideal for use in seismic regions or cases with repeated loading. The controlled manufacturing system allows the material to be designed and manufactured using QC measures comparable with metal-forming processes.

In the area of characterization, several research test methods have been developed for mechanical testing. Tests include properties measured by tension, flexure, compression, shear, and high-speed tension testing for fibers, composites, matrix, and, finally, characterization of the interface using both analytical and experimental techniques.

In the theoretical segments, effect of interface properties on the response of the composite materials is addressed in detail. Finite element modeling and nonlinear fracture mechanics approaches based on the R-curves were used to study the toughening mechanisms in brittle matrix composites.

TEXTILE REINFORCED AND HIGH-VOLUME CONTENT CEMENT COMPOSITES

Fabric reinforced cement-based composites are a new class of sustainable construction materials with superior tensile strength and ductility. These materials have the potential for becoming load-bearing structural members; therefore, a wide array of structural and nonstructural applications are possible. The constitutive response that entails damage evolution under tensile loading is the primary and fundamental component of mechanical response in these systems. The tensile response is presented in detail. Composites manufactured by means of the pultrusion technique have been documented to show tensile strengths in the range of 20–30 MPa and strain capacity of up to 6% to 8%. Such mechanical properties allow structural designers and architects to develop many exotic lightweight and stiff structures. This enhanced behavior in these compliant systems is primarily governed by interfacial bond characteristics between fabrics and matrix. Models on the basis of composite laminate theory, nonlinear homogenized constitutive response, and fracture mechanics are presented to simulate the tensile, shear, and flexural behavior. These models relate the properties of the matrix, fabric, interface, and the damage parameters to the overall mechanical response. Application areas of new structural panels, impact, blast resistance, repair and retrofit, earthquake remediation, strengthening of unreinforced masonry walls, and beam–column connections are presented.

DEVELOPMENT OF DESIGN METHODOLOGIES FOR FIBER REINFORCED COMPOSITES

Guidelines for the design of fiber reinforced concrete materials are needed to incorporate the cost-effectiveness of using new materials in the context of reduction in labor, materials, and equipment. Analytical closed form solutions are presented for design and analysis of composite systems such as beams, slabs, retaining walls, and buried structures. Examples are directly related to new methods to design elevated slabs with fiber reinforced concrete instead of continuous reinforcement. The procedures proposed use the recent developments in the rheological modifications of the fresh matrix, such as viscosity modifiers for the production of SCC as one of the design parameters that allow for mixing, transporting, and placing of short-length, high-volume fiber reinforced concrete. Properly designed fiber reinforced concrete structures can economically compete with traditional reinforced concrete structures. Various chapters directly address the following tools:

- Closed form solutions for analysis and design based on fracture mechanics and stress crack width approach to incorporate the contribution of fibers
- Constitutive modeling of the tensile performance

- Closed form solutions for tensile and flexural response
- Computation of load deformation in closed form solutions
- Finite element implementation of the design tools
- Design for elevated slabs
- Design for slabs on ground and retaining walls
- Design for flexural beams
- Design for minimum reinforcement design

SUSTAINABILITY—THE MAIN DRIVER FOR NEW MATERIALS AND DESIGN METHODS IS THE ECONOMY OF CONSTRUCTION SYSTEM

Life cycle cost modeling has enabled many construction alternatives to be attractive, simply because the total cost of the project is reduced by choosing fiber reinforced concrete as opposed to traditional reinforced concrete systems. During the past 4 years, we have developed several technical tools that would improve the production process. The QC can be monitored using statistical process control tools. The book will present applications in statistical process control to improve the quality of production processes.

From a sustainability point of view, we address material shortages along with technically feasible production technologies and life cycle maintenance costs of structural systems. This book requires the user to rethink and reevaluate the way we traditionally design and manufacture reinforced concrete structures. Our traditional design methodologies ignore tensile capacity of concrete altogether, treat the cracking and associated durability problems as an afterthought, and are inherently inefficient, wasteful, and therefore expensive. By using innovative lightweight materials designed using fundamental aspects of mechanics of composite materials, we can design efficient structural systems. The proposed method of approach in this book is to optimize the structural performance by

(1) Developing new materials such as fabric and fiber reinforced concrete using novel reinforcing systems in conjunction with woven, knitted, polymeric, natural, and/or alkali resistant glass
(2) Developing new analysis and design guidelines so that the material models can be directly integrated within structural analysis software; both material and structural design can be accomplished concurrently
(3) Developing alternative solutions that allow for sustainable development of infrastructure systems using blended cements, innovative reinforcing systems, natural fibers, statistical process control, and optimized designs

REFERENCES

1. Garau, P., Sclar, E., and Carolini, G. (2005), "A Home in the City—The Report of the UN Millennium Project Task Force on Improving the Lives of Slum Dwellers," London: Earthscan.
2. Coburn, A. W., and Spence, R. J. S. (2002), *Earthquake Protection*, 2nd ed. Chichester, UK: Wiley.
3. Roy, D. M. (1999), "Alkali-activated cements: Opportunities and challenges," *Cement and Concrete Research*, 29, 249–254.
4. Mobasher, B., and Ferraris, C., "Simulation of Expansion in Cement Based Materials Subjected to External Sulfate Attack," RILEM International Symposium: Advances in Concrete through Science and Engineering, March 2004, comp. Weiss, Jason, and Shah, Surendra P. (on CD).
5. Portland Cement Association. (2007), *Economics of the U.S. Cement Industry*. Skokie, IL: PCA.
6. *Optimal Procedures for Quality Assurance Specifications*. Publication No. FHWA-RD-02-095. http://www.cement.org/basics/cementindustry.asp
7. Aïtcin, P. C. (1998), *High-Performance Concrete*. London: E & FN Spon, p. 591.

8. Zia, P., "State-of-the-Art of HPC: An International Perspective," Proceedings PCI/FHWA International Symposium on High Performance Concrete, New Orleans, October 1997, 49–59.

9. Malvar, L. J., Cline, G. D., Burke, D. F., Rollings, R., Sherman, T. W., and Greene, J. (2002, September–October), "Alkali–silica reaction mitigation: State-of-the-art and recommendations," *ACI Materials Journal*, 99(5), 480–489.

10. Yunovich, M., Thompson, N. G., and Virmani Y. P. "Corrosion of Highway Bridges: Economic Impact and Life-Cycle Cost Analysis," 2004 Concrete Bridge Conference: Building a New Generation of Bridges, May 17–18, 2004, Charlotte, NC.

11. Ozyildirim, C., "Effects of Temperature Development of Low Permeability in Concretes," VTRC 98–R14, Virginia Transportation Research Council, Charlottesville, 1998.

12. ASTM C1556, "Standard Test Method for Determining the Apparent Chloride Diffusion Coefficient of Cementitious Mixtures by Bulk Diffusion," in *Annual Book of ASTM Standards*, Vol. 4.02.

13. ACI Committee 214, ACI 214R-02, "Evaluation of Strength Test Results of Concrete, Part 1," in *ACI Manual of Concrete Practice*. American Concrete Institute, 2002.

14. *An Introduction to Self-Consolidating Concrete (SCC)*. Technical Bulletin No. tb-1500. W. R. Grace & Co.-Conn.

15. Roy, D. M. and Asaga, K. (1979), "Rheological properties of cement mixes: III. The effects of mixing procedures on viscometric properties of mixes containing superplasticizers," *Cement and Concrete Research*, 9, 731–739.

16. Khayat, K. H. (1999), "Workability, testing and performance of self-consolidating concrete," *ACI Materials Journal*, 96(3), 346–353.

17. Bonen, D., and Shah, S.P. (2005), "Fresh and hardened properties of self-consolidating concrete," *Progress in Structural Engineering and Materials*, 7(1), 14–26.

18. Ferraris, C. F. (1999), "Measurement of the rheological properties of high performance concrete: State of the art report," *Journal of Research of the National Institute of Standards and Technology*, 104(5), 461–478.

19. Proceedings of the 2nd North American Conference on the Design and Use of Self-Consolidating Concrete Westin Michigan Avenue, RILEM, Chicago, October 30–November 2, 2005.

2 Historical Aspects of Conventional Fiber Reinforced Concrete Systems

INTRODUCTION

Modern-day development of fiber-reinforced concrete (FRC) started more than 100 years ago, soon after the commercialization of concrete and reinforced concrete. Having said that, the manufacture and use of composite materials for construction goes back to more than 3000 years (Bentur). The cost-effectiveness of using cement-based materials for residential buildings is evident; however, ways to increase the damage tolerance and ductility of cement products have been the limiting factor. Next to steel, cement-based composites have demonstrated the highest strength-to-weight ratio of any building material. They are noncombustible and will not contribute fuel to the spread of a fire. Cement composites will not rot, warp, split, crack, or creep and are 100% recyclable. Furthermore, they are not vulnerable to termites, fungi, or organisms. The dimensional stability in the form of expansion or contraction with moisture movement can be considered in design equations. The consistent material quality of cement products developed under strict manufacturing protocol in a precast operation allows them to meet strict standards. Finally, there is a strong labor pool that can be trained to work with the production of precast cement-based products.

The interest in the use of FRC is continuously growing. This chapter addresses some of the main developmental aspects of cement composites. Steel FRC is already widely used in structures as the main reinforcement in structures under bending [1]. Early examples of this are the square slabs of the Heathrow Airport car park in London [2] or the foundation slab of Postdamer Platz in Berlin [3]. One of the main advantages of using composites as compared to the conventional materials is that design loads, called live and dead loads, can be carried by sections that are much smaller in size. This results in reductions in weight, service loads, deflection criteria, and resistances to wind and earthquake loads. Thin section FRC cladding panels are widely used in the construction industry. Compared with conventional precast concrete, the production, transportation, and installation costs of these materials are significantly lower [4]. The following sections review a series of product developments in the field of FRC with special attention being paid to high volume content materials that are primarily used for thin section applications.

PREHISTORIC DEVELOPMENTS

The development of new and innovative materials throughout the last millennia has been instrumental at every stage of societal development. An overview of human evolution shows us that making and using tools were among the first priority of early societies, a fact that even today differentiates developed and developing societies.

Handling and understanding tools was perhaps the first requirement for survival for early humans. Although it is generally accepted that sharp objects made out of rocks, bones, and wood were the first engineering materials, we extend these categories to the next generation of tools and man-made materials today. A unique finding of wild flax fibers from a series of Paleolithic

FIGURE 2.1   A funerary model of a weaver's loom found in an Egyptian tomb. (Adapted from "The industrial uses of asbestos," *Scientific American*, 258–259, April 22, 1876.)

layers located in the foothills of the Caucasus, Georgia, indicates that prehistoric hunter-gatherers were making cords for stone tools, weaving baskets, or sewing garments. The earliest evidence for humans using fibers is the discovery of wool and dyed flax fibers found in a prehistoric cave in the Republic of Georgia that dates back 36,000 years [5]. With the development of shelter, fire, and agriculture, the development of weaving items out of twines was a means of converting one-dimensional branches into two-dimensional flat and curved surfaces so that they could be used for storage, transportation, or a variety of other purposes. Fabrics and textiles offered early humans a multitude of tools for covering, clothing, shelter, and security from natural elements. The development of weaving technology dates back to 8000 BC as there is evidence of cloth being made in Mesopotamia and in Turkey as far back as 7000–8000 BC. They were therefore, among the first engineering products made out of natural materials. These technologies were followed by basket weaving and also the development of tools and machinery for weaving. A funerary model of a weaver's workshop found in an Egyptian tomb is shown in Figure 2.1. It served as the necessary blueprint for reconstruction of the trade. This model contains a horizontal loom, warping devices and other tools, and weavers in action.

The development of ceramics, metals, and copper all followed next. Our contemporary society is identified as the electronics age. All of these categories measure human development in terms of materials categories developed specific to our era.

ASBESTOS CEMENT

Asbestos cement was the first diverse, popular, and industrial level cement composite material. It consisted of portland cement reinforced with asbestos fibers. Asbestos is a fibrous silicate mineral that maintains chemical resistance, especially to alkalis. Fire resistance, mechanical strength (due to the fibers' high length-to-diameter ratio), flexibility, and resistances to friction and wear are also characteristics of this kind of cement [6]. Its inexpensive processing and special chemical and physical properties that make it virtually indestructible helped asbestos' popularity in the building industry starting in the 1880s [7]. The portland cement matrix ultimately binds the fibers of asbestos into a hard mass, which is a durable material, mechanically and chemically compatible with the fibers. In addition, good bond to the cement matrix with no evidence of an interfacial transition zone is attributed to the hydrophilic surface of the asbestos fibers [8]. The proportion of cement to asbestos fibers varied over a range of 10% to 75% by weight, depending on the desired characteristic [9].

Approximately 95% of asbestos used in building products was in the form of chrysotile (white asbestos), although amosite (brown asbestos) was occasionally used [10]. Other forms of asbestos, namely crocidolite, anthophylite, tremolite, and actinolite, were not used, as their brittle and thin fibers were considered to be excessively hazardous [11]. Asbestos cement was first manufactured in the United States in 1905, introduced first in the form of a coating for pipes and boilers as shown in Figure 2.2 [12]. In 1907, an invention by an Austrian engineer, Ludwid Hatschek, enabled the industrial manufacturing of preformed asbestos cement products [2]. The Hatschek machine, a wet transfer roller, was used to produce the initial asbestos cement sheets (Figure 2.3), whereas two other

FIGURE 2.2   Manufacture of nonconducting asbestos cement coverings for steam-pipes and boilers. ("The industrial uses of asbestos," *Scientific American*, April 22, 1876, 258–259.)

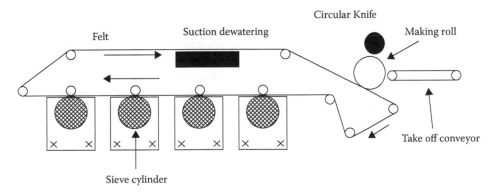

FIGURE 2.3   Simplified representation of a Hatschek machine for asbestos cement production. (Adapted from Wilden, J. E. (1986), *A Guide to the Art of Asbestos Cement*, Winchester, England: Taylor & Partners Translations, 108.)

manufacturing processes included the Mazza process for pipes and the Magnani semidry process for corrugated sheets [13]. After being formed, most products were steam cured to achieve the optimum microstructure for strength and durability. These asbestos cement building products have many desirable material characteristics, such as being lightweight, impermeable to water, durable, tough, resistant to rot. These building materials were also resistant to termites, soiling, corrosion, warping, and fire, and easy to clean and maintain [14]. Because of its low thermal conductivity, asbestos cement is a good electrical insulator. These highly desirable material characteristics, apparent in the new found material, sparked growth in the manufacture of a plethora of forms and styles to suit different needs [15].

With the refinement of the asbestos, cement mixture, and the forming and curing procedures, the market soon developed major commercial products of synthetic roof and wall shingles, corrugated wall and roof panels, flat millboard, and decorative wall and ceiling moldings. Additional manufactured products included water pipes, ceramic tiles, electrical switchboard panels, tabletops, electrical conduits. Even smaller diameter pipers were constructed for use in purlins and trusses during wartime construction to conserve steel and lumber [10].

The main building construction materials (in order of their production volume) were siding and roofing shingles and flat and corrugated sheets. These asbestos cement products lent themselves to rapid construction techniques and, therefore, were particularly useful for lightweight housing and industrial buildings.

Large flat asbestos cement sheets were available in sizes of 12-ft. long × 4-ft. wide and in thicknesses ranging from approximately an eighth of an inch to one inch [16]. By 1950, approximately 1 billion square feet of asbestos cement products had been produced for use in the building industry [3]. Use of asbestos cement in the United States continued to climb for another three years before reaching the peak of its popularity, only to plummet to a quick death in 1973 when the EPA implemented the initial ban on asbestos [17]. Some asbestos fibers, when inhaled, constitute a health hazard leading to asbestosis, a form of lung cancer. These health risks prompted the establishment of strict environmental regulations on working with asbestos. Health risks were shown to be greatest during mining and production processes, but minimal during installation and use of asbestos cement products [4].

HATSCHECK PROCESS

Industrial development of FRC materials is attributed to Ludwig Hatschek who, in 1900, developed a process to produce asbestos cement. This process is based on the movement of an endless felt through a slurry containing mixtures of cement and fiber with a water solids ratio in excess of 10. The felt collects the solids in the form of thin layers that become thicker every time it passes through the slurry. The Hatschek process is shown in Figure 2.3. As the desired thickness is achieved, the board is cut and pressed to dewater. Normal or autoclave curing is followed next. This is an efficient production method yielding up to 500 to 600 m²/h for boards measuring 6 mm thick [18].

Despite its worldwide use, there are disadvantages associated with the Hatschek method of production. Various fibers types are used as the filtering media in the retention of cement particles during manufacture. Thus, the fiber length used is shorter than the optimum length, resulting in brittle products, because the fiber pullout mechanism is not present. Furthermore, because the production is based on laminating thin layers, the board performance in areas subjected to freeze–thaw is poor, with the delamination failure being a common mode of failure.

Because of the health hazards associated with asbestos, the majority of Hatschek machines have been converted to use wood fibers. The development of wood and polymeric-based fibers such as polyethylene reinforcing pulps has resulted in a successful changeover of the equipment and production methods [19]. However wood is deleterious to the setting (and curing) of cement. The sugars in wood significantly inhibit the setting of the cement [20–22]. One of the ways to mitigate such deleterious effects is to remove the water and sodium hydroxide-soluble wood chemicals (e.g., Kraft process). The use of coupling agents in the surface treatment of discrete wood fibers enhances the mechanical performance, as well [23]. Having said that, degradation mechanisms such as alkali attacks and/or fiber embrittlement are not removed completely [24]. Loss of strength and ductility has been observed for composites subjected to cycles of heat and humidity, although the use of calcium chloride, calcium hydroxide, fly ash, and other admixtures have been somewhat helpful in the setting and initial curing [25]. The strength and toughness of the composites still decrease with time, however [26, 27].

FERROCEMENT

The most traditional form of continuous fiber cement composites are ferrocement products. In contrast to thin section composites that are reinforced with randomly distributed short fibers, ferrocement [28] uses small wire mesh as a main reinforcement and mortar as the matrix. Small scales of ingredients, reinforcement, and sand minimize the flaw sizes in material, consequently leading to the increase in overall strength. As alternatives to steel, other materials such as polypropylene (PP) [29], asbestos, and glass have also been shown to improve the tension capacity and ductility.

The widespread use of ferrocement in the construction industry has occurred in many developing countries during the last 50 years. The use of this construction material in the United States, Canada, and Europe is quite limited, however, perhaps because of the labor costs. The main worldwide applications of ferrocement construction to date have been for silos, tanks, roofs, and, most of all, boats [30]. The construction of ferrocement can be divided into four phases that include (a) fabricating the steel rods to form a skeletal framing system, (b) tying or fastening rods and mesh to the skeletal framing, (c) plastering, and (d) curing. Most of these steps are labor intensive.

Ferrocement has a very high tensile strength-to-weight ratio and superior cracking behavior in comparison with conventional reinforced concrete. This means that thin ferrocement structures can be made relatively light and watertight and are hence an attractive material for the construction of boats, barges, prefabricated housing units, and other portable structures. Although for these application ferrocement is more efficient on a weight basis, it is frequently more economical to build with conventionally reinforced concrete.

Even though construction with ferrocement may not be cost-effective in many applications, this material competes favorably with fiberglass laminates or steel when used in special structures. With the development of new mesh reinforcing systems and more efficient production techniques, ferrocement-based approaches have become competitive in a wide range of applications requiring thin structural elements.

CEMENT COMPOSITES IN MODULAR AND PANELIZED CONSTRUCTION SYSTEMS

The National Association of Home Builders NAHB [31, 32] identified the development of alternatives to lumber and wood framing for residential construction as one of the main areas of interest. To propose an alternative system to wood-based construction, one must consider several aspects, including material innovativeness and durability, analysis and design methodology, and constructability issues. Cement composites are attractive because they can be formed into thin sections, panels, beams, and slender columns. Furthermore, the same thin section fiber reinforced cement composite technology can be used to repair columns, reinforced concrete members, and masonry walls. The design methodology for composites can be developed on the basis of the fundamentals of structural analysis.

The benefits of panelized concrete construction include: a light framing material, homogeneous sections (there is less of a need to cull, resulting in less scrap and waste that is at 20% for lumber), and price stability (>50% increase in steel prices in three years). Benefits to the consumer include: higher strength, safer structures, less maintenance, slower aging, fire safety, and stronger connections (e.g., screwed versus nailed). From a design perspective, the analysis and fabrication of the truss and wall systems can be easily automated—the cutting of the sections and the jointing of the chord and Web members are the only two major required operations [33].

The use of modern structural analysis design tools in the home building industry is absent because the majority of designs are conducted by an adherence to a minimum set of specifications using a prescriptive approach. Motivated by the market forces to reduce cost, structural design methodologies are developed to better address the available economical alternatives. Yet these techniques are practiced today as a secondary approach to verify the designer's experience. Other industries such as the aerospace or automotive businesses, have recognized the benefits of using a systems design approach in which the analysis tool (e.g., finite element method) and the design optimization tool (e.g., nonlinear programming) are closely linked to design requirements, manufacturing, and cost issues. Such tools not only reduce the cost and increase the safety of the system but also help in reducing the design cycle time.

GLASS FIBER REINFORCED CONCRETE

Alkali-resistant (AR) glass fibers are used extensively in the manufacturing of thin (10 mm or 1/2 in.), concrete composites with greatly reduced weights (as little as one-eighth of the weight of the equivalent concrete product). AR fiberglass is formed into continuous strands or rovings or pre-chopped strands, either loose or bonded into a mat for hand layup. Glass fiber reinforced concrete (GFRC) is a composite material consisting of a mortar of cement and fine aggregate reinforced with AR glass fibers. It is manufactured by the simultaneous spraying of chopped glass fibers and port-land cement mortar onto a mold [34]. Major developmental efforts took place during the 1970s and 1980s with the invention of AR glass fibers containing up to 16% zirconia [35–38]. GFRC materials have been widely used, and their properties studied extensively worldwide. GFRC may be thought of as a thin-section concrete, with a typical thickness of 10 to 15 mm.

The reduced spacing of fibers strengthens the composite at the microlevel by bridging the micro-cracks before they reach the critical flaw size [39]. In comparison with steel fibers, the small diameter of the individual glass fibers ensures a better and more uniform dispersion. In addition, the high surface area and relatively small size of glass fiber bundles offers a significant distribution capability and crack bridging potential as compared with steel fibers. The fibers are also randomly distributed, offering efficiency in load transfer. Furthermore, due to the small size of the individual fibers, as shown in Figure 2.4, their bond strength also is far superior to the polymeric fibers, thus increasing the efficiency of fiber length so that there is limited debonding and fiber pullout. Finally, because of the resilience and highly compliant nature of the fiber bundles that bridge the matrix cracks at a random orientation, they are able to orient so as to carry the load across the crack faces. This factor is expected to substantially increase the durability of composite laminates made with glass fibers.

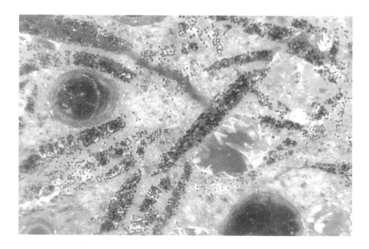

FIGURE 2.4   AR glass fiber dispersion in cement paste.

GFRC-based products lend themselves to a wide variety of applications such as cladding panels, small enclosures, noise barriers, drain channels, formwork, and the construction of many architectural details. Generally, for these products, GFRC is a factory-produced material where the composite performance is obtained with a fiber content of 2% to 5% by weight, depending on the product application and production method used. Ingredients of the matrix include ordinary or rapid setting portland cement, dry-graded sand, acrylic polymer in emulsion form for curing (optional for many products), superplasticizer, and water. Metakaolin, fly ash, and Silica fume may be also added as partial cement replacement materials to improve the long-term properties of the matrix. The fiber dosage is typically in the 2% to 5% range as mixed into a simple 1:1 cement–sand and water matrix. GFRC has a high compressive strength along with good tensile and bending properties so products can be designed in thin (10 mm or 1/2 in.) sections [40].

The mechanical properties of GFRC composites depend on the fiber content, water–cement ratio, density, sand content, fiber orientation, fiber length, and polymer content, if used. Typical properties for traditional spray-up GFRC containing 5% by weight of glass fibers are shown in Table 2.1 [36], which details the most widely accepted design procedure in the industry.

In addition to the traditional spray-up GFRC, the use of premixed GFRC is growing for the production of certain thin cementitious products. Typical properties for premixed GFRC are shown in Table 2.2 [36]. Generally, premixed GFRC has a lower fiber content, uses shorter fibers, and has significantly greater three-dimensional fiber orientation than the largely two-dimensional orientation obtained with spray-up GFRC, which all contribute to it having lower mechanical performance than the spray-up GFRC.

TABLE 2.1
Typical Range of Traditional GFRC Properties[a]

Property	28 Days	Aged[b,c]
Density (dry)	1900–2000 kg/m^3	1900–2000 kg/m^3
Compressive strength	50–80 Mpa	70–80 Mpa
Flexural		
Yield (FY)	6–10 Mpa	7–10 Mpa
Ultimate strength (FU)	14–24 Mpa	9–17 Mpa
Modulus of elasticity	10–20 Gpa	10–20 Gpa
Direct tension		
Yield (TY)	5–7 Mpa	5–8 Mpa
Ultimate strength (TU)	7–11 Mpa	5–8 Mpa
Strain to failure	0.6%–1.2%	0.03%–0.08%
Shear		
Interlaminar	3–5.5 Mpa	3–5.5 Mpa
In plane	7–11 Mpa	5–8 Mpa
Coefficient of thermal expansion	~20 × 10^{-6} (mm/mm/°C)	~20 × 10^{-6} (mm/mm/°C)
Thermal conductivity	0.5–0.6 (W/m °C)	0.5–0.6 (W/m °C)

Source: Majumdar, A. J., *Proceedings of Symposium L, Advances in Cement–Matrix Composites*, Materials Research Society, University Park, PA, 37–59, November 1980.

[a] Typical values and not intended for design or control purposes. Cement–sand ratio in the above composites ranges between 1:1 and 3:1.

[b] Developed from accelerated testing programs on GFRC immersed in 50°C to 80°C water.

[c] Commercially available modified cementitious matrices specially developed for GFRC yield substantial improvements in long-term properties, particularly the tensile strain capacity.

TABLE 2.2
Typical Range of Premix GFRC Properties[a]

Property	28 Days
Density (dry)	1800–2000 kg/m³
Compressive strength	40–60 Mpa
Flexural	
Yield (FY)	5–8 Mpa
Ultimate strength (FU)	10–14 MPa
Modulus of elasticity	10–20 GPa
Direct Tension	
Yield (TY)	4–6 Mpa
Ultimate strength (TU)	4–7 MPa
Strain to failure	0.1%–0.2%
In-plane Shear	4–7 Mpa
Coefficient of thermal expansion	~20 × 10⁻⁶ (mm/mm/°C)
Thermal conductivity	0.5–0.6 (W/m °C)

Source: Majumdar, A. J., *Proceedings of Symposium L, Advances in Cement–Matrix Composites*, Materials Research Society, University Park, PA, 37–59, November 1980.

[a] Typical values and not intended for design or control purposes.

The appearance of the finished product is directly related to the form material and the quality of the mold itself. The mold can be made from various material such as steel, plywood, plastic, rubber molding compounds, glass fiber reinforced polyester resin, or GFRC itself. Various manufacturing equipment are available. Depending on the production process used, three basic techniques for fiber incorporation in GFRC applications include:

- Premix, using prechopped 12- to 40-mm (1/2- to 1 5/8-in.) fiber
- Spray process with purpose built equipment and continuous rovings of fiber
- Hand layup using a chopped strand mat or woven filament net

In the premix process, 12 to 40 mm (1/2–1 5/8 in.) of chopped strands are stirred into the mix at low speed until they are dispersed. In the spray technique, a continuous strand is fed into a chopping gun, cut to 25 to 37mm (1–1 1/2 in.), and simultaneously sprayed out with the cement mortar. The cement mortar is fed to the spray gun by pump and is sprayed using compressed air. The glass fibers and cement mortar are sprayed together onto a prepared mold where they are compacted to produce the composite GRC using rollers. The strands are orientated in two dimensions: In the hand layup process, the mortar mix is sprayed or painted onto the mold or form and the chopped strand mat or net laid on. The mortar is then forced through with the brush or a compaction roller.

The GFRC composites are not immune to durability problems [41]. Because the glass fibers are manufactured in bundles of several hundred fibers, each with a diameter in the range of 12 μm, the filling up of interstitial pores in the glass fibers increases the bond strength and rigidity of the interfacial zone, thereby resulting in a range of responses depending on the fiber length, bond, and interface conditions [42–46]. This results in an increase in the bond strength and a loss in the flexibility of the fibers [36, 47]. Use of pozzolans could also delay such processes to a certain degree [48, 49]. The increase in the bond strength causes a loss of flexibility among the fibers within a strand [49]. Once the strength of the bond is improved, the load-carrying capacity becomes dominated by the strength of the fiber. For fibers of sufficient length, an effective bridging mechanism forms across the matrix cracks. If the crack is opened because of environmental or loading conditions, the bridging force of the fiber may reach as high as the tensile strength of the fiber. This indicates that for a fully bonded fiber, the strength of the composite may be governed by the strength of the fibers. Ductility loss is measured by means of tensile and flexural tests [50]. The effect of aging on the interfacial properties of glass fiber strands can be measured using pullout or strand-in-cement tests [50].

The "strand in cement" test was developed to characterize aging, in which a single strand was encased in a cement paste prism and then aged in hot water at various temperatures before testing to failure [51]. The accelerated aging tests were then compared with 10-year real-life weathering data and were based on the comparison of empirical relationships between accelerated aging regimes using a range of temperatures between 5°C and 80°C and real weathering acceleration factors [52–54]. On the basis of the model, a table, including the relationship between time in accelerated aging (different temperatures) and exposure to real weathering, was proposed [41]. It has been used as a basis for testing GFRC products ever since.

One of the challenging problems in the modeling and testing of glass fiber composites is the determination of the bond parameters of the glass fibers with the matrix. Tests have been developed to characterize the pullout response of glass fibers from a portland cement matrix [55]. New methods have also been developed to measure the tensile response and the stress–strain diagrams of glass fiber bundles under normal and saturated $Ca(OH)_2$ solutions [56, 57]. Improved results have also been obtained using amorphous metal glass fibers under alkaline conditions [58]. Several approaches are becoming available to reduce the pore water solution's alkalinity, thus providing a less aggressive environment for the fibers [59]. Techniques are also being developed to characterize the strand effect in glass fiber composites. By determining strand perimeters using digital analyses of images captured from petrology of thin sections, one can obtain the effective perimeter parameter and use this to characterize the bond [60].

CELLULOSE FIBERS

Cellulose or vegetable fibers, including sisal, coconut, jute, bamboo, pulp (cellulose), hemp, flax, jute, kenaf, and wood fibers, are prospective reinforcing materials. Their use until now has been more empirical than technical and are therefore a very generic term for a vast range of materials and can refer to fibers of any size, from particulate to long strands, which trace their beginning from a natural feedstock, rather than being fibers using a man-made material, or animal by-products such as silk or horse hair. Cellulose fibers require only a low degree of industrialization for their processing, and, in comparison with an equivalent weight of the most common synthetic reinforcing fibers, the energy required for their production is small. Hence the cost of fabricating these composites is also low [61, 62]. They have been tried as reinforcement for cement matrices in developing countries mainly to produce low-cost thin elements. The sisal plant leaf is a functionally graded composite structure that is reinforced by three types of fibers: structural, arch, and xylem fibers. The first occurs in the periphery of the leaf, providing resistance to tensile loads. The others present secondary reinforcement, occurring in the middle of the leaf as well as a path for nutrients (see Figure 2.5).

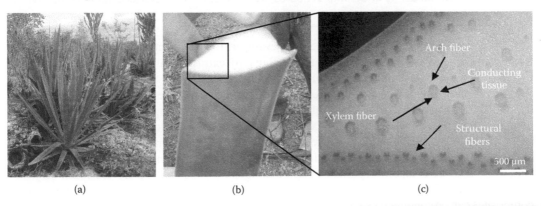

(a)   (b)   (c)

FIGURE 2.5   The sisal (a) plant, (b) leaf, and (c) leaf cross section showing different fiber types.

As a substitute of asbestos fibers, natural fibers have been traditionally used in the form of chopped, short, or pulp for the production of thin roofing and cladding elements [63–65]. An increased use of these materials in applications such as the internal and external partitioning walls may lead to the development of low cost-sustainable materials [66]. Even though natural fiber cement composites have been mainly reinforced by short or pulp cellulose fibers, their application is still limited because of two main challenges: lack of durability and lack of ductility.

From an economical development and sustainability perspective, natural fiber reinforced cement composites provide an opportunity to develop an agriculture-based economy in arid environments making materials for the housing construction. The economic incentives for developing countries with low labor costs are enormous because of the availability and production of natural fiber farms, which require a small capital investment and a low degree of industrialization. Furthermore, in comparison with the most common synthetic reinforcing fibers, natural fibers require less energy to produce and are the ultimate green products.

In North America, cellulose-based FRCs have found increasing applications in residential housing nonstructural components (such as siding, roofing, flat panel applications including underlayment, and tile backer board) and lumber substitutes (such as trim, fascia, and corner boards). Cellulose-based FRC products have been limited in exterior applications because of degradation from ambient wetting and drying.

CONTINUOUS FIBER SYSTEMS

A major portion of the theoretical methods concentrate on the toughening mechanisms in discrete fiber reinforced systems. However, both experimental and theoretical results show that increasing the aspect ratio of fibers significantly improves the mechanical performance of these composites. Despite the facts, new processing, manufacturing, and analysis techniques for continuous fiber cement-based composite laminates have received little or no attention. Because of their significant strength and ductility improvement, these composites may prove to be the ideal choice for materials under severe loading conditions.

Standard processing techniques such as conventional mixing have been primarily used for the development of cement-based composites (FRC materials). This results in composites with a relatively low volume fraction of short, discrete fibers distributed in a random two- or three-dimensional orientation. Three major tasks addressed in the development of cement-based composites are defined in terms of the following:

- Development of manufacturing technique for various composite types
- Development of theoretical and experimental characterization techniques
- Theoretical modeling of the interaction between fiber, matrix, interface, and manufacturing process

Several novel techniques have been recently developed for manufacturing composite laminates. Some of these processes include high energy mixing [55], extrusion, pultrusion [67], filament winding [68], compression molding, cross ply and sandwich lamination techniques [69], and hybrid reinforcement [70]. These materials offer a significant degree of strength, ductility, and versatility, and can be easily be used as a new class of structural, load bearing members. Computer-aided manufacturing provides the computational and control power needed to develop economic and versatile materials. Cement-based composite laminates prepared with continuous fibers allow the full potential of fibers in reinforcing the matrix to be used, because the manufacturing technique is fully controlled and the composite laminates can be designed for the specific service loads they may encounter. The effect of ply orientation and stacking sequence on the mechanical properties of these composites can be controlled using stochastic or deterministic models that combine the principles of classical laminate theory, fracture mechanics, and the micromechanics of distributed damage.

THIN SECTION COMPOSITES USING TEXTILES

Woven meshes have been used in the manufacture of thin sheet concrete products [71, 72]. Placement of fiber reinforcement at the surface of the composites results in higher matrix and composite strengths. This is partly due to the length effect of the continuous fibers, which allows for full stiffness utilization. In comparison, large volume fractions (>3%) of chopped fibers are required to effectively reinforce concrete products. A woven or bonded textile facilitates panel production in a continuous manufacturing line. Because the primary mode of loading is due to bending (especially during installation), the fibrous mesh provides a continuous reinforcement in the tensile region. The composites under study consist of woven or fibrillated mesh of both ductile and brittle continuous fibers. The use of short fibers in addition to surface reinforcement is advantageous for both short-term handling installation and long-term performance.

Because of panel weight considerations, lightweight concrete mixtures with densities around 1280 kg/m^3 (80 pcf) are desirable, which result in a matrix with relatively low bending strength. Fiber reinforcement also increases the ductility several-fold. Use of a low fiber volume fraction moderately increases the first-crack strength, ultimate tensile, and flexural strengths in addition to granting a better postpeak response. Fibers control the cracking process by preventing localization while also generating a homogeneous microcracking state to dissipate energy over the entire volume [73]. This book will address the development and various areas of application for cement-based textile composites.

REFERENCES

1. Walraven, J. (1999), "The evolution of concrete," *Structural Concrete*, 1, 3–11.
2. Di Prisco, M., and Toniolo, G., "Structural applications of steel fibre reinforced concrete," Proceedings of the International Workshop, Milan, Italy, CTE Publ., Milan, Italy, April 4, 2000.
3. ACI 544.4R (1988), "Design considerations for steel fiber reinforced concrete" (Reported by ACI Committee 544), *ACI Structural Journal*, 85(5), 563–580.
4. Mobasher, B., Sheppard, T., Slager, R., and Krapf, W. (1991), "Direct Exterior Finish Systems (DEFS) Using Fiber Reinforced Cement Based Substrates," American Concrete Institute, Spring Convention, Boston, MA, March 1991.
5. Kvavadze, E., Bar-Yosef, O., Belfer-Cohen, A., Boaretto, E., Jakeli, N., Matskevich, Z., and Meshveliani, T. (2009), "30,000-year-old wild flax fibers," *Science*, 325(5946), 1359.
6. Michaels, L., and Chissick, S. S. (eds.) (1979), *Asbestos, Properties, Applications, and Hazards*, New York: Wiley, 1–2.
7. Amy Lamb Woods. (2000), *Keeping a Lid on It: Asbestos-Cement Building Materials*. Philadelphia: Technical Preservation Services, National Park Service. Available online at http://www.nps.gov/history/nps/tps/recentpast/index.htm. Accessed 30 June 2008.
8. Bentur, A., and Mindess, S. (1990), *Fibre Reinforced Cementitious Composites*, London: Elsevier Applied Science, 288–304.
9. Rosato, D. V. (1959), *Asbestos: Its Industrial Applications*, New York: Reinhold Publishing Corp., 1, 62.
10. Michaels, L., and Chissick, S. S. (eds.) (1979), *Asbestos, Properties, Applications, and Hazards*, New York: Wiley, 533.
11. United States Department of the Interior Bureau of Mines (1952), *Materials Survey: Asbestos*. Washington, DC: U.S. Government Printing Office, I-1–I-4.
12. Hornbostel, C. (1978), *Construction Materials: Types, Uses and Applications*, New York: John Wiley & Sons, 82.
13. St John, D. A., Poole, A. B., and Sims, I. (1998), *Concrete Petrography: A Handbook of Investigative Techniques*, London: Arnold Publishers, 320–322.
14. "The industrial uses of asbestos" (1876, April 22), *Scientific American*, 258–259.
15. Wilden, J. E. (1986), *A Guide to the Art of Asbestos Cement*, Winchester, England: Taylor & Partners Translations, 108.
16. Lechner, E. (1934, June), "Recent innovations in the manufacture of asbestos-cement," *Cement and Cement Manufacture*, 7(6), 180–181.

17. National Trust for Historic Preservation (1993), "Coping with contamination: A primer for preservationists," Information Bulletin No. 70, 12.
18. Hiendl, H. (1900), *Asbestos Cement Machinery,* Straubing, Germany: CL. Attenkofer, Ludwigsplatz 30, 8440, 128 pp.
19. Pehanich, J. L., Blankenhorn, P. R., and Silsbee, Michael R. (2004), "Wood fiber surface treatment level effects on selected mechanical properties of wood fiber–cement composites," *Cement and Concrete Research*, 34(1), 59–65.
20. Simaputang, M. H., Lange, H., Kasim, A., and Seddiq, N. Proceedings of the 3rd International Conference on Inorganic Bonded Wood and Fiber Composite Materials, Forest Products Research Society, Madison, Wisconsin, 1989, 33–42.
21. Clark, K. E., and Sherwood, P. T. (1956), "Further studies on the effect of organic matter on the setting of soil-cement mixtures," *Journal of Applied Chemistry*, 6, 317–324.
22. Soroushian, P., Marikunte, S., and Won, J. P. (1994, November), "Wood fiber reinforced cement composites under wetting–drying and freezing–thawing cycles," *Journal of Materials in Civil Engineering*, 6(4), 595–611.
23. Couts, R. S. P., and Campbell, M. D. (1979, October), "Coupling agents in wood fibre-reinforced cement composites," *Composites*, 10(4), 228–232.
24. Vinson, K. I., and Daniel, J. I. (1990), "Specialty cellulose fibers for cement reinforcement," *ACI Special Publication*, 124, 1–18.
25. Sargaphuti, M., Shah, S. P., and Vinson, K. D. (1993), "Shrinkage cracking and durability characteristics of cellulose fiber reinforced concrete," *ACI Materials Journal*, 90(4), 309–318.
26. Hofstrand, A. D., Moslemi, A. A., and Garcia, J. F. (1984), "Curing characteristics of wood particles from nine northern Rocky Mountain species mixes with Portland cement," *Forest Products Journal*, 34(2), 57–61.
27. Hachmi, M., and Moslemi, A. A. (1989), "Correlation between wood-cement compatibility and wood extractives," *Forest Products Journal*, 39(6), 55–58.
28. Naaman A. E., and Shah S. P. (1971), "Tensile test of ferrocement," *Journal of the American Concrete Institute*, 68, 693–698.
29. Naaman A. E., Shah, S. P., and Throne J. L. (1984), *Some Developments in Polypropylene Fiber for Concrete*, Special publication. Farmington Hills, MI: American Concrete Institute, 375–396.
30. Naaman, A. E. (2000), *Ferrocement and Laminated Cementitious Composites*. Ann Arbor, MI: Techno Press 3000, 372 pp., ISBN 0-9674030-0-0.
31. NAHB. (1994), *Alternative Framing Materials in Residential Construction: Three Case Studies*. Upper Marlboro, MD: NAHB Research Center.
32. NAHB. (1996), *Prescriptive Method for Residential Cold Formed Steel Framing*, 1st ed. Upper Marlboro, MD: NAHB Research Center.
33. Mobasher, B., Chen, S.-Y., Young, C., and Rajan, S. D. (1999, August), "Cost-based design of residential steel roof systems: A case study," *Structural Engineering and Mechanics*, 8(2), 165–180.
34. Proctor, B.A., "Past Development and Future Prospect for GRC Materials," Proceedings, International Congress on Glass Fibre Reinforced Cement, The Glass Fibre Reinforced Cement Association, Gerrards Cross, Paris, November 1981, 50–67.
35. Majumdar, A. J., and Ryder, J. F. (1968), "Glass fibre reinforcement of cement products," *Glass Technology*, 9(2), 78–84.
36. Majumdar, A. J., *Proceedings of Symposium L, Advances in Cement–Matrix Composites*, Materials Research Society, University Park, PA, 1980, 37–59.
37. Majumdar, A. J. (1970), "Glass fibre reinforced cement and gypsum products," *Proceedings of the Royal Society of London, A*, 319, 69–78.
38. Laws, V., Langley, A. A., and West, J. M. (1986), "The glass fibre/cement bond," *Journal of Materials Science*, 21(1), 289–296.
39. Mobasher, B., and Li, C. Y. (1996), "Mechanical properties of hybrid cement based composites," *ACI Materials Journal*, 93(3), 284–293.
40. Frondistouyannas, S. (1977), "Flexural strength of concrete with randomly oriented glass fibers," *Magazine of Concrete Research*, 29(100), 142–146, 1977.
41. Majumdar, A. J., and Laws, V. (1991), *Glass Fibre Reinforced Cement*. Oxford: BSP Professional Books, p. 197.
42. Bentur, A., and Diamond, S. (1986, November), "Effect of aging of glass fibre reinforced cement on the response of an advancing crack on intersecting a glass fibre strand," *International Journal of Cement Composites and Lightweight Concrete*, 8(4), 213–222.

43. Curiger, P., (1995), "Glass fibre reinforced concrete: Practical design and structural analysis" Düsseldorf: Germany Publisher Beton-Verlag.

44. Bentur, A., and Diamond, S. (1984), "Fracture of glass fiber reinforced cement," *Cement and Concrete Research*, 14, 31–34.

45. Bentur, A. "Mechanisms of potential embrittlement and strength loss of glass fiber reinforced cement composites," Proceedings of the Durability of Glass Fiber Reinforced Concrete Symposium organized by PCI, Concrete Symposium organized by PCI, November 12–15, 1985, Chicago, ed. S. Diamond, 1986, 109–123.

46. Shah, S. P., Ludirdja, D., Daniel J. I., and Mobasher, B. (1988, September–October), "Toughness-durability of glass fiber reinforced concrete systems," *ACI Materials Journal*, 352–360. Discussion 85-M39, *ACI Materials Journal*, July–August 1989, 425.

47. Mobasher, B., and Shah, S. P. (1989, September–October), "Test methods for evaluation of toughness of GFRC systems," *ACI Materials Journal*, 448–458.

48. Leonard, S., and Bentur, A. (1984, September), "Improvement of the durability of glass fiber reinforced cement using blended cement matrix," *Cement and Concrete Research*, 14(5), 717–728.

49. Diamond, S. (1981, May), "Effects of two danish fly ashes on alkali contents of pore solutions of cement–flyash pastes," *Cement and Concrete Research*, 11(3), 383–394.

50. Li, Z., Mobasher, B., and Shah, S. P. (1991), "Characterization of interfacial properties in fiber-reinforced cement based composites," *Journal of the American Ceramic Society*, 74(9), 2156-2164.

51. Litherland, K. L., Oakley, D. R., and Proctor, B. A. (1981), "The use of accelerated aging procedures to predict the long term strength of GRC composites," *Cement and Concrete Research*, 11, 455–466.

52. Aindow, A. J., Oakley, D. R., and Proctor, B. A. (1984), "Comparison of the weathering behaviour of GRC with predictions made from accelerated aging tests," *Cement and Concrete Research*, 14, 271–274.

53. Litherland, K. L., "Test Methods for Evaluating the Long Term Behaviour of GFRC," Proceedings of the Durability of Glass Fiber Reinforced Concrete Symposium organized by PCI, ed. S. Diamond, November 12–15, 1985, Chicago, 1986, 210–221.

54. Proctor, B. A., Oakley, D. R., and Litherland, K. L. (1982), "Developments in the assessment and performance of GRC over 10 years," *Composites*, 13, 173–179.

55. Mobasher, B., and Li, C. Y. (1996), "Effect of interfacial properties on the crack propagation in cementitious composites," *Advanced Cement Based Materials*, 4, 93–105.

56. Rabinovich, F. N., and Lemysh, L. L. (1996, November–December), "Stress–strain state of fiber bundles in glass-fiber-reinforced concrete structural elements," *Glass and Ceramics*, 53(11–12), 373–376.

57. Pribyl, M., Sevcik, V., and Kutzendorfer, J. (1995, September), "Modeling the elastic behavior of refractory concrete reinforced with glass-fibers," *Ceramics-Silikaty*, 39(2), 41–44.

58. Alonso, C., Acha, M., and Andrade, C. (1993, September), "Reactivity of metallic-glass fibers in simulated concrete pore solutions," *Solid State Ionics*, 63-5, 639–643.

59. Purnell, P., Short, N. R., Page, C. L., Majumdar, A. J., and Walton, P. L. (1999), "Accelerated ageing characteristics of glass-fibre reinforced cement made with new cementitious matrices," *Composites Part A: Applied Science and Manufacturing*, 30(9), 1073–1080.

60. Purnell, P., Buchanan, A. J., Short, N. R., Page, C. L., and Majumdar, A. J. (2000, September), "Determination of bond strength in glass fibre reinforced cement using petrography and image analysis," *Journal of Materials Science*, 35(18), 4653–4659.

61. Aziz, M. A., Paramasivam, P., and Lee, S. L. (1984), "Concrete reinforced with natural fibres," in Swamy, R. N. (ed), *Natural Fibre Reinforced Cement and Concrete*. London: Blackie and Son Ltd., vol. 5, 106–140.

62. Tolêdo Filho, R. D., Kuruvilla J., Ghavami, K., and England, G. L. (1999), "The use of sisal fibre as reinforcement in cement based composites," *Revista Brasileira de Engenharia Agrícola e Ambiental*, 3(2), 245–256.

63. Swift, D. F., and Smith, R. B. L. (1979), "The flexural strength of cement-based composites using low modulus (sisal) fibers," *Composites*, 6(3), 145–148.

64. Coutts R. S. P., and Warden P. G. (1992), "Sisal pulp reinforced cement mortar," *Cement and Concrete Composites*, 14(1), 17–21.

65. Toledo Filho, R. D., Ghavami, K., England, G. L., and Scrivener, K. (2003), "Development of vegetable fiber-mortar composites of improved durability," *Cement and Concrete Composites*, 25(2), 185–196.

66. Toledo Filho, R. D., Scrivener, K., England, G. L. (2000), Ghavami K. Durability of Alkali-Sensitive Sisal and Coconut Fibers in Cement Mortar Composites. *Cement and Concrete Composites*, 22(2), 127–143.

67. Pivacek A., and Mobasher, B. (1997), "A Filament Winding Technique for Manufacturing Cement Based Cross-Ply Laminates," Innovations Forum, *ASCE Journal of Materials Engineering*, 55–58.
68. Mobasher, B., Pivacek A., and Haupt, G. J. (1997), "Cement based cross-ply laminates," *Advanced Cement Based Materials*, 6, 144–152.
69. Mobasher, B., and Pivacek, A. (1998), "A filament winding technique for manufacturing cement based cross-ply laminates," *Journal of Cement and Concrete Composites*, 20, 405–415.
70. Perez-Pena, M., Mobasher, B., and Alfrejd, M. A. (1991), "Influence of Pozzolans on the Tensile Behavior of Reinforced Lightweight Concrete," Materials Research Society, Symposium "O," Innovations in the Development and Characterization of Materials for Infrastructure, December 1991, Boston, MA.
71. Balaguru, P. N., and Shah, S. P. (1992), *Fiber-Reinforced Cement Composites*. New York: McGraw-Hill, 365–412.
72. Odler, I. (1991), "Fiber-Reinforced Cementitious Materials," Materials Research Society Symposium Proceedings, 211, Eds. S. Mindess and J. Skalny, Pittsburgh, PA, 265–273.
73. Mobasher, B., Castro-Montero, A., and Shah, S. P., "A study of fracture in fiber reinforced cement-based composites using laser holographic interferometry," *Experimental Mechanics*, 30(90), 286–294.

3 Ductile Cement Composite Systems

INTRODUCTION

Application of fiber reinforced concrete (FRC) materials in the area of thin precast boards is becoming increasingly popular because the production, transportation, and installation costs are significantly lower than the conventional precast concrete. During the past 40 years, the need to replace the asbestos cement-based products necessitated the development of new classes of cementitious materials. These developments are in several areas of matrix systems, reinforcing materials, and also manufacturing techniques. Inherent differences in the properties of fibers make hybrid reinforcement a design variable in the design of FRC composites. Areas of application for these materials include exterior and interior of buildings, roofing products, floor underlayment, water and wastewater pipes, cast in place forms, rebar free applications, and various retrofit projects.

The use of strong fibers in the tension-weak brittle cementitious matrix increases the ductility of composite. In FRC applications, low fiber volume fractions are used, which result in a moderate increase in tensile or flexural strength. The ultimate strain capacity of the matrix is lower than the strain capacity of the fibers; hence, the matrix fails before the full capacity of fibers is achieved. The fibers bridge the cracks and contribute to the energy dissipation through the process of debonding and pullout. Mineral fibers such as glass have a high modulus and fail in a relatively brittle manner. Polymeric fibers are strong and ductile, but with a lower modulus. In addition, bond characteristics of the organic fibers are comparatively inferior to metallic or mineral fibers. Various alternatives are presented to use the reinforcing capacity of fibers. These mechanisms are addressed in subsequent chapters.

In a uniaxial tensile test, the first major crack occurs at a stress level referred to as the bend over point (BOP). BOP is a function of the fiber type, volume fraction, and interfacial properties [1]. By properly selecting fiber types and matrix formulations, the BOP level can be increased to as much as 15 MPa. Beyond the BOP, the composite response is further affected by the interface characteristics and the ability of fibers to carry additional load, leading to strong and ductile composites. Durability is another key performance factor. Long-term durability may cause a ductile to brittle transition response. For an optimum composite performance, parameters such as the interfacial stiffness and adhesional and frictional bond strength must be considered.

Recent advances in textile manufacturing for construction systems have introduced new opportunities. Woven fabric reinforced composites have been recognized as more competitive than unidirectional composites. This is due to their stability and deformation characteristics that result from coupling of reinforcement in transverse and longitudinal directions, an area change due to the trellising of the yarns, and finally from the stretching of undulated fibers in the load direction. When the textile geometry possesses a double curvature, the forming process usually results in in-plane shear deformation, which consequently leads to complex redistribution and reorientation of fibers in the woven fabric reinforced composite. The change of fiber orientation during large deformation induces additional anisotropy into the composite. Although the existing material models involving this nonorthogonality are very limited, Chapter 5 focuses in methodology to show the broad geometrical changes that textiles undergo to realign themselves with the direction of the load to carry the forces.

MECHANICS OF TOUGHENING

In the area of short fiber composites, if the volume fraction of fibers is increased beyond a critical level, the fiber dispersion will result in reduced specific spacing and reduction of critical flaw size. Romauldi and Batson [2] were among the first to suggest that the strength cement paste mortar and concrete could be significantly increased by using closely spaced fibers. Their experimental studies showed that the stress at which a brittle matrix will crack can be increased by using high modulus fibers. Most importantly, as the wire spacing was reduced for the same volume fraction, the first-crack strength of the matrix was increased. Considerable modification in the behavior of the material is observed as the matrix cracks. The fibers bridge across the cracks and so provide postcracking ductility. These mechanisms will be discussed in detail in this book.

Aveston et al. [3] addressed the mechanics of toughening in brittle matrix composites. Using energy balance, they showed the analytical foundations of increase strain capacity in composite systems when the fiber volume fraction exceeds a critical level. Figure 3.1a represents the distribution of flaws in a solid. Figure 3.1b represents the same solid, but now containing fibers, indicating that the initial flaw distribution is influenced by the introduction of fibers. Figure 3.1c indicates that the initial flaw distribution is not influenced by the introduction of fibers but that the fibers straddle the initial flaws, $s < 2c$. Figure 3.1d shows a schematic representation of a cement paste that has undergone multiple fracture of the matrix. The inset illustrates that the longitudinal stress in the fibers is not uniform [4]. Various theoretical approaches have demonstrated the mechanisms operative in increasing the strength, toughness, and fracture properties of the composites. The critical volume fraction of fibers necessary for distributed cracking has been studied by Li and Wu [5],

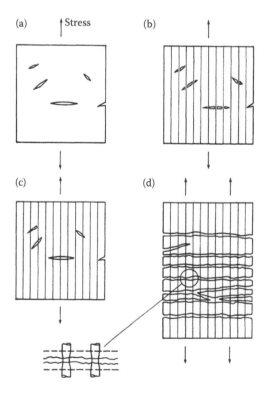

FIGURE 3.1   Schematic representation of flaws in a brittle solid. (From Hannant, D. J., Hughes, D. C., Kelly, A., Alford, N. McN., and Bailey, J. E., "Toughening of cement and other brittle solids with fibres [and discussion]," *Philosophical Transactions of the Royal Society of London. Series A, Mathematical and Physical Sciences,* 310(1511), 175–190, 1983. With permission.)

Visalvanich and Naaman [6], Tjiptobroto and Hansen [7], Yang et al. [8], Li et al. [9], and Mobasher and Li [10].

Fibers strengthen these composites at the microlevel by bridging the microcracks before they reach the critical flaw size. After the formation of the first crack, fibers contribute to the toughness of the composite through dissipative means. Beyond the first cracking level, it is assumed that the matrix does not significantly contribute to the strength of the composite material; however, because of the interlock mechanism, the matrix phase still carries significant force. With a higher volume of fibers, the microcracks can be controlled by arrest and bridging mechanisms. This allows the stresses to be transferred back into the matrix forming more microcracks in the matrix material and increasing the toughness considerably. Once a certain volume fraction of fibers is exceeded, the ultimate strength and ultimate strain increase considerably due to parallel microcracking [11].

Because of their very high specific surface area, the small-sized fibers such as whiskers result in excellent strengthening mechanisms. Using high-energy shear mixing technique, rheology modifying agents or hybrid reinforcement systems using several fiber types can also developed to optimize composite performance for strength and ductility. The matrix formulations developed can then be subjected to an extrusion and/or compression molding process for forming. The compression molding procedure is an ideal method for reducing the water–cement ratio and porosity of the mix while increasing the molding ability of short fiber composites [12]. This process however has its weaknesses because the rheology of a two-phase suspension such as cement paste is affected by the shear stresses applied during the extrusion or injection molding process. On the basis of these principals, discrete carbon and alumina-based whiskers of approximately a few microns in diameter were developed, which are discussed in Chapter 12 [39].

In a series of papers, Krenchel and Stang [13], Mobasher et al. [14], and Stang et al. [15] showed the correlation between processing and mechanism of distributed cracking in continuous fiber cement-based systems. The nature of microcracking is shown in Figure 3.2. They showed that by using a high-volume fraction of continuous fibers, the tensile strength of the composites may be increased significantly. By developing a manual pultrusion technique, they were able to develop composites containing up to 13% volume fraction of polypropylene (PP) fibers and a BOP of 15 MPa. The discussions on the basis of the mechanics of continuous fiber systems and textiles are addressed in Chapters 10 and 11.

MACRO-DEFECT-FREE CEMENTS

Macro-defect-free (MDF) cements are composites of a water-soluble polymer and a cement, characterized by its extremely low porosity and unusual mechanical properties [16]. Flexural strength, for example, is at least one order of magnitude higher than that of plain concrete. The

(a) 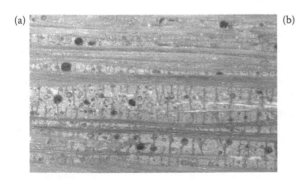 (b)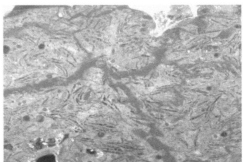

FIGURE 3.2 Microcracking development in a unidirectional cement composite with (a) PP fibers, and (b) Alkali resistant glass fibers. (Adapted from Pivacek, A., "Development of a filament winding technique for manufacturing cement based materials," MS thesis, Arizona State University, 2001.)

possible applications of these materials are extensive, varying from electronic substrate, armor ceramics, and substitution of fiber reinforced plastics [17, 18]. One of the most prominent features of these composites is the rheology of the fresh paste that can be processed as polymers, using calendering, extrusion, molding, pressing, and so forth. A typical MDF cement preparation method was developed by Birchall [19]. However, there are restrictions associated to these materials because the major limitation is the decay of mechanical properties at high relative humidity [20, 21].

Hasegawa et al. [22] have developed composites on the basis of the MDF cement concepts using hydrophobic phenol resin precursor through essentially water-free cement paste. The key points of this innovation are (i) to adhere cement particles by strong bonding technique, (ii) to decrease stress concentrated defects from the body, and (iii) to avoid using water and obtain hydration by water released from the phenol resin precursor during heat curing. This precursor acts as both processing aid and the hydration resource. To reduce the porosity, cement paste was mixed and calendered through the twin roll mill. As a result, they succeeded in establishing the composition and the process for a new class of high flexural strength cement-based material (phenol resin-cement composite) with remarkably enhanced moisture and thermal stabilities.

The flexural strength of the composite was higher than 120 MPa. Once the sheets were further densified under heat pressing with 6 MPa at 80°C before heat curing, flexural strength was improved to higher than 200 MPa. This achievement of strength may be due to the reduction of porosity. Bending modulus of elasticity in the range between 32 and 45 GPa, which depends on fabrication system was obtained.

DUCTILE COMPOSITES WITH HIGH-VOLUME FIBER CONTENTS

Many theoretical methods concentrate on the toughening mechanisms in discrete fiber reinforced systems. Both experimental and theoretical results show that increasing the aspect ratio of fibers significantly improves the mechanical performance. Despite the facts, new processing, manufacturing, and analysis techniques for continuous fiber cement-based composite laminates has received little or no attention. Because of their significant strength and ductility improvement, these composites may prove to be the ideal choice for materials for sustainability or materials under severe loading conditions.

Standard processing techniques such as conventional mixing have been traditionally used for the development of cement-based composites. This results in composites with a relatively low-volume fraction of short discrete fibers distributed in a random two- or three-dimensional orientation. Three major tasks addressed in the development of cement-based composites include the following:

- Manufacturing techniques for various composite types using short and long fiber systems
- Theoretical and experimental characterization techniques in terms of analysis and design
- Interaction between different fiber, textile, matrix, and interface systems

Several novel techniques have been recently developed for manufacturing composite laminates. Innovative manufacturing provides the computational and control power needed to develop economic and versatile materials. For example, cement-based composites laminates prepared with continuous fibers allow the full potential of fibers in reinforcing the matrix to be used because the manufacturing technique is fully controlled and the composite laminates can be designed for the specific service loads they may encounter. Some of these processes include high-energy mixing [55], extrusion, pultrusion [21], filament winding [23], compression molding, cross-ply and sandwich lamination techniques [24], and hybrid reinforcement [25]. These materials collectively offer a significant degree of strength, ductility, and versatility and can be easily be used as structural load bearing members. The effect of ply orientation and stacking sequence on the mechanical properties of these composites can be controlled using stochastic or deterministic models that combine the principles of classical laminate theory, fracture mechanics, and micromechanics of distributed damage.

EXTRUSION

Extrusion of cement composites has been only moderately investigated and described in the literature [26–30]. Contrary to that of colloidal materials like clays and other fine mineral particles, extrusion of granular materials is intrinsically a difficult process for several reasons. The size of the particles is relatively large, and quite often irregular particle shapes allow for direct frictional particle to particle contacts to occur. The large size of the voids results in filtration of the liquid through the solid skeleton, thereby increasing the local solid volume fraction. Both friction and filtration may lead to flow blocking through arch formation or jamming. Addition of water-soluble polymers dramatically improves the reliability of the process. Kneading is the essential step before extrusion, and the available information is even scarcer than for extrusion [31]. Lombois-Burger et al. [32] investigated the role of polymers on both kneading and extrusion. Both the polymer-to-cement ratio and the water-to-cement ratio varied using both adsorbing and nonadsorbing. The kneading and extrusion parameters (dispersion energy, torque plateau value, extrusion pressure, paste ageing) were determined using instrumented kneader and ram extruder.

An injection molding setup for manufacture short fiber reinforced cement composites is shown in Figures 3.3a and 3.3b [33]. The configuration is based on a 1HP inductive motor operating at 1750 RPM. The motor is connected to a 30:1 gear drive to reduce the RPM and increase the torque. The power is transmitted to a single screw auger that rotates inside a 76.2-mm-diameter (3-in.) steel tube. The spacing between the augur teeth decreases along the length of the augur; hence, a constant volume of the material is compressed as moves along the length of the augur. The die section

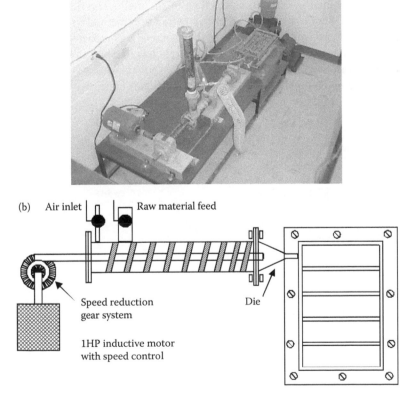

FIGURE 3.3 (a) The extrusion system for manufacturing cement composites containing short fibers. (b) Schematics of the extrusion- injection molding setup.

is a conical shape that reduces in volume, further pressurizing the slurry. This section is connected to the mold that is filled up with the material exiting the die. Air pressure is maintained at 40 psi at the inlet to help create the pressure difference between the internal and the external portion of the system. Vacuum is applied at the vent holes in the mold to reduce the hydrostatic back pressure developed because of compaction of the material in the die and the mold. The mold is vibrated during the filling up process. The system allows the manufacture FRC materials with several volume fractions of hybrid PP, ceramic, and carbon fibers. The properties of these composites are addressed in Chapter 12.

COMPRESSION MOLDING

A compression molding setup uses normal pressure to press the fresh paste into the desired shape. A prismatic shape used is shown in Figure 3.4 [12]. At the bottom of the mold, a fine mesh is used to allow for draining the excess water as the specimen is consolidated under the static pressure. After a known weight of the slurry is placed in the mold, pressure is applied from the top plunger using a hydraulic ram at the level of 1 MPa. The pressure is maintained for up to 2 h while the excess porosity and water in the paste is drained. The sample was removed from the mold after the final set, and the differential density before and after compression molding determines the degree of compaction.

SPIN CASTING

Spin casting, also known as centrifugal casting, is a method of using centrifugal force to produce castings from a mold. Typically, a disc-shaped mold is spun along its central axis at a set speed. The casting mortar is poured in through an opening of the mold. The filled mold then continues to spin as the paste sets. Spin casting is an effective fabrication method to produce concrete poles, masts, or pipes. Through the centrifugal process, the concrete is compacted and the desired shape, mostly round or ellipsoidal, is obtained. Because concrete has to be reinforced with steel bars that are susceptible to corrosion, textile-based systems offer an advantageous system [34]. Furthermore, the placement of the reinforcement is time consuming and hence expensive and leads to rather thick

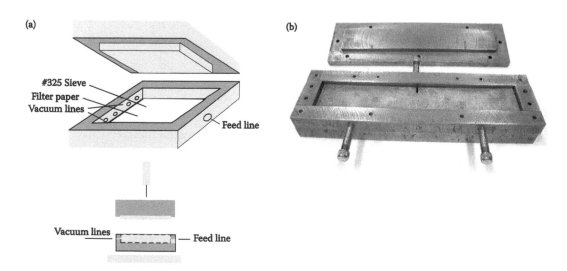

FIGURE 3.4 Schematic presentation of the compression molding setup and a picture of the mold before sample preparation. (Adapted from Haupt, G. J., "Study of cement based composites manufactured by extrusion, compression molding and filament winding," MS thesis, Arizona State University, 1997.)

and heavy structural elements. The application of short fiber reinforced cement pastes or mortar or textiles is a suitable alternative.

Generally, the casting materials used for competing processes like extrusion, injection molding, or spin casting are similar, but more control on the rheology is required for spin casting. For example, any formulation that sets too rapidly when cast with centrifugal force may result in incomplete filling of the mold as well as a rough surface finish.

Spin casting is a low-temperature process. Depending on the mix formulation, the product may become overly soft or hard. Slow setting and low viscosity in addition to nonuniform and unrestricted flow of the casting material may substantially affect the quality and finish of the final products. Using conventional manufacturing methods, no special curing or mixing techniques were necessary to obtain a very high strength of more than 30 MPa in bending.

Workable, high-strength, ductile, and durable fiber reinforced composites have been obtained in spin-cast elements [35]. The influence of different fibers (type and dimensions) and of different dosage of a latex dispersion on the workability and strength of the composites with respect to the spin casting process was characterized. The bending strength of carbon and PVA fiber reinforced composites of conventionally cast specimens and of spun-cast prototype pipes were reported in the range of 28 to 30 MPa with as much as 4.5 vol% PVA fibers. Fabrication of ductile, high-strength, and durable composites under real production conditions is possible.

MIXING HIGH-VOLUME FRACTION COMPOSITES

Fiber reinforcement increases the ductility severalfold. The use of a low fiber volume fraction moderately increases the first-crack strength, ultimate tensile, and flexural strengths in addition to better postpeak response. Fibers control the cracking process by preventing localization and generate a homogeneous microcracking state to dissipate energy over the entire volume [36].

Because of the high fiber content of the mixes, a conventional drum mixer would not be able to adequately disperse the fibers. Omni mixers have shown to address the agglomeration and fiber balling effects quite well [37]. This mixer uses a rubber casing to allow for flexibility along the walls. The flexible rubber tub with a wobble disk mounted at its center, which acts as the mixing blade, is shown in Figure 3.5. Rotation of the wobble plate contributes to the mixing action. The flexible sides, a shorter mixing time, and the absence of blades reduce fiber damage normally observed in conventional drum mixers. The motion of the casing as the mixer is rotating allows for breakup of the fiber clumps and results in a better distribution of fibers.

High-energy shear mixer was investigated for mixing of high-volume fraction of short fibers in a cement-based matrix. It was shown that fiber volume fractions as high as 8% PP-based fibers

(a)  (b)

FIGURE 3.5 Side view and the internal section of the Omni mixer.

and 16% chopped carbon fibers can be easily incorporated in the matrix. The mechanical response of the composites manufactured using this techniques is presented in the following publications [10, 38–41] and discussed in detail in Chapter 12.

COMPOSITES USING CONTINUOUS FIBERS AND TEXTILES

MESH REINFORCED CEMENTITIOUS SHEETS

Woven meshes have been used in the manufacture of thin sheet concrete products [42–44]. This is partly due to the length effect of the continuous fibers that allow for full stiffness utilization. In comparison, large volume fractions (>3%) of chopped fibers are required to effectively reinforce concrete products. A woven or bonded textile facilitates panel production in a continuous manufacturing line. Because the primary mode of loading is due to bending especially during installation, the fibrous mesh provides a continuous reinforcement in the tensile region. Placement of fiber reinforcement at the surface of the composites results in higher composite strengths. Composites may use woven or fibrillated mesh of both ductile and brittle continuous fibers. The use of short fibers in addition to surface reinforcement as a hybrid composite is also advantageous for both short-term handling and long-term performance. Fiber reinforced composites with significant postpeak toughness can be fabricated by using PP fiber contents of around 1% or higher.

Polymeric and glass meshes consisting of polymer-coated woven E glass, woven PP, or polyester are being used in the industry. The mesh is effective in stabilization of shrinkage and mechanical cracks that originate at the surface. The flexural and tensile results indicate higher first-crack strengths and better postpeak response.

The penetration of cement paste in between the opening of the fabrics is a controlling factor in the ability of the system to develop and maintain the bond. Such penetration is dependent on the size of the fabric opening and the viscosity (rheology) of the matrix. Also, the viscosity of the fresh mix is an important factor in the pultrusion process discussed next. If the mixture is too stiff, the fabric may be damaged as it goes through the cement bath; however, if the mixture is too soft, not enough matrix will adhere to the fabric as leaves the cement bath. The use of additives such as fly ash can influence the viscosity of the fresh cement mixture.

Figure 3.6 presents the interlocking mechanism offered by a knitted polymeric fiber within a textile. Note that the penetration of matrix is essential in the development of geometrical bond. Figure 3.6b represents the nature of a fibrillated PP fiber and indicates that the opening the fiber film results in anchorage development during mixing and casting. Figure 3.6c shows that at the microstructural level, the adhesional bond between the fiber surface and matrix is essentially quite low; hence, it is the geometric interlock that contributes to the majority of the bond adhesions. The geometric interlock combined with the continuous length of polymeric fibers in a textile system makes the textile reinforcement using polymeric fibers a viable option.

PULTRUSION

An efficient production method for fabric–cement composites is the pultrusion process. Assuring uniform production, pultrusion has been demonstrated to produce cement composites with continuous filaments (filament winding technique), exhibiting significantly improved performance. Cement composites containing 5% alkali-resistant (AR) unidirectional glass fibers produced by pultrusion achieved tensile strength of 50 MPa [45, 46]. The behavior of fabric–cement composites exposed to tensile loads is characterized by multiple cracking. The nature of multiple cracking and the resulting stress–strain curve, toughness, and strength are dependent on the properties of the reinforcing fabrics, the cement matrix, and the interface bond and the anchorage of the fabrics developed [47]. Microstructural features such as crack spacing, width, and density allow formulation of the damage evolution as a function of macroscopically applied strain [48, 49] and are discussed in subsequent chapters.

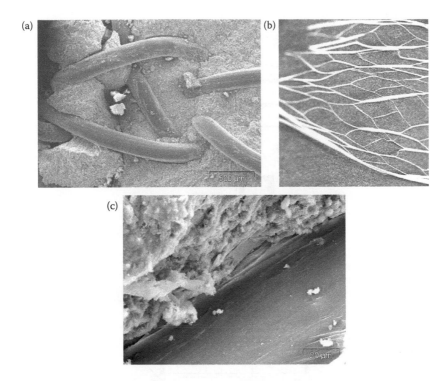

FIGURE 3.6 (a) Interlocking mechanism of a knitted polyethylene fiber. (b) Nature of a fibrillated PP fiber. (c) The limited adhesional bond at the fiber–matrix interface: surface and adhesions.

The penetration of cement paste in between the opening of the fabrics is a controlling factor in the ability of the system to develop and to maintain the bond. Such penetration depends on the size of the fabric opening and the rheology of the matrix because if the mixture is too stiff, the fabric may be damaged during production; however, if the mixture is too fluid, not enough matrix will adhere to the fabric. The use of additives such as fly ash can influence the viscosity of the fresh cement mixture.

In the pultrusion process, continuous reinforcements are passed through a slurry infiltration chamber and then pulled through a set of rollers to squeeze the paste in the openings of the fabric, to remove excessive paste, and to form composite laminates. Additional pressure is then applied on top of the laminates to improve penetration of the matrix in between the opening of the fabrics and the bundle filaments. The pultrusion setup is presented in Figures 3.7a and 3.7b showing the pulling of a slurry infiltrated textile through a set of rollers. The sample is collected on a rotating mandrel. Pictures of the preliminary setup of the equipment while preparing specimens are shown in Figure 3.8.

Using a pilot plant fabric–cement sheets with width of up to 75 cm and thickness of 1 to 3 cm can be produced. Composites with multiple layers of fabric can be manufactured, resulting in reinforcement content of up to 9.5% by volume of AR glass and or polyethylene (PE) fibers. Cross-ply laminates can also be manufactured by rotating the mandrel after each layer is placed. After forming the sample, pressure is needed to be applied on top of the composite laminates to improve penetration of the matrix in between the opening of the fabrics. A constant pressure of 15 kPa is typically applied. Most of this pressure is removed with in 1 h after casting.

Several parameters related to the processing effects have been addressed, which include the influence of pressure applied during the process either by the cylinders (shear stresses) or at the latter stage (normal or hydrostatic pressure). Effects of mix formulation, direction of casting, matrix formulation, and textile type have also been addressed. Fabrics used include bonded, knitted, and plain weave

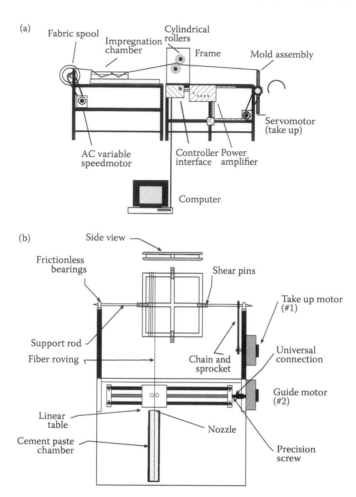

FIGURE 3.7 Schematics of the pultrusion using a filament winding setup. (After Mobasher, B., and Pivacek, A. (1998), "A filament winding technique for manufacturing cement based cross-ply laminates," *Cement and Concrete Composites*, 20, 405–415.)

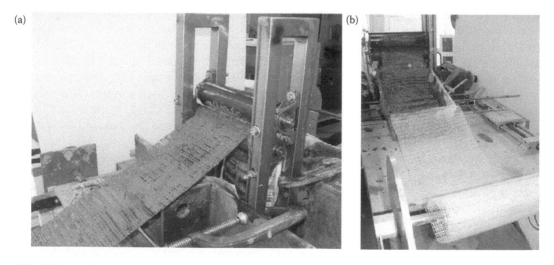

FIGURE 3.8 Manufacturing of fabric cement composites with the pultrusion process. (a) AR glass fabric. (b) Polyethylene fabric.

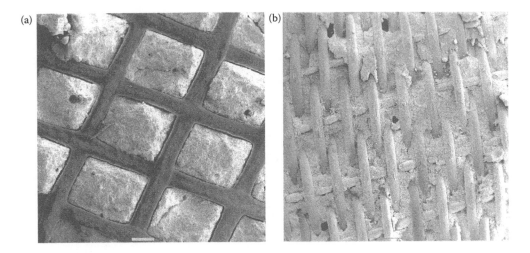

FIGURE 3.9   SEM micrograph of matrix embedment in two types of fabrics (scale bar = 500 μm).

woven fabrics. In bounded fabrics, a perpendicular set of yarns (warp and weft) are glued together at the junction points (see Figure 3.9a). In woven fabrics, the warp and the fill (weft) yarns pass over and under each other (see Figure 3.9b). Multifilament AR glass bonded fabrics use a yarn density of two yarns per centimeter in both directions of the fabric. The AR glass fibers were with tensile strength of 1270 to 2450 MPa, elasticity modulus of 78,000 MPa, and filament diameter of 13.5 μm. The woven fabrics were made from monofilament PE with 22 yarns per centimeter in the reinforcimg direction (warp yarns) and 5 yarns per centimeter in the perpendicuar direction (fill yarns). The PE fibers were with tensile strength of 260 MPa, 1760 MPa modulus of elasticity, and 0.25 mm diameter.

MATRIX PHASE MODIFICATIONS

RAPID SETTING

The necessity to increase the speed of board production may impose various modifications to the cement chemistry. Some of these approaches include addition of carbon dioxide to form calcium carbonate and help in setting the boards, that is, a forced carbonation of the hydrating portland cement microstructure in the green stage [50]. The durability of these systems is of concern although as a partial decomposition of monosulfate to form ettringite is possible during the long-term life of the products [51]. Ettringite is commonly used as a setting agent in the manufacturing of composites [52, 53]. The durability of these systems is also questionable. Addition of carbonates such as potassium, sodium, or ammonium carbonate as much as 7% to 15% by weight of cement can reduce the final set to approximately 2 to 5 min. This could however introduce deleterious alkalis into the microstructure, while the ammonia released by the ammonium carbonate present a health hazard.

FLY ASH

The use of pozzolans in blended cements is becoming a widely accepted. This is partly due to the cost saving potentials associated with the utilization of the industrial by-product waste materials and additional benefits of reactions that lower the free calcium hydroxide, leading to a lower alkalinity and higher ultimate mechanical strength. Potential benefits include reduced segregation, bleeding, heat of hydration, drying shrinkage, permeability, alkali–silica reaction, and sulfate attack potential, while increasing ultimate tensile and compressive strength [54]. Alternatively, the disadvantages of

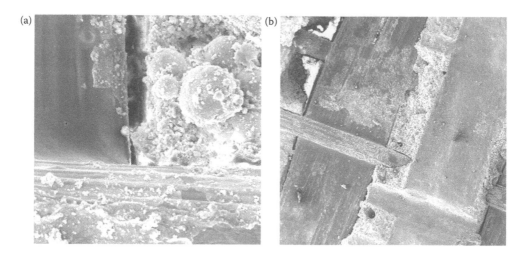

FIGURE 3.10 (a) Fly ash particles in the vicinity of the yarn junction. (b) The matrix penetration in between the two adjacent fabric layers. (Adapted from Mobasher, B., Peled, A., and Pahilajani, J., "Pultrusion of fabric reinforced high flyash blended cement composites," *Proceedings of the RILEM Technical Meeting, BEFIB*, 1473–1482, 2004.)

using pozzolanic admixtures such as fly ash may be observed in lowered early strength, lowered freeze–thaw durability, and increased air entraining mixture dosage requirement.

Pozzolanic reactions reduce the calcium hydroxide level to produce more calcium silicate hydrates (CSH) gel and thus enhance strength. The reactions are strongly influenced by the pH of the pore water [55, 56]. Cements with a lower pH such as blends with silica fume or granulated blast furnace slags reduce the reactivity of fly ash. Higher strength levels, although desirable, are accompanied by an increased brittleness [57]. Fiber reinforcement can increase the toughness of composite, whereas reduction in alkalinity due to the pozzolan could increase the performance level of fibers that would otherwise be susceptible to the alkaline attack [58].

The microstructure reveals much about the use of additives such as fly ash. Specimens observed under scanning electron microscope show that fly ash addition improves the microstructure of the fabric paste bond as shown in Figure 3.10a. By occupying the capillary voids and the transition zone in the vicinity of the inter yarn spacing, strong bonding in mixtures with high levels of fly ash obtained while due to lower viscosity and better filling of the matrix at the interfacial region as shown in Figure 3.10b.

CALCIUM HYDROXIDE REDUCTION

Many matrix modifications are aimed at reducing the amount of free calcium hydroxide produced during cement hydration either by using ingredients/additives such as ground granulated blast furnace slag, silica fume [60], or fly ash [61]. Alternatively, non–ordinary portland cement (OPC) matrices, in particular those based on calcium aluminates or sulfoaluminates [62] are used.

More matrix ingredients, such as metakaolin [63–65] and acrylic polymer [66–68], and cement types, such as sulfoaluminate cement [69–72], inorganic phosphate cement [73–76], and calcium aluminate cement [77, 78], were further investigated in terms of improving glass FRC (GFRC) long-term durability. Results of the comparison of flexural strength decay are normally very positive. Ambroise and Pera [70] concluded that GFRC using AR glass and calcium sulfoaluminate cements outperform those using OPC. Cuypers et al. [74, pp. xv, xvi] reported that inorganic phosphate cement GFRC property retention is better for both E and AR glass compared with AR glass OPC GFRC. Marikunte et al. [63] studied the hot-water durability of AR glass fiber reinforced composites in blended cement matrix and were rated for their flexural and tensile performance. The different matrices selected were (a) cement, (b) cement + 25% metakaolin, and (c) cement + 25% silica fume. Specimens after normal curing of 28 days were immersed in a hot water bath at 50°C for up to 84 days and then tested in flexure and tension. Results indicate that the blended cement consisting of metakaolin significantly improves the durability of GFRC composite. Beddows and Purnell [64] and Purnell and Beddows [65] used for the development of a durability model more than 1100 aged GFRC samples including OPC I and OPC II and GFRC with metakaolin, polymers, and sulfoaluminates (44% of the data available from literature, 25% from industry, 28% generated), and they concluded that AR glass composites with OPC + 5% polymer + a proprietary cement with sulfoaluminate-based additives (trade name Nashrin) were the most durable of the GFRCs tested. Additional discussion is provided in a recent ACI publication [79].

RHEOLOGY

The difference in the mechanical behavior of the samples due to the use of fly ash can be attributed to the rheology of the matrix. Tests were conducted on the fresh paste mixtures using a Brookfield rheometer. The shear rate was varied during the tests ranging from a higher rpm value (20) at the start to a lower value (10). Figure 3.11a shows a plot of viscosity with the shear rate. The results in the plot indicate that the viscosity of the control samples increases much faster as compared with the three different levels of fly ash, indicating that the presence of fly ash causes a slower rate of stiffness gain. The yield strength of the sample decreases with the addition of fly ash as well. This may explain the ability of high fly ash content mixtures to develop a better bond strength due to the ability of the fresh paste to infiltrate the fabric openings. Figure 3.11b represents the shear strength versus the shear rate. It is observed that the shear rate increases with time, but with the increase in the amount of fly ash, strength gain is slow.

HYBRID SHORT FIBER REINFORCEMENT

The strengthening and toughening mechanisms for cement-based composites should be viewed at several size scales. To strengthen the matrix, the specific fiber spacing must be sufficiently low to reduce the allowable flaw size. This may be achieved by using short whiskers of approximately

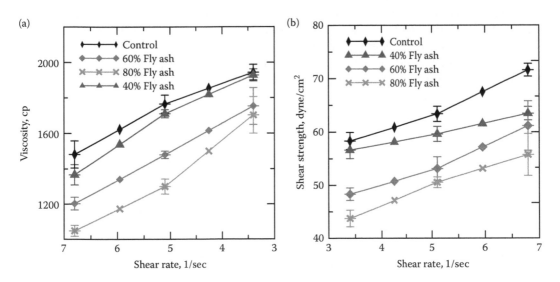

FIGURE 3.11 (a) Viscosity versus shear rate plots for various levels of fly ash. (b) Shear strength versus shear rate plots for various levels of fly ash in matrix. (After Mobasher, B., Peled, A., and Pahilajani, J., "Pultrusion of fabric reinforced high flyash blended cement composites," *Proceedings of the RILEM Technical Meeting, BEFIB*, 1473–1482, 2004.)

few microns in diameter. Some typical examples include carbon and alumina-based whiskers. Carbon whiskers have been used in FRC materials for more than 20 years [80, 81]. Alumina-based fibers in a cement-based matrix are a relatively recent development. These fibers, which average approximately 2.5 μm in diameter, provide microcrack bridging before a critical flaw size has reached [33].

Whiskers increase the ultimate strength by stabilizing the microcracks that occur before the peak load. However, mechanisms such as fiber debonding and pullout seem to be limited because the short length of fibers cannot provide significant bridging length across the crack faces.

HYBRID REINFORCEMENT: WOVEN MESH AND DISCRETE FIBERS

Various combinations of mesh and matrix reinforcement can be manufactured. The design objectives may be based on parameters such as strength, stiffness, toughness, and durability [82]. Perez-Pena and Mobasher [82] studied the response of short fibers (glass, PP, and nylon) used in lightweight aggregate core matrix with glass or PP meshes. Note that when a polymeric textile such as PP is used, matrix cracking load may not surpass the first cracking zone, although the toughness is significantly increased. By using approximately 1.0% by volume of AR glass fibers in a PP mesh or glass mesh composite, an increase of approximately 50% in first-crack flexural strength is observed. For the case of glass textiles, the increase in flexural strength was accompanied by a decrease in both the crack spacing and the maximum deflection. Note that addition of AR fibers to the lightweight aggregate core matrix increases the composite stiffness.

CONCLUSIONS

A study on the use of various fiber composite systems was presented. Various polymeric and synthetic fibers can be used to alter the properties of the composites for specific functions. PP mesh was shown to be an effective reinforcing means of thin concrete products. The detrimental effects of aging in the alkaline cement environment are thus avoided. Glass mesh and AR glass fiber composites showed sufficient load carrying capacity.

The use of high-energy mixing in the manufacture of cement-based composites with high volume fraction of short fibers was demonstrated. Microstructural analysis of the fracture surfaces indicated a uniform fiber distribution. The use of alumina-based fibers in cementitious composites was also investigated. Alumina and carbon fibers are shown to stabilize the microcracks and increase the strength of FRC composites. The toughening effect after peak load is not significant because of the short length of the fibers. Polypropylene fibers, when added to the hybrid composites, increase the toughness significantly.

REFERENCES

1. Mobasher, B., Ouyang, C. S., and Shah, S. P. (1991), "Modeling of fiber toughening in cementitious composites using an R-curve approach," *International Journal of Fracture*, 50, 199–219.
2. Romauldi, J. P., and Batson, G. B. (1963, June), "Mechanics of crack arrest in concrete," *Journal of the Engineering Mechanics Division*, 89(3), 147–168.
3. Aveston, J., Cooper, G. A., and Kelly, A. (1971). "The Properties Fibre Composites," *Conference Proceedings of the National Physical Laboratory*, IPC Science and Technology Press Ltd., Guildford, 15–26.
4. Hannant, D. J., Hughes, D. C., Kelly, A., Alford, N. McN., and Bailey, J. E. (1983, September 24), "Toughening of cement and other brittle solids with fibres [and Discussion]," *Philosophical Transactions of the Royal Society of London. Series A, Mathematical and Physical Sciences*, 310(1511), 175–190.
5. Li., V. C., and Wu, H. C. (1992), "Conditions for pseudo strain hardening in fiber reinforced brittle matrix composites," *Applied Mechanics Reviews*, 45(8), 390–398.
6. Visalvanich, K., and Naaman, A. E. (1983), "A fracture model for fiber reinforced concrete," *Journal of the American Concrete Institute*, 80(2), 128–138.

7. Tjiptobroto, P., and Hansen, W. (1993), "Tensile strain hardening and multiple cracking in high performance cement based composites," *ACI Materials Journal*, 90(1), 16–25

8. Yang, C. C., Mura, T., and Shah, S. P. (1991), "Micromechanical theory and uniaxial tests of fiber reinforced cement composites," *Journal of Materials Research*, 6(11), 2463–2473.

9. Li, S. H., Shah, S. P., Li, Z., and Mura, T. "Micromechanical analysis of multiple fracture and evaluation of debonding behavior for fiber reinforced composites," *International Journal of Solids and Structures*, 30(11), 1429–1459.

10. Mobasher, B., and Li, C. Y. (1996, May–June), "Mechanical properties of hybrid cement based composites," *ACI Materials Journal*, 93(3), 284–293.

11. Pivacek A., and Mobasher, B. (1997), "A filament winding technique for manufacturing cement based cross-ply laminates," Innovations Forum, *ASCE Journal of Materials Engineering*, 9(2), 55–58.

12. Haupt, G. J. (1997), "Study of cement based composites manufactured by extrusion, compression molding and filament winding," MS thesis, Arizona State University.

13. Krenchel, H., and Stang, H. (1994), "Stable microcracking in cementitious materials," in Brandt, A. M., Li, V. C., and Marshall, I. H. (eds.), *Brittle Matrix Composites 1*. Cambridge: Woodhead Publishing Limited, 20–33.

14. Mobasher, B., Stang, H., and Shah, S. P. (1990), "Microcracking in fiber reinforced concrete," *Cement and Concrete Research*, 20, No. 5, 665–676.

15. Stang, H., Mobasher, B., and Shah, S. P. (1990), "Quantitative damage characterization in polypropylene fiber reinforced concrete," *Cement and Concrete Research*, 20(4), 540–558.

16. Roy, D. M. (1987), "New strong cement materials: chemically bonded ceramics," *Science*, 235, 651–658.

17. Birchall, J. D., Majid, K. I., Staynes, B. W., Rahman, A. A., Dave, N. J., Taylor, H. F. W., Tamas, F., Majumdar, A. J., Champion, A., Roy, D. M., and Bensted, J. (1983), "Cement in the context of new materials for an energy-expensive future," *Philosophical Transactions of the Royal Society of London. Series A, Mathematical and Physical Sciences*, 310(139), 31–42.

18. Rodrigues F. A., and Joekesa, I. (1998). "Macro-defect free cements: a new approach," *Cement and Concrete Research*, 28(6), 877–885.

19. Birchall, J. D., Howard, A. J., and Kendall, K. (1981), "Flexural strength and porosity of cements," *Nature*, 289, 388.

20. Kendall, K., Howard, A. J., Birchall, J. D., Pratt, P. L., Proctor, B. A., and Jefferis, S. A. (1983), "The relation between porosity, microstructure and strength, and the approach to advanced cement-based materials," *Philosophical Transactions of the Royal Society of London. Series A, Mathematical and Physical Sciences*, 310(1511), 139–153.

21. Mallick, P. K. (1988), *Fiber-Reinforced Composites: Materials, Manufacturing and Design*. New York: Marcel Dekker Inc.

22. Hasegawa, M., Kobayashi, T., and Dinilprem Pushpalal, G. K. (1995, August), "A new class of high strength, water and heat resistant polymer—cement composite solidified by an essentially anhydrous phenol resin precursor," *Cement and Concrete Research*, 25(6), 1191–1198.

23. Mobasher, B., Pivacek A., and Haupt, G. J. (1997), "Cement based cross-ply laminates," *Journal of Advanced Cement Based Materials*, 6, 144–152.

24. Mobasher, B., and Pivacek, A. (1998), "A filament winding technique for manufacturing cement based cross-ply laminates," *Cement and Concrete Composites*, 20, 405–415.

25. Perez-Pena, M., Mobasher, B., and Alfrejd, M. A. "Influence of Pozzolans on the Tensile Behavior of Reinforced Lightweight Concrete," *Materials Research Society, Symposium "O," Innovations in the Development and Characterization of Materials for Infrastructure*, December 1991, Boston, MA.

26. Srinivasan, R., Deford, D., and Shah, S. P. (1999), "The use of extrusion rheometry in the development of extruded fiber-reinforced cement composites," *Concrete Science and Engineering*, 1, 26–36.

27. Shao, Y., Qiu, J., and Shah, S. P. (2001), "Microstructure of extruded cement-bonded fibreboard," *Cement and Concrete Research*, 31, 1153–1161.

28. Mu, B., Li, Z., Chui, S. N. C., and Peng, J. (1999), "Cementitious composite manufactured by extrusion technique," *Cement and Concrete Research*, 29, 237–240.

29. Peled, A., and Shah, S. P. (2003), "Processing effects in cementitious composites: extrusion and casting," *Journal of Materials in Civil Engineering*, 15, 192–199.

30. Kendall, K. (1987), "Interparticle friction in slurries," in Briscoe, J. and Adams J. (eds.), *Tribology in Particulate Technology*. Bristol: AdamHilger, 91–102.

31. Yang, M., and Jennings, H. M. (1995), "Influence of mixing methods on the microstructure and rheological behavior of cement paste," *Advanced Cement Based Materilas*, 2, 70–78.

32. Lombois-Burger, H., Colombet, P., Halary, J. L., and Van Damme, H. (2006), "Kneading and extrusion of dense polymer–cement pastes," *Cement and Concrete Research*, 36, 2086–2097.

33. Mobasher, B., and Li, C. Y., "Processing Techniques for Manufacturing High Volume Fraction Cement Based Composites," *Proceedings of the First International Conference for Composites in Infrastructure*, ICCI '96, ed. H. Saadatmanesh and M. R. Ehsani, 1996, 123–136.

34. Dilger W. H., Ghali A., and Rao S. V. K. M. (1996), "Improving the durability and performance of spun-cast concrete poles," *PCI Journal*, 41(4), 68–90.

35. Kaufmann, J., and Hesselbarth, D. (2007, November), "High performance composites in spun-cast elements," *Cement and Concrete Composites*, 29(10), 713–722.

36. Mobasher, B., Castro-Montero, A., and Shah, S. P. (1990), "A Study of Fracture in Fiber Reinforced Cement-Based Composites Using Laser Holographic Interferometry," *Experimental Mechanics*, 30(90), 286–294.

37. Garlinghouse, L. H., and Garlinghouse, R. E. (1972, April), "The Omni mixer—A new approach to mixing concrete," *ACI Journal*, 69(4), 220–223.

38. Mobasher, B., and Li, C. Y. "Fracture of Whisker Reinforced Cement Based Composites," Proceedings of the International Symposium on Brittle Matrix Composites 4, ed. A. M. Brandt, V. C. Li., and I. H., Marshall, Warsaw, September 2003, 116–124.

39. Mobasher, B. "Micromechanics of Fracture in Fiber Reinforced Cement Based Composites," Department of Construction Engineering, Technical University of Catalunia, Barcelona, Spain, May 1994.

40. Mobasher, B., and Li, C. Y. "Tensile Fracture of Carbon Whisker Reinforced Cement Based Composites," *ASCE Materials Engineering Conference, Infrastructure: New Materials and Methods for Repair*, ed. K. Basham, November 13–16, 1994, 551–554.

41. Haupt, G. J. (1997), "Study of cement based composites manufactured by extrusion, compression molding and filament winding," MS thesis, Arizona State University.

42. Balaguru, P. N., and Shah, S. P. (1992), *Fiber-Reinforced Cement Composites*. New York: McGraw-Hill, 365–412.

43. Odler, I., "Fiber-reinforced Cementitious Materials," *Materials Research Society Symposium Proceedings*, ed. S. Mindess and J. Skalny, Pittsburgh, PA, 1991, Vol. 211, 265–273.

44. Hannant, D. J., Zonsveld, J. J., and Hughes, D. C. (1978, April), "Polypropylene film in cement based materials," *Composites*, 9, 83–88.

45. Mobasher, B., Pivacek, A., and Haupt, G. J. (1997), "Cement based cross-ply laminates," *Journal of Advanced Cement Based Materials*, 6, 144–152.

46. Peled, A., and Mobasher, B. (2004), "Pultruded fabric–cement composites," *ACI Materials Journal*, 102(1), 15–23.

47. Peled, A., Sueki, S., and Mobasher, B. (2006), "Bonding in fabric–cement systems: Effects of fabrication methods," *Cement and Concrete Research*, 36(9), 1661–1671.

48. Mobasher, B., Peled, A., and Pahilajani, J. (2006), "Distributed cracking and stiffness degradation in fabric–cement composites," *Materials and Structures*, 39(3), 17–331.

49. Peled, A., and Mobasher, B. (2006), "Properties of fabric–cement composites made by pultrusion," *Materials and Structures*, 39(8), 787–797.

50. Lahtinen, P. K., *Proceedings of the 2nd International Inorganic Bonded Wood and Fiber Composite Materials*, Moscow, ID, 1990, 32–34.

51. Kuzel, H. J., and Strohbauch, E. (1989), "Reactions associated with the action of CO_2 on heat treated hardened cement pastes," *ZKG*, 42(8), 413–418.

52. Ramachandran V. S. (1995). *Concrete admixtures handbook: properties, science, and technology*. Technology & Engineering Noyes Publications, 1153.

53. Aoki, Yukio, Proceedings of the 2nd International Inorganic Bonded Wood and Fiber Composite Materials, Moscow, ID, 1990, 35–44.

54. Halstead, W. J., "Quality Control of Highway Concrete Containing Fly Ash," Virginia Highway and Transportation Research Council, Report VHTRG 81-R38. Charlottesville, VA, February 1981.

55. Fraay, L. A., Bijen, J. M., and de Haan, Y. M. (1989), "The reaction of fly ash in concrete: A critical examination," *Cement and Concrete Research*, 19, 235–146.

56. McCarthy, G. J., and Lauf, R. J. (1985), "Fly ash and coal conversion by-products: Characterization, utilization and disposal," *Materials Research Society Symposia Proceedings*, 43, 53–60.

57. Gettu R., Bazant, Z. P., and Karr, M. E. (1990, November–December), "Fracture properties and brittleness of high-strength concrete," *ACI Materials Journal*, 87(6), 608–618.

58. Kohno, K., and Horri, K. (1985), "Use of by-products for glass fiber reinforced concrete," *Transactions of the Japan Concrete Institute*, 7, 25–32.

59. Mobasher, B., Peled, A., and Pahilajani, J., "Pultrusion of fabric reinforced high flyash blended cement composites," *Proceedings of the RILEM Technical Meeting, BEFIB*, 2004, 1473–1482.

60. Kumar, A., and Roy, D. M., "Microstructure of glass fiber/cement paste interface in admixture blended Portland cement samples," *Proceedings of the Durability of Glass Fiber Reinforced Concrete Symposium organized by PCI*, November 12–15, 1985, ed. S. Diamond, Chicago, 1986, 147–156.

61. Leonard, S., and Bentur, A. (1984), "Improvement of the durability of glass fiber reinforced cement using blended cement matrix," *Cement and Concrete Research*, 14, 717–728.

62. Litherland, K. L., and Proctor, B. A., "The effect of matrix formulation, fibre content and fibre composition on the durability of glassfibre reinforced cement," *Proceedings of the Durability of Glass Fiber Reinforced Concrete Symposium organized by PCI*, ed. S. Diamond, Chicago, November 12–15, 1985, 124–135.

63. Marikunte, S., Aldea, C., and Shah, S. (1997), "Durability of glass fiber reinforced cement composites: effect o silica fume and metakaolin," *Journal of Advanced Cement Based Composites*, 5, 100–108.

64. Beddows, J., and Purnell, P., "Durability of new matrix glassfibre reinforced concrete," GRC 2003 *Proceedings of 12th Congress of the GRCA*, ed. J. N. Clarke and R. Ferry, Barcelona, Spain, October 2003, paper 16.

65. Purnell, P., and Beddows, J. (2005, October–November), "Durability and simulated ageing of new matrix glass fibre reinforced concrete," *Cement and Concrete Composites*, 27(9–10), 875–884.

66. Bijen, J., "A survey of new developments in glass composition, coatings and matrices to extend service life of GFRC," *Proceedings of the Durability of Glass Fiber Reinforced Concrete Symposium organized by PCI*, ed. S. Diamond, Chicago, November 12–15, 1985, 251–269.

67. Bijen, J. (1983, July–August), "Durability of some glass fiber reinforced cement composites," *ACI Materials Journal*, 305–311.

68. Ball, H., "Durability of naturally aged, GFRC mixes containing Forton polymer and SEM analysis of the facture interface," *GRC 2003 Proceedings of 12th Congress of the GRCA*, ed. J. N. Clarke and R. Ferry, Barcelona, Spain, October 2003.

69. Qi, C., and Tianyou, B., "A review of the development of the GRC industry in China, GRC 2003," *Proceedings of the 12th Congress of the GRCA*, ed. J. N. Clarke and R. Ferry, Barcelona, Spain, October 2003, paper 37.

70. Ambroise, J., and Péra, J., "Durability of glass-fibre cement composites: Comparison between normal, portland and calcium sulphoaluminate cements," *Proceedings Composites in Construction 2005—Third International Conference*, Lyon, July 11–13, 2005, 1197–1204.

71. Litherland, K. L., Oakley, D. R., and Proctor, B. A. (1981), The use of accelerated aging procedures to predict the long term strength of GRC composites, *Cement and Concrete Research*, 11, 455–466.

72. Gartshore, G. C., Kempster, E., and Tallentire, A. G., "A new high durability cement for GRC products," *The Glassfibre Reinforced Cement Association 6th Biennial Congress, Maastricht, Netherlands Proceedings*, October 21–24, 1991, 3–12.

73. Cuypers, H., Wastiels, J., Orlowsky, J., and Raupach, M., "Measurement of the durability of glass fibre reinforced concrete and influence of matrix alkalinity," *Proceedings of Brittle Matrix Composites 7*, ed. A. M. Brandt, V. V. Li, and I. H. Marshall, Warsaw, October 13–15, 2003, 163–172.

74. Cuypers, H., Van Itterbeeck, P., De Bolster, E., and Wastiels, J., "Durability of cementitious composites," *Composites in Construction 2005—Proceedings of the Third International Conference, Lyon, France*, July 11–13, 2005, 1205–1212.

75. Cuypers, H., Gu, J., Croes, K, Dumortier, S., and Wastiels, J., "Evaluation of fatigue and durability properties of E-glass fibre reinforced phosphate cementitious composite," Proceedings of Brittle Matrix Composites 6, ed. A. M. Brandt, V. C. Li, and I. H. Marshall, Warshaw, October 9–11, 2000, 127–136.

76. Cuypers, H., Wastiels, J., Van Itterbeeck, P., De Bolster, E., Orlowsky, J., and Raupach, M. (2006), "Durability of glass fibre reinforced composites experimental methods and results," *Composites. Part A*, 37, 207–215.

77. Brameshuber, W., and Brockmann, T., "Calcium aluminate cement as binder for textile reinforced concrete," International Conference on Calcium Aluminate Cements (CAC), Edinburgh, Scotland, July 16–19, 2001, 659–666.

78. Brameshuber, W., and Brockmann, T., "Development and optimization of cementitious matrices for textile reinforced concrete, GRC 2001," *Proceedings of the 12th International Congress of the International Glassfibre Reinforced Concrete Association*, ed. N. Clarke and R. Ferry, Dublin, Ireland, May 14–16, 2001, 237–249.

79. American Concrete Institute (2010). *ACI 544.5R-10 Report on the Physical Properties and Durability of Fiber-reinforced Concrete*. Farmington Hills, MI: American Concrete Institute.

80. Nishioka, K., Yamakawa, S., and Shirakawa, K., "Properties and application of carbon fiber reinforced cement based composites," *Proceedings of the Third International Symposium on Developments in Fiber Reinforced Cement and Concrete*, RLEM Symposium FRC 86, 1. RILEM Technical Committee 49-TFR, July 1986.
81. Ali, M. A., Majumdar, A. J., and Rayment, D. L. (1972), "Carbon-fibre reinforcement of cement," *Cement and Concrete Research*, 2, 201–212.
82. Perez-Pena, M., and Mobasher, B. (1994), "Mechanical properties of hybrid fiber reinforced cementitious composites," *Cement and Concrete Research*, 24(6), 1121–1132.

4 Textile Reinforcement in Composite Materials

INTRODUCTION

There are many alternatives to the classification of material systems based on their physical, chemical, microstructural, and mechanical properties. For example, one can characterize them based on their ingredients, processing, and applications areas. A majority of man-made materials can be divided into four classes: metals, polymers, ceramics, and composites. Fiber and textile reinforced composites are part of the general class of engineering materials called composites. A rigorous definition of composite materials is difficult to achieve because the first three categories of homogeneous materials are sometimes heterogeneous at submicron dimensions (e.g., precipitates in metals). Composites are characterized by being multiphase materials within which the phase distribution and geometry have been deliberately tailored to optimize one or more properties [1]. This is clearly an appropriate definition for fiber, whisker, and platelet reinforced composites for which there is one phase, called the matrix, reinforced by a fibrous reinforcement. This combination however results in orthotropic properties described in the following sections.

Unidirectional laminated composites exhibit excellent in-plane but poor interlaminar properties. This is due to the lack of reinforcements in the thickness direction and leads to poor damage tolerance in the presence of interlaminar stresses. In order to reduce the degree of orthotropy, textiles such as plain weave fabrics are used as reinforcements in composites to obtain balanced ply properties and improved interlaminar properties. In the textile reinforced composites, the fibers used are first conformed into the shape of a textile. Despite the reduced stiffness and strength in the in-plane directions, fabrics are advantageous because of improved cost of manufacturing. It is therefore important to study the mechanical behavior of such composites to fully realize their potential.

In principle, there are many combinations of fiber and matrix available for textile reinforced composites. A large range of materials choices are available that can be used with a range of manufacturing techniques. One can analyze and design materials at the same time that the manufacturing is considered. This flexibility is advantageous in comparison to other classes of engineering materials, where the material is produced first and then machined and formed into the desired shape. The full range of possibilities for composite materials is also very large. Reinforcements may consist of arrange of fibers such as S-glass, R-glass, a wide range of carbon, boron, ceramic (e.g., alumina, silicon carbide), polymeric, natural, and aramid fibers. The reinforcement can come in the form of long (continuous) or short fibers, disks or plates, spheres, or ellipsoids. Matrices include a wide range of polymers (epoxides, polyesters, nylons, etc.), metals (aluminum alloys, magnesium alloys, titanium, etc.), cements, and ceramics (SiC, glass ceramics, etc.).

Processing methods include hand layup, autoclave, resin transfer molding, injection molding for polymer matrices, squeeze casting and powder metallurgy routes for metals, chemical vapor infiltration, and prepregging routes for ceramics. Those interested in a general introduction to composite materials should consult one of a number of texts including: Matthews and Rawlings [2], Hull and Clyne [3], and Ogin [4]. (a good introduction to the fabrication of polymer matrix composites is provided by Bader et al. [5]).

The market for composite materials can be loosely divided into two categories: "reinforced plastics" based on short fiber E-glass reinforced unsaturated polyester resins, which account for more than 95% of the volume, and "advanced composites," which make use of high-performance fibers (carbon, boron, aramid, SiC, etc.), matrices (e.g., high-temperature polymer matrices, metallic or ceramic matrices), or advanced design or processing techniques [1]. Even within these loosely defined categories, it is clear that textile composites are "advanced composites" by virtue of the manufacturing techniques required to produce the textile reinforcement. This chapter will be mostly concerned with the development and terminologies used for textile reinforced composites consisting of both ductile matrix and brittle matrix systems [6, 7].

TERMINOLOGY AND CLASSIFICATIONS SYSTEMS

High-performance fibers made of AR glass, carbon, polyethylene, polypropylene, polyvinyl alcohol, and aramid (trade name Kevlar) can be made up as filament or twisted yarns. For reinforcing purposes, filament yarns are the better choice because they offer strength, forming, and handling efficiency. The following list defines the terminology used throughout this text in addressing the various technical aspects of characterization and modeling of cement composite systems [8, 9].

Fiber and Fabric Terminology

- *Filament*—a single long and continuous fiber with a predominantly circular shape.
- *Yarns (or filament yarns)*—a bundle of elementary fibers or filaments. Yarns consist of several hundreds up to thousands of single filaments. The fineness of the yarn, indicated as tex (gram per 1000 m), depends on the number of filaments, the average filament diameter, and the fiber density.
- *Tex*—a unit of fiber fineness defined by the weight in grams of 1000 m of yarn; the lower the number, the finer the yarn (weight per unit length, i.e., 1 g/1000 m = 1 tex).
- *Denier*—another term for fiber fineness such that 1 denier = 1 g/9000 m.
- *Roving*—process where filaments are spun into larger diameter threads. These threads are then commonly used for woven reinforcing glass fabrics and mats and in spray applications.
- *Fiber fabric*—web-form fabric reinforcing material that has both warp and weft directions.
- *Fiber mats*—web-form nonwoven mats of glass fibers. Mats are manufactured in cut dimensions with chopped fibers, or in continuous mats using continuous fibers.
- *Chopped fiber glass*—processes where lengths of glass threads are cut between 3 and 26 mm, threads are then used in plastics most commonly intended for molding processes.
- *Glass fiber short strands*—short 0.2–0.3 mm strands of glass fibers that are used to reinforce thermoplastics most commonly for injection molding.
- *Woven fabric*—manufactured by intertwining two rectangular crossing thread systems, defined as warp and weft (or fill). The kind and crossing of warp and weft is called the "weaving pattern." The weaving pattern influences the fabric properties and design. One of the variables in the process is how often a thread is crossing threads of the other system on a certain length. This length is called floating.
- *Woven*—a cloth formed by weaving. It only stretches in the bias directions (at an angle between the warp and weft directions). There are three basic weaving patterns—plain, twill, and atlas weave.

Composites

- *Anisotropic*—material properties that are different in all directions.
- *Orthotropic*—material properties that are different in three mutually perpendicular directions.
- *Isotropic*—material properties that are the same in all directions. Almost all engineering alloys such as aluminum and steel in the annealed condition are isotropic.

- *Preferred orientation*—a material that has some degree of anisotropy is said to have preferred orientation. Highly wrought alloys, including aluminum and steel, will have different properties in the direction of elongation. Drawn wire and extrusions are well-known examples.
- *Random orientation*—a material that is said to have random orientation possesses some degree of isotropy. Engineering alloys are made up of an assembly of crystals, each of which may be orthotropic, but on a macroscale exhibit apparent isotropy.
- *Homogeneous*—properties are the same from point to point in the material. This effect is very scale dependent. Even a material that consists of two or more phases or constituents can be considered homogenous if the representative volume element (RVE) is sufficiently large. Most engineering alloys are considered homogeneous by this standard. Cement composites that have macrosized fibers and coarse aggregates would require a larger sample size to appear homogeneous.
- *Heterogeneous*—properties are different from point to point in a material. On a fine enough scale, almost cement composites are heterogeneous.
- *Lamina*—a single layer containing reinforcements in the plane of the layer. This is the usual building block for design and fabrication of composite structures.
- *Laminate*—a laminate is a stack of lamina (usually with alternating or varying principal directions). The laminas are arranged in a particular way to achieve some desired effect.
- *Micromechanics*—predicts composite behavior from the interaction of constituents on the appropriate scale. In whisker composites, the appropriate scale is a fraction of a micrometer. For reinforced concrete, the appropriate scale is several centimeters.
- *Macromechanics*—predicts composite behavior by presuming homogeneous material. The effects of constituents are detected only as averaged apparent properties.

Composites are classified by various criteria based on their physical, mechanical, and long-term properties. Matrix type is often used to describe composites. The most common designations based on ductile or brittle nature matrix type are polymer matrix composites, metal matrix composites, and Ceramic Matrix Composites (CMC). Table 4.2 lists some of the mechanical characteristics

TABLE 4.1
Stiffness and Strength of a Range of Fiber Materials Used in General Composites Area

Material	Specific Gravity	Modulus psi ×10^6	Strength psi ×10^3
AISI 1010 steel	7.87	30.0	53
AISI 4340 steel	7.87	30.0	75
6061-T6 aluminum alloy	2.70	10.1	45
7178-T6 aluminum alloy	2.70	10.1	88
17-7 PH stainless steel	7.87	28.5	235
Inco 718 nickel alloy	8.20	30.0	203
HS carbon fiber-epoxy (unidirectional)	1.55	20.0	225
HM carbon fiber-epoxy (unidirectional)	1.63	31.2	180
E-glass fiber-epoxy (unidirectional)	1.85	5.7	140
Kevlar 49 fiber-epoxy (unidirectional)	1.38	11.0	200
Boron fiber-6061 Al alloy (unidirectional)	2.35	32.0	161
Carbon fiber-epoxy (quasi-isotropic)	1.55	6.6	84
HM carbon fiber-epoxy (unidirectional)	1.63	31.2	180

of these three types of composites. Cement composites fall under the general area of CMCs, also referred to as brittle matrix composites. Alternatively, composites are classified by the shape or continuity of the reinforcement as shown in Figure 4.1, in which they are described as continuous or discontinuous relative to their continuity and as particulate or fiber relative to the shape of the reinforcement. Illustrations of these types of composites and some of their characteristic properties as determined by the failure modes are presented in the following chapters.

AR GLASS FIBERS

AR glass filament yarns were designed especially for their high alkalinity resistance in the reinforcement of portland-cement-based materials. AR glass contains more than 15% by mass of zirconia. The basic materials, including silica sand, clay, and limestone, are melted at temperatures up to 1350°C and pulled off the spinning nozzle with a speed between 25 and 150 m/s and diameters ranging from 9 to 27 μm. After spinning a coating material defined as sizing (0.5–1.5 mass% of the fiber) consisting of organic polymers dispersed in water is applied on the filaments. Approximately 400 to 6600 of filaments are combined to form a yarn. The sizing is important because it protects and improves the yarn properties as well as its adhesion with the matrix material. Table 4.2 presents a variety of yarns produced by two main manufacturers of AR glass fibers.

TABLE 4.2
Selected Commercially Available Alkali-Resistant Glass Fibers

Product Type	Producer	Tex	Diameter (μm)	No. Filaments
AR310S-800/DB	Nippon Electric Glass, Japan	310	13,5	800
AR620S-800/TM		620	13,5	1600
AR1100S-800/TM		1100	16	2000
AR2500S-800/DB		2500	24	2000
LTR ARC 320 5325	Saint-Gobain Vetrotex, Spain	320	14	800
LTR ARC 640 5325		640	14	1600
LTR ARC 1200 5325		1200	19	1600
LTR ARC 2400 5325		2400	27	1600

Source: Brameshuber, W., "State of the art report of RILEM Technical Committee," TC 201-TRC: Textile Reinforced Concrete—State-of-the-Art Report of RILEM TC 201-TRC, p. 292, 2006. With permission.

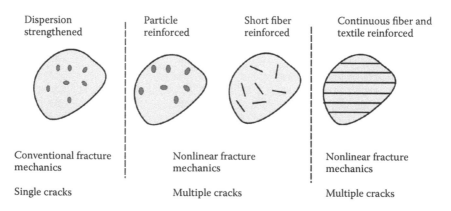

FIGURE 4.1 Composites classified by reinforcement shape of continuity.

KEVLAR

Kevlar fabrics are popular wraps that are commonly used in the fan housing of an aircraft engine or bulletproof vests. Kevlar, which is golden brown in color, is a synthetic fiber made by EI du Pont de Nemours & Co., Inc. [11]. It is a polymer formed by combining a large number of smaller molecules, called monomers, in a regular pattern. It is synthesized from the monomers 1,4-phenyl-diamine (para-phenylenediamine) and terephthaloyl chloride. The result is a polymeric aromatic amide (aramid) with alternating benzene rings and amide groups as shown in Figure 4.2. When they are produced, these polymer strands are aligned randomly. To make Kevlar, these strands are dissolved and spun, causing the polymer chains to orient in the direction of the fiber. Its chemical name and formula is poly-paraphenylene terephthalamide and $C_{14}H_{10}N_2O_2$, respectively. Kevlar is approximately five times stronger than steel on an equal weight basis yet is lightweight, flexible, and comfortable. Table 4.3 shows the properties of a plain-woven Kevlar fabric.

CARBON FILAMENTS AND YARNS

The carbon atoms in a fiber are bonded together in microscopic crystals that are more or less aligned parallel to the long axis of the fiber. The crystal alignment makes the fiber very strong for its size. Basic materials are polyacrylnitrile and mesophase pitch. They can be made spinnable by

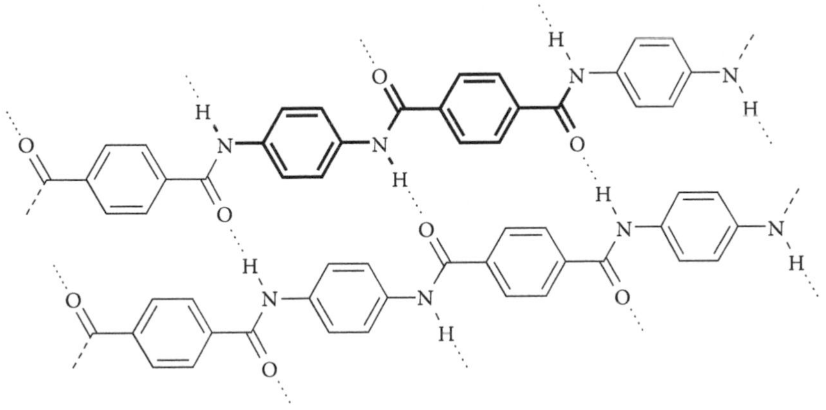

FIGURE 4.2 Chemical structure of Kevlar. (From Registered Trademark, E. I. du Pont de Nemours & Co., Inc. retrieved from http://www.dupont.com/kevlar, 2003. With permission.)

TABLE 4.3
Properties of a Plain Woven Kevlar Fabric

Material	Ply Count	Bulk density (lb/in³)	Linear density (lb/in)	c/s area per ply (in²)	Specimen Size (in)
Kevlar AS-49	17 × 17	0.00530516	9.457(10⁻⁷)	1.78(10⁻⁴)	2.5 × 12

Source: Rajan, S. D., B. Mobasher, J. Sharda, V. Yanna, C. Deenadaylu, D. Lau, and D. Shah, "Explicit finite element analysis modeling of multi-layer composite fabric for gas turbine engines containment systems: Part 1, Static tests and modeling," Final report, DOT/FAA/AR-04/40,P1 Office of Aviation Research, Washington, DC, November 2004, 115 pp.; Rajan, S. D., B. Mobasher, S. Sankaran, D. Naik, and Z. Stahlecker, "Explicit finite element modeling of multilayer composite fabric for gas turbine engine containment systems: Phase II, Part 1, Fabric material tests and modeling," DOT/FAA/AR-08/37, P1, Office of Aviation Research, Washington, DC, February 2009.

polymerization and thermal treatment. The anisotropic PAN or pitch fiber from the spinning process becomes unmeltable in an oxidation procedure at temperatures of 200°C to 300°C. Fibers of high strength (HT fibers) are created from PAN at 1500°C to 1700°C and fibers of high modulus of elasticity at 2200°C to 3000°C by graphitization incandescence. The strength of HT fibers is between 3000 and 5000 MPa, the modulus of elasticity between 200 and 250 GPa. HM fibers range from 2000 to 4500 MPa in tensile strength and from 350 to 450 GPa in modulus. Tests on commercial carbon fiber yarns have shown yarn strengths of more than 2000 MPa, depending on the fineness.

TEXTILE REINFORCED COMPOSITES

Textile reinforced composites have been in service in engineering applications for many years in low-profile, relatively low-cost applications. Although there has been a continual interest in textile reinforcement since the early 1970s, there has been a recent surge in our understanding of the mechanics and use of textile reinforcement. Although there is the possibility of a range of new applications for which textile reinforcement may replace current technologies, textile reinforcement is in competition with relatively mature composite technologies, such as fiber reinforced concrete, that uses the more traditional methods of blending with ordinary concrete. Textile reinforced concrete materials can potentially reduce manufacturing costs and enhance processing, with a clear and confirmed improvement in mechanical properties. Several textile techniques are likely to be combined for some applications. For example, a combination of braiding and knitting can be used to produce an I-shaped structure [14].

The properties that are important for structural applications include stiffness, strength, strain capacity, and resistance to damage/crack growth. The range of textiles under development for composite reinforcement is indicated in the schematic diagram shown in Figure 4.3 from Ramakrishna [15]. The following sections provide an introduction to textile reinforced composite materials using woven, braided, knitted, or stitched textiles. For more information, refer to the relevant papers listed. However, we must appreciate the complexity of the mechanical properties of textile reinforced composites as compared with the more traditional continuous fiber reinforcement of laminated composites.

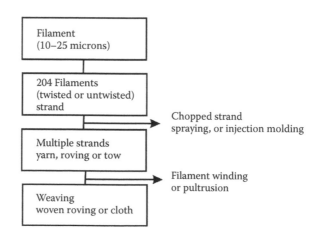

FIGURE 4.3   Use of filaments in composites. (From Ramakrishna, S., "Characterization and modeling of the tensile properties of plain weft-knit fabric reinforced composites," *Composites Science and Technology*, 57, 1–22, 1997. With permission.)

TEXTILE FIBERS

Textile fibers can be woven, braided, or knitted into precursor forms before infiltration by a matrix. Generally, the continuous fibers are monofilament or multifilament, with a range of diameters. Fibers can be polymers, carbons, or ceramics, examples of which are given in the following list:

- Polymers—polyethylene, Kevlar®, nylon, polypropylene, poly vinyl alcohol
- Carbons—graphite, carbon
- Ceramics—glass, alumina
- Natural—sisal fibers
- Metallic—steel fibers

TEXTILE FORMS

The fiber precursor is drawn, extruded, or spun as a group of individual fibers. These groups are referred to as *filaments*, *strands*, *ends*, or *tows*, depending on the industry. Glass fiber groups are called "strands," whereas carbon fiber groups are called "tows." For example, 3k tows means 3000 carbon fibers per tow. Strands or tows are collected into larger groups called *rovings*, which can be woven into heavy coarse fabrics or chopped and pressed to produce a mat or felt. To produce a fine textile, the strands or tows can be given a slight twist (less than one turn per inch). Twisted strands are called *yarns*. Twisting single-strand yarns together produces heavy yarns. The order of manufacturing shown in Figures 4.4 and 4.5 represent the multi-scale nature of transition from filaments, strands, yarns, and to fabrics. Filaments and yarns are the first elements of manufacturing composites. A typical ventricular yarn cross section is shown in Figure 4.5.

MONOFILAMENTS (25–200 μM, CONTINUOUS)

Monofilament fibers are produced as a single strand by chemical vapor deposition or single-strand drawings and are generally more expensive than textile fibers. Typical monofilaments are boron on tungsten, boron on carbon, silicon carbide on tungsten, Silicon carbide on Carbon, and alumina.

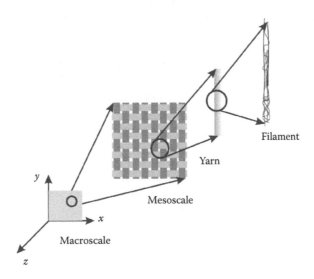

FIGURE 4.4 Discretization in textile composites representing scales of textile, yarn, and filament. (From Roye, A., and Th. Gries, "Design by application—Maßgeschneiderte Abstandskettengewirke für den Einsatz als Betonbewehrungen," *Kettenwirk-Praxis* 39, H. 4, S. 20–21, 2004. With permission.)

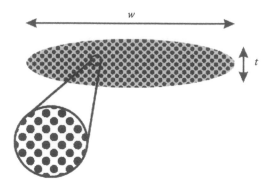

FIGURE 4.5 Lenticular yarn cross section.

WHISKERS (<1 μM, DISCONTINUOUS)

Whiskers are very fine, nearly perfect single crystals that have near theoretical strength and stiffness. They are generally produced by vapor deposition and are commonly used in the dispersion strengthening of composites as shown in Figure 4.1. Common metal whiskers include Fe, Cu, and Ni. Common ceramic whiskers include silicon carbide, alumina, and boron carbide. They are used for metal matrices and CMCs.

TEXTILE TERMINOLOGY

A number of properties make woven fabrics more attractive than their unidirectional fiber and non-woven alternatives. They have very good drapability, allowing complex shapes to be formed with no gaps. Manufacturing costs are reduced, because a single biaxial fabric replaces two nonwoven plies and the ease of handling lends itself more readily to automation. Woven-fabric composites also show an increased resistance to impact damage compared with nonwoven composites, with significant improvements in compressive strengths after impact. These advantages are gained, however, at the expense of lower stiffness and strength than equivalent nonwoven composites.

A variety of fabrics can be woven from yarns. Yarns that lie in lengthwise or machine direction are called *warp* yarns. The yarns that lie in the crosswise direction are called *fill* yarns. There are five commonly used weave patterns used in composite construction:

- *Plain weave*—one warp yarn passes over and under one filling yarn. This pattern has maximum stability with respect to slippage and fabric distortion.
- *Basket weave*—two or more warp yarns interlock over two or more filling yarns. This pattern has greater pliability than plain weave.
- *Twill weave*—one or more warp yarns over and under two or more filling yarns. This produces a very drapeable fabric.
- *Crowfoot satin weave*—one warp yarn interlocking over three and under one filling yarn. This pattern can conform to very complex contours.
- *Eight-harness satin weave*—one warp yarn interlock over seven and under one filling yarn. This pattern produces the highest strength composite.

A fabric is a collection of fiber tows or filaments arranged in a given pattern. Both fibers and matrix are responsible for bearing the mechanical loads [16]. Fabrics are classified as woven, nonwoven, knitted, or braided [17]. Furthermore, they can also be classified into two-dimensional reinforcement and three-dimensional reinforcement fabrics. The geometry of the woven composite is complex, with unlimited architectures such that even the simplest plain weave structures are quite complicated to characterize and represent by geometrical models. Plain weave fabrics are formed

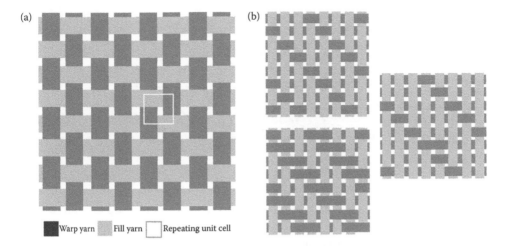

FIGURE 4.6 (a) Definition of warp and fill yarns. (b) Basic weave construction (a) plain, 3 × 1 twill, 2 × 2 twill, and 5H satin weave.

by interlacing or weaving two sets of orthogonal tows. As shown in Figure 4.6a, the longitudinal direction tows are known as the warp, whereas the tows in the transverse direction are known as the fill tows or weft. The interlacing causes bending in the tows, called *tow crimp*. Some examples of fabrics are plain weave, satin weave, weft knitted, warp knitted, and orthogonal fabrics. The stiffness and strength of fabric reinforced composites are controlled by the fabric architecture and material properties of the fiber and matrix. The fabric architecture depends upon the undulation, crimp, and density of the fiber tows, which are untwisted strands of fibers. The undulation or waviness of the tows causes crimps (bending) in the tows, which significantly reduces the stiffness and strength of the composite.

Woven fabrics are characterized by the interlacing of two or more yarn systems and are among the most widely used textile reinforcements in a variety of applications. Woven reinforcement exhibits good stability in the warp and weft directions and highest cover or yarn packing density in relation to fabric thickness [18]. Figure 4.6b represents several basic weave construction, including plain weave, 3 × 1 twill, 2 × 2 twill, and 5H satin weave. The weave patterns are determined by the number of warp and fill yarns that are crossed over during the weaving process. Two- or three-dimensional textile performs may be woven using biaxial or triaxial weaving. Braiding in two-dimensional systems is flat braiding or circular braiding, whereas knitting in two-dimensional textiles is accomplished by warp knitting or weft knitting (see Figure 4.7). Nonwoven fabrics are manufactured by mechanical or chemical processes. There are combinations of knitting and weaving for two- and three-dimensional textiles, with Figure 4.8 showing a three dimensional knitted fabric [9, 19].

SCRIMS

Scrims are defined as textile fabrics produced by superposing thread systems with or without fixing the crossing points. If the reinforcing threads are exactly positioned and drawn in the load direction, the highest stiffness and tenacity can be reached. Two thread systems form a biaxial scrim, three or more systems a multiaxial scrim.

Multiaxial scrims offer a broad variety of properties: drawn threads, any combination of angles between the layers, any design of layers, and a free choice of the weight per unit area. The structure of a multiaxial scrim consists of several layers of reinforcing thread systems with different orientations and a knitting thread structure—the warp knit. The knitting thread bonds the thread layers

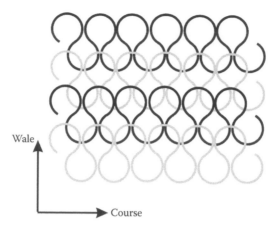

FIGURE 4.7   Schematic diagram of a plain weft-knit fabric.

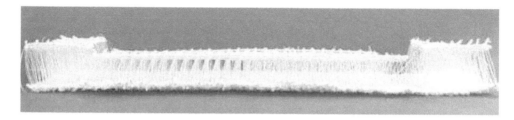

FIGURE 4.8   Three-dimensional knitted fabric for variable thickness composites. (From Roye, A., and Gries, Th., "Design by application—Maßgeschneiderte Abstandskettengewirke für den Einsatz als Betonbewehrungen," *Kettenwirk-Praxis*, 39, H. 4, S. 20–21, 2004. With permission.)

and forms a warp knitted scrim. By varying the reinforcing thread systems and the mesh structure (fringe, tricot, etc.), the character of the scrim can be adapted to its final shape. Figure 4.9 shows a lay-up of a multiaxial warp knit. It consists of up to eight layers which can be adjusted in almost any direction (e.g., 0°, 90°, +45°, –45°). The 0° direction is the machine direction, the so-called "warp direction." Threads laid in other angles are called weft thread systems [20, 21].

Only a few textile fabrication processes are adaptable for cement composites. The most important criterion is the possibility to create open structures for proper impregnation. A good permeability and complete envelope with the concrete is provided by an open grid structure. To ensure satisfactory handling, there must be no displacement of the threads. Certain manufacturing methods (e.g., warp knitting with insertion of reinforcing threads) are therefore better suited than others (e.g., braiding) [6].

STITCH-BONDED FABRICS

The stitch-bonding technique is an established production process used to manufacture textiles for the reinforcement of concrete. Stitch-bonded fabrics or multi-plies are multiple layers of yarn sheets connected by a stitching yarn as shown in Figure 4.10. A stitch-bonded multiaxial fabric is composed of multiple layers of differently aligned reinforcing yarn sheets and a threaded mesh structure. Some of the obvious advantages of this method are in the use of noncrimped yarns and combination of several layers at specified angles (multiaxial). Stitch-bonding fabrics also allows for user-defined arrangements and combinations in each layer. These fabrics contain up to eight layers with each layer's reinforcing yarns being laid at various angles (i.e., 0°, 90°, +45°, –45°) to one another. These layers can be user defined, as mentioned earlier [21, 22].

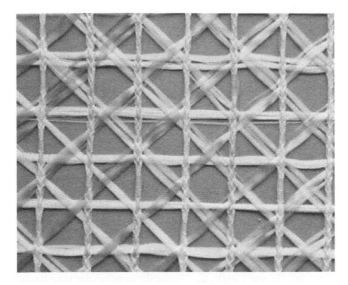

FIGURE 4.9 Multiaxial lay-up of a warp-knit scrim. (From Roye, A., and Gries, Th., "Design by application—Maßgeschneiderte Abstandskettengewirke für den Einsatz als Betonbewehrungen," *Kettenwirk-Praxis*, 39, H. 4, S. 20–21, 2004. With permission.)

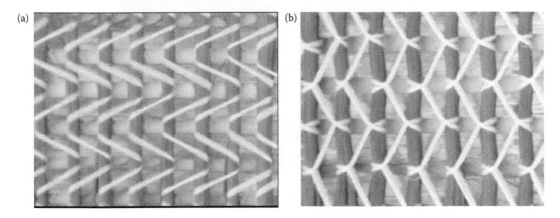

FIGURE 4.10 Symmetrically stitch-bonded fabric made of carbon filament threads—(a) front and (b) back view.

LENO WEAVE TECHNIQUE

In a leno weave, the warp and the weft are not held together by interweaving, but rather by a separate binding thread system. The binder threads wrap around the warp and weft threads a half rotation each time. After each weft thread change, their rotational direction crosses to the opposite side. This is shown in Figure 4.11. The binder thread can run simultaneously with the warp or can be fed separately [23]. Figure 4.12 shows a multiaxial weaving machine.

ANALYSIS OF WOVEN TEXTILE COMPOSITES

The application of the finite element method to textile composites requires the division of the composite into a number of unit cells interconnected at nodal points. The discretization of a composite at the fiber level, however, results in extremely large mathematical models for woven

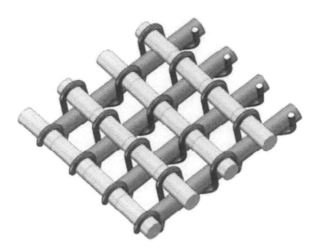

FIGURE 4.11 Model of Leno Weave with stretched warp and weft thread and binder threads. (From Roye, A., Gries, T., Engler, T., Franzke, G., and Cherif, C., "Possibilities of textile manufacturing for load-adapted concrete reinforcement," in Dubey, A. (ed.), *Textile-Reinforced Concrete*, Proceedings of ACI Fall Convention, Kansas City, November 2005, SP-250CD-3, CD-ROM, 2008. With permission.)

FIGURE 4.12 Coating process on the multiaxial knitting machine. (From Roye, A., Gries, T., Engler, T., Franzke, G., and Cherif, C., "Possibilities of textile manufacturing for load-adapted concrete reinforcement," in Dubey, A. (ed.), *Textile-Reinforced Concrete*, Proceedings of ACI Fall Convention. Kansas City, November 2005, SP-250CD-3, CD-ROM, 2008. With permission.)

reinforcements. The individual phases within the geometry result in a number of interfaces, sliding surfaces, and nonlinear contact formulations, particularly when orthogonal adjacent layers have a great degree of lateral slip and rotation during loading. These mechanisms are computationally expensive for direct inclusion in finite element models. That is perhaps the reason why so many elastically equivalent material models have been developed to homogenize the material properties into an orthotropic unit cell material model.

The majority of closed-form analyses of woven fabric composites have been based on laminated plate theory [24, 25], (as will be discussed in the following chapters) with special emphasis being placed on characterization of the RVE. Numerical methods rely on the finite element method as shown by Chou [26] who presented models to evaluate the thermomechanical properties of woven fabric composites. The mosaic model treats the woven composite as an assemblage of asymmetric cross-ply laminates, ignoring the fiber continuity and undulation. The fiber undulation model for plain and twill weave composites takes these complexities into account by considering a slice of the crimped region and by averaging the properties with the aid of the laminate plate theory. Woo and Whitcomb [27] were also among the first in this area who developed one-dimensional models that were extended to two dimensions by Naik and Shembekar [28].

COMPOSITE MODULI IN TEXTILE REINFORCEMENTS

One of the simplest laminate configurations for continuous unidirectional fiber reinforced composites is the cross-ply laminate, for example, $(0/90)_s$, which is 0/90/90/0. For such a laminate, the Young's moduli parallel to the 0° and 90° directions, E_x and E_y, are equal and, to a good approximation, are just the average of E_1 and E_2. Yang and Chou [29] and Chou and Ito [30] have schematically shown the change in these moduli, E_x and E_y, for a carbon fiber reinforced epoxy laminate with a range of fiber architectures but the same fiber volume fraction of 60% (see Figure 4.13). This diagram provides a good discussion point by showing the lower degree of orthotropic response in textile reinforced composites as compared to cross ply laminates. The cross-ply composite has E_x and E_y moduli of approximately 75 GPa. In the biaxial weaves of the eight-harness satin and the plain weave, the moduli both fall to approximately 58 and 50 GPa, respectively. These reductions reflect the crimps in the interlaced woven structure, with more crimps per unit length in the plain weave producing a smaller modulus.

The triaxial fabric, with three sets of yarns interlaced at 60° angles, behaves similarly to a $(0/\pm60)_s$ angle-ply laminate. Such a configuration is quasi-isotropic for in-plane loading; that is, it has the same Young's modulus for any direction in the plane of the laminate. The triaxial fabric shows a further reduction in E_x and E_y to approximately 42 GPa, but this fabric benefits from a higher in-plane shear modulus than the biaxial fabrics. The range of properties for a multiaxial warp-knit fabric (or multilayer multidirectional warp-knit fabric) reinforced composite is also shown, lying between the triaxial fabric and above the cross-ply laminate (at least for the modulus E_x), depending on the precise geometry. In this case, warp, weft, and bias yarns (usually ±45) are held together by

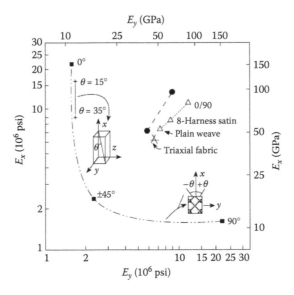

FIGURE 4.13 Predicted E_x and E_y moduli for a range of reinforcement architectures; $\pm\theta$ angle ply (for $\theta = 0$ to ±45 to 90), cross-ply (0/90), eight-harness satin and plain woven, triaxial woven fabric, braided ($\theta = 35°$ to 15°), and multiaxial warp knit (•--•) for the same fiber volume fraction of 60%. (From Horrocks, A. R., and Anand, S. C., *Handbook of Technical Textiles*, Textile Institute, CRC Press, 2000; Yang, and Chou, T. W., *Proceedings of ICCM6/ECCM2*, ed. F. L. Matthews et al., 5.579–5.588, 1987. With permission.)

"through-the-thickness" chain or tricot stitching. Finally, a three-dimensional braided composite is shown, with braiding angles in the range 15° to 35°. This type of fiber architecture gives very aniso-tropic elastic properties as shown by the very high E_x moduli (which are fiber dominated) and the low E_y moduli which are matrix dominated. In the following sections, the properties of these textile reinforcements (woven, braided, knitted, stitched) will be discussed in more detail.

MODELING OF TEXTILE COMPOSITES AT THE REPRESENTATIVE VOLUME LEVEL

Several approaches are used in computing the effective properties of composites. These include the method of cells (MOC), variations of the MOC, finite element modeling with virtual testing, and classical lamination theory (CLT). With each method, only a representative unit of material is con-sidered because of the repetitive pattern in the composite material. The terms "representative unit cell (RUC)," "unit cell," or "RVE" will be used interchangeably. An example of a repetitive unit cell for a plain weave fabric [31] is shown in Figure 4.14. Symmetric conditions are used to improve the computational efficiency—one quarter of the unit cell model is shown in the same figure. Analytical methods, including CLT and MOC, have been successful in determining effective material proper-ties. Some of the earliest CLT models have been used to determine the elastic modulus of woven fabric composites [5, 32–34]. One of the more recent CLT models referred to as Mesotex [35] is general enough to capture the three-dimensional elastic properties and the ultimate failure strengths of several types of fabric composites and is very computationally efficient. Models using the MOCs have shown a favorable correlation with experimental results [36, 37].

Most analytical models of the elastic behavior of plain weave fabrics are based on CLT and numeri-cal methods. Huang [38] developed a micromechanics bridging model to predict the elastic proper-ties of woven fabric composites. The geometric models of the fabrics (RVE) were well described in his studies. The tow cross section was assumed to be elliptical, and a tow undulation was described by a sinusoidal function in their model.

The discretization procedure applied by Huang and coworkers to the RVE of the fabric composite divided it into a number of subelements [22]. Each subelement consists of the tow segments and the pure matrix. The tow segments were considered as unidirectional composites in their material coor-dinate system. The elastic response (compliance) of the tow segments and the matrix were assem-bled to achieve the effective stiffness of the subelement using classical laminate theory (isostrain condition). The overall elastic property of the RVE was calculated by assembling the compliance matrix of the subelements under isostress assumption [39].

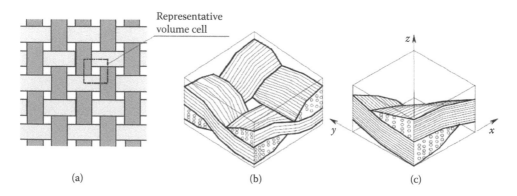

FIGURE 4.14 (a) Weave architecture (b) RVE (c) quarter of the RVE. (From Tabiei, A., and Yi, W., "Comparative study of predictive methods for woven fabric composite elastic properties," *Journal of Composite Structures*, 58, 149–164, 2002. With permission.)

There are two main categories: analytical models and numerical models [10, 40–45]. Chou and Ito [30] developed one-dimensional analytical models of the plain weave laminated composites in order to determine their mechanical properties. The undulation or wave nature of the fill tow was not considered, whereas the undulation of warp tow was assumed to be sinusoidal. Two types of cross section were assumed for the fill tows: sinusoidal and elliptical. The isostrain condition was used in evaluating the stiffness of the plain weave laminates.

Ishikawa and Chou [46–48] developed three models to predict the elastic properties of woven fabric laminates, each of which has its own set of limitations. The mosaic model [27] was used to predict the stiffness of satin weave fabric composites. The model neglects the tow crimp and idealizes the composite as an assemblage of asymmetric cross-ply laminates. Then, an isostress or an isostrain condition was used to predict the stiffness of the laminate depending on whether the laminates are assembled in series or parallel. Because the model neglects the tow crimp, the prediction of stiffness is not accurate. The fiber undulation model [27] or the one-dimensional model considers fiber undulation in the longitudinal direction but neglects it in the transverse direction. The bridging model [28], a combination of mosaic and fiber undulation model, was developed for satin weave fabrics. The model reduces to the crimp model [27] for plain weave fabrics and hence the stiffness prediction is not accurate.

One of the computationally efficient approaches used by Tabiei and Yi is referred to as a four-cell model in which the quarter cell RVE is divided into four subcells as shown in Figure 4.15 [49]. An example of a finite element mesh of a woven fabric unit cell [50] is shown in Figure 4.16. Finite elements are more computationally expensive than closed-form methods. However, they do provide detailed stress-strain distributions.

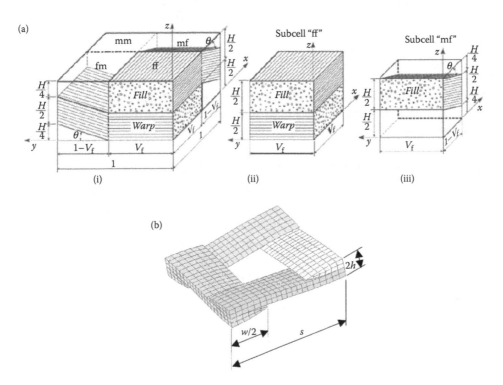

FIGURE 4.15 (a) Division of quarter RVE into subcells using the four-cell method. (From Tabiei, A., and Yi, W., "Comparative study of predictive methods for woven fabric composite elastic properties," *Journal of Composite Structures*, 58, 149–164, 2002. With permission.) (b) Example of FE mesh of a plain woven fabric unit cell. (From Tabiei, A., and Yi, W., "Comparative study of predictive methods for woven fabric composite elastic properties," *Journal of Composite Structures*, 58, 149–164, 2002. With permission.)

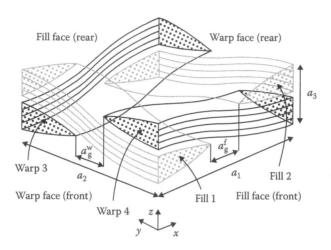

FIGURE 4.16 Schematic representation of the fabric geometry. (From Peng, X. Q., and Cao, J., "Numerical determination of mechanical elastic constants of textile composites," *15th Annual Technical Conference of the American Society for Composites*, College Station, TX, 2000. With permission.)

Naik and Ganesh [51] developed two-dimensional micromechanical models of plain weave fabrics to determine the elastic properties of the fabrics, taking the warp and weft tow undulation into consideration. The elastic constants of the warp and fill tows were calculated within each element (considering the undulation angle), and then the stiffness of the element was calculated using the classical laminate theory. The compliance of the slices was calculated from the element stiffness matrix using isostress conditions.

Other models for the prediction of the elastic models for two-dimensional "hybrid" woven fabrics generally represented as Figure 4.16, including those of Vandeurzen et al. [52, 53], Hahn and Tsai [54], Hahn and Pandey [55], Averill et al. [56], Blackletter et al. [57], and others [58], who developed a three-dimensional finite element model of a plain weave fabric. The models work well in predicting the elastic modulus, but the prediction of in-plane shear modulus is not accurate.

MECHANICAL STRENGTH AND DAMAGE ACCUMULATION

The mechanical properties of woven fabric reinforced composites are dominated by the type of fiber used, the weaving parameters, and the stacking and orientation of the various layers [59]. Glass reinforced woven fabrics yield composites with lower mechanical properties because of the much lower value of the glass fiber modulus (as compared with carbon). The potential advantages of woven fabrics over equivalent nonwoven carbon/epoxy laminate are that the modulus of the biaxial (0/90) woven laminate was slightly lower than the nonwoven cross-ply laminate (50 GPa compared with 60 GPa, respectively). The compressive strength after the impact event was increased by more than 30%. Bishop and Curtis found that tensile strength was 23% lower than in UD equivalent laminates [60].

Damage under tensile loading in woven composites is characterized by the development of matrix cracking in the off-axis tows at strains above 0.3% to 0.4%. Most investigations of damage have considered biaxial fabrics loaded in the warp direction. Cracks initiate in the weft bundles, and an increasing density of cracks develops with increasing load (or strain).

The accumulation of cracks is accompanied by a gradual decrease in the Young's modulus of the composite. In woven systems, the matrix cracking can lead to considerable delamination in the region of the crimps in adjacent tows, which further reduces the mechanical properties [61]. Damage modeling has been attempted using finite element methods or closed-form models [25].

It is clear that in the future, additional models will continue to be developed to better address many of the complicated mechanics and kinematic aspects of the textile materials.

REFERENCES

1. Bader, M. G. (1997). *An Introduction to Composite Materials*. Short course notes. Surrey, UK: University of Surrey.
2. Matthews, F. L., and Rawlings R. (1994). *Composite Materials: Engineering and Science*. London: Chapman and Hall.
3. Hull, D., and Clyne, T. W. (1996). *An Introduction to Composite Materials*. Cambridge: Cambridge University Press.
4. Ogin, S. L. (2000). "Textile reinforced composite materials," in Horrocks, A. R., and Anand, S. (eds.), *Handbook of Technical Textiles*. Boca Raton, FL: CRC Press.
5. Bader, M. G., Smith, W., Isham, A. B., Rolston, J. A., and Metzner, A. B. (1990). *Delaware Composites Design Encyclopedia: Volume 3. Processing and Fabrication Technology*. Lancaster, PA: Technomic Publishing.
6. Kuo, W.-S., and Chou, T.-W. (1995), "Elastic response and effect of transverse cracking in woven fabric brittle matrix composites," *Journal of the American Ceramic Society*, 78(3), 783–792.
7. Pryce, A. W., and Smith P. A. (1992), "Behaviour of unidirectional and crossply ceramic matrix composites under quasi-static tensile loading," *Journal of Materials Science*, 27, 2695–2704.
8. Hausding, J., Lieboldt, M., Franzke, G., Helbig, U., and Cherif, C. (2006), "Compression-loaded multi-layer composite tubes," *Composite Structures*, 7(1–2), 47–51.
9. Hausding, J., Engler, T., Kleicke, R., and Cherif, C., "High Productivity and Near-Net Shape Manufacture of Textile Reinforcements for Concrete," Proceedings of the GRC 2008, Prague, 2008, 113–122.
10. Brameshuber, W., "State of the art report of RILEM Technical Committee," TC 201-TRC: Textile Reinforced Concrete—State-of-the-Art Report of RILEM TC 201-TRC, 2006, 292.
11. Registered Trademark, E.I. du Pont de Nemours & Co., Inc.
12. Rajan, S. D., Mobasher, B., Sharda, J., Yanna, V., Deenadaylu, C., Lau, D., and Shah, D., "Explicit finite element analysis modeling of multi-layer composite fabric for gas turbine engines containment systems: Part 1. Static tests and modeling," Final report, DOT/FAA/AR-04/40,P1 Office of Aviation Research, Washington, DC, November 2004, 115.
13. Rajan, S. D., Mobasher, B., Sankaran, S., Naik, D., and Stahlecker, Z., "Explicit finite element modeling of multilayer composite fabric for gas turbine engine containment systems: Phase II, Part 1. Fabric material tests and modeling," DOT/FAA/AR-08/37, P1, Office of Aviation Research, Washington, DC, February 2009.
14. Nakai, A., Masui M., and Hamada, H., "Fabrication of large-scale braided composite with I-shaped structure," *Proceedings of the 11th International Conference on Composite Materials (ICCM-11)*, Gold Coast, Queensland, Australia, Australian Composites Structures Society and Woodhead Publishing, 1997, 3830–3837.
15. Ramakrishna, S. (1997), "Characterization and modeling of the tensile properties of plain weft-knit fabric reinforced composites," *Composites Science and Technology*, 57, 1–22.
16. Barbero E. J. (1999). *Introduction to Composite Materials Design*. Philadelphia: Taylor & Francis.
17. Pandey R. (1995). "Micromechanics based computer-aided design and analysis of two-dimensional and three-dimensional fabric composites," Dissertation, Pennsylvania State University, PA.
18. Scardino, F. (1989). "An introduction to textile structures and their behaviour," in Chou, T. W., and Ko, F. K. (Eds.), *Textile Structural Composites, Chapter 1, Composite Materials Series*, Vol. 3. Oxford: Elsevier.
19. Badawi, S. S. (2007). "Development of the weaving machine and 3D woven spacer fabric structures for lightweight composites materials," PhD thesis, Technischen Universität Dresden, 186.
20. Roye, A., and Gries, Th. (2004), "Vom Haustraum zum Traumhaus," *Kettenwirk-Praxis*, 38, H. 3, S. 26–27.
21. Roye, A., and Gries, Th. (2004), "Design by application—Maßgeschneiderte Abstandskettengewirke für den Einsatz als Betonbewehrungen," *Kettenwirk-Praxis*, 39, H. 4, S. 20–21.
22. Hausding, J., and Cherif, C. (2010), "Improvements in the warp knitting process and new patterning techniques for stitch-bonded textiles," *Journal of the Textile Institute*, 101.
23. Roye, A., Gries, T., Engler, T., Franzke, G., and Cherif, C. "Possibilities of textile manufacturing for load-adapted concrete reinforcement," *Textile-Reinforced Concrete. Proceedings of ACI Fall Convention*, ed. A. Dubey, Kansas City, November 2005, SP-250CD-3, CD-ROM 2008.

24. Agarwal, B. D., and Broutman, L. J. (1980). *Analysis and Performance of Fiber Composites*. New York: John Wiley and Sons.
25. Jones, R. M. (1975). *Mechanics of Composite Materials*. Washington DC: Scripta (McGraw-Hill).
26. Chou, T. W. (1992). *Microstructural Design of Fiber Composites*. Cambridge: Cambridge University Press.
27. Woo K., and Whitcomb, J. (1994), "Global/local finite element analysis for textile composites," *Journal of Composite Materials*, 28, 1305–1321.
28. Naik, N. K., and Shembekar, P. S. (1992), "Elastic behaviour of woven fabric composites: I. Lamina analysis," *Journal of Composite Materials*, 26, 2196–2225.
29. Yang, J. M., and Chou, T. W., Proceedings of ICCM6/ECCM2, ed. F. L. Matthews et al., 1987, 5.579–5.588.
30. Chou, T. W., and Ito, M. (1998), "An analytical and experimental study of strength and failure behavior of plain weave composites," *Journal of Composite Materials*, 32, 2–30.
31. Tabiei, A., and Yi, W. (2002), "Comparative study of predictive methods for woven fabric composite elastic properties," *Composite Structures*, 58, 149–164.
32. Ishikawa, T., and Chou, T. W. (1982), "Stiffness and strength behavior of woven fabric composites," *Journal of Composite Materials*, 17, 3211–3220.
33. Ishikawa, T., and Chou, T. W. (1982), "Elastic behavior of woven hybrid composites," *Journal of Composite Materials*, 16, 2–19.
34. Ishikawa, T., and Chou, T. W. (1983), "One-dimensional micromechanical analysis of woven fabric composites," *AIAA Journal*, 21(12), 1714–1721.
35. Scida, D. (1999), "A micromechanics model for 3D elasticity and failure of woven-fibre composite materials," *Composite Science and Technology*, 59, 505–517.
36. Jiang, Y., Tabiei, A., and Simitses, G. J. (2000), "A novel micromechanics-based approach to the derivation of constitutive equations for local/global analysis of plain-weave fabric composite," *Composite Science and Technology*, 60, 1825–1833.
37. Tanov, R., and Tabiei, A. (2001), "Computationally efficient micro-mechanical woven fabric constitutive," *Journal of Applied Mechanics*, 68(4), 553–560.
38. Huang, Z. M. (1999), "The mechanical properties of composites reinforced with woven and braided fabrics," *Composites Science and Technology*, 479–498, Vol. 60.
39. Barbero, E. J., Trovillion, J., Mayugo, J. A., and Sikkil, K. K. (May 2006), "Finite element modeling of plain weave fabrics from photomicrograph measurements," *Composite Structures*, 73(1), 41–52.
40. Pandey, R. (1995). "Micromechanics based computer-aided design and analysis of two-dimensional and three-dimensional fabric composites," Dissertation, Pennsylvania State University, PA.
41. Barbero E. J., and Luciano, R. (1995), "Micromechanics formulas for the relaxation tensor of linear viscoelastic composites with transversely isotropic fibers," *International Journal of Solid Structures*, 32, 1859–1872.
42. Luciano, R., and Sacco, E. (1998), Variational methods for the homogenization of periodic heterogeneous media, *European Journal of Mechanics. A, Solids*, 17(4), 599–617.
43. Kollegal, M. G., and Sridharan, S. (1998), "A simplified model for plain woven fabrics," *Journal of Composite Materials*, 34, 1757–1785.
44. Barbero, E. J., Damiani, T. M., and Trovillion, J. (2005), "Micromechanics of fabric reinforced composites with periodic microstructure," *International Journal of Solid Structures*, 42, 2489–2504.
45. Barbero, E. J., and Luciano, R. (1994), "Formulas for the stiffness of composites with periodic microstructure," *International Journal of Solid Structures*, 31, 2933–2943.
46. Ishikawa, T., and Chou, T. W. (1983), "One-dimensional micromechanics analysis of woven fabric composites," *AIAA Journal*, 21, 1714–1721.
47. Ishikawa, T., and Chou, T. W. (1982), "Stiffness and strength behavior of woven fabric composites," *Journal of Materials Science*, 17, 3211–3220.
48. Ishikawa, T, and Chou, T. W. (1983), "Nonlinear behavior of woven fabric composites," *Journal of Composite Materials*, 17, 399–413.
49. Tabiei, A., and Yi, W. (2002), "Comparative study of predictive methods for woven fabric composite elastic properties," *Composite Structures*, 58, 149–164.
50. Peng, X. Q., and Cao, J. "Numerical determination of mechanical elastic constants of textile composites," 15th Annual Technical Conference of the American Society for Composites, College Station, TX, 2000.
51. Naik, N. K., and Ganesh, V. K. (1992), "Prediction of on-axes elastic properties of plain weave fabric composites," *Composites Science and Technology*, 45, 135–152.

52. Vandeurzen, Ph., Ivens, J., and Verpoest, I. (1996), "A three-dimensional micromechanics analysis of woven-fabric composites: I. Geometric analysis," *Composites Science and Technology*, 56, 1303–1315.

53. Vandeurzen, Ph., Ivens, J., and Verpoest. I. (1996), "A three-dimensional micromechanics analysis of woven-fabric composites: II. Elastic analysis," *Composites Science and Technology*, 56, 1317–1327.

54. Hahn, H. T., and Tsai, S. W. (1973), "Nonlinear elastic behavior of unidirectional composite laminae," *Journal of Composite Materials*, 7, 102–118.

55. Hahn, H. T., and Pandey, R. (1994), "A micromechanics model for thermo-elastic properties of plain weave fabric composites," *Journal of Engineering Materials and Technology*, 116, 517–523.

56. Aitharaju, V. R., and Averill, R. C. (1999), "Three-dimensional properties of woven-fabric composites," *Composites Science and Technology*, 59, 1901–1911.

57. Blackletter, D. M., Walrath, D. E., and Hansen, A. C. (1993), "Modeling continuum damage in a plain weave fabric-reinforced composite material," *Journal of Composites Technology and Research*, 15, 136–142.

58. Scida, D., Aboura, Z., Benzeggagh, M. L., and Bocherens, E. (1999), "A micromechanics model for 3D elasticity and failure of woven-fiber composite materials," *Composites Science and Technology*, 59, 505–517.

59. Kollegal, M. G., and Sridharan, S. (1998), "Strength prediction of plain woven fabrics," *Journal of Composite Materials*, 34, 241–257.

60. Bishop, S. M., and Curtis, P. T. (1984), "An assessment of the potential of woven carbon fibre reinforced plastics for high performance applications," *Composites*, 15, 259–265.

61. Gao, F., Boniface, L., Ogin, S. L., Smith, P. A., and Greaves, R. P. (1999), "Damage accumulation in woven fabric CFRP laminates under tensile loading. Part 1: Observations of damage; Part 2: Modelling the effect of damage on macromechanical properties," *Composites Science and Technology*, 59, 123–136.

62. Horrocks, A. R., and Anand, S. C., *Handbook of Technical Textiles*, The Textile Institute, CRC Press, 2000.

5 Single Yarns in Woven Textiles: Characterization of Geometry and Length Effects

INTRODUCTION

To properly characterize the deformation patterns in textile materials and model individual yarn behavior in a woven fabric, the size, shape, and meso-level deformation locations of the yarns need to be verified as a function of applied strain. As an exercise to exhibit the correlation between the applied external deformations and meso-level changes, the case of a plain-woven Kevlar textile will be described in the next section to demonstrate the changes in the internal structure of the system subjected to tensile loading in the primary direction. This approach has been extended to cases of shear and biaxial loading and points to the complexities that are involved in modeling and testing of textiles, especially when they are subjected to any loads in directions other than the principal materials direction.

KEVLAR FABRIC

Kevlar® 49 is a registered trademark of E. I. du Pont de Nemours & Co. and has been used in many engineering applications due to its high strength, modulus, and strength-to-weight ratio. Research on the quasi-static strength of yarns, along with studies of dynamic strength, have been reported by several authors.

Kevlar 49 fabric is manufactured using a plain weave of 17 × 17 yarns (per linear inch) each consisting of Kevlar filaments. Figure 5.1a shows the detail of the woven structure of Kevlar 49 fabric with typical properties being shown in Table 5.1. The yarns in the woven structure consist of hundreds of filaments, as shown in Figure 5.1b. The total cross-sectional area for each ply was calculated using the values of the linear density and bulk density of the material. The cross-sectional area of each yarn was calculated by taking into account the linear density of the material and dividing it by its bulk density. The total cross-sectional area of the specimen was defined as the cross-sectional area per yarn multiplied by the number of yarns of the specimen.

The undeformed geometry of the woven Kevlar fabric is shown in Figure 5.1. The detail of these steps is described in an FAA report by Rajan et al. [1] who used an experimental procedure to capture the geometry under various states of deformation. The procedure involves loading a fabric swath in tension and freezing the displacement by gluing the yarns while under an applied strain. The schematic diagram of the experimental procedure is shown in Figure 5.2. Samples were loaded to three different strain levels: 1.0%, 1.5%, and 2.0%. The stress–strain response of the textile up to the designated point of loading is shown in Figure 5.3. By clamping through the thickness to restrain the geometry in this condition, the displacement was frozen in the sample. Thin sectioning and polishing is conducted in both the warp and fill directions and the cross sections are observed under both optical microscopy and also scanning electron microscopy as shown in Figure 5.4. By applying image processing techniques, quantitative measures of the geometry of the yarns are obtained as a function of the applied strain.

Figures 5.5 and 5.6 show images of longitudinal cross sections of fill and warp yarns under different strain levels. It is evident that both the fill and the warp yarns can be represented as having

TABLE 5.1
Basic Material Property of Kevlar 49® Fabric

Material	Yarn Count (yarns/cm)	Bulk Density (g/cm³)	Linear Density (g/cm)	Cross-Sectional Area per Yarn (cm²)
Kevlar® 49	6.7×6.7	1.44	$1.656\ (10^{-3})$	$1.15\ (10^{-3})$

(a)   (b)

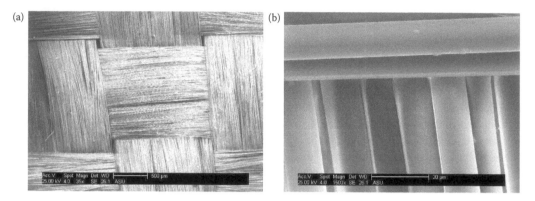

FIGURE 5.1 Scanning electron microscope images: (a) plain woven Kevlar fabric, (b) individual filaments within each yarn.

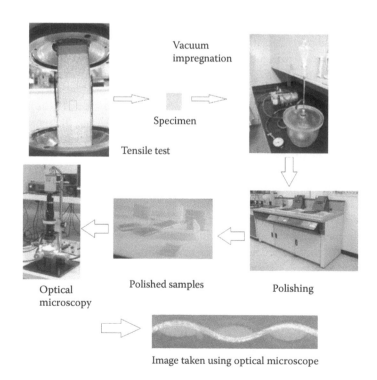

FIGURE 5.2 Schematic of overall experimental procedure.

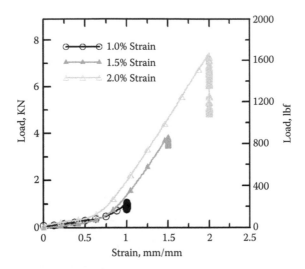

FIGURE 5.3 Stress–strain curve of Kevlar during test.

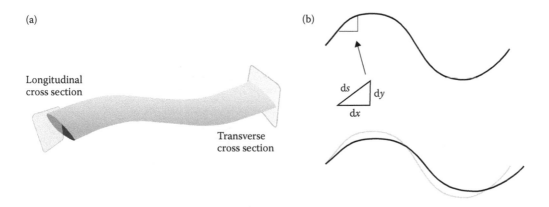

FIGURE 5.4 Fiber length and shape change due to applied strain.

elliptical cross sections along their transverse cross section and sinusoidal curves along the longitudinal cross section.

Figures 5.5 and 5.6 show the cross sections of fill and warp yarns at 0%, 1.0%, 1.5%, and 2% strain levels. As is evident, the cross sections of yarns undergo significant changes as the strain level is increased. Images at increasing strain levels indicate that the fibers along the loaded direction (warp) straighten out, whereas the yarns in perpendicular direction (fill) become wavier. The waviness of fill yarn increases as the strain level increases, whereas the warp yarns assume a sawtooth shape with an increasing strain level. The cross section of the warp yarns converts from an elliptical shape to a more circular form, whereas the cross section of fill yarns becomes flatter and the individual filaments distribute over the entire length of the warp yarns. At a strain of 2%, the warp yarns are completely covered by the fill yarns in a periodic support structure as shown in Figure 5.6d. It is clear that these deformations are representative of an increasing contact pressure between the two perpendicular yarn systems. The nature of these pressure distributions is unknown; however, it is expected that through proper calibration procedures and the use of finite element modeling, a better understanding of these interyarn pressures can be obtained.

A qualitative analysis image analysis procedure to analyze the images and characterize the deformation patterns is presented next. The initial yarn cross sections in both warp and fill

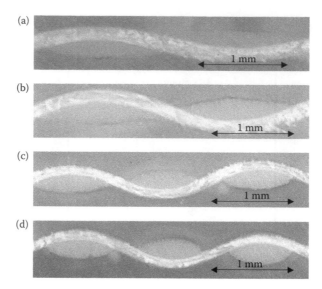

FIGURE 5.5   Fill yarn longitudinal cross section: (a) 0%, (b) 1.0%, (c) 1.5%, and (d) 2% strain level (fill yarns are in plane, and warp yarns are out of plane).

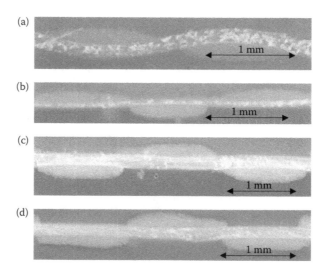

FIGURE 5.6   Warp yarn longitudinal cross section: (a) undeformed, (b) 1.0%, (c) 1.5%, and (d) 2% strain level (warp yarns are in plane, and fill yarns are out of plane).

directions were approximated as elliptical in cross section, whereas the length of the yarn was projected to follow a sinusoidal curve (see Figure 5.7). The parameters associated with these curves have been estimated by fitting sine and elliptical curves to the cross sections using the least square method.

As the samples are loaded, the warp yarns straighten out by removal of crimps, and they can no longer be approximated using a sinusoidal curve. It should be noted that the interfacial pressure between fill and warp yarns does not allow warp yarns to become completely straight; hence, they have dips/valleys with dips at the location of fill yarns and valley in between the fill yarns. This characteristic response was approximated using a periodic step function as a fitting parameter. Therefore, warp yarns under high strain levels were fitted using step functions instead of sinusoidal curves. Parameters were computed separately for each sample at each strain level. The overall or average parameters were computed by fitting a single curve through all the selected points from different samples at a strain level. Figure 5.8 shows all the selected points and fitted curve for undeformed geometry of fill yarn.

The warp yarns under load can be approximated using a square wave function as shown in Figure 5.8. Because the square wave function is small and gets smaller with an increasing load/strain level, it becomes increasingly difficult to fit step functions through the selected curves. Instead of fitting a step function as a single element in the curve, two straight lines representing the edges can be applied through selected points on the warp yarns. The distance between these two lines would be taken as the step height.

Figure 5.9 compares the undeformed geometry of fill and warp cross sections. Note that the fill yarns have a higher degree of waviness than the warp yarns. The effect of strain levels on the transverse geometry of the fill yarns is shown in Figure 5.10. Note that as the strain is increased, the fill yarns curvature increases and the wavelength of the curves decreases.

One can use the images to estimate the geometrical parameters for fill and warp yarns for undeformed samples as well as for 1.0%, 1.5%, and 2.0% strained samples. Figure 5.11 represents a typical distribution of points in the cross section of a fill yarn in an undeformed stage. The parameters of the curve fit can be obtained and used to generate a best fit curve as shown. The parameters for undeformed samples can be further used to generate micromechanical model of the fabric. Parameters associated with transverse cross section (elliptical) of fill yarns are presented by Rajan et al. [1].

Using the major and minor axis (a and b) parameters associated with the yarns for various loading conditions, one can represent the shape of the cross section at different strain levels. As the strain level increases, the ratio of major to minor axis also increases. This means that the yarns become flatter with increasing strain. This phenomenon can also be observed in the images shown in Figures 5.12 and 5.13.

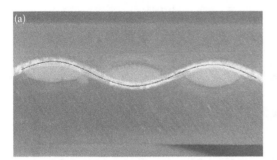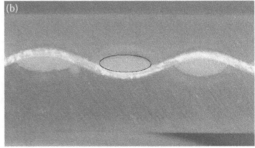

FIGURE 5.7 (a) Sinusoidal curve parameter estimation. (b) Cross-sectional approximation.

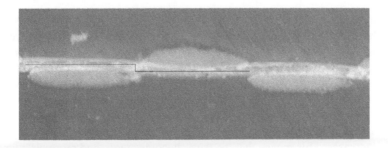

FIGURE 5.8 Cross-sectional approximation using square wave function.

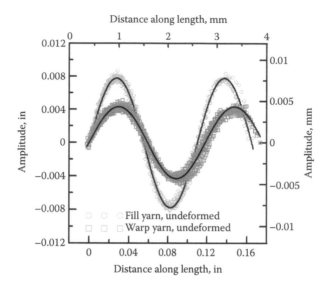

FIGURE 5.9 Comparison between the fill and the warp directions in the preloaded conditions.

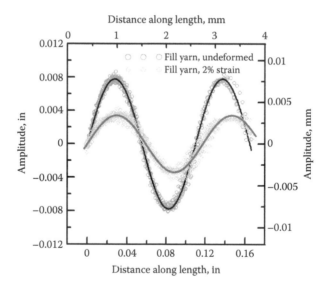

FIGURE 5.10 Comparison of fill sections at two different strain levels.

Figure 5.12 shows the effect of strain on the major and minor axis of the fill and warp yarns. Both the major and the minor axes of fill yarn increases with an increase in the strain level. The increase in the major axis is up to approximately $\varepsilon = 1.5\%$ and remains relatively constant beyond this level. The minor axis of the fill yarns increases as the strain level increases, indicating that the increase in waviness results in the contraction of the yarn and the bulging of the fill yarns. Also note that the density of the warp yarns is reduced or the voids between fibers decrease as they are stretched and become more circular. Figures 5.13a and 5.13b show the estimation of warp yarn transverse cross section under loading. As the strain level increases, the major axis of warp yarns decreases while the minor axis increases. This means that the yarns becomes more and more circular with increasing strain. This phenomenon can also be observed in the images shown in Figure 5.8 as well.

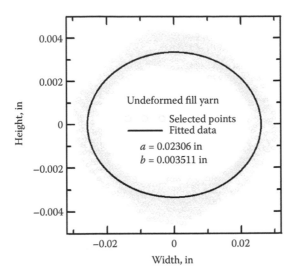

FIGURE 5.11 Fill cross-sectional (undeformed) ellipse curve fit through point cloud.

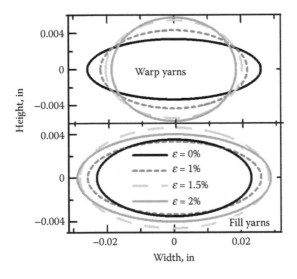

FIGURE 5.12 Comparison of warp and fill cross section (ellipse) at different strain levels.

Figure 5.14 shows the parameters of the geometry of fill yarn under 0%, 1.0%, 1.5%, and 2.0% strain levels. The amplitude indicates that the waviness of the shape of the sine curve, and the decreasing period corresponds to the increasing strain in the warp direction. As the warp yarns are loaded, the fill yarns become wavier, indicated by the increase in the amplitude of the sine curve. The change in the period of the sine curve correlates directly with the average strain in the warp direction as shown in Figure 5.10, which compares to the undeformed shape of the fill yarns after a strain of 2% has been applied in the warp direction.

Figure 5.14 shows the effect that warp yarn loading has on the shape of the sine curve in a longitudinal cross section of fill yarn. A slight change in the period is attributed to the reorientation of yarns under stress. The fill yarn becomes wavier as the warp yarns are loaded, as shown by the increase in the amplitude of the sine curve. This phenomenon represents the transfer of the slack from the warp

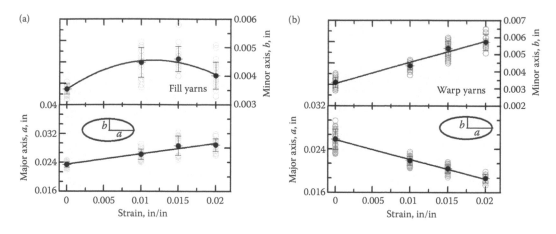

FIGURE 5.13 (a) Comparison of cross section (ellipse) at different strain levels: (a) fill yarns and (b) warp cross sections.

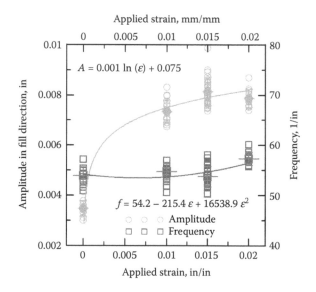

FIGURE 5.14 Comparison of fill cross section (sine curve) at different strain levels.

yarns to the fill yarns. Furthermore, an increase in the amplitude is much more evident during the initial phase of loading than it is later.

The longitudinal geometry of warp yarns follows a sinusoidal curve, and as these yarns are loaded, they become flatter and represent (approximately) a step function. This can be seen as removal of the crimping effect due to the application of the load. The image analysis of longitudinal and transverse cross sections of the warp and fill of Kevlar has shown that there is a basic difference in the geometry of these two yarns. These differences can be attributed to the weave pattern and the process of waving.

The geometrical parameters computed in previous section can be used to build a finite element model of the Kevlar yarns as shown in Figure 5.4a; however, to simulate fabrics using explicit analysis, it is important to verify the single yarn properties initially. Using the data generated in this section, the elliptical and sinusoidal representation of a yarn has been used to build a model of Kevlar warp yarn and has been analyzed using the finite element method [2].

SINGLE YARN TENSILE TESTS

To estimate the average material properties at the yarn level, statistical parameters for strength, strain capacity, and stiffness were gathered from single yarn tensile tests conducted on yarns extracted from fabric swaths. The objectives of single yarn tensile tests are as follows:

(1) To measure the tensile response of Kevlar single yarns under uniaxial tensile loading
(2) To provide estimation of strain capacity and longitudinal stiffness of the yarn
(3) To capture the effect of gage length on failure stress

Samples for single yarn tensile tests were obtained by removing warp direction yarns from the woven fabric (as shown in Figure 5.15). A universal joint was connected to the testing frame to allow rotation of the grip and remove any potential bending moment. The universal joint also helped in the alignment of yarn during test. To avoid any slipping of test samples during testing, the yarn was fed and wrapped around the upper and lower mandrels and aligned using a laser beam level. The single yarn tests were conducted and tested on a servohydraulic closed loop control Materials Test Systems, Inc, MTS-810 test frame under a displacement control at a strain rate of $4.167E^{-4}$ 1/s. Ten replicate samples of six different lengths of specimen were also tested. The first step was to document the effect of gage length on the peak stress, Young's modulus, and strain at ultimate strength.

The stress-strain curve obtained from the test with gage lengths of 2, 5, and 8 in. are shown in Figure 5.16a. The variation among the tests results is small, which ensures their repeatability. The Young's modulus computed for the 2-in. long samples is 22780 MPa, the average peak stress is 1899 MPa, and the strain at peak stress is estimated to be 0.12 mm/mm. The Young's modulus increases as the gage length increases. As the gage length increases from 5 to 8 in., the modulus increases from 43725 to 57913 MPa. The average ultimate strength decreases from 1792 to 1768 MPa, and the strain at this level decreases from 0.054 to 0.042 mm/mm. The stress-strain curve obtained from the test with gage length of 11, 14, and 17 in. is shown in Figure 5.16b. The Young's modulus computed for these gage lengths range from 67161 to 77440 MPa, while the average peak stress is 1698 to 1587 MPa, and the strain at peak stress is estimated to be 0.034 to 0.026 mm/mm. Note that as the yarn length is increasing, the stiffness increases, but the ultimate strength and strain capacity decrease marginally.

The effect of gage length on Kevlar yarn properties is prominent. Figure 5.17 shows the stress-strain curve for gage lengths tested. Note that as the gage length increases, the Young's modulus of

FIGURE 5.15   Single yarn test setup.

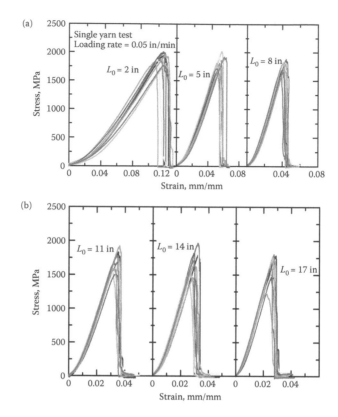

FIGURE 5.16   Stress–strain curves for (a) gage length = 2, 5, and 8 in. and (b) gage length 11, 14, and 17 in. Kevlar yarns.

the yarns increases while peak stress is reduced and ultimate strain capacity both decrease. These variations are attributed to two factors, being the Weibull strength effect, which due to the fiber length and the probability distribution of the flaws in larger lengths, leads to a decrease in strength as the fiber length increases. The second factor is the initial waviness in the length of the filaments, which leads to unequal stress distribution in the filament length. As the fiber length increases, there is more uniformity in the length, and the initial stretching and fiber alignment decreases. The effect of filament waviness on the strength of the fibers has been addressed by Zohdi et al. [3–5]. A quantitative analysis of gage length effect allows for the determination of Weibull parameters.

Figure 5.17 shows the effect of gage length on the ultimate yarn strength. A linear relation between the peak stress and gage length is demonstrated later. The effects of gage length on Young's modulus and strain at ultimate strength are also shown. We observed that as gage length increases, Young's modulus of the yarns increases. On the other hand, strain as peak stress decreases. The trend of material properties discussed earlier in this chapter is further shown in Figure 5.18. In this case, a second-order polynomial curve was used to empirically relate the effect of gage length on the Young's modulus, whereas a log curve was used for the strain at peak stress. Empirical equations governing these effects are shown as follows:

$$\sigma_{uts} = -19L_0 + 1916 \tag{5.1}$$

$$E = 8590 + 7950L_0 - 231L_0^2 \tag{5.2}$$

$$\ln\varepsilon_{uts} = -0.7 \times \ln L_0 - 1.7 \tag{5.3}$$

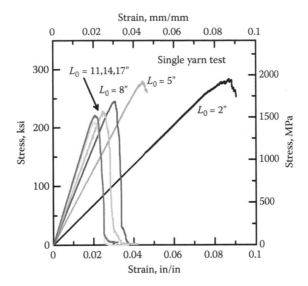

FIGURE 5.17   Stress–strain response of Kevlar yarns versus gauge length.

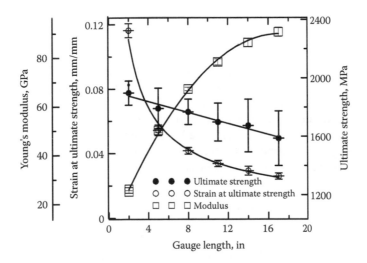

FIGURE 5.18   Strength, modulus, and strain at ultimate strength versus gauge length.

WEIBULL ANALYSIS

The classical approach to the strength of the materials is a deterministic route in which a single value that is characteristic of the material is supposed to exist. In experiments undertaken to determine the single valued strength, considerable scatter in the results is usually observed. If the scatter is not considered to be a feature of the material, it is usually attributed to uncontrollable experimental variables. As a consequence, the second central moment of the experimental data, the standard deviation, is interpreted as an indicator of the success of standardizing the experimental setup and procedures. Therefore, standard deviation can be considered to be an indicator of the quality of an experiment or testing method.

The aforementioned deterministic view is less popular in the technical sciences because, if it is valid, then identical experiments performed on specimens of different sizes should yield the same results for failure stress. However, it has been shown that even for uniformly produced Kevlar, glass, or ceramic yarns, larger specimens have a lower failure stress when compared with smaller ones. These systematic differences cannot be explained by random variations in experimental procedures, but by imperfections included in the yarn structure. The existence of microscopic flaws can cause a material to fail long before its ideal strength is reached. Although this ideal strength might be interpreted as true strength, for practical purposes it is more important to understand the actual strength of the material than true strength. The strength distribution function for a brittle material needs to take into consideration the inhomogeneous flaw distribution. The probability of failure initiation depends on the mean number of critical defects in a specimen of a given size. These flaws may then be categorized as surface or volume flaw densities.

Logically, we can accept imperfections as an integral part of a material and account for their presence when describing its strength. The distribution of imperfections and hence the strength is of a probabilistic nature. By assuming that the initial failure is governed by the flaw size distribution and the weakest linking yarn, we can use the well-known Weibull failure criteria (the most widely used approach for explaining the variation in fiber strength) as a function of gage length. The Weibull statistics apply for materials in which there is a uniformly defined flaw size distribution. The aspects of the relationships between the Weibull statistics and material and loading structure are analyzed by Danzer [6]. The basic form of the Weibull equation for cumulative probability density is

$$F(\sigma) = 1 - e^{-(\sigma/\sigma_0)^m} \tag{5.4}$$

where σ is the failure stress and σ_0 is the reference/scaling value related to the mean and m is the Weibull modulus/shape parameter of a two-parameter Weibull model. Modifications to the Weibull theory apply to the assumption of a uniform flaw distribution and a uniformly stressed volume. Once these two parameters are taken into account, it has been shown that the ratio of the strength of ceramic materials in flexure to their tensile strength can be determined on the basis of the Weibull parameter [7–9]. In order to include length in the Weibull model, parameters related to gage length or volume of the yarns can be introduced in the model. The modified three-parameter Weibull equation for cumulative probability density is given as

$$F(\sigma) = 1 - e^{-(L/L_0)(\sigma/\sigma_0)^m} \tag{5.5}$$

where L represents the volume of the yarn and L_0 represents the scaling value for the volume.

A total of six different lengths were used to estimate the length effect. The strain rate during these tests was maintained at 0.025/min. Hence, the variation in the peak stress was not dependent on the strain rate. Ten replicates of each gage length were also tested for strength and. Figure 5.19 shows the significant dependence ultimate strength on gage length. Cumulative probability distribution of the test data was obtained, and the parameters are presented in Table 5.2. Figure 5.19 also shows the Weibull curve fitting to experimental data using the least square method. As the gage length increases, the cumulative probability plot shifts toward lower stress values, which is a clear indicator of the dependence of peak stress on the gage length.

In a general three-dimensional model, one can represent the effective volume V as a characteristic parameter, such that for a specimen under uniform tension, it is the tested specimen volume, and for other geometries, it is defined using Weibull parameters [7, 10]:

$$F(\sigma) = 1 - e^{-V(\sigma/\sigma_0)^m} \tag{5.6}$$

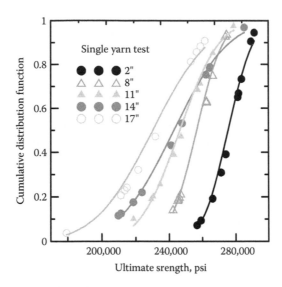

FIGURE 5.19 Cumulative probability of failure versus peak stress for different gage lengths.

TABLE 5.2
Weibull Parameters for Kevlar Yarns

	Gage Length (in.)					
	2	**5**	**8**	**11**	**14**	**17**
σ_0 (psi)	280130	265140	260160	249790	250160	239370
m	21.7	11.5	14.3	11.9	7.9	8.8

The Weibul representation of the mean and coefficient of variation (CV) of the strength of a volume of material is given as

$$\bar{\sigma} = \sigma_0 V^{-1/m} \Gamma\left(1 + \frac{1}{m}\right) \tag{5.7}$$

$$CV = \left[\frac{\Gamma\left(1 + (2/m)\right)}{\Gamma^2\left(1 + (1/m)\right)} - 1\right]^{1/2} \times 100 \ (\%) \tag{5.8}$$

It is postulated that the differences between tensile and flexural strengths can be best explained using the derivations for a Weibull-based material tested under different effective volume and stress gradients. In the case of the bend test, a stress gradient exists in the specimen. The effective volume, V, for the three-point bend strength is [11]

$$V = \frac{1}{2(m+1)^2 V_0} \tag{5.9}$$

where V_0 is the volume of the specimen between two supports. The effective volume of the four-point bend strength tested with a load span of one-third of a support span is

$$V = \frac{m+3}{6(m+1)} V_0 \tag{5.10}$$

where V_0 is the volume of a specimen between two supports. These models are applicable to a range of cement composites. Examples include materials with several different flaw distributions or rising crack resistance. The minimum number of test specimens necessary to guarantee a reliable prediction of Weibull's theory depends on the loaded (effective) volumes of the test specimens. The number of critical defects increases if the stress amplitude increases. Therefore, as shown in Equation 5.7, the most significant statistical features of the strength of brittle materials are the increase in the probability of failure with increasing specimen size and with increasing load amplitude.

REFERENCES

1. Rajan, S. D., Mobasher, B., Stahlecker, Z., Bansal, S., Zhu, D., Morea, M., and Dhandapani, K. (2010), "Explicit finite element modeling of multilayer composite fabric for gas turbine engine containment systems—Phase III Part 2," Arizona State University Fabric Material Tests, DOT/FAA/AR-10/23, Final Report, November 2010, 112.
2. Zhu, D., Soranakom, C., Mobasher, B., and Rajan, S. D. (2011), "Experimental study and modeling of single yarn pull-out behavior of Kevlar® 49 fabric," *Journal of Composite Materials*, 42, 868–879.
3. Zohdi, T. I., and Powell, D. (2006), "Multiscale construction and large-scale simulation of structural fabric undergoing ballistic impact," *Computer Methods in Applied Mechanics and Engineering*, 195, 94–109.
4. Lightweight Ballistic Protection of Flight-Critical Components on Commercial Aircraft Part 3: Zylon Yarn Tests. Report number: DOT/FAA/AR-04/45, P3.
5. Zohdi, T. I., and Steigmann, D. J. (2002), "The toughening effect of microscopic filament misalignment on macroscopic fabric response," *International Journal of Fracture*, 115, L9–L14.
6. Danzer, R. (1992), "A general strength distribution function for brittle materials," *Journal of the European Ceramic Society*, 10(6), 461–472.
7. Noguchi, K., Matsuda, Y., Oishi, M., Masaki, T., Nakayama S., and Mizushina, M. (1990), "Strength analysis of yttria stabilized tetragonal zirconia polycrystals," *Journal of the American Ceramic Society*, 73, 2667.
8. Weil, N. A., and Daniel, I. M. D. (1964), "Analysis of fracture probabilities in nonuniformly stressed brittle materials," *Journal of the American Ceramic Society*, 47(6), 268–274.
9. McCartney, L. N. (October 1979), "Extensions of a statistical approach to fracture," *International Journal of Fracture*, 15(5), 477–487.
10. Davies, D. G. S. (1973), "The statistical approach to engineering design in ceramics," *Proceedings of the British Ceramic Society*, 22, 429–452.
11. Weil N. A., and Daniel, I. M. D. (1964), "Analysis of fracture probabilities in nonuniformly stressed brittle materials," *Journal of the American Ceramic Society*, 47(6), 268–274.

6 Introduction to Mechanics of Composite Materials

INTRODUCTION

This chapter covers basic ideas concerning how an applied mechanical load is shared between the matrix and fibers. The mathematical idealization process starts with assumptions regarding the simple case of one-dimensional formulation for a two-phase composite using continuous unidirectional fibers reinforcing a homogeneous elastic matrix. The case of loading parallel to the fiber axis represented as a single laminate model is derived next. Under the assumption that there is a strong bond between matrix and fiber, one is able to allow a majority of the load transfer from the matrix into the fiber over a short transfer distance. This is a general simplification from a practical starting point and originates from the composite laminates, extending to reinforced concrete members under axial or flexural loads. A composite laminate is a sheet of reinforcing fibers or textile that is saturated with a matrix paste. One or several of these laminates may be stacked up on top of each other at various orientations to form a lamina. Further assumptions made here include imposing the equal strain condition that leads to the rule of mixtures expression for the Young's modulus. This is followed by the cases of transverse loading of a continuous fiber composite and axial loading with discontinuous fibers. Before the development of the modeling aspects, a set of parameter definitions are used to define the matrix, fibers, and interface properties and characteristics. The properties of the interface region between fiber and bulk matrix will greatly affect the efficiency of the reinforcing system and modeling aspects. The role and function of the interface will, however, be addressed in detail in other chapters.

VOLUME FRACTION

The volume fraction of the fiber component V_f and the matrix component V_m is defined as

$$V_f = \frac{v_f}{v_c} \quad V_m = \frac{v_m}{v_c} \tag{6.1}$$

where v_f is the volume of the fiber, v_m is the volume of the matrix, and v_c is the volume of the composite. Assuming that there are no pores, the sum of the volume fractions of all constituents in a composite must equal 1. In a two-component system consisting of one fiber and one matrix, then the total volume of the composite is

$$v_c = v_f + v_m \quad V_m = 1 - V_f \tag{6.2}$$

Similarly, the weight fractions W_f and W_m of the fiber and matrix can be defined in terms of the fiber weight, w_f, the matrix weight, w_m, and the composite weight, w_c. Hence,

$$W_f = \frac{w_f}{w_c} \quad W_m = \frac{w_m}{w_c} \quad w_c = w_f + w_m \quad W_m = 1 - W_f \tag{6.3}$$

COMPOSITE DENSITY

The density of the composite in terms of volume fraction can be found by calculating the weight of the composite that will be composed of the weights of their constituent,

$$\rho_c = \rho_m V_m + \rho_f V_f \tag{6.4}$$

or expressed in terms of the weight fractions as

$$\rho_c = \frac{1}{(W_m / \rho_m) + (W_f / \rho_f)} \tag{6.5}$$

After fabrication, composites often contain voids. Voids must be treated as a constituent with volume but with no weight. To determine the volume fraction of voids, V_v, express the void volume as $V_v = V_{actual} - V_{theoretical}$, where $V_{theoretical}$ is the volume of the solid constituents that contribute to the weight of the composite and V_{actual} is the measured or actual volume of the composite including the voids. The volume fraction is then found to be

$$V_v = \frac{\rho_{theoretical} - \rho_{actual}}{\rho_{theoretical}} \tag{6.6}$$

NATURE OF LOAD SHARING AND LOAD TRANSFER

The best way to increase the fiber content is to align them in a specific direction and bond them using matrix under a term defined as a lamina or ply. A single ply is defined as a lamina and is modeled as an orthotropic sheet in-plane stress. The three principal material axes of the orthotropic lamina are longitudinal and transverse to the fiber direction and normal to the lamina surface, which are denoted as 1, 2, and 3. An orthotropic material is defined using its elastic properties with respect to three orthogonal directions. There are nine independent elastic constants for a general orthotropic material. If a uniaxial tensile force is applied to an orthotropic material along one of its orthotropic symmetry axes, no angular distortion will result, since it is assumed that normal stresses produce only normal strains and no shear strains. The application of a tensile force along an axis that is not one of the material's special orthotropic axes produces angular distortions; that is, the material no longer behaves in an orthotropic manner.

The concept of load sharing between the matrix and the reinforcing fiber is the basis for the mechanical behavior of a composite. Consider loading a composite parallel to the fibers. Because they are bonded together, both fiber and matrix will stretch by the same amount in that direction, that is, they will have equal strains, ε (Figure 6.1). If the fibers are stiffer than in the matrix phase, because of a higher Young modulus, E, then they will be carrying a larger stress. On the other hand, if the fibers have a lower modulus (like polymeric fibers), then the stress will be higher in the matrix phase. This illustrates the concept of load sharing or distribution between matrix and fiber. By putting the sum of the contributions from each phase equal to the overall load, the weighted average Young modulus of the composite can be obtained using what is referred to as the "rule of mixtures." This is also referred to as "equal strain" or "Voigt" case.

As an external force is applied to a composite, the load is shared by the matrix and reinforcement (m and f, respectively). Equilibrium requires that the load carried by each phase is the product of the average stress in that phase (σ_m, and σ_f) and its sectional area (A_m and A_f). Compatibility is defined in terms of a constant strain applied to both systems of fiber and matrix in a parallel manner. The geometry of the problem allows us to relate the volume fraction V_f of reinforcement and matrix to their cross-sectional area. Then, equating the externally imposed force "P" to the sum of

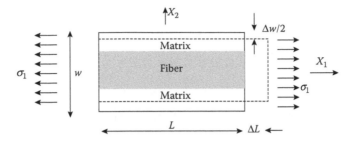

FIGURE 6.1 Isostrain model, also known as Voigt or parallel model.

these two contributions and dividing through by the total area, the "rule of mixtures" as a basic and important equation of composite theory is obtained. The applied stress σ_c is due to the applied P is the sum of the forces in both phases.

Geometrical definition:

$$V_i = \frac{A_i}{A} \quad i = m, f; \quad A = A_m + A_f \quad 1 = V_m + V_f \tag{6.7}$$

Constitutive relationships:

$$\sigma_i = E_i \varepsilon_i \quad i = c, m, f \tag{6.8}$$

where E_c, E_m, and E_f are the Young modulus of composite, matrix, and fibers. V_m and V_f and are the volume fraction of matrix and fibers, respectively:

Compatibility:

$$\varepsilon = \varepsilon_m = \varepsilon_f \tag{6.9}$$

Equilibrium:

$$P = \sigma_c A = \sigma_m A_m + \sigma_f A_f \tag{6.10}$$

Division by the total area results in

$$\sigma_c = \sigma_m V_m + \sigma_f V_f \tag{6.11}$$

Substitution of Equation 6.8 in Equation 6.11 results in the definition of the composite stiffness E_c as estimated by the rule of mixtures, and relates the volume-averaged matrix and fiber stiffness in a composite:

$$E_c = E_m V_m + E_f V_f \tag{6.12}$$

A certain proportion of an imposed load will be carried by the fiber and the remainder by the matrix. During the elastic range of response of the composite, this proportion is independent of the applied load. Instead, it depends on the volume fraction, shape, and orientation of the reinforcement and on the elastic properties of both constituents.

A lower-bound approach to the problem is obtained by calculating the transverse direction using an isostress method such that the continuity of the stress is imposed, whereas the total strain is

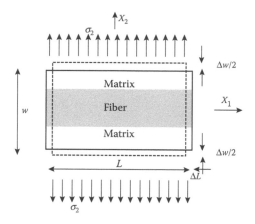

FIGURE 6.2 Isostress model or Kelvin model, series model.

defined by the addition of strains in each phase. As shown in Figure 6.2, in the series or Kelvin approach, the external force is applied to a composite, the load is uniformly applied to both matrix and reinforcement. The strain in by each phase is the ratio of the average stress in that phase by its Young's modulus. By defining the overall strain as the sum of these two contributions, the "rule of mixtures" is obtained.

Compatibility:

$$\Delta_c = \Delta_m + \Delta_f = w_m \varepsilon_m + w_f \varepsilon_f = w_c \varepsilon_c \tag{6.13}$$

$$\frac{w \sigma_c}{E_c} = \frac{w_m \sigma_m}{E_m} + \frac{w_f \sigma_f}{E_f} \tag{6.14}$$

Equilibrium:

$$\frac{P}{A} = \sigma_c = \sigma_m = \sigma_f \tag{6.15}$$

The composite stiffness E_c is estimated by the rule of mixtures,

$$\frac{1}{E_c} = \frac{V_m}{E_m} + \frac{V_f}{E_f}, \tag{6.16}$$

and shear modulus is obtained similarly,

$$\frac{1}{G_c} = \frac{V_m}{G_m} + \frac{V_f}{G_f}. \tag{6.17}$$

COMPUTATION OF TRANSVERSE STIFFNESS

The stiffness and strength of the composite to a load applied transverse to the fiber direction is expected to be much lower because the matrix phase is not shielded from carrying stress to the same degree as for axial loading. Because of this, the prediction of the transverse stiffness of a composite from the elastic properties of the constituents is more difficult than the axial value.

A lower bound on the stiffness of composite is obtained from the "equal stress" (or "Reuss") assumption shown in Figure 6.2 [1]. The value is underestimated, because in practice there are parts of the matrix that are effectively "in parallel" with the fibers (as in the equal strain model) rather than "in series" as is assumed. Empirical expressions such as "Halpin–Tsai" give much better approximations than the "equal stress" method [2]. The Halpin–Tsai expression for transverse stiffness is as follows:

$$E_2 = \frac{E_m(1+\xi\, \eta V_f)}{1-\eta V_f}$$ (6.18)

in which

$$\eta = \frac{E_f - E_m}{E_f + \xi E_m}$$ (6.19)

The value of ξ may be taken as an adjustable parameter, but its magnitude is generally of the order of unity. The expression gives the correct values in the limits of $V_f = 0$ and $V_f = 1$ and in general gives an accurate representation with experiment over the complete range of fiber content. A general conclusion is that the transverse stiffness (and strength) of an aligned composite is poor. The modulus perpendicular to the fibers is given by E_2, which, for a given fiber volume fraction, is much lower than the rule of mixtures expression. This is because the longitudinal modulus is fiber dominated and the transverse modulus is matrix dominated. The Halpin–Tsai equations can be generalized in the following form:

$$\frac{P_c}{P_m} = \frac{1+\xi\, \eta V_f}{1-\eta V_f}$$ (6.20)

$$P = \frac{1}{V_f + \eta(1-V_f)}(V_f P_f + \eta(1-V_f)P_m)$$ (6.21)

The parameters in Table 6.1 can be applied to compute all orthotropic elastic properties, such as G_{12} and υ_{23}. Parameter η is a measure of the relative magnitude of average stress in the fiber and matrix. If these values are set to 1, the equations converge to the rule of mixtures. Ideally, these values should be determined by experimental means. Parameter ξ reflects the amount of reinforcement and depends on fiber geometry, packing, and loading. For circular fibers, $\xi = 2$ (for E_2) and $\xi = 1$ (for G_{12}), whereas for Shear modulus, ξ is often related to V_f by $\xi = 1 + 40\,(V_f)$, which gives slightly better agreement than $\xi = 1$. Note that as $\xi \to 0$, equations revert to the simple model, $\eta \to 1 - E_m/E_f$ and $1/E_2 \to V_f/E_f + V_m/E_m$. Finally, as $\xi \to \infty$, equations approach the rule of mixtures (Figure 6.3).

TABLE 6.1
Parameters of the Halpin–Tsai Equations for Average Composite Properties

	P	P_f	P_m	η
E_1	E_1	E_f	E_m	1
ν_{12}	ν_{12}	ν_f	ν_m	1
E_2	$1/E_2$	$1/E_f$	$1/E_m$	½
K	$1/k$	$1/k_f$	$1/k_m$	$\frac{1}{2}(1-\nu_m)$
G_{12}	$1/G_{12}$	$1/G_f$	$1/G_m$	½

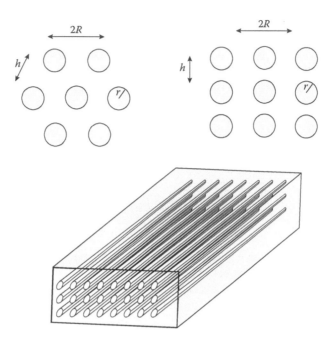

FIGURE 6.3   Closed packing of the fibers within a ply.

Using the various methods of cells [3–5], the lower and the upper bounds on elastic properties for axial shear and bulk moduli are computed and represented as shown in Table 6.2.

The packing of the fibers in composites is limited by the clustering limits such that ideal distributions do not normally occur in practice and volume fractions from 50% to 65% are typical with maximum levels of approximately 70%. When the fibers are modeled as parallel cylindrical inclusions with varying degrees of contact, as expressed by a contiguity factor, C is defined as 0 when fibers are isolated and $C = 1$ when all the fibers are in contact with real values somewhere in between [6]. In these cases, the stiffness relationships are a function of the packing due to fiber contiguity and expressed as

$$E_2 = A\left[(1-C)\frac{K_f(2K_m+G_m)-G_m(K_f-K_m)(1-V_f)}{(2K_m+G_m)+2(K_f-K_m)(1-V_f)}\right]+ \\ C\left[\frac{K_f(2K_m+G_f)+Gf(K_m-K_f)(1-V_f)}{(2K_m+G_m)-2(K_m-K_f)(1-V_f)}\right] \qquad A = 2[1-v_f+(v_f-v_m)v_m] \tag{6.22}$$

$$v_{12} = (1-C)\frac{K_f v_f(2K_m+G_m)V_f + K_m v_m(2K_f+G_m)(1-V_f)}{K_f(2K_m+G_m)-G_m(K_f-K_m)(1-V_f)}+ \\ C\frac{K_m v_m(2K_f+G_f)(1-V_f)+K_f v_f(2K_m+G_f)V_f}{K_f(2K_m+G_f)+G_f(K_m-K_f)(1-V_f)} \tag{6.23}$$

$$G_{12} = (1-C)G_m\frac{2G_f-(G_f-G_m)(1-V_f)}{2G_m+(G_f-G_m)(1-V_f)}+CG_f\frac{(G_f+G_m)-(G_f-G_m)(1-V_f)}{(G_f+G_m)+2(G_f-G_m)(1-V_f)} \tag{6.24}$$

Figure 6.4 shows the close packing of the carbon fibers in an epoxy system before and after the composite is exposed to elevated temperatures. It is clear from this that the formation of damage due to aging can be modeled using the contiguity factor.

TABLE 6.2
Upper and Lower Limits Using the Various Methods of Cells

Property	Lower Bound	Upper Bound
Longitudinal modulus E_1	$E_f V_f + (1-V_f)E_m + \dfrac{4V_f(1-V_f)(v_f - v_m)^2}{V_f/k_m + (1-V_f)/k_f + 1/G_f}$	$E_f V_f + (1-V_f)E_m + \dfrac{4V_f(1-V_f)(v_f - v_m)^2}{V_f/k_m + (1-V_f)/k_f + 1/G_m}$
Poisson ratio, v_{12}	$v_f V_f + (1-V_f)v_m + \dfrac{V_f(1-V_f)(v_f - v_m)(1/k_m - 1/k_f)}{V_f/k_m + (1-V_f)/k_f + 1/G_f}$	$v_f V_f + (1-V_f)v_m + \dfrac{V_f(1-V_f)(v_f - v_m)(1/k_m - 1/k_f)}{V_f/k_m + (1-V_f)/k_f + 1/G_m}$
Shear modulus, G_{12}	$G_{12} = G_m + \dfrac{V_f}{1/(G_f - G_m) + V_f/2G_m}$ $G_{21} = G_m + \dfrac{V_f}{1/(G_f - G_m) + (1-V_f)(k_m + 2G_m)/2G_m(k_m + G_m)}$	$G_{12} = G_f + \dfrac{(1-V_f)}{1/(G_m - G_f) + V_f/2G_f}$ $G_{21} = G_f + \dfrac{(1-V_f)}{1/(G_m - G_f) + V_f(k_f + 2G_f)/2G_f(k_f + G_f)}$
Bulk modulus	$k = k_f + \dfrac{1-V_f}{1/(k_m - k_f) + V_f/(k_f + G_f)}$	$k = k_m + \dfrac{V_f}{1/(k_f - k_m) + (1-V_f)/(k_m + G_m)}$

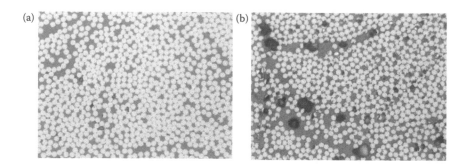

FIGURE 6.4   Carbon–epoxy core used for high-temperature–low sag (HTLS) conductors before and after heat exposure 120 h at 210°C. (From Gorur, R., Mobasher, B., Olsen, R., Shinde, S., and Erni, J., "Characterization of composite cores for high temperature-low sag (HTLS) conductors," *PSERC, IAB Meeting*, Cornell University, May 27–29, 2009. With permission.)

For uniaxially aligned fibers of finite length, the rule of mixtures can be modified by the inclusion of a length correction factor [8, 9]:

$$E_c = \eta_L E_f V_f + E_m V_m \tag{6.25}$$

It can be shown that [10],

$$\eta_L = 1 - \frac{\tanh(\beta L/2)}{\beta L/2}, \quad \text{where} \quad \beta = \frac{G_m 2\pi}{E_f A_f \ln(R/R_f)} \tag{6.26}$$

where G_m is the shear modulus of the matrix, L is the fiber length, A_f is the fiber cross-sectional area, R_f is the radius of the fiber, and R is the mean separation of the fibers. The effect of fiber length distribution can be described as,

$$E_1 = \sum_{j=1}^{M} E_f V_{fj} \left[\frac{\tanh(\beta L/2)}{\beta L/2} \right] + E_m(1 - V_f) \tag{6.27}$$

STRENGTH OF A LAMINA

According to the law of mixtures, in a simple aligned fiber composite loaded parallel to the fibers, both the matrix and the fiber experience the same strain. Any deviation from this linear behavior takes place when one of the two components fails or exhibits a nonlinear response such as creep, fatigue, and so forth. This initial failure takes place at the lower value of the matrix fracture strain or the fiber fracture strain. There are two types of failure to consider. Either the matrix fails or the fiber fails first. In cement-based composite systems, the former situation is more common because the matrix has a strain capacity lower than the fibers. The latter case is observed in metal matrix composites or thermoplastic polymer composites in which, because of a plastic deformation in the matrix, the failure strain of the fiber is the smaller value.

The stresses at which the fibers and the matrix fail determines the strength of a composite and can be calculated using a few simplifying assumptions. These assumptions include continuous, aligned fibers, with the load applied along the direction of the fibers. The bond between the fiber and the matrix, preventing interfacial slip. The longitudinal strength of a composite lamina can be obtained as an extension of the rule of mixtures. The limiting condition is the smaller value of two parameters, strain to failure of matrix or fibers, which determine the nature of load transfer after the first cracking is observed. In the case that the strain to failure of the matrix is assumed too much lower than the

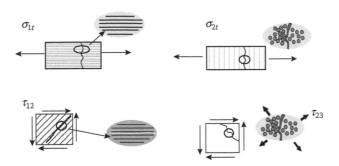

FIGURE 6.5   Tensile and shear modes of failure in fiber composite systems.

fibers and the fiber volume fraction is relatively large in the range of engineering composite materials (i.e., more than 10% and up to approximately 60%), the composite strength, σ_c, is given as

$$\sigma_{cu} = \sigma_{fu} V_f \tag{6.28}$$

where σ_{fu} is the fiber strength (Figure 6.5).

CASE STUDY 1: MATRIX FAILS FIRST, σ_{MU} GOVERNS

In cement composites that use a relatively low volume fraction of fibers, the matrix constitutes the major load-bearing section. The addition of fibers gradually increases the strength and post cracking ductility as the applied load is partitioned between the fibers and the matrix. When the strain in the composite reaches the fracture strain of the matrix, the matrix phase will fail. The load will then transfer gradually to the fibers, which, due to their relatively low volume fraction, will experience a large jump in stress and undergo debonding and gradual pullout, leading to their ultimate failure as the strain on the composite is further increased. When the composite is deformed, the elastic modulus is linear. At the strain at which the matrix is about to fracture, ε_{mu}, the stress in the composite can be determined using Hookes' law, because both the fiber and the matrix are still behaving elastically. That is,

$$F = \sigma A = (V_f E_f \varepsilon_f + (1 - V_f)\sigma_{mu})A \tag{6.29}$$

As the stress in the matrix reaches the fracture stress, σ_{mu}, the fiber stress is still much lower than its fracture stress ($\varepsilon_{fu} > \varepsilon_{mu}$). What follows next is a function of the mode of loading (either a load control or a deformation control). Either a series of cracks occur sequentially, or they form immediately, depending on whether or not the load is dropped as soon as the cracks occur. Ultimately, the distinction is irrelevant to the overall strength of the composite, but it does affect the shape of the stress–strain curve. In this case, we are only considering the load control which means that the sample is not allowed to unload as its compliance increases due to cracking. Before matrix cracking, the load on the composite is

$$F = \sigma A = \left(V_f \frac{E_f}{E_m} \sigma_{mu} + (1 - V_f)\sigma_{mu} \right) A \tag{6.30}$$

$$F = \sigma A = \left(V_f (1 - \frac{E_f}{E_m}) + 1 \right) \sigma_{mu} A \tag{6.31}$$

After the matrix breaks, only the fibers remain to carry the load and the portion of the load that needs to be redistributed is $(1-V_f)\sigma_{mu}A$. Stress in the fiber proportionately increases by

$$\Delta\sigma_f = \frac{(1-V_f)E_m}{V_f E_f} \tag{6.32}$$

If this increase elevates the stress in the fiber to its fracture strength, then the fibers will also fail instantaneously. Such conditions occur when the volume fraction of fibers is sufficiently small and the matrix portion of the load is large. This condition is called a matrix-dominated fracture. Fiber-dominated fractures takes place only if the jump in stress is not sufficient to break the fibers so that the load can then be increased with a much lower stiffness until the fibers fail at their ultimate strength, that is, at

$$F = \sigma A = V_f E_f \varepsilon_{fu} A = \sigma_{fu} A \tag{6.33}$$

Figure 6.6a represents the mode of cracking when the strain capacity of matrix is higher than the strain capacity of the fiber. The first cracking strength can therefore be defined as,

$$\sigma_1 = \begin{cases} V_f\sigma_{fu} + (1-V_f)E_m\varepsilon_{fu} & \varepsilon_{mu} > \varepsilon_{fu} \\ V_f\sigma_{fu} + (1-V_f)\sigma_{mu} & \varepsilon_{mu} = \varepsilon_{fu} \\ V_f E_f\varepsilon_{mu} + (1-V_f)\sigma_{mu} & \varepsilon_{mu} < \varepsilon_{fu} \end{cases} \tag{6.34}$$

If the fiber and the matrix have the same ultimate strain, then the ultimate strength is defined as

$$\sigma_{cu} = V_f\sigma_{fu} + (1-V_f)\sigma_{mu} \tag{6.35}$$

Figure 6.7 shows the condition and the effect of fiber volume fraction on the response of the composite under two different cases of matrix-dominated and fiber-dominated failure modes. Figure 6.7b shows the effect of fiber volume content in the fiber-dominated failure range. As the strain level is increased, the post cracking stiffness of the composite also increases.

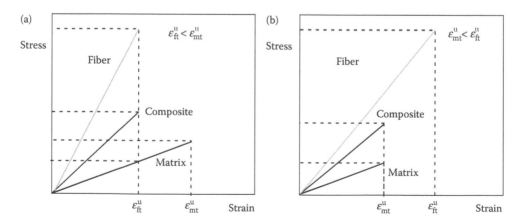

FIGURE 6.6 Mode of cracking when the strain capacity of matrix is higher than the fiber: (a) $\varepsilon_{ft} < \varepsilon_{mt}$ and (b) when $\varepsilon_{ft} > \varepsilon_{mt}$.

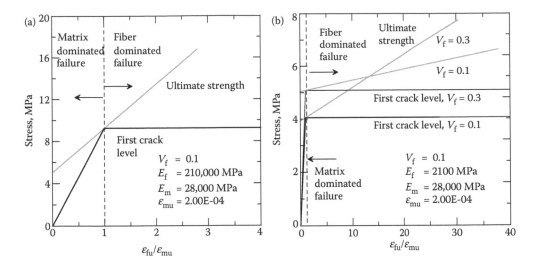

FIGURE 6.7 Effect of strain capacity of fiber with respect to matrix on the composite's nominal stress strain response. (a) Separation of matrix-dominated failure and fiber-dominated failure. (b) Effect of fiber volume fraction.

CASE STUDY 2: FOUR STAGES OF CRACKING

A simplified way of seeing the four stages of cracking in a brittle matrix composite is presented in this section. During Stage 1, both the fiber and the matrix are elastic, and no mismatch in strain between the two phases exists. There is therefore no shear stress between the two phases. This assumption results in stress in the fiber at the moment of cracking to be obtained from the rule of mixtures:

$$\sigma_{f0} = E_f \varepsilon_{mu} = E_f \frac{P}{A_c E_c} = E_f \frac{P}{A_c (E_m V_m + E_f V_f)} = \frac{1}{(E_m / E_f)(1 - V_f) + V_f} \frac{P}{A_c} \tag{6.36}$$

where

    P = the load applied on the unit cross section of the composite
    A_f = the area of cross section of each fiber
    V_f = volume fraction of fibers in the composite
    V_m = volume fraction of matrix in the composite
    A_c = unit area of the cross section of the composite
    E_m = elastic modulus of cement matrix before cracking under tension
    E_f = elastic modulus of fiber under tension

During Stage 2, the initiation of matrix cracking results in the transfer of the load to the fiber at the crack plane. The stress in the fiber right after cracking at the level of matrix crack is,

$$\sigma'_{f0} = \frac{P}{\sum A_f} = \frac{P}{A_c V_f} = \frac{1}{V_f} \frac{P}{A_c} \tag{6.37}$$

During Stage 3, which is defined by the initiation of multiple cracks, three important stress distributions are defined: the interfacial shear stress, $\tau(x)$; the stress in fiber, $\sigma_f'(x)$; and the stress in the matrix $\sigma_m'(x)$, as shown in Figure 6.8. Greszczuk's [11] equation of interfacial shear stress is defined as

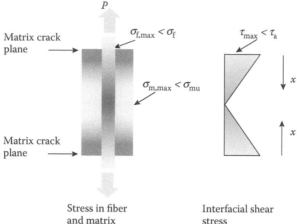

FIGURE 6.8 Stress distribution in a composite with a bonded interface between two matrix cracks.

$$\tau_x = \frac{P_f \beta_2}{2\pi r}[\sinh(\beta_2 x) - \coth(\beta_2 l)\cosh(\beta_2 x)] \tag{6.38}$$

$$\beta_2 = \left(\frac{2G_m}{b_i r E_f}\right)^{(1/2)} \qquad b_i = \frac{1 - \sqrt{V_f}}{\sqrt{V_f}} r \tag{6.39}$$

$$P_f = \sigma'_{f_0} A_f = \frac{1}{V_f}\frac{P}{A_c} A_f = \frac{P}{V_f}\frac{A_f}{A_c} \tag{6.40}$$

where
   P_f = load applied on each fiber
   r = radius of the fiber
   b_i = effective width of the interface
   G_m = shear modulus of the matrix at the interface
   l = half length of the crack spacing 2 s.
The shear stress distribution and the maximum shear stress are obtained as

$$\tau(\max) = \tau(x=0) = \frac{P_f \beta_2}{2\pi r}\coth\left(\beta_2 l\right) \tag{6.41}$$

$$\tau(x) = \tau_{max}\left(1 - \frac{x}{l}\right) \tag{6.42}$$

Assuming a linear shear stress distribution from the maximum shear stress, the stress in the fiber along the cracking surface is obtained by

$$\sigma'_f(x) = \sigma'_{f0} - \frac{1}{\pi r^2}\int_0^x \tau(\xi)2\pi r\, d\xi = \sigma'_{f0}\frac{1}{\pi r^2}\int_0^x\left(\tau_{max}\left(1 - \frac{\xi}{l}\right)\right)(2\pi r)\,d\xi$$
$$= \sigma'_{f0} - \frac{P_f \beta_2}{\pi r^2}\coth(\beta_2 l)\left(x - \frac{x^2}{2l}\right) \tag{6.43}$$

$$\sigma_f'\,|_{max} = \sigma_f'(x=0) = \sigma_{f0}' \tag{6.44}$$

$$\sigma_f'\,|_{min} = \sigma_f'\,(x=l) = \sigma_{f0}' - \frac{P_f\beta_2 l}{2\pi r^2}\coth(\beta_2 l) \tag{6.45}$$

whereas the stress in matrix is

$$\sigma_m(x) = \frac{1}{A_m}\int_0^x \tau(\xi)2\pi r d\xi = \frac{V_f}{(1-V_f)A_f}\int_0^x \tau(\xi)2\pi r d\xi = \frac{V_f}{(1-V_f)A_f}2\pi r\tau_{max}\left[x-\frac{x^2}{2l}\right] \tag{6.46}$$

$$\sigma_m|_{min} = \sigma_m(x=l) = 0 \tag{6.47}$$

$$\sigma_m\Big|_{max} = \sigma_m(x=l) = \frac{V_f}{(1-V_f)A_f}2\pi r\tau_{max}\frac{l}{2} = \frac{V_f}{(1-V_f)A_f}\frac{1}{2}P_f\beta_2 l\coth(\beta_2 l) \tag{6.48}$$

Therefore, the stress in both matrix and fiber oscillate between the maximum and the minimum levels such that the total force remains constant throughout the sample length. Average strain in laminates can be defined as

$$\bar{\varepsilon} = \bar{\varepsilon}_f = \frac{1}{E_f l_0}\int_0^l \sigma_f'(x)dx = \frac{1}{E_f l_0}\int_0^l\left(\sigma_{f0}' - \frac{P_f\beta_2}{\pi r^2}\coth(\beta_2 l_0)\left(x-\frac{x^2}{2l_0}\right)\right)dx$$

$$= \frac{\sigma_{f0}'}{E_f}\left[1-\beta_2\coth(\beta_2 l_0)\frac{l_0}{3}\right] = \varepsilon_{mu}\left[\frac{E_m}{E_f}\left(\frac{1}{V_f}-1\right)+1\right]\left[1-\beta_2\coth(\beta_2 l_0)\frac{l_0}{3}\right] \tag{6.49}$$

Therefore, the theoretical crack spacing is

$$\Delta_c = 2(\bar{\varepsilon}\, l_0 - \int_0^l \frac{\sigma_m(x)}{E_m}dx)$$

$$= 2\left(\varepsilon_{mu}\left[\frac{E_m}{E_f}\left(\frac{1}{V_f}-1\right)+1\right]\left[1-\beta_2\coth(\beta_2 l_0)\frac{l_0^2}{3}\right] - \varepsilon_{mu}\left[1+\frac{E_f}{E_m}\frac{V_f}{1-V_f}\right]\beta_2\coth(\beta_2 l_0)\frac{l_0^2}{3}\right) \tag{6.50}$$

Multiple cracking up to crack saturation point can continue under steady-state conditions or increasing load levels. The cracking process continues in a composite until a failure criteria for one or all of the stress measures are exhausted. The three stress measures are σ_f, σ_m, and τ_{max}. As long as there is a sufficient amount of shear stress to transfer the load to the matrix and cause it to crack ($\tau_{max} < \tau_{au}$, $\sigma_{m,max} < \varepsilon_{mu}E_m$), and the fiber strength has not been violated ($\sigma_{f,max} < \sigma_{fu}$), then more the load can be applied to the fiber until the matrix cracks in the middle section again or there is debonding between fiber and matrix.

It is also possible for frictional sliding to occur under these conditions; the stress distribution along the interface is nonlinear, and one would have to use numerical methods for a solution to the problem as shown by Sueki et al. [12] and Soranakom and Mobasher [13] and its further applications to model tension stiffening [14, 15]. The debonding mode would inherently slow down the level of matrix cracking toward a saturation level where no more matrix cracks could occur. The final stage

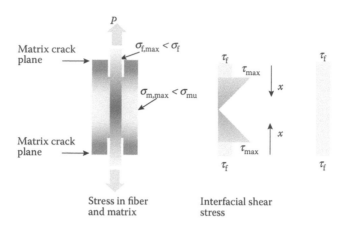

FIGURE 6.9 Stress distributions in a composite with a debonded interface between two matrix cracks.

is when a fully debonded interface is achieved, as shown in Figure 6.9. At that point, only the frictional shear between the fiber and matrix can transfer the load to the matrix.

When the interface is fully debonded, fiber slipping dominates the response, and the cracking remains saturated and evenly distributed. Because no more cracking takes place, a lower-bound estimate of the composite modulus is due to the contribution of fibers only, that is, $E_c = E_f V_f$, and no more force can be transferred into the matrix through the interface.

LAMINATED COMPOSITES

The calculations of previous section show that to obtain the highest stiffness and strength of the composite, a high proportion of aligned and continuous fibers should be used. Such an aligned system, however, is highly anisotropic, weak, and compliant in the transverse direction. Because high strength and stiffness are required in various directions within a plane, one can stack together a number of sheets, each having the fibers oriented in different directions. Such a stack is called a laminate. An example is shown in Figure 6.8. The procedures to incorporate the stiffness of single plies into the stiffness of cross ply composites is presented in this section (Figure 6.10).

Although unidirectional plies are high in axial stiffness and strength, their anisotropic properties make them impractical for multiaxial loading conditions. With a laminate, one can tailor the properties in different directions such that the material design and the component design are carried out simultaneously. Both elastic and strength properties can be computed using the stresses on the individual plies. The procedures in this section address the stiffness of ply as a function of the orientation of the materials, principal direction with respect to the loading direction and the response of multiple stacked layers. By imposing a strain or curvature distribution on all the individual plies, the stresses that are generated in the individual plies are added to correlate the internal and external forces and the displacements.

STIFFNESS OF AN OFF-AXIS PLY

The elastic response of the ply to stresses parallel and normal to the fiber axis has been shown in previous sections. The stiffness of a single ply, in either axial or transverse directions (E_1 and E_2), was obtained using the slab model or the Halpin–Tsai expressions (see the Nature of Load Sharing and Load Transfer section). Other elastic constants, such as the shear modulus (G_{12}) and Poisson's ratios, are readily calculated in a way similar to these two methods.

Laminated composites will usually combine lamina with fibers at different orientations. To predict the laminate properties, stress–strain relations are required for loading a lamina at an angle θ to the fiber direction, and for loading both in plane and in bending.

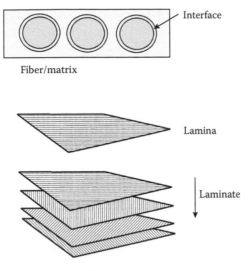

FIGURE 6.10 Definition of interfacial region and lamina and laminates. (Adapted from Bader, M. G., Short course notes for "An Introduction of Composite Material," University of Surrey, 1997; Ogin, S. L., Textile Reinforced Composite Materials, *Handbook of Technical Textiles*, Horrocks, A. R., and Anand, S. (eds.), Textile Institute, CRC Press, 2000; and from Bader, M. G., Smith, W., Isham, A. B., Rolston, J. A., and Metzger, A. B., *Delaware Composites Design Encyclopedia*, Vol. 3, *Processing and Fabrication Technology*, Technomic Publishing, Lancaster, PA, 1990. With permission.)

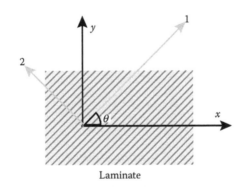

FIGURE 6.11 Orientation of local and material coordinate systems. Material axes (1–2) oriented an angle θ from the global axes (x–y).

First, it is necessary to establish the stiffness of an oriented ply so that the fibers lie at some arbitrary angle to the stress axis. Second, further calculations are needed to find the stiffness of a given stack. Consider first a single ply. The stiffness for any loading angle is evaluated by considering only the stresses in the plane of the ply. The applied stress is then transformed to get its components parallel and perpendicular to the fibers. The strains generated in these directions can be calculated from the (known) stiffness of the ply when referred to these axes. Finally, these strains are transformed into values relative to the loading direction, giving the stiffness (Figure 6.11).

Composite mechanics for laminated composites are well developed. Many textbooks deal with the subject (e.g., [16, 17]). Using the stiffness of a unidirectional composite ply from measurable engineering constants, one can determine how that stiffness varies with orientation. The transformation of the stiffness of a lamina between an arbitrary orientation (x–y) and material coordinates (1–2), as is illustrated in Figure 6.9. Figure 6.10 shows the representation of the stress both in the arbitrary x–y direction and its corresponding stresses equivalent measures in the principal materials directions. The

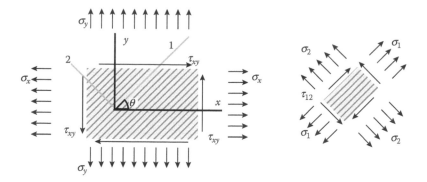

FIGURE 6.12   Material axes (1–2) oriented an angle θ from the global axes (x–y).

objective is to predict the stiffness and the stress distribution of a general laminate loaded in the arbitrary direction, θ. Starting with the stress–strain relationship in the material direction (Figure 6.12),

$$\varepsilon_1 = \frac{\sigma_1}{E_1} - \frac{v_{12}\sigma_2}{E_1}$$

$$\varepsilon_2 = \frac{\sigma_2}{E_2} - \frac{v_{21}\sigma_1}{E_2} \tag{6.51}$$

$$\gamma_{12} = \frac{\tau_{12}}{G_{21}}$$

Assuming that the only load applied is the stress in the x direction, one can write the stress transformation as,

$$\sigma_1 = \sigma_x\cos^2\theta$$

$$\sigma_2 = \sigma_x\cos^2\theta \tag{6.52}$$

$$\tau_{12} = -\sigma_x\cos\theta\cos\theta$$

One can also express the strain in terms of local coordinates using transformation:

$$\varepsilon_x = \varepsilon_1\cos^2\theta + \varepsilon_2\sin^2\theta - \gamma_{12}\sin\theta\cos\theta$$

$$\varepsilon_y = \varepsilon_1\sin^2\theta + \varepsilon_2\cos^2\theta + \gamma_{12}\sin\theta\cos\theta \tag{6.53}$$

$$\gamma_{xy} = 2\varepsilon_1\sin\theta\cos\theta - 2\varepsilon_2\sin\theta\cos\theta + \gamma_{12}(\cos^2\theta - \sin^2\theta)$$

Substitution of Equations 6.29 and 6.30 in Equation 6.31 expresses the strain and stresses in local coordinates:

$$\varepsilon_x = \sigma_x\left[\frac{\cos^4\theta}{E_1} + \frac{\sin^4\theta}{E_2} + \frac{1}{4}\left(\frac{1}{G_{12}} - \frac{2v_{12}}{E_1}\right)\sin^2 2\theta\right]$$

$$\varepsilon_y = -\sigma_x\left[\frac{v_{12}}{E_1} - \frac{1}{4}\left(\frac{1}{E_1} + \frac{2v_{12}}{E_1} + \frac{1}{E_2} - \frac{1}{G_{12}}\right)\sin^2 2\theta\right] \tag{6.54}$$

$$\gamma_{xy} = -\sigma_x\sin2\theta\left[\frac{v_{12}}{E_1} + \frac{1}{E_2} - \frac{1}{2G_{12}} - \cos^2\theta\left(\frac{1}{E_1} + \frac{2v_{12}}{E_1} + \frac{1}{E_2} - \frac{1}{G_{12}}\right)\right]$$

For example, the modulus, E_x, of a ply loaded at an angle θ to the fiber direction is given as

$$E_x = \frac{\sigma_x}{\varepsilon_x} = \left[\frac{\cos^4\theta}{E_1} + \frac{\sin^4\theta}{E_2} + \frac{1}{4}\left(\frac{1}{G_{12}} - \frac{2v_{12}}{E_1}\right)\sin^2 2\theta \right]^{-1} \qquad (6.55)$$

These operations can be expressed mathematically in tensor notation. Because we are concerned with stresses and strains within the plane of the ply, only two normal and one shear stressors are involved. The first step of resolving the applied stresses, σ_x, σ_y, and τ_{xy}, into components parallel and normal to the fiber axis, σ_1, σ_2, and τ_{12} (see Figure 6.10), depends on the angle, θ, between the loading direction (x) and the fiber axis (1). The stress transformation rule is defined as

$$\begin{bmatrix} \sigma_1 \\ \sigma_2 \\ \tau_{12} \end{bmatrix} = [T] \begin{bmatrix} \sigma_x \\ \sigma_y \\ \tau_{xy} \end{bmatrix} \qquad (6.56)$$

where the transformation matrix is given as

$$[T] = \begin{bmatrix} \cos^2\theta & \sin^2\theta & 2\sin\theta\cos\theta \\ \sin^2\theta & \cos^2\theta & -2\sin\theta\cos\theta \\ -\sin\theta\cos\theta & \sin\theta\cos\theta & \cos^2\theta - \sin^2\theta \end{bmatrix} \qquad (6.57)$$

or

$$\begin{bmatrix} \sigma_1 \\ \sigma_1 \\ \tau_1 \end{bmatrix} = T_{ij} \begin{bmatrix} \sigma_x \\ \sigma_y \\ \sigma_{xy} \end{bmatrix} = \begin{bmatrix} \cos^2\theta & \sin^2\theta & 2\sin\theta\cos\theta \\ \sin^2\theta & \cos^2\theta & -2\sin\theta\cos\theta \\ -\sin\theta\cos\theta & \sin\theta\cos\theta & \cos^2\theta - \sin^2\theta \end{bmatrix} \begin{bmatrix} \sigma_x \\ \sigma_y \\ \sigma_{xy} \end{bmatrix} \qquad (6.58)$$

The strain transformation can also be expressed using the same matrix, T; however, using the Reuter matrix, the ½ factors will disappear, and the transformation will look as follows:

$$\begin{bmatrix} \varepsilon_1 \\ \varepsilon_2 \\ \frac{1}{2}\gamma_{12} \end{bmatrix} = [T] \begin{bmatrix} \varepsilon_x \\ \varepsilon_y \\ \frac{1}{2}\gamma_{xy} \end{bmatrix} \qquad \begin{bmatrix} \varepsilon_1 \\ \varepsilon_2 \\ \gamma_{12} \end{bmatrix} = [R][T][R]^{-1} \begin{bmatrix} \varepsilon_x \\ \varepsilon_y \\ \gamma_{xy} \end{bmatrix} \qquad (6.59)$$

where

$$R = \begin{bmatrix} 1 & 0 & 0 \\ 0 & 1 & 0 \\ 0 & 0 & 2 \end{bmatrix} \qquad (6.60)$$

The stress–strain relationship expressed in the matrix notation for a general orthotropic material is expressed using the compliance matrix, S, which relates the stress to strain within a lamina loaded in its principal directions [16]:

$$\varepsilon_j = S_{ij}\sigma_i \tag{6.61}$$

$$\begin{bmatrix} \varepsilon_1 \\ \varepsilon_2 \\ \gamma_{12} \end{bmatrix} = \begin{bmatrix} S_{11} & S_{12} & 0 \\ S_{21} & S_{22} & 0 \\ 0 & 0 & S_{66} \end{bmatrix} \begin{bmatrix} \sigma_1 \\ \sigma_2 \\ \tau_{12} \end{bmatrix} \quad S_{11} = \frac{1}{E_1}; \quad S_{12} = -\frac{v_{12}}{E_1}; \quad S_{22} = \frac{1}{E_2}; \quad S_{66} = \frac{1}{G_{12}} \tag{6.62}$$

Also, we know that, because of symmetry in material behavior,

$$\frac{v_{12}}{E_1} = \frac{v_{21}}{E_2} \tag{6.63}$$

Knowing the material properties obtained from elementary tests, we can generate the parameters for stress–strain relations for a general state of stress as described in Equation 6.40. Alternatively, one can express the relationship in terms of the inverse of compliance, S, or the stiffness matrix, Q, defined as

$$\begin{bmatrix} \sigma_1 \\ \sigma_2 \\ \tau_{12} \end{bmatrix} = S_{ij}^{-1} \varepsilon_j = Q_{ij} \varepsilon_j = \begin{bmatrix} Q_{11} & Q_{12} & 0 \\ Q_{21} & Q_{22} & 0 \\ 0 & 0 & Q_{66} \end{bmatrix} \begin{bmatrix} \varepsilon_1 \\ \varepsilon_2 \\ \gamma_{12} \end{bmatrix} \tag{6.64}$$

with its components

$$Q_{11} = \frac{E_1}{1 - v_{12}v_{21}} \quad Q_{22} = \frac{E_2}{1 - v_{12}v_{21}} \quad Q_{12} = \frac{v_{12}E_2}{1 - v_{12}v_{21}} \quad Q_{66} = G_{12} \tag{6.65}$$

Therefore, a unidirectional composite, which is isotropic in the second and third planes, has four independent elastic constants and four independent stiffness coefficients. Substituting Equations 6.36 and 6.37 in Equation 6.42 results in the mapping in-between stresses and strain in an arbitrary direction for a lamina:

$$\begin{bmatrix} \sigma_x \\ \sigma_y \\ \tau_{xy} \end{bmatrix} = [T]^{-1} \begin{bmatrix} \sigma_1 \\ \sigma_2 \\ \tau_{12} \end{bmatrix} = [T]^{-1}[S]^{-1} \begin{bmatrix} \varepsilon_1 \\ \varepsilon_2 \\ \gamma_{12} \end{bmatrix} = [T]^{-1}[Q] \begin{bmatrix} \varepsilon_1 \\ \varepsilon_2 \\ \gamma_{12} \end{bmatrix} = [T]^{-1}[Q][R][T][R]^{-1} \begin{bmatrix} \varepsilon_x \\ \varepsilon_y \\ \gamma_{xy} \end{bmatrix} = [\bar{Q}] \begin{bmatrix} \varepsilon_x \\ \varepsilon_y \\ \gamma_{xy} \end{bmatrix} \tag{6.66}$$

The compliance of an arbitrary orientation angle is denoted as $\bar{Q}_{ij}$ and can be written as,

$$\bar{Q}_{ij} = T^{-1}\bar{S}^{-1}R\,T\,R^{-1} \quad \text{or} \quad [\bar{Q}] = [T]^{-1}[Q][R][T][R]^{-1} = \begin{bmatrix} \bar{Q}_{11} & \bar{Q}_{12} & \bar{Q}_{16} \\ \bar{Q}_{12} & \bar{Q}_{22} & \bar{Q}_{26} \\ \bar{Q}_{16} & \bar{Q}_{26} & \bar{Q}_{66} \end{bmatrix} \tag{6.67}$$

Or, in algebraic terms,

$$\bar{Q}_{11} = Q_{11}\cos^4\theta + 2(Q_{12} + 2Q_{66})\sin^2\theta\cos^2\theta + Q_{22}\sin^4\theta$$

$$\bar{Q}_{22} = Q_{11}\sin^4\theta + 2(Q_{12} + 2Q_{66})\sin^2\theta\cos^2\theta + Q_{22}\cos^4\theta$$

$$\bar{Q}_{12} = (Q_{11} + Q_{22} - 4Q_{66})\sin^2\theta\cos^2\theta + Q_{12}(\sin^4\theta + \cos^4\theta)$$

$$\bar{Q}_{16} = (Q_{11} - Q_{12} - 2Q_{66})\sin\theta\cos^3\theta + (Q_{12} - Q_{22} + 2Q_{66})\sin^3\theta\cos\theta \qquad (6.68)$$

$$\bar{Q}_{26} = (Q_{11} - Q_{12} - 2Q_{66})\sin^3\theta\cos\theta + (Q_{12} - Q_{22} + 2Q_{66})\sin\theta\cos^3\theta$$

$$\bar{Q}_{66} = (Q_{11} + Q_{12} - 2Q_{12} - 2Q_{66})\sin^2\theta\cos^2\theta + 2Q_{66}(\sin^4\theta + \cos^4\theta)$$

This approach can be used to update the amount of elastic stiffness in the matrix with the changes due to cracking (Figure 6.13).

Classical laminate theory assumes that all layers are perfectly bonded together, and that the strains are continuous throughout the laminate thickness. In the absence of an applied moment (for a symmetric laminate), the strains in any given direction are equal, for example, $(\varepsilon_x)_k = (\varepsilon_x)_{k+1}$. If there is an applied moment, strains through the thickness can be calculated by the curvature, as in the case of engineering beam theory as shown in Figure 6.11. Thin plate theory assumes that any line initially straight and normal to the midplane of the plate remains straight and normal to the midplane after extension and flexure (this is equivalent to ignoring shear strain measures γ_{xz} and γ_{yx}) (Figure 6.14).

In Figure 6.12, the line AD is normal to the middle surface. Intersecting the middle surface at B. u_0 is the displacement of point B on the middle surface in the x direction, and β is the slope of the

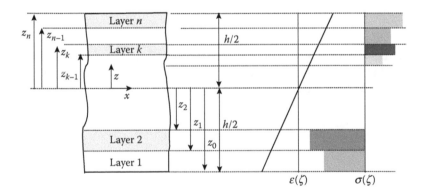

FIGURE 6.13 Schematics of laminate geometry, the strain and stress distribution.

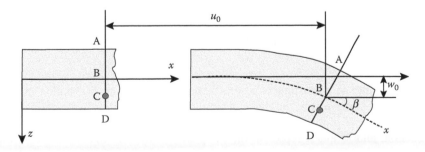

FIGURE 6.14 Bending of a laminate illustrating curvature of middle surface and displacement of point C.

middle surface. z_c is the distance from B on the middle surface to C in the z direction. Therefore, C is displaced in the x direction. On the basis of the assumptions of thin plate theory,

$$\beta = \frac{\partial w_0}{\partial x} \tag{6.69}$$

$$u_c = u_0 - z_c \frac{\partial w_0}{\partial x} \tag{6.70}$$

If v is the displacement in the y direction,

$$v = v_0 - z \frac{\partial w_0}{\partial y} \tag{6.71}$$

$$\varepsilon_x = \frac{\partial u}{\partial x} = \frac{\partial u_0}{\partial x} - z \frac{\partial^2 w_0}{\partial x^2} \tag{6.72}$$

$$\varepsilon_y = \frac{\partial v}{\partial y} = \frac{\partial v_0}{\partial y} - z \frac{\partial^2 w_0}{\partial y^2} \tag{6.73}$$

$$\gamma_{xy} = \frac{\partial u}{\partial y} + \frac{\partial v}{\partial x} = \frac{\partial u_0}{\partial y} + \frac{\partial v_0}{\partial x} - z \frac{\partial^2 w_0}{\partial x \partial y} \tag{6.74}$$

Therefore, stress at a point in the kth layer of the laminate can be located from the stress–strain relation described in Equation 6.53, where ε_x^0, ε_y^0, γ_{xy}^0 are the in-plane strain at the midplane, and κ_x, κ_y and κ_{xy} are the plate curvatures of the midplane surface,

$$\begin{bmatrix} \varepsilon_x \\ \varepsilon_y \\ \gamma_{xy} \end{bmatrix} = \begin{bmatrix} \varepsilon_x^0 \\ \varepsilon_y^0 \\ \gamma_{xy}^0 \end{bmatrix} + z \begin{bmatrix} \kappa_x \\ \kappa_y \\ \kappa_{xy} \end{bmatrix} \tag{6.75}$$

$$\begin{vmatrix} \sigma_x^k \\ \sigma_y^k \\ \sigma_{xy}^k \end{vmatrix} = \begin{bmatrix} QT_{11}^k & QT_{12}^k & QT_{16}^k \\ QT_{12}^k & QT_{22}^k & QT_{26}^k \\ QT_{16}^k & QT_{26}^k & QT_{66}^k \end{bmatrix} \begin{bmatrix} \begin{vmatrix} \varepsilon_x^0 \\ \varepsilon_y^0 \\ \gamma_{xy}^0 \end{vmatrix} + z \begin{vmatrix} k_x \\ k_y \\ k_{xy} \end{vmatrix} \end{bmatrix} \tag{6.76}$$

For a composite laminate consisting of several lamina each with an orientation of θ^m, where m ranges from the first to the nth ply, the classical lamination theory results in the derivation of lamina stiffness components. M represents the moment per unit length, N is the force per unit length of the cross section, and ε^0 and κ represent the midplane strains and the curvature of the section, respectively (Figure 6.15):

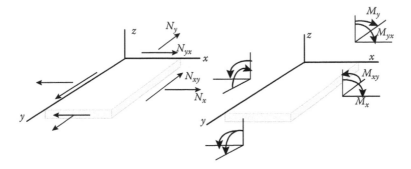

FIGURE 6.15   Force and moment distribution.

$$\begin{bmatrix} N_x \\ N_y \\ N_{xy} \end{bmatrix} = \int_{-h/2}^{h/2} \begin{bmatrix} \sigma_x \\ \sigma_y \\ \sigma_{xy} \end{bmatrix} dz \qquad (6.77)$$

$$\begin{bmatrix} M_x \\ M_y \\ M_{xy} \end{bmatrix} = \int_{-h/2}^{h/2} \begin{bmatrix} \sigma_x \\ \sigma_y \\ \sigma_{xy} \end{bmatrix} z\,dz \qquad (6.78)$$

$$\begin{bmatrix} [N] \\ [M] \end{bmatrix} = \begin{bmatrix} [A] & [B] \\ [B] & [D] \end{bmatrix} \begin{bmatrix} [\varepsilon^0] \\ [\kappa] \end{bmatrix} \qquad (6.79)$$

where A, D, and B represent the extensional, the bending, and the coupling stiffness, respectively. Using this approach, the stresses and strains in each lamina may be obtained by inverting the laminate constitutive relationship as discussed by Agarwal and Broutman [17]. If we know the laminate strains and curvatures, the stress distribution per lamina is computed for each loading step either in direct or in an incremental fashion as,

$$A_{ij} = \sum_{m=1}^{n} \bar{Q}_{ij}^m (z_m - z_{m-1}), \quad B_{ij} = \frac{1}{2}\sum_{m=1}^{n} \bar{Q}_{ij}^m (z_m^2 - z_{m-1}^2), \quad D_{ij} = \frac{1}{3}\sum_{m=1}^{n} \bar{Q}_{ij}^m (z_m^3 - z_{m-1}^3) \quad (6.80)$$

In balanced composites where $B = 0$, an uncoupled formulation can be used by separating axial and flexural stresses. The problem can be solved incrementally if nonlinear material properties are specified. For each iteration, the incremental loads and strains are determined, which updates the loads and strains of previous increments,

$$\Delta N_x = [\bar{A}] \ [\Delta \varepsilon^0] \qquad (6.81)$$

Similarly,

$$\Delta M_x = [\bar{D}][\Delta \kappa] \qquad (6.82)$$

PLY DISCOUNT METHOD

This procedure is used to analyze the failure of these composites using the first ply failure approach [18]. An incremental procedure is used, and the strains and forces are updated incrementally according to a piecewise linear assumption. The total strains, stresses, and forces are updated and checked against the failure criterion for either the matrix phase or the lamina phase. As a layer meets the failure criteria, the stiffness of the layer is reduced in accordance to the damage law used. Assuming that N represents the nominal applied load, at each increment,

$$N_i^j = N_{i-1}^j + \Delta N_i^j \tag{6.83}$$

$$\varepsilon_i^j = \varepsilon_{i-1}^j + \Delta \varepsilon_i^j \tag{6.84}$$

$$\begin{bmatrix} \Delta N \\ \Delta M \end{bmatrix} = \begin{bmatrix} \bar{A} & \bar{B} \\ \bar{B} & \bar{D} \end{bmatrix} \begin{bmatrix} \Delta \varepsilon^0 \\ \Delta \kappa \end{bmatrix} \tag{6.85}$$

where the $\bar{A}$ is modified at each increment to take into account the fact that some of the layers have fractured. After each failure, the incremental loads and strains are determined, and the results are added to the loads and strains at the previous ply. In a symmetric laminate, there is no shear coupling effects, and the submatrix $\bar{B}$ reduces to zero. For a state of uniaxial tension ($\Delta \kappa = 0$), Equation 6.61 reduces to

$$\begin{bmatrix} \Delta N \end{bmatrix} = \begin{bmatrix} \bar{A} \end{bmatrix} \begin{bmatrix} \Delta \varepsilon^0 \end{bmatrix} \tag{6.86}$$

In the present calculations, the matrix and fiber elastic properties were used in the rule of mixtures to calculate the E_1. Calculation of the transverse modulus E_2 and ν_{12} was achieved using the Halpin–Tsai equations [2]. The strength of the matrix in the presence of fibers depends on the mechanical properties and volume fraction of the fibers, matrix, and the interface. Theoretical justifications for the evaluation of matrix strength has been provided in an earlier work [10].

FAILURE CRITERIA

Several methods are available to predict the strength of a composite. If the stresses in the two constituents are known, as in the case of the long fiber under axial loading, then these values can be compared with the corresponding strengths of each phase to determine which will fail. This ultimate strength approach is a logical development from the analysis of axial stiffness, with additional input from the variable of the ratio between the strengths of fiber and matrix.

Such predictions are, in practice, complicated by uncertainties about *in situ* strengths, interfacial properties, residual stresses, and so forth. Instead of relying on predictions such as those outlined earlier, it is often necessary to measure the strength of the composite, usually by loading the fibers parallel, transverse, and in shear. This provides a basis for the prediction of whether a component will fail when a given set of stresses is generated, although other factors such as environmental degradation or the effect of failure modes on toughness may require attention.

Failure criteria are used to compare the actual state of stress with the failure stresses measured in the tests. Failure criteria can be developed for a unidirectional composite from a macro mechanical point of view. Once failure criteria have been developed for individual plies, they can be incorporated into a failure analysis procedure for a laminate.

MAXIMUM STRESS THEORY

The maximum stress theory, when applied to an orthotropic material, assumes that the failure of a ply occurs when any of the stress components in the principal material directions (1–2) reaches its corresponding strength property. In other words, failure happens under any of the following conditions:

$$\sigma_1 \geq \sigma_{1t}^{fu}, \quad \sigma_2 \geq \sigma_{2t}^{fu}, \quad \sigma_1 \geq \sigma_{1c}^{fu}, \quad \sigma_2 \geq \sigma_{2c}^{fu}, \quad \tau_{12} \geq \tau_{12}^{fu} \tag{6.87}$$

The theory considers each component independently, ignoring any interaction. Each criterion must also be applied in the appropriate principal material direction. Actual failure stress measured in uniaxial tensile testing of off-axis specimens differs from the predictions of the criteria. Therefore, a more comprehensive failure criterion, which considers the interactions among stress components, will be used for the multidirectional models.

Interactive Failure Criterion, Tsai–Hill

Because rolled metals have slightly different properties in the roll direction than in the other two directions, R. Hill developed an extension of the Von Mises–Hencky distortion energy theory as a yield criterion for orthotropic materials. S.W. Tsai utilized these constants for the failure strengths for a unidirectional composite. In his procedure, it is assumed that individual stresses are applied one at a time along principal material directions, as in the tests to measure strengths.

Now, in our case, we can include that, in the plane stress condition, in which unidirectional composites are normally characterized, the only stress components are σ_1, σ_2, and τ_{12}. Setting the other components to zero, we have the following equation (the Tsai–Hill criterion):

$$\frac{\sigma_1^2}{S_1^2} - \frac{\sigma_1 \sigma_2}{S_1^2} + \frac{\sigma_2^2}{S_2^2} + \frac{\tau_{12}^2}{S_{12}^2} = 1 \tag{6.88}$$

$$F(\sigma_1, \sigma_2, \tau_{12}) = 1 \tag{6.89}$$

$$F(\sigma_1, \sigma_2, \tau_{12}) = F_{11}\sigma_1^2 + 2F_{11}\sigma_1\sigma_2 + F_{22}\sigma_2^2 + F_{66}\sigma_{12}^2 + F_1\sigma_1 + F_2\sigma_2 = 1 \tag{6.90}$$

$$F_1 = \frac{1}{|\sigma_{1t}|} - \frac{1}{|\sigma_{1c}|} \quad F_2 = \frac{1}{|\sigma_{2t}|} - \frac{1}{|\sigma_{2c}|} \quad F_{11} = \frac{1}{|\sigma_{1t}\sigma_{1c}|} \quad F_{22} = \frac{1}{|\sigma_{2t}\sigma_{2c}|} \quad F_{66} = \frac{1}{\sigma_{6u}^2} \tag{6.91}$$

REFERENCES

1. Jones, R. M. (1975). *Mechanics of Composites Materials*. New York: McGraw-Hill Book Co.
2. Halpin, J. C., and Tsai, S. W. "Environmental factors in composite materials design," Air Force Materials Research Laboratory, Technical Report, AFML-TR-67-423, 1967.
3. Adams, D. F., and Tsai, S. W. (1969), "The influence of random filament packing on the transverse stiffness of unidirectional composites," *Journal of Composite Materials*, 3(3), 368–381.
4. Hashin, Z., "On elastic behaviour of fibre reinforced materials of arbitrary transverse phase geometry," *Journal of the Mechanics and Physics of Solids*, 13(3), 119–134.
5. Whitney, J. M., and Riley, M. B. (1966), "Elastic properties of fiber reinforced composite materials," *AIAA Journal*, 5, 1537–1542.
6. Tsai, S. W., "Structural Behavior of Composite Materials," NASA CR71, July 1964.
7. Gorur, R., Mobasher, B., Olsen, R., Shinde, S., and Erni, J., "Characterization of composite cores for high temperature-low sag (HTLS) conductors," PSERC, IAB Meeting, Cornell University, May 27–29, 2009.

8. Darlington, M. W., McGinley, P. L., and Smith, G. R. (May 1976), "Structure and anisotropy of stiffness in glass fibre-reinforced thermoplastics," *Journal of Materials Science*, 11(5), 877–886.
9. Ekvall, J. C. (1961), "Elastic properties of orthotropic monofilament laminates," ASME, Paper No. 61-AV-56, Los Angeles.
10. Cox, H. L. (1952), "The elasticity and strength of paper and other fibrous materials," *British Journal of Applied Physics*, 3, 72–79.
11. Greszczuk, L. B. "Thermoelastic Properties of Filamentary Composites," Proceedings of the AIAA 6th Structural Materials Conference, vol. 285, 1965.
12. Sueki, S., Soranakom, C., Peled, A., and Mobasher, B. (2007), "Pullout-slip response of fabrics embedded in a cement paste matrix," *Journal of Materials in Civil Engineering*, 19, 9.
13. Soranakom, C., and Mobasher, B. (2009), "Geometrical and mechanical aspects of fabric bonding and pullout in cement composites," *Materials and Structures*, 42, 765–777.
14. Soranakom, C., and Mobasher, B. (2010), "Modeling of tension stiffening in reinforced cement composites: Part I. Theoretical modeling," *Materials and Structures*, 43, 1217–1230.
15. Soranakom, C., and Mobasher, B. ((2010)), "Modeling of tension stiffening in reinforced cement composites: Part II. Simulations vs. experimental results," *Materials and Structures*, 43, 1231–1243
16. Jones, R. M. (1975). *Mechanics of Composites Materials*. New York: McGraw-Hill.
17. Agarwal, B. D., and Broutman, L. J. (1990). *Analysis and Performance of Fiber Composites*, 2nd ed. New York: Wiley.
18. Kwon, Y. W., Allen, D. H., and Talreja, R. (2007). *Multiscale Modeling and Simulation of Composite Materials and Structures*. Springer, 630 pp.

7 Mechanical Testing and Characteristic Responses

INTRODUCTION

Macrostructure properties such as strength, stiffness, and ductility are determined through standard mechanical tests. Much more information is obtained about the interaction of various internal phases when tests measure the entire specimen response under controlled conditions. By documenting the nature of cracking and the mechanisms that contribute to enhanced ductility in composites, proper modeling can be conducted. This is achieved through procedures to detect the incipient failure as the load is being applied and through the development and control of the test in response to that failure. The development of closed-loop testing procedures has led to the identification of the phenomenon known as strain softening. This characteristic response is attributed to the gradual decrease of load carrying capacity with an increase in the strain or widening of cracks in a tensile stress field. The interpretation of the stress–strain curve thus obtained is not straightforward, because its postpeak portion of the response is influenced significantly by the specimen size, geometry, control mode of test, and loading setup. Moreover, the postpeak deformation is not homogenous, but is localized within a narrow zone that undergoes progressive damage and cracking. Nonetheless, obtaining the postpeak response demonstrates the behavior of the specimen in terms of its ductility, energy absorption capacity, and the parameters that contribute to it. A series of tests, including the mechanical properties developed, that address tension, flexure, compression, in-plane shear, interlaminar shear, bond, and connection. Characteristic material responses obtained in this chapter include elastic properties, stiffness, toughness, stress–strain curves, fracture toughness, ultimate tensile strength, crack width, crack spacing, and the moment–curvature relationship.

CONCEPTS OF CLOSED-LOOP TESTING

The ability of a mechanical testing system to accurately respond to a wide range of input parameters depends on the functionality of the instrumentation in monitoring and controlling the loading actuators. This is commonly known as the control parameter. In general, the control can be classified as an open or closed loop, in which the loop signifies the use of the system output as feedback by the control process. In an open-loop control, the system response and output is not used by the controller, and the process depends only on the system input (see Figure 7.1a). The variables that can be controlled in such systems are usually the actuator (piston) displacement and the applied load (or pressure), both of which are not significantly affected by the behavior of the test specimen. This is analogous to other automated systems, such as programmable washing machines and toasters. In closed-loop controls (CLC), the output of the controlled variable is directly monitored by the controller (see Figure 7.1b). This can be any parameter that is accessible to the controller, such as the specimen displacement, strain, and crack opening. Its actual and desired (reference input) values are equalized indirectly by the controller by manipulating the movement of the actuator. Analogies for this system include the autopilots of ships and planes and the cruise control in cars. CLC has also been applied in the active control structures [1], where the process is similar to testing systems.

In closed-loop controlled systems, as shown in Figure 7.1b, the current value of the controlled variable is fed back to the controller and compared with the reference input signal. The difference between the two signals (i.e., the error) is used to manipulate the actuator, and therefore the process is also known as negative feedback control [2]. The reference input in testing machines is provided by a function generator. The feedback signal is normally the output of a transducer, which is monitored continuously in analog controllers and sampled at discrete increments using digital controllers. Obviously, the scope of CLC is greater than in open-loop control systems, because the range of controlled variables is much wider. Even for the same controlled variable, say, piston displacement, the closed-loop system produces a more accurate output than the open-loop system. CLC drawbacks include a higher initial cost and the need for more operator skill. In addition, the lag between the actual response and the corrective action of the controller may result in loss of control, overcorrection, or undercorrection. Because of these considerations, closed-loop controllers have to be properly designed through modeling and analysis. The theories and techniques used in the analysis of CLC systems, as well as a more mathematical treatment of control systems, can be found in several publications [1, 3–6].

CLC is most useful when there is a rapid and unpredictable change in the system input or specimen behavior, such as when a crack becomes unstable and propagates rapidly. Therefore, the transient response in the time domain is important and is normally evaluated by the response to a step input as described by the parameters shown in Figure 7.2. These parameters are strongly interrelated and have to be optimized for the best transient performance. On the other hand, the performance of the system under dynamic cyclic input is characterized by the response in the frequency domain, which is characterized mainly by the maximum frequency sustained by the system and the phase lag between the input and the output signals.

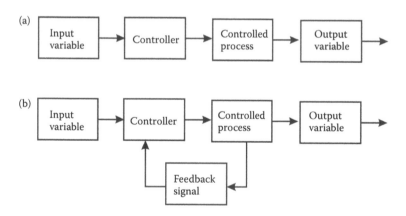

FIGURE 7.1   Schematics of (a) open-loop and (b) closed-loop testing.

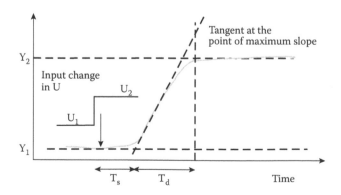

FIGURE 7.2   Response of a CLC system to an input impulse.

COMPONENTS AND PARAMETERS OF CLC

THE PROPORTIONAL-INTEGRAL-DERIVATIVE (PID) CONTROLLER

The most common testing machine configuration is shown in Figure 7.1, where the controller is in series with the controlled process. In this setup, called a series compensation, the negative feedback controller generates a control signal that, in its simplest form, is given as

$$u(t) = K_p e(t); \quad e(t) = r(t) - c(t) \tag{7.1}$$

where $r(t)$ is the reference input, $c(t)$ is the output of the controlled variable (i.e., the feedback signal), $e(t)$ is the error signal, $u(t)$ is the control signal, t represents time, and K_p is a constant. This type of control, in which the control signal is obtained by simply amplifying the error, is called proportional control. The parameter K_p is consequently called the proportional gain. Although the proportional element is the critical component of the controller, other complementary elements are needed to make it more versatile. A commonly used configuration is the PID controller, where the letters stand for the Proportional, Integral, and Derivative actions generated by the controller. The corresponding control signal is of the form

$$u(t) = K_p e(t) + K_I \int e(t) dt + K_D \frac{de(t)}{dt} \tag{7.2}$$

where the second and the third terms are the integral and the derivative elements, respectively, and parameters K_I and K_D are the integral and the derivative gains. Each element of the PID controller performs a specific function. In other words, the proportional element governs the dynamic behavior of the system. A sluggish system response, characterized by a long rise time (Figure 7.2; T_d), is improved by boosting the control signal, that is, by increasing K_p. However, a very large K_p tends to make the system unstable or decreases the damping of the oscillations (i.e., settling time). The integral element reduces the steady-state error, because the integration over time makes it sensitive to the presence of even a small error. In stable systems, integral control improves the steady-state error by one order. For example, if the error is constant for a certain input, the integral element reduces it to zero. This is especially useful for increasing system accuracy during slow and low-frequency tests and for maintaining the mean level of high-frequency input signals. Additionally, an increase in K_I leads to an increase in the damping (i.e., decrease in the oscillations in the transient response). However, this occurs at the expense of the rise and settling times, which also increase.

The response of a system to an input step function is shown in Figure 7.3. Derivative control primarily improves the system performance in high-frequency operations. By using the slope of the error, the derivative element anticipates overshoots and takes corrective action before they actually

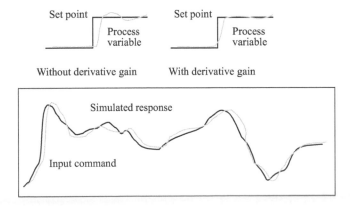

FIGURE 7.3 Comparison of input command and simulated response and the effect of derivative gain.

occur. Therefore, this element is used mainly for decreasing the maximum overshoot and for damping the oscillations in the transient response (see Figures 7.2 and 7.3). Obviously, it affects the system only when there is a significant change in the error and therefore does not improve a constant steady-state error. Because it uses the slope of the error signal, the derivative control accentuates any high-frequency noise that enters the system (e.g., from transducers). The use of the PID controller with a proper choice of parameters produces satisfactory results. Nevertheless, modifications are sometimes made for specific purposes [6, 7]. For example, the velocity of the controlled variable is used by the controller, instead of the derivative element, in systems where it can be measured directly (i.e., without differentiating the output with respect to time). This process, known as rate feedback control, improves the damping and suppresses the occurrence of large overshoots in the initial transient response. Another improvement of the PID controller is the inclusion of feed-forward compensation [7]. This provides an additional degree of freedom and increases the system response to sudden changes in input, especially during high-frequency loading. The PID controller also improves system fidelity when working with soft specimens in load control, as well as with large actuators, heavy fixtures, and moving load cells. An example of PID loops that incorporate feed-forward control is shown in Figure 7.6 [8].

ACTUATORS AND SERVOMECHANISM

Two types of actuators are normally used to apply loads: those with helicoidal screws driven by electric motors (called electromechanical), and those driven by hydraulic actuators. When the actuator is part of a closed-loop testing system, it is manipulated by the controller through a servomechanism. The function of the servomechanism is to drive the actuator so that its movement is in a direction to minimize the error function, which is defined as the difference between the programmed function and the feedback signal's response. Consequently, closed-loop controlled systems are also known as servo-controlled systems.

HYDRAULIC ACTUATORS AND SERVOVALVES

Hydraulic actuators are of two classes: single acting (Figure 7.4a), where the load is proportional to the applied pressure, and double acting (double ended as in Figure 7.4b or single ended), where the load is proportional to the difference between the pressures in the two actuator chambers. Single-acting actuators are normally used in open-loop systems, in which the pump pressure is controlled directly. Although it is possible to conduct closed-loop tests on single-acting actuators, these hydraulic systems are unable to reclaim system energy in order to respond to sudden changes in command under a dynamic condition. Double-acting actuators are governed by electrohydraulic servovalves under CLC. A typical two-stage servovalve is shown schematically in Figure 7.5.

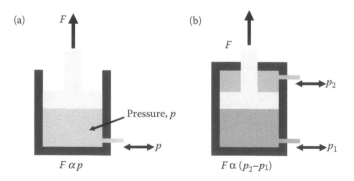

FIGURE 7.4 Single- and double-acting actuators. (a) Single-acting actuator with force proportional to pressure. (b) Double-acting actuators with force proportional to differential pressure.

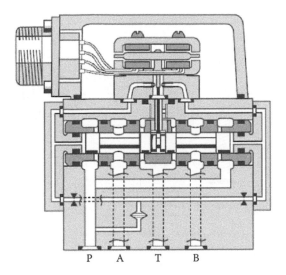

FIGURE 7.5 A schematic diagram of the internal structure of a two-stage servovalve. (Courtesy of MOOG Corporation.)

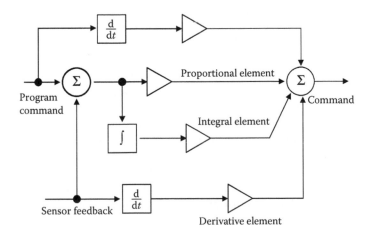

FIGURE 7.6 Schematics of a PID controller setup.

Its function is to provide the actuator with oil at a flow rate that is proportional to the error signal. Two of its ports are connected to a pump: one to the pressure outlet, which provides oil at a constant pressure (normally 21 MPa), and another to the return inlet. The electric signal creates a differential opening in the valve that controls the flow rate between the two chambers. The net difference in pressure results in movement of the floating actuator and an application of the load.

SERVOHYDRAULIC TESTING MACHINES

THE ELECTRONICS

The basic components and the configuration of a typical PID controller closed-loop testing are represented in Figure 7.6. The first component is the function generator, which produces an electrical signal that matches the desired testing protocol in terms of a ramp, sinusoidal, sawtooth, or random function. This reference input is fed into the controller that, furthermore, collects the output of the

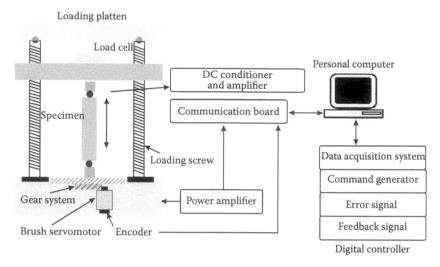

FIGURE 7.7 Closed-loop testing system.

controlled variable. Another important aspect is the measurement of the controlled variable [2]. One characteristic of the specimen behavior, such as a load, displacement, deformation, velocity, or acceleration, is chosen as the controlled variable. Usually, the output of the controlled variable is measured by a transducer that is in contact with the specimen, such as a load cell, displacement gauge, or an extensometer. These two signals are added in order to generate the control signal that governs the servovalve. The controller can be of two basic types: analog, where, theoretically, the loop is always closed, or digital, where the loop is closed in the order of 5000 times every second. Such a high loop-closure rate is sufficient for conducting stable tests for brittle materials like rock and concrete [2, 9]. In general, noise should be maintained at less than 0.25% of the full transducer range.

Figure 7.7 shows the schematics of the testing setup with computer data acquisition and control parameters for a complete CLC test. Such an apparatus can be used to conduct tension, compression, fatigue, fracture, and flexure tests. To ensure a stable control, the controlled variable should be sensitive to the failure of the specimen. For brittle materials, this corresponds to the direction of maximum tensile stress and the direction perpendicular to crack propagation. The criteria for selecting the controlled variable and other practical considerations are discussed in the following subsections when the different configurations are described.

COMPRESSION TEST

The compression test is the most common tool used for characterizing concrete. Although it is conventionally used only to obtain the maximum stress (i.e., strength) and the modulus of elasticity, the test can be extended into the postpeak regime to determine the entire stress–strain response. Standard ASTM tests are conducted under a load or actuator displacement control conditions. In these cases, once the maximum load is attained, the resistance of the specimen to an increasing load level is exhausted. If the test machine is unable to adjust the load by moving energy away from the specimen, the specimen shatters suddenly, due to the stored strain energy in the sample and system. If the stiffness of the testing system is high enough to ensure that the energy released by the machine is lower than that consumed by the specimen during deformation, then it is possible to maintain stability during the unloading phase.

The test configuration needed for obtaining the stable postpeak response depends on the behavioral class and the brittleness of the material. Available options for a controlled variable include the following: load, stroke, axial displacement, transverse, circumferential displacement, and their combinations. Load control is excluded after peak load, because it will not permit a decrease in load after the peak. Postpeak response can be obtained by an extensometer placed using a chain-like fixture around the specimen used to measure the circumferential displacement as the control. A typical example includes a special ring-like fixture for measuring the specimen axial, and circumferential strain, as shown in Figure 7.8. Two linear variable differential transducers (LVDTs) are used to measure the axial strain in the specimen using a gage length that encompasses a majority of the axial length. A chain (with rollers for minimizing friction) can also be wrapped around the specimen to support the extensometer measuring the circumferential dilatation and relating it to the transverse strain. This test initially starts with the load control and then the control is transferred to extensometer that controls the test before the specimen experiences the peak load, through the peak load, and through the postpeak response until the end of the test. Figure 7.9a shows the response of the axial and circumferential transducers. The switch over point is also identified.

Two classes of behavior are observed here. Under normal conditions, the axial strain increases as the circumferential displacement increases; however, under a snap-back condition, a decrease in axial displacement during the descending part of load strain response takes place as the circumferential strain increases. Both these modes have been observed in the analysis of the data. This

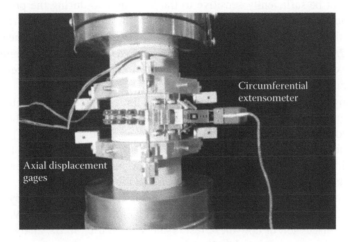

FIGURE 7.8 A compression test sample with axial and circumferential transducers.

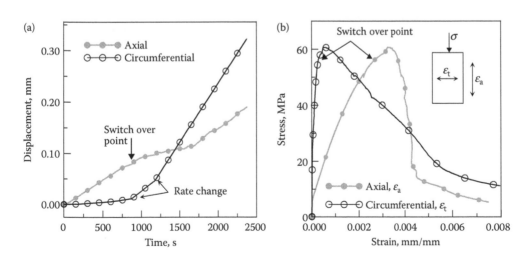

FIGURE 7.9 Closed-loop testing in compression using (a) an axial deformation followed by circumferential deformation as the controlled variable and (b) a compressive stress versus strain response. (After Gettu, R., Mobasher, B., Carmona, S., and Jansen, D. (1996), "Testing of concrete under closed-loop control," *Invited Article, Journal of Advanced Cement Based Materials*, 3(2), 54–71.)

---

is partly explained by the fact that the postpeak deformation is not homogenous, but is localized within a narrow zone that undergoes progressive damage and cracking. The energy released due to unloading in the bulk of the specimen dominates the displacement response.

The interpretation of the stress–strain curve in the postpeak region is not straightforward, because it represents an average displacement in both elastic and inelastic zones and is further influenced by the specimen geometry and loading setup. Moreover, the postpeak deformation is not homogeneous, but is localized within a narrow zone that undergoes progressive damage and cracking. Therefore, the stress–strain curve beyond the peak, as shown in Figure 7.9b, represents a combination of structural and material responses [27].

Figures 7.9a and 7.9b show the typical curves obtained using circumferential deformation as the controlled variable in tests of concretes with compressive strength. These plots show that, while the sample exhibits strain softening (class I behavior), the tendency toward snap back (class II behavior) increases with the choice of axial strain as the control variable. More importantly, the circumferential deformation always increases throughout the test. In some cases, the increase in circumferential deformation may not be sufficiently sensitive to the load applied during the prepeak regime [31]. This can be handled by initially using a load or axial deformation as the controlled variable and then switching to the circumferential deformation control when the specimen begins to dilate significantly. Figure 7.9b shows the stable response obtained in a test in which the control was switched from a constant axial displacement rate to a constant circumferential displacement rate when a certain circumferential displacement was reached.

Figure 7.10 shows the effect of AR glass fibers on the compressive strength and ductility of concrete in compression. Figure 7.10 shows the stress versus the circumferential strain for concrete containing a hybrid combination of two fiber lengths 12 and 24 mm (50%–50%). The stress–strain results are shown for various ages at 3, 7, and 28 days. The increase in strength from 3 to 7 days is more significant than from 7 to 28 days. There is also an increase of 35% in strength from 3 to 7 days. The postpeak energy absorption at 28 days is much higher than at 3 days. The hybrid system of concrete containing of 12 to 24 mm (50%–50%) fibers is more effective in early age strength than strength at 28 days. At the same time, a better toughening of the system is observed after 28 days of curing. A complete discussion of how these test results can be used in the analysis of various fiber combinations and systems is presented by Desai et al. [10, 11].

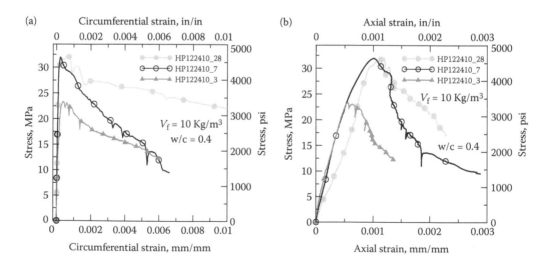

FIGURE 7.10 Effect of AR glass fibers on the compressive strength and ductility of concrete. (After Desai, T., Shah, R., Peled, A., and Mobasher, B., "Mechanical properties of concrete reinforced with AR-glass fibers," *Proceedings of the 7th International Symposium on Brittle Matrix Composites* (BMC7), Warsaw, October 13–15, 2003, 223–232.)

UNIAXIAL TENSION TEST

The entire load-displacement curve of concrete under uniaxial tension was first discovered in the 1960s. Concrete plates were loaded in parallel with steel bars [34] to avoid unstable failures after the peak. However, such passive methods of control do not always yield accurate results, because the deformation of concrete increases homogeneously at first, but near the peak load, it localizes within a planar region that develops as a crack. The region outside the crack unloads while the crack continues to open. Therefore, the total load-displacement response normally exhibits snap-back behavior because of the large unloading region. Unlike in compression, the localized zone in tension is narrow and does not necessarily pass through the middle of the specimen. This makes the test quite difficult to control. The use of notched sections or "dog-bone" specimens ensures that the crack occurs within the zone of the reduced section. When the displacement over this zone is measured, the corresponding load-displacement curve is often free from snap back (because most of the unloading material lies outside the gauge length). This displacement can then be used as the controlled variable. Another problem in tension tests of plate specimens is that the crack rarely propagates simultaneously from both the edges. This implies that the displacement should be monitored on both sides of the specimen with two transducersm, and that the controlled variable has to be either the dominant sensor or the sum of the absolute values of their outputs to eliminate bending effects.

Tension tests are performed using different types of grip configurations that simulate fixed or pinned conditions. The most convenient fixed-type connections use hydraulic grips. Pin-type grips, on the other hand, are attached to thin plates that are glued onto the specimens or fixtures with wedge-type frictional grips and universal joints that allow the specimen to rotate at the support. These fixtures have been used successfully on cement composites systems [12, 13]. Figure 7.11 represents the schematics of the tensile test setup. Notched specimens are tested in tension using two LVDTs of a sufficiently small range (1.5 mm) mounted on both sides of the notched specimen at a gage length of approximately 10 mm to measure the crack opening. The use of notched sections or "dog-bone" specimens ensures that the crack occurs within the zone of the reduced section. When the displacement over this zone is measured, the corresponding load-displacement curve is often free from snap back and provides stability at the postpeak [14].

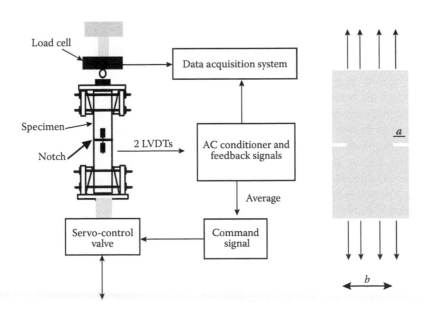

FIGURE 7.11 The schematics of the tensile test setup.

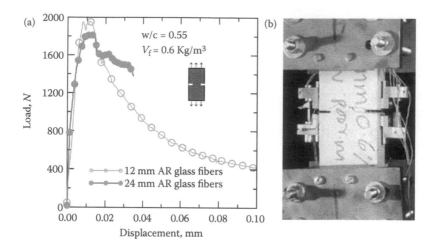

FIGURE 7.12 Tensile test on AR glass fiber composites containing 12- and 24-mm-long fibers.

Two displacement sensors are placed across the potential failure plane on one of the flat faces, and their output is averaged and used as the controlled variable. In this manner, stable pre- and post-cracking responses of high-strength concrete specimens, with and without fibers, have been calculated. The response of the LVDTs are conditioned and used as the command signal to control the servovalve. Mobasher and Li [13] used a similar procedure for determining the tensile response of fiber reinforced concrete (FRC) under tensile loading.

Figure 7.12 represents the results of a tensile test on fiber composites containing 12- and 24-mm-long AR glass fibers. Note that the response of the specimen may be uniform in some cases, whereas in other cases, averaging the signal may result in spurious results if the sample experiences some level of rotation due to fiber distribution. Under these conditions, the response may result in the sudden failure of the sample, as shown in the case of the 24-mm-long samples tested [10,11]. Figure 7.12b shows the path of growth of a tensile crack, which may deviate from the plane of the notches. As the fiber content increases, the potential for a homogenized deformation across the notch diminishes.

FLEXURE TEST

The behavior in a flexural test is initially linearly elastic. However, because of the tensile stress field, microcracks form and begin to propagate. The location of the crack is not predetermined if the sample is not notched, but cracking along the centerline of the specimen initiates and grows along the depth of the beam, because of flexural stress distribution. As the deformation localizes at the main crack, the response is dominated by crack propagation and, at some stage, the maximum load is reached. As the load drops, the crack continues to open while the remainder of the specimen undergoes elastic unloading. Flexure tests conducted under crack openings or tensile displacements are called Mode I tests. Tests that involve crack sliding or shear displacements are Mixed Mode Tests. The most popular Mode I test configuration for concrete is the loading of the notched beam at midspan [15]. The test is performed under CLC with Crack Mouth Opening Deformation (CMOD) as the controlled variable. Measurement of the sample deflection allows one to compute the energy absorbed.

Figure 7.13 represents the flexural test arrangement. Typical specimens tested in a three-point bending configuration use a 406 × 101 × 101 or a 152 × 152 × 609 mm flexural beam specimens with an initial notch. A three-point bending loading fixture should eliminate extraneous deformations such as support settlements and specimen rotations. The CMOD can be measured across the face of the notch using extensometers. The deflection of the beam should also measure the centerline deflection with respect to support settlements.

The first crack load is defined as the load at which the deformation response deviates from linearity, a parameter that, at many times, is defined subjectively. The stress measure derived from this is referred to as the nominal flexural stress. Fracture energy, G_f, is defined as the area under entire load-deflection curve normalized with respect to the crack ligament area. The area under the load-deflection curve can be calculated by numerical integration of the load-deflection response using Simpson's algorithm.

CLC direct tensile tests performed on a variety of cement composites like AR glass and fabric composites are reported throughout this book. The presence of fibers reduces the potential for localization. The typical load versus deflection relationships of FRC materials are shown Figure 7.14, representing a low volume fraction fiber loading of 0.6 kg/m³ using 24-mm-long AR glass fibers. Metal plates with the dimensions of 25 × 30 mm were glued on the surface of the specimen at the grips to minimize damage. The tensile load and elongation were recorded and converted to the stress–strain response. The tensile strength, stiffness, ductility, strain capacity, and cracking parameters are reported [11]. The shapes of these curves are similar and show the increase in the load

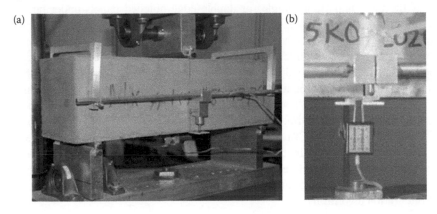

FIGURE 7.13 The flexural test setup showing the hinges in the top and bottom in addition to deflection measuring jig and CMOD extensometer gage.

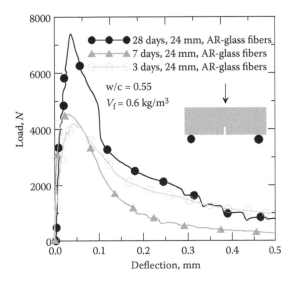

FIGURE 7.14 Load versus deflection relationship of FRC materials representing a relatively low-volume fraction for fiber loading. (After Desai, T., Shah, R., Peled, A., and Mobasher, B., "Mechanical properties of concrete reinforced with AR-glass fibers," *Proceedings of the 7th International Symposium on Brittle Matrix Composites* (BMC7), Warsaw, October 13-15, 2003, 223–232.)

carrying capacity of the composite with age. The response is quasi-linear up to the peak load, and the specimens retain a major portion of their tangential stiffness. It is evident from the figure that the effect of low fibers is more pronounced in the early ages when the strength of the matrix material had not increased significantly. The postpeak deformation is relatively substantial, but decreases significantly as the age of specimen increases.

FRACTURE TESTS

Fracture tests are performed on specimens with notches or initial cracks so that the behavior is governed exclusively by crack initiation and its subsequent propagation. As such a test progresses, the deformation localizes at the notch. Hence, the best controlled variable is the crack opening displacement. Fracture tests are conducted under several loading configurations, with those that only involve an opening under a tensile mode are called Mode I tests, which are generally easier to perform and interpret. Moreover, because the tensile strength of concrete materials is relatively low, the Mode I response is the most important. Single- or double-edge notched panels under pure tension are preferred, but this is a difficult test to design and to conduct [16]. Gopalaratnam and Shah [17] performed double-edge notched tests where the controlled variable was the average of the two notched (or crack) mouth opening displacements [17]. The most popular Mode I test configuration for concrete is the one in which the notched beam is loaded at midspan. The test is best performed under CLC with the crack mouth opening as the controlled variable [18]. Two RILEM recommendations [19, 20] for determining material fracture parameters are based on the stable postpeak response obtained using crack opening control. Another similar application of crack opening control is in the toughness characterization of FRCs using the load–CMOD response as proposed by Gopalaratnam and Gettu [21]. Mixed mode fracture tests are normally conducted with parameters similar to Mode I tests. Fracture energy (G_f) is defined as the area under entire load-deflection curve, normalized with respect to crack ligament. The area can be calculated from the load-deflection response using numerical integration algorithms.

CYCLIC TEST

Loading/unloading tests are also conducted using the CLC method with an extensometer to measure the Crack Mouth Opening Displacement (CMOD) as the feedback signal. Another displacement transducer may be used to measure the vertical deflection at the midspan of the beam along the direction of the applied load. The mechanism is the same as the monotonic flexural notched beam test as shown in Figure 7.13. Several cycles may be defined within the magnitude of CMOD at the onset of each unloading cycle. As the CMOD reading reaches a set value for each cycle, the control will transfer to load signal and a reduction in the load magnitude is imposed to reach to nearly a minimal level. At this minimum point, the control mode will switch back to CMOD control, and the process will be repeated. The unloading is conducted under load-control. After the applied load is reduced to zero, the sample is reloaded to the next level of CMOD prescribed by the program. The first crack load may be defined as the load at which the deformation response deviates from linearity. The envelope of the cyclic loading/unloading response can be used for the calculation of fracture parameters. Figure 7.15 represents the load–CMOD response of a cyclic test conducted on a beam in three-point bending.

Loading/unloading tests can help measure the R-curves or other fracture parameters, such as the critical stress intensity factor, $K_{IC}{}^S$, and the critical Crack Tip Opening Displacement CTOD$_C$. The RILEM Committee on Fracture Mechanics provides guidelines for determining the critical stress intensity factor $K_{IC}{}^S$ and the critical crack tip opening displacement CTOD$_C$ of mortar and concrete beams. The critical stress intensity is defined as the stress–intensity as calculated at the effective crack tip. The critical crack tip opening displacement is defined as the crack tip opening displacement, which is calculated at the original notch tip of the specimen using maximum load and

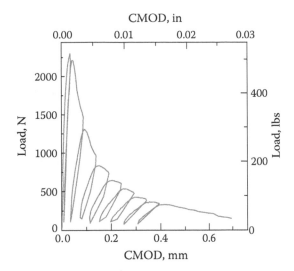

FIGURE 7.15 Load versus CMOD obtained from a loading–unloading test in three-point bending.

effective crack length. Using $K_{IC}{}^S$ and $CTOD_C$, along with the modulus of elasticity, E, the fracture resistance and the energy dissipation of concrete can be calculated.

In comparison with the RILEM test methods, the R-curve determination uses the entire loading history as opposed to only the peak load and the effective compliance at that point. Both methods use a compliance calibration technique to derive the materials and structural properties. The R-curve is dependent on the size and geometry of the specimen and represents material's resistance to the initiation and propagation of cracks [22].

COMPLIANCE-BASED APPROACH

The compliance approach is based on the assumption that, as stable crack propagates, the original flaw or notch opening is resisted by the fibers. Crack growth leads to an increase in the compliance, which, if measured, can be used in a back-calculation to obtain an elastically equivalent traction-free crack length.

Using the crack mouth opening as a measure, compliance is defined as the rate of increase in displacement as a function of load. When a notched specimen exhibits infinitesimal crack growth under constant load or displacement conditions, changes in the load deformation response before and after the crack propagation are observed. The CMOD is used as a displacement component of the compliance, C, as defined by the inverse of the slope of the load/CMOD curve as shown in Figure 7.16. Compliance change as a function of crack length can be used to calculate the increase in the fracture toughness.

A notched beam is subjected to loading/unloading cycles under a three-point bending action as illustrated in Figure 7.16. The CMOD is measured using a LVDT mounted at the crack tip. The load applied to the specimen is controlled to obtain a constant rate of CMOD. The increase in load for a given crack opening correlates with the fracture energy of a beam.

The unloading compliance of the specimen at various loading points can be used to calculate the effective crack length using the linear elastic fracture mechanics (LEFM) equations described as follows. The strain energy release rate for a three-point bending specimen is obtained from the stress intensity factor handbook [2] . K_I is given as

$$K_I = \frac{3s\,P\sqrt{\pi a}}{b}\,g(\alpha) \quad \alpha = \frac{a}{b} \tag{7.3}$$

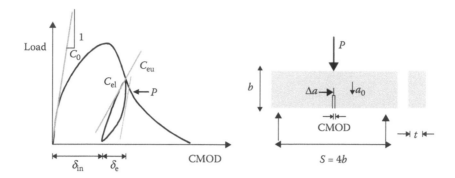

FIGURE 7.16 Definition of compliance parameters obtained from loading/unloading tests under three-point bending.

$$g(\alpha) = \begin{cases} \dfrac{1.99 - \alpha(1-\alpha)(2.15 - 3.93\alpha + 2.7\alpha^2)}{2\,t\sqrt{\pi b}\,\,(1+2\alpha)(1-\alpha)^{3/2}} & s/b = 4 \\[4mm] \dfrac{\left(1.0 - 2.5\alpha + 4.49\alpha^2 - 3.98\alpha^3 + 1.33\alpha^4\right)}{2tb(1-\alpha)^{3/2}} & s/b = 2.5 \end{cases} \tag{7.4}$$

where S represents the span, P represents the far field load, b represents the beam depth, α represents the matrix crack length, and t represents the beam thickness. The fiber pullout mechanism and the closing pressure (due to fibers) are primarily smeared and considered in the compliance term.

In applying this method to FRC materials, the effect of closing pressure due to fiber reinforcement on the stress crack width is smeared in the overall compliance of the composite, which is treated as a homogeneous material. Using the LEFM formulas, the CMOD can be calculated as a function of the applied load, the specimen geometry, and the elastic properties of the material. The CMOD is given as

$$\text{CMOD}(\alpha) = \frac{6\,Sa_0\,P(\alpha)\,V(\alpha)}{Eb^2 t} \qquad C(\alpha) = \frac{\text{CMOD}(\alpha)}{P(\alpha)} \tag{7.5}$$

$$V(\alpha) = 0.76 - 2.28\alpha + 3.87\alpha^2 - 2.04\alpha^3 + \frac{0.66}{(1-\alpha)^2} \qquad \alpha = \frac{a}{b} \tag{7.6}$$

where a_0 is the initial crack length. On the basis of this response, the relationship between the compliance and the crack length is defined, and the initial compliance of the specimen is used to calculate the Young's modulus. For subsequent loading and unloading cycles, the unloading compliance at a given crack length, C_{eU}, is used in a nonlinear equation to solve for an effective crack length, a, according to Equation 5:

$$f(\alpha) = E - \frac{6\,S\,(a_0 + \Delta a)\,V(\alpha)}{Cb^2 t} = 0, \qquad \alpha = \frac{a_0 + \Delta a}{b} \tag{7.7}$$

The stress intensity at the tip of the effective crack may be obtained using Equation 6 and reported as the R-curve, K^R. This definition of the R-curve is based on a modified LEFM approach, because only the elastic component of the compliance change is considered,

$$K^R(a) = \frac{3PS\sqrt{\pi(a_0 + \Delta a)}}{2\,b^2 t}\,g(\alpha) \tag{7.8}$$

R-curves can be further extended by the use of the strain energy release rate, G, to include inelastic effects such as nonlinear displacements. By assuming that crack growth is under constant load, the energy release rate due to the incremental crack growth is defined as

$$G(a) = \frac{1}{2t}\frac{\partial C}{\partial a}P^2 \tag{7.9}$$

For specimens that exhibit residual displacements by not unloading to the origin, additional terms are needed to account for the rate of change of inelastic displacement with respect to crack growth. Several researchers, including Sakai and Bradt [23] and Wecharatana and Shah, [24] in addition to Mai and Hakeem [25], have proposed this additional term as

$$G(a) = \frac{P^2}{2t}\frac{\partial C}{\partial a} + \frac{P}{t}\frac{\partial \delta_r}{\partial a} \tag{7.10}$$

This expression can be computed from the experimental data. By measuring the load (P) and the crack length ($a = a_0 + \Delta a$) values at successive intervals of crack growth, the compliance/crack length relationship can be constructed. For every loading/unloading cycle, four parameters are recorded. These include the effective crack length obtained from the previous procedure, the compliance, the load at the initiation of unloading, and the total inelastic CMOD, which is the residual displacement as the sample is completely unloaded. The first step involves plotting both the compliance and the inelastic displacement as a function of the effective crack length. These two curves are subjected to a second- or third-order polynomial curve fit. The coefficients of the polynomial are used to obtain the rate of change of compliance and the rate of change of inelastic CMOD as a function of crack length. Differentiation can also be achieved using other algorithms, such as the cubic spline method. Equation 8 is calculated at every incremental crack length and used in Equation 9 to define an elastically equivalent effective toughness, K^R,

$$K^R(a) = \sqrt{E'_c G(a)} \tag{7.11}$$

where $E' = E/(1 - v^2)$ for plane strain and E for plane stress. E and v represent the elastic modulus and Poisson's ratio of the material, respectively. Results of this analysis can be used to compute properties such as the effective modulus of elasticity, E, the effective stress intensity factor at crack lengths corresponding to maximum load (hereby referred to as the critical stress intensity factor, K_{IC}^S), and the associated crack opening at the initial notch tip, which is defined as the critical crack tip opening displacement, $CTOD_c$, and fracture toughness, G_f.

MECHANICAL PERFORMANCE—TEST METHODS FOR MEASUREMENT OF TOUGHNESS OF FRC

There are several code-based flexural tests available. The Japanese standard, JCI-SF4 and the *Methods of Tests for Flexural Strength and Flexural Toughness of Fiber Reinforced Concrete* standard specifies a method of measuring the flexural strength and flexural toughness of FRC unnotched beam specimens [26, 27]. This standard dictates that a 100 × 100 × 350 mm (4 × 4 × 14 in.) unnotched beam specimen with a test span of 300 mm (12 in.) and a third-point loading configuration must be used. The test procedure for determining the flexural strength and flexural toughness of fiber reinforced concrete is very similar to ASTM C1609/C1609M [28] and other flexural tests [29–32]. This method provides for the determination of the residual flexural strength ratio and the equivalent flexural strength.

ASTM C1399-04 [33] and 1609 demonstrate the post–first crack behavior of FRC. Both tests are modifications to ASTM C1018, which was marred by inaccuracies because of the inaccurate definition of terms, such as toughness indices, as material parameters. These material parameters were normalized with respect to the first-cracking point, which, itself, is a difficult property to measure using non-scientific equipment.

ASTM C1399 was developed to eliminate or to minimize the shock load that affected data recovery at the time of first crack in the ASTM C1018 test method. The Averaged Residual Strength (ARS) was proposed by Banthia and Dubey [34–36]. The postcrack strength in the Banthia and Dubey method, can be obtained by a simple open-loop testing machine. First, a steel plate is placed underneath a concrete beam and the specimen is loaded under four-point bending apparatus until the concrete cracks. Then, the steel plate is removed and the cracked specimen is reloaded in order to obtain postcrack flexural strengths at the deflection levels, 0.02, 0.03, 0.04, and 0.05 in. Finally, the equivalent stress results are averaged to represent an ARS value. This parameter has been used to compare different material formulations; however, many designers have been persuaded to use them as a tensile strength measure. It is very important to mention that the ARS value is not in any way an equivalently elastic stress and is not to be associated with the postcrack tensile strength or the tensile residual strength parameter.

The average residual strength (ARS) (ASTM C1399-04) [33] is computed for specified beam deflections after the beam has been cracked under separate loading conditions. The test obtains the portion of the load-deflection curve beyond which significant cracking damage has occurred, and provides a measure of post-crack load carrying capacity. However, the interpretation of the load carried by the specimen at a cracked stage into an apparent strength could result in a significant overestimation of the tensile or flexural strength. The ARS method, therefore, does not represent the average stress–carrying ability of a cracked beam. Using the residual strength at the four deflections specified in ASTM C1399 is also therefore, erroneous, because the use of equations in calculating the elastic section modulus of a cracked section are invalid. Under flexural loads, the neutral axis and stress distribution of the idealized linear elastic model are quite different than the experience of the cracked material in reality.

Several case studies and examples [37] point to the inadequacies of the inverse analysis techniques that have been used to obtain residual tensile capacity, such as the average residual stress method. Designers should be aware of the shortcomings of these methods and approaches when determining member capacity.

ROUND PANEL TESTS

FRC materials with a low volume fraction of fibers are particularly suitable for structures with a high degree of redundancy where a stress redistribution may occur. Because of this redistribution, large fracture areas are involved (with a high number of fibers crossing them), and structural behavior is mainly governed by the mean value of the material properties. Furthermore, because of the large fracture areas, the scatter of experimental results from structural tests is remarkably lower than those obtained from the beam tests. A round determinate panel (RDP) test was proposed by ASTM C-1550 [38] and is shown in Figure 7.17. The round plate is supported at three points along its perimeter and is a statically determinate test (a round slab, $\varphi = 800$ mm, thickness = 75 mm, with three supports at 120°). The crack pattern is predictable and the postcracking material properties can be determined using back-calculation. However, handling and placing such a specimen is quite complicated because of the large size and, consequently, high weight. In addition, standard servo-controlled loading machines may not fit with the geometry of the panel, which is too big for most of them.

Measurements of toughness from this test in accordance with basic assumptions regarding the mode and nature of crack opening leads to results that are still being evaluated using fundamental mechanics approaches [39]. Additional analysis of the load carrying capacity of the section is required with tools like the finite element method in order to extract material behavior.

Figure 7.18 shows the test results of a round slab containing macropolymeric fibers at a loading rate of 12 lb./yd.[3] The sample diameter of $S = 910$ mm and thicknesses of $t = 76$ mm. The load-deflection diagram where the first cracks occur at 15 kN is recorded during the testing, but the total cracking

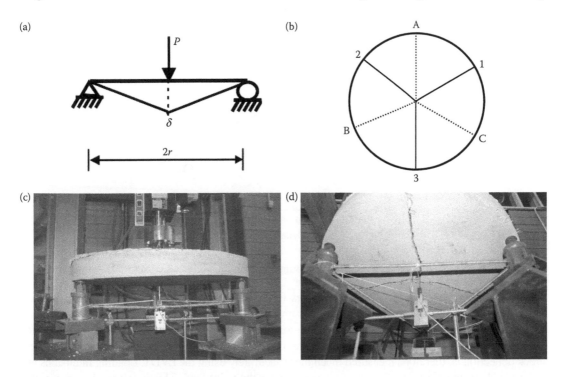

FIGURE 7.17 Round panel determinate panel test apparatus.

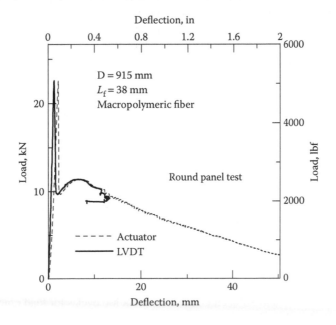

FIGURE 7.18 Load-deflection result of a round slab containing macropolymeric fibers.

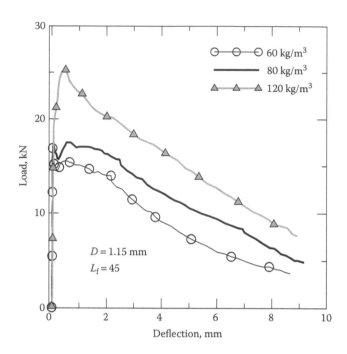

FIGURE 7.19 Round panel tests with continuous support along the perimeter on steel FRC samples.

occurs at 22 kN. Beyond this level, yield hinges form, and the member continues to carry the load beyond the first crack point. The response measured by the cross head and also the LVDT mounted underneath the sample are quite similar, but the compliance at the initial part of the test measured for the sample is quite different than the compliance of the sample and test assembly combined. The first crack deflection is, therefore, measured much more accurately using the LVDT. Loading intensities are quite visible (minute cracks of 0.1 mm opening) under the point load. This is followed by a pseudo-elastic linear stage with a slight increase in the rate of deflection where we can observe the formation of yield lines. Note that the sample is continuously resting along its perimeter as shown in Figure 7.17a.

Figure 7.19 represents the round panel experiments conducted on samples with continuous support along the entire perimeter. These experiments, which are different than the determinate round panel tests due to support conditions, are also applicable as a structural approach to back-calculate the materials properties and to represent quasi-full scale testing for steel FRC round slabs with a net diameter of $S = 1.15$ m and thickness of $t = 150$ mm [40]. Procedures for the back-calculation of stress crack width relationships and materials properties using finite element methods have been proposed [41].

FATIGUE TESTS

Cement composites are expected to endure microcracking during their service life. Fatigue tests establish the basis for the threshold stress levels for such damage tolerant composites. Most of the fatigue tests in concrete or FRC have been performed under bending loads [42]. Jun and Stang [43] used a constant ratio R between the minimum and the maximum load amplitude. Naaman and Hammoud [44] found that FRC mixtures containing 2% of hooked steel fibers can sustain bending fatigue stresses more than twice that of plain concrete. Precracked FRC specimens presented average fatigue lives on the order of 10 cycles for loads within 10% to 90% of static strength, 8000 cycles for a load range of between 10% and 80%, and more than 2.7×10^6 cycles for load range of 10% to 70%. Parant et al. [45] tested multiscale steel fiber cement composite under bending fatigue and observed that, below a loading ratio of 0.88 (maximum fatigue stress ranging from 35.9 to 40.8 MPa for a

Modulus of Rupture [MOR] of 61.5 MPa), specimens survived up to 2 million fatigue cycles. Only a few publications address fatigue in uniaxial compression [46, 19] and tension loads [47, 48].

A uniaxial tension fatigue test was developed to study the effects of microcracking on the mechanical properties of continuous sisal fiber cement composites [49]. The development of these composites is addressed in several papers and also in Chapter 19 [50, 51]. The fatigue behavior is normally examined in terms of the stress versus cycles and the stress–strain hysteresis behavior of the composites. Composites were tested at stress levels ranging from 30% to 80% of the monotonic ultimate tensile strength. Monotonic tensile testing was performed for composites that survived 10^6 tests in order to determine the residual strength. Crack spacing was measured by image analysis. There was no observed loss in strength, but there was a decrease in Young's modulus and an increase in the first crack strength with increasing fatigue stress. Fluorescent optical microscopy can be used to investigate the microcrack formation in composites subject to fatigue loading. During the cyclic fatigue test, an effective laminate stiffness as a function of the progressive cracking in the matrix can be measured to assess the damage. Using compliance calibration, the effective laminate compliance is correlated with the number of cycles. Results are then used to assess the degradation and its associated effects such as crack spacing and density [49].

Figure 7.20 shows the stress versus cycles behavior of a sisal reinforced cement composite tested at various maximum stresses (4–9.8 MPa). It can be seen from this test that the composites can survive 10^6 cycles up to 6 MPa, representing 50% of the Ultimate Tensile Strength (UTS). The stress level of 6 MPa can be considered as the threshold limit at which composites may present fatigue failure at cycles close to 10^6. Beyond 6 MPa, all the composites failed below 10^3 cycles. It was observed that for high stress levels (i.e., >6 MPa), all the cracks were formed during the first few cycles. Examination of failed samples indicates that the number of cracks (12) in fatigue specimens was the same as the ones observed in monotonic tensile tests. After their formation, cracks started to exercise under fatigue load. The cycles at these high stress levels caused a degradation of the fiber–matrix interface that increased the rate of crack opening, ultimately leading to a complete composite failure at low cycles (i.e., <10^3).

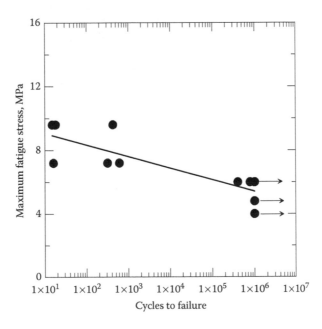

FIGURE 7.20 Fatigue response of sisal reinforced cement composites. (After Silva, F., Mobasher, B., Toledo Filho, R. (2010), "Fatigue behavior of sisal fiber reinforced cement composites," *Materials Science and Engineering: A*, 527(21–22), 5507–5513.)

IMPACT RESISTANCE

The impact resistance of FRC has been measured by several test methods: Charpy, Izod, drop weight, split Hopkinson bar, explosive, and ballistic impact [52, 53]. These tests can be either instrumented or heuristically based, and the resistance can be measured by means of fracture energy, damage accumulation, and/or measurement of the number of drops needed to achieve a determined damage or stress level. The results depend on many variables, such as the size of the specimen, machine compliance, strain rates, the type of instrumentation, and test setup.

Instrumented pendulum-type impact tests were developed by Gopalaratnam et al. [54]. This apparatus was used to characterize the properties of glass FRC (GFRC) specimens. It was observed that the first-cracking load (as determined from deflection measurements) was higher than that determined from the extreme fiber–tensile strain of the centerline measurements. Thus, an accelerated aging environment reduces toughness more severely than the modulus of rupture.

An impact test based on the free-fall drop of an instrumented hammer on a three-point bending configuration test was recently developed [55]. The schematics of this test are shown in Figure 7.21. The impact energy can be investigated by using different drop heights and weights. The hammer and specimen accelerations are recorded continuously during the impact event so that the relative energy imparted to the specimen can be differentiated from the total energy of the hammer. The impact load can be measured by conventional load cells, but piezoelectric load cells offer a much better response spectrum. Damage mechanisms can be investigated by characterizing the cracking patterns using a high-speed digital camera. Mechanical properties obtained under impact loading are normally compared with static three-point bending tests. Using this method, the impact response of unidirectional continuous sisal fiber reinforced cement composites, as well as pultruded cement composites are studied.

Properties of cement composite systems have been investigated under three-point bending conditions using an instrumented drop weight impact system [55]. AR glass cement composites samples were studied based on the impact load, deflection response, acceleration, and absorbed energy. It

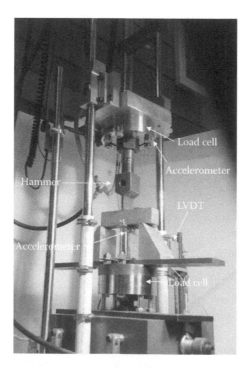

FIGURE 7.21   Setup of the drop weight impact testing method.

was concluded that the maximum flexural stress and absorbed energy of composites increase with drop height. In beam specimens, complete fracture did not take place as cracks formed and closed due to rebound. Significant microcracking in the form of radial fan cracking was also observed, whereas interlaminar shear was the dominant failure mode in the plate specimens. These mechanisms are discussed in Chapter 21.

The potential energy of the hammer is the input energy. It depends on its drop height, mass, and the amount of energy lost during the free fall drop of the hammer due to friction. Some of the input energy is absorbed by the test specimen, whereas the remaining energy is either dissipated by friction or transferred to the test apparatus through the supports after the impact event.

The input potential energy of the hammer, U_i, is defined as the follows:

$$U_i = mgH = \frac{1}{2}mv_0^2 + U_d = U_k + U_f + U_d \qquad (7.12)$$

where m is mass as above, g is the acceleration of gravity, H is the drop height of hammer, v_0 represents the hammer velocity before impact, U_d represents the frictional dissipated energy between the time of release of the hammer until just before the impact event, U_k represents the absorbed energy by the specimen, and U_f represents the energy remaining in the system after the failure of the specimen has taken place. This energy may be elastically stored in the sample and may result in its rebound or transmit through the specimen to the support. The total absorbed energy (kinetic energy dissipated in the specimen), U_k, was defined as the follows:

$$U_k = \int_{t=0}^{t=t^*} P(t)v(t)dt \approx \sum P(t)\Delta d(t) \qquad (7.13)$$

where $P(t)$ and $v(t)$ represent the force and velocity history of the impact event, t^* represents the impact event duration, and $\Delta d(t)$ represents the deflection increment history of test specimen. The total absorbed energy was evaluated using the area under the load-deflection curves. Peak absorbed energy was computed as the area under the load-deflection curve up to the peak load. The computed individual absorbed energies are normalized with the respective samples cross-sectional areas. The load-deflection curves were then analyzed to measure the flexural stress and the absorbed energy for all samples.

RESTRAINED SHRINKAGE

If a prismatic cement-based specimen is restrained on the length direction, uniaxial tensile stresses, which are similar to a uniaxial tensile test, are produced. The linear specimens have the advantage of a relatively straight-forward data interpretation; however, it is difficult to provide sufficient restraint to produce cracking with linear specimens, especially when cross-sectional dimensions are large [56]. Unfortunately, because of difficulties associated with providing sufficient end restraint, these test methods are generally not used for quality control procedures [57].

Different types of loadings and shapes, such as plate-type specimens, have been used to simulate the cracking of concrete due to restrained shrinkage [58]. When a restraint to shrinkage is provided in two directions, a biaxial state of stress is produced. Consequently, the results obtained from plate-type specimens may depend on specimen geometry in addition to the material properties [56].

A restrained shrinkage test using a steel ring was done as early as 1939 to 1942 by Carlson and Reading [59]. As a result of drying, the concrete ring would tend to shrink, but the steel ring would prevent this so that cracking occured. More recently, to better quantify early-age cracking tendency of cementitious material, instrumented rings have been used by researchers to measure the magnitude of the tensile stresses that develop inside the material [60–63]. Because of its simplicity and economy, the ring test has been developed to meet both AASHTO [64] and ASTM [65] standards. The main difference between these standards is the ratio of the concrete to steel ring thickness, which influences the degree of restraint provided to the concrete.

AGING AND WEATHERING

The primary concern for GFRC composites is the durability of glass fibers in the alkaline environment of cement. Despite the use of improved alkaline-resistant glass fibers (AR glass) and pozzolanic materials like silica fume and fly ash, durability concerns still exist [66]. Marikunte et al. [67] and Shah et al. [66] immersed cured GFRC specimens in a hot water bath at 50°C for a long period of time and then tested the samples in flexure and tension. The concept behind the accelerated aging test is that the strength loss at high temperatures can be used to predict the strength loss at lower temperatures that would be expected to occur over a long period of time [67]. The results indicated that the blended cement consisting of pozzolanic materials significantly improves the durability of GFRC composite.

REFERENCES

1. Leipholz, H. H. E., and Abdel-Rohman, M. (1986), *Control of Structures*, The Netherlands: Martinus Nijhoff Dordrecht.
2. Gettu, R., Mobasher, B., Carmona, S., and Jansen D. C. (March 1996), "Testing of concrete under closed-loop control," *Advanced Cement Based Materials*, 3(2), 54–71.
3. Schwarzenbach, J., and Gill K. F. (1984). *System Modelling and Control*. London: Edward Arnold.
4. Franklin, G. F., Powell, J. D., and Emami-Naeini, A. (1991). *Feedback Control of Dynamic Systems*. Reading, MA: Addison-Wesley.
5. Kuo, B. C. (1991). *Automatic Control Systems*. London: Prentice-Hall.
6. Hinton, C. E. (1995). "Dynamic Control Methods", *in Materials Metrology and Standards for Structures Performance*. In B. F. Dyson, M. S. Loveday, and M. G. Gee (eds.), London: Chapman and Hall, 28.
7. Stephanopoulos, G. (1984). *Chemical Process Control: An Introduction to Theory and Practice*. Englewood Cliffs, NJ: Prentice Hall.
8. Miller, R.W. (1977). *Servomechanisms: Devices and Fundamentals*. Reston, VA: Publishing Co. (Prentice-Hall).
9. Hudson, J. A., Crouch S. L., and Fairhurst, C. (1972), "Soft, stiff and servo-controlled testing machines: a review with reference to rock failure," *Engineering Geology*, 6(3), 155–189.
10. Desai, T., Shah, R., Peled, A., and Mobasher, B. "Mechanical properties of concrete reinforced with AR-glass fibers," *Proceedings of the 7th International Symposium on Brittle Matrix Composites (BMC7)*, Warsaw, October 13–15, 2003, 223–232.
11. Desai, T. (2001), "Mechanical properties of conventional concrete reinforced with alkali-resistant glass fibers," MS thesis, Arizona State University.
12. Mobasher, B., and Shah, S. P. (September–October 1989), "Test parameters in toughness evaluation of glass fiber reinforced concrete panels," *ACI Materials Journal*, 86(5), 448–458.
13. Mobasher, B., and Li, C. Y. (1996), "Mechanical properties of hybrid cement based composites," *ACI Materials Journal*, 93(3), 284–293.
14. Perez-Pena, M., and Mobasher, B. (1994), "Mechanical properties of fiber reinforced lightweight concrete composites," *Cement and Concrete Research*, 24(6), 1121–113.
15. Hillerborg, A. (1989). In "Fracture of concrete and rock: recent developments," in Shah, S. P., Swartz, S. E., and Barr, B. (eds.), Amsterdam, Netherlands: Elsevier, 369–378.
16. Hillerborg A. (1989), in Shah, S. P. , Swartz, S. E., and Barr, B. (eds.), *Fracture of Concrete and Rock: Recent Developments*. Amsterdam, The Netherlands: Elsevier, 369–378.
17. Gopalaratnam, V. S., and Shah, S. P. (1985), "Softening response of plain concrete in direct tension," *ACI Journal*, 82, 310–323.
18. Swartz, S. E., Hu K-. K., and Jones, G. L. (1978), "Compliance monitoring of crack growth in concrete," *Journal of the Engineering Mechanics Division*, 104, 789–800.
19. RILEM 89-FMT draft recommendation. (1990). "Determination of fracture parameters (K^S_{Ic} and $CTOD_c$) of plain concrete using three-point bend tests," *Materials and Structures*, 23, 457–460.

20. RILEM 50-FMC draft recommendation. (1985). "Determination of fracture energy of mortar and concrete by means of three-point bend tests on notched beams," *Materials and Structures*, 18, 285–290.

21. Gopalaratnam, V. S., and Gettu, R. (1995), "On the characterization of flexural toughness in fiber reinforced concretes," *Cement and Concrete Composites*, 17, 239–254.

22. Mobasher, B., and Li, C. Y., and Arino, A. (1995). *Experimental R-Curves for Assessment of Toughening in Micro-fiber Reinforced Hybrid Composites*. American Concrete Institute, ACI SP-155-5, 93–114.

23. Sakai, M., and Bradt, R. C. (1986), "Graphical methods for determining the nonlinear fracture parameters of silica and graphite refractory composites," *Fracture Mechanics of Ceramics*, 7, 127–142.

24. Wecharatana, M., and Shah, S. (1983), "A model for predicting fracture resistance of fiber reinforced concrete," *Journal of Engineering Mechanics*, 109(5), 819–829.

25. Mai, Y. W., and Hakeem, M. I. (1984), "Slow crack growth in cellulose fibre cements," *Journal of Materials Science*, 19(2), 501–508.

26. Japan Concrete Institute, Japanese standard JCI-SF4, Methods of tests for flexural strength and flexural toughness of fiber reinforced concrete, JCI standard Test Methods for Fiber Reinforced Concrete, 45–51, JCI: Tokyo, 1983.

27. European Committee for Standardization, CEN/TC 229, Standard Reference EN 14651:2005+A1:2007 Test method for metallic fibre concrete—Measuring the flexural tensile strength (limit of proportionality (LOP), residual), 2005.

28. ASTM Standard C1609 / C1609M-10 Standard test method for flexural performance of fiber-reinforced concrete (Using Beam With Third-Point Loading) ASTM International, West Conshohocken, PA, 2010, DOI: 10.1520/C1609_C1609M-10. www.astm.org.

29. C1581/C1581M-09a, "Standard test method for determining age at cracking and induced tensile stress characteristics of mortar and concrete under restrained shrinkage."

30. ASTM C293 ASTM Committee C09 on Concrete and Concrete Aggregates, "Standard test method for flexural strength of concrete (using simple beam with center-point loading)," ASTM International, 2007.

31. ASTM C1399 ASTM Committee C09 on Concrete and Concrete Aggregates, "Standard test method for obtaining average residual-strength of fiber-reinforced concrete," ASTM International, 2007.

32. ASTM C78 ASTM Committee C09 on Concrete and Concrete Aggregates, "Standard test method for flexural strength of concrete (using simple beam with third-point loading)," ASTM International, 2007.

33. ASTM C1399-04, "Standard tests method for obtaining average residual-strength of fiber-reinforced concrete," ASTM International, PA, 2004, 6 pp.

34. Banthia, N., and Dubey, A. (November–December 1999), "Measurement of flexural toughness of fiber-reinforced concrete using a novel technique—Part 1: Assessment and calibration," *ACI Materials Journal*, 96(6), 651–656.

35. Banthia, N., and Dubey, A. (January/February 2000), "Measurement of flexural toughness of fiber-reinforced concrete using a novel technique—Part 2: Performance of various composites," *ACI Materials Journal*, 97(1), 3–11.

36. ASTM Standard C1399 / C1399M-10 "Standard test method for obtaining average residual-strength of fiber-reinforced concrete" ASTM International, West Conshohocken, PA, 2010, DOI: 10.1520/C1399_C1399M-10. www.astm.org.

37. Soronakom, C., and Mobasher B., "Flexural analysis and design of strain softening fiber reinforced concrete," ACI proceedings, ACI-SP-272, Ed. G.J. Para-Montesinos, G. J., Balaguru, P. 173–187, 2010.

38. ASTM Standard C1550-10a "Standard test method for flexural toughness of fiber reinforced concrete (Using Centrally Loaded Round Panel)" ASTM International, West Conshohocken, PA, 2010, DOI: 10.1520/C1550-10A. www.astm.org.

39. Tran, V. N. G., Bernard, E. S., and Beasley, A. J. (May 2005), "Constitutive modeling of fiber reinforced shotcrete panels," *Journal of Engineering Mechanics*, 131, (5), 512–521.

40. Destrée, X., Planchers structurels en béton de fibres métalliques: Justification du modèle de calcul. Université Laval, CRIB, 11 Juin 1998, Troisième Colloque International Francophone sur les Bétons Renforcés de Fibres Métalliques.

41. Soranakom, C., Mobasher, B., and Destreé, X. "Numerical simulation of FRC round panel tests and full scale elevated slabs," Deflection and Stiffness Issues in FRC and Thin Structural Elements, ACI SP-248-3, 2008, 31–40.

42. Ramakrishnan, V., and Lokvik, B. J. (1992). "Flexural fatigue strength of fiber reinforced concretes," in Reinhardt, H. W., and Naaman, A. E. (eds.), *High Performance Fiber Reinforced Cement Composites: Proceedings of the International RILEM/ACI Workshop*. London: E&FN SPON, 271–287.

43. Jun, Z., and Stang, H. (January–February 1998), "Fatigue performance in flexure of fiber reinforced concrete," *ACI Materials Journal*, 95(1), 58–67.
44. Naaman, A. E., and Hammoud, H. (1998), "Fatigue characteristics of high performance fiber-reinforced concrete," *Cement and Concrete Composites*, 20, 353–363.
45. Parant, E., Rossi P., and Boulay, C. (2007), "Fatigue behavior of a multi-scale cement composite" *Cement and Concrete Research*, 37, 264–269.
46. Gao, L., and Hsu, T. C. C. (1988), "Fatigue of concrete under uniaxial compression cyclic loading," *ACI Materials Journal*, 95, 575–581.
47. Xianhong M., and Yupu, S. (2007), "Residual tensile strength of plain concrete under tensile fatigue loading," *Journal of Wuhan University of Technology, Materials Science Edition*, 22, 564–568.
48. Saito, M. (1987), "Characteristics of microcracking in concrete under static and repeated tensile loading," *Cement and Concrete Research*, 17, 211–218.
49. Silva, F., Mobasher, B., and Filho, R. D. T. (2010), "Fatigue behavior of sisal fiber reinforced cement composites," *Materials Science and Engineering, A*, in press.
50. Silva, F., Zhu, D., Mobasher, B., and Filho, R. D. T. (2010), "Impact behavior of sisal fiber cement composites under flexural load" *ACI Materials Journal*, in press.
51. Silva, F., Mobasher, B., and Filho, R. D. T. (2009), "Cracking mechanisms in durable sisal fiber reinforced cement composites," *Cement and Concrete Composites*, 31, 721–730.
52. Balaguru, P. N., and Shah, S. P. (1992). *Fiber-Reinforced Cement Composites*. New York: McGraw-Hill.
53. Wang, N., Mindess, S., and Ko, K. (1996), "Fibre reinforced concrete beams under impact loading," *Cement and Concrete Research*, 26, 363–376.
54. Gopalaratnam, V. S., Shah, S. P., and John, R., "A modified instrumented Charpy test for cement based composites," *Experimental Mechanics*, 24(2), 102–111.
55. Zhu, D., Gencoglu, M., and Mobasher, B. (2009), "Low velocity flexural impact behavior of AR glass fabric reinforced cement composites," *Cement and Concrete Composites*, 31, 379–387.
56. Grzybowski, M., and Shah S. P. (1990), "Shrinkage cracking of fiber reinforced concrete," *ACI Materials Journal*, 87(2), 138–148.
57. Weiss, W. J., and Shah, S. P. (1997), "Recent trends to reduce shrinkage cracking in concrete pavements," Aircraft/Pavement Technology: In the Midst of Change, Seattle, WA, 217–228.
58. Padron I, and Zollo R. F. (1990), "Effect of synthetic fibers on volume stability and cracking of portland cement concrete and mortar," *ACI Materials Journal*, 87(4), 327–332.
59. Carlson, R. W., and Reading, T. J. (1988), "Model study of shrinkage cracking in concrete building walls," *ACI Structural Journal*, 85(4), 395–404.
60. Shah, S. P., Karaguler, M. E., and Sarigaphuti, M. (1992), "Effects of shrinkage reducing admixture on restrained shrinkage cracking of concrete," *ACI Materials Journal*, 89(3), 289–295.
61. Hossain, A. B., and Weiss, J. (2004), "Assessing residual stress development and stress relaxation in restrained concrete ring specimens," *Cement and Concrete Composites*, 26, 531–540.
62. Mane, S. A., Desai, T. K., Kingsbury D., and Mobasher, B. (2002), "Modeling of restrained shrinkage cracking in concrete materials," *ACI Special Publications*, 206(14), 219–242.
63. Soranakom C, Bakhshi, M., and Mobasher, B. "Role of alkali-resistant glassfiber in suppression of restrained shrinkage cracking of concrete materials," 15th Glassfibre Reinforced Concrete Association Congress, Prague, Czech Republic, April 20–23, 2008.
64. AASHTO PP 34-99, "Standard practice for estimating the crack tendency of concrete," 2004. Washington, DC: American Association of State and Highway Transportation Officials.
65. ASTM Standard C1581 / C1581M - 09a "Standard test method for determining age at cracking and induced tensile stress characteristics of mortar and concrete under restrained shrinkage," ASTM International, West Conshohocken, PA, 2010, DOI: 10.1520/C1581_C1581M-09A. www. astm.org.
66. Shah, S. P., Ludirdja, D., Daniel, J. I., and Mobasher, B. (1988), "Toughness-durability of glass fiber reinforced concrete systems," *ACI Materials Journal*, 85(5), 352–360.
67. Marikunte, S., Aldea, C., and Shah, S. P. (1997), "Durability of glass fiber reinforced cement composites: Effect of silica fume and metakaolin," *Advanced Cement Based Materials*, 5(3–4), 100–108.

8 Fiber Pullout and Interfacial Characterization

INTRODUCTION

The load-carrying capacity of cement-based composites depends on the performance of three components: the fiber, the matrix, and the interface. The brittle matrix cracks when subjected to tensile stresses greater than its tensile strength. The occurrence of cracking is unavoidable but controllable by mean of a force transfer mechanism. A well-developed bond between matrix and fiber increases the force transfer between the two phases. Continuous reinforcement systems, such as textiles, transmit localized forces to regions with lower stress, leading to multiple cracking. The net result is that the structural stiffness slowly degrades. On the contrary, poor bonding and discontinuous fiber reinforcement result in a fewer cracks, which widen without offering much resistance to load. The stiffness degenerates rapidly and crack widths become visible. Proper modeling of the load transfer between the two materials is a crucial tool in the development of high-performance discrete fibers and textile reinforced cement composites.

SIGNIFICANCE OF INTERFACIAL MODELING

The stiffness of fiber–interface–matrix systems directly affects the toughening mechanisms. In brittle matrix composites, if the ultimate strain capacity of the fibers exceeds that of the matrix, fibers will bridge the matrix cracks. The force transferred by the bridging reduces the stress intensity at the tip of the matrix crack. The constitutive response of the debonding phase depends on the length of cylindrical or planar shear microcracks that form at the interface and the ability of the interface to transmit traction across a matrix crack. For properly designed systems, the matrix's tolerance to crack propagation and, thus, the composite strength may increase significantly as shown by Aveston et al. [1] and Kelly and Tyson [2]. The fibers bridge across the matrix cracks, while processes of debonding and pullout lead to toughening and energy dissipation.

Several fiber pullout models have been able to characterize the bond properties of single fibers with a cementitious matrix. One of the main justifications for the modeling of pullout mechanisms is that it will better address closing pressure formulations. The distribution of traction force over the crack length depends on the interaction of the debonding fibers with the matrix crack in an opening mode. Closing pressure formulations have been modeled on several approaches, including closing pressure forms that are square root dependant on the fiber slip [3, 4] and that correlate the fiber, matrix, and interface properties to the composite response (see the work of McCartney [5], Budiansky et al. [6], and Marshall et al. [7]).

Scholars, working through a variety of analytical solutions [8–10] assume that, although fiber and interface behave elastically in bonded regions, there is a constant residual shear strength in the debonded region. Bond strength models for rebars also address the pre- and postpeak response [11, 12]; however, the more detailed the bond strength model, the more complex the analytical solution. Currently, most bond properties of textile structures are obtained from straight fiber pullouts that treat the textile (grid reinforcement) as an equivalent smooth longitudinal fiber. The mode of load transfer in textile composites, however, is much more complicated because of the function of transverse yarns that provide anchorage. As shown in Figure 8.1, glass fiber transverse yarns are fully embedded in the

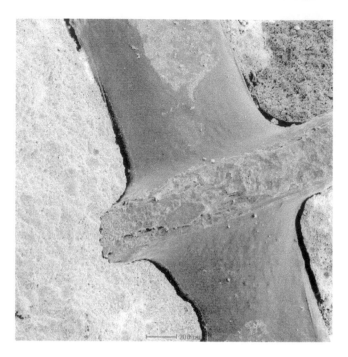

FIGURE 8.1   Role of transverse yarns in anchorage of longitudinal yarns in AR glass fiber composites.

matrix. Therefore, the equivalent bond properties computed on the basis of a one-dimensional bond pullout model are relatively high and include the effect of the mechanical anchorage of transverse yarn and slack existing in fibers.

Modeling the failure process at the interface requires a numerical simulation of the stiffness degradation before the peak load using a formulation for the stable propagation of the debonded zone. Interface toughness and frictional sliding resistance can be measured by using micromechanical models to curve fitting the load–slip response. Theoretical load–slip responses can be obtained using stress-based or crack growth fracture criteria. The available modeling schemes are based on a variety of techniques based on analytical, finite, and finite difference approach. Two approaches toward measuring shear strength and fracture mechanics are commonly used in the analyses of pullout tests [13]. The simplest models are reasonably accurate and assume that sliding along a debonded interface is governed by a constant shear stress, τ [14–16]. Coulomb's Law of Friction has been used to study the effects of residual and thermal stresses [17–19]. Ballarini et al. [20] and Mital and Chamis [21] carried out finite element studies using the Coulomb friction law. Analytical approaches, however, are quite versatile in understanding the various operating mechanisms during the debonding and pullout process.

The mechanism of pullout of straight yarns is based on the shear lag, which includes frictional and adhesion bond. The relationship between the pullout force and its associated slip displacement is obtained using a sliding contact surface to model the debonding and slip. The debonding criterion is defined using the interfacial strength. After debonding, the Coulomb friction is introduced in the debonded zone to account for interlocking effects. This simulation provides the theoretical pullout response as affected by the parameters of the interface as it is calibrated with experimental pullout results.

In a fracture mechanics, the debonding zone propagates by requiring that the energy release into the interface to be equal to its resistance energy [11, 22–24]. The shear strength and fracture criteria have been combined as a Mode 2 crack growth coupled with a Coulomb-type debonded interface [25] or a Dugdale–Barenblatt-like model [26]. Additional approaches that include an R-curve

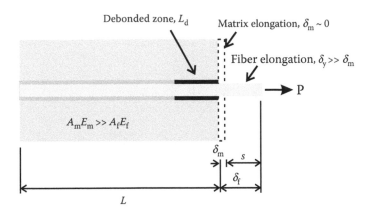

FIGURE 8.2 Pullout slip and the various zones of bonded, debonded, and sliding interface.

formulation for the formation and growth of the delamination damage zone [27] and the finite element implementation of the interface models are also available [28].

A variety of pullout tests have been developed. Figure 8.2 shows the schematics of test parameters during a pullout–slip test. The Various zones of the interface include the bonded zone, the debonded zone, and the sliding zone. Using the complete load–slip response, the stiffness of the interfacial zone and shear strength parameters can be obtained by using a standard fiber pullout model. In a pullout test by Li et al. [29], fiber debonding under slip-controlled closed-loop tests was measured. This approach allowed for the measurement of the aging in the alkali-resistant glass (AR glass) fiber composites, as well as characterization of the differences in the failure mechanism. Moire interferometry has alternatively been used by Shao et al. to observe the growth of debonding microcracks at the interface of fiber reinforced concrete (FRC) materials [30, 31]. Test methods have also been developed to characterize the pullout response of glass fibers from a portland cement matrix [32].

Interfaces play an important role in the long-term durability of FRC. Their characterization is important for service life modeling and prediction. Interfaces in FRC materials are quite porous with low strengths and stiffnesses. The microstructure of these interfaces contains up to 50 μm from the fiber surface and consists of a duplex film, a zone rich in calcium hydroxide crystals, and a porous zone [33]. The durability of materials such as natural fibers or glass-based composites is associated with an increased bond strength and a loss of the flexibility of fiber bundles. Fiber fracture precedes complete debonding and results in loss of ductility and strength [34]. Several methods are available in reducing the pore water solution's alkalinity, thus providing a less aggressive environment for the fibers [35]. Methods have also been developed to characterize the strand effect in glass fiber composites. By determining strand perimeters using digital analyses of images captured from thin sections, one can obtain the effective perimeter parameter and use this to characterize/optimize the bond and fiber length parameters [36].

ANALYTICAL DERIVATION FOR FIBER PULLOUT FIBER AND TEXTILE COMPOSITES

A pullout model that can be used to characterize the parameters of fiber and textile cement systems is presented in this section. The derivation for this model is similar to the single-fiber analogy created by Naaman et al. [37, 38]. The first set of solutions allow for the comparison with fiber composites. In the case of fabrics, the effect of transverse yarns is included using additional terms to address the bond contributions of orthogonal fill yarns. But, the geometrical interlock of a textile

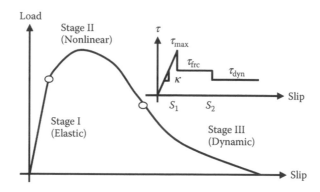

FIGURE 8.3   Pullout–slip response and shear strength diagram.

system into obvious chemical and mechanical bond characteristics results in the overestimation of bond properties.

The characteristic of the curve as shown in Figure 8.3 can be divided into various stages of stress distribution in the fiber. Initially, the linear response corresponds to the perfect bonding of the fiber and the cement paste. At certain point on the ascending curve, the response becomes nonlinear, because some parts of the fiber start debonding and propagate along the embedded length. This nonlinearity on the ascending curve is observed until, reaches the maximum strength. The debonding continues in the postpeak region and until the entire length becomes debonded and the fiber start to slide out dynamically. There are, therefore, three stages: elastic, nonlinear, and dynamic. The derivations, along with the assumed shear stress and force distribution for each stage, will be explained in the next section.

The static equilibrium of the specimen requires that the forces in the fiber and matrix, where F is the force in fiber and M is the force in matrix, are equal. According to Hooke's law, the relationship of the force can be express as:

$$A_f E_f \varepsilon_f = -A_m E_m \varepsilon_m \tag{8.1}$$

where A is the cross-sectional area, E is the Young's modulus, and ε is the axial strain. The subscripts "f" and "m" refer to the fiber and matrix, respectively. Along the embedded length of fiber in matrix, the tension force from the fiber gradually transfers to the matrix by mean of shear at the interface. This mechanism can be described in the differential form:

$$\frac{dF}{dx} = -\frac{dM}{dx}\psi\tau \tag{8.2}$$

where ψ is the circumference of the fiber and τ is the shear at the fiber–matrix interface. During the elastic stage, when a small load is applied, the local shear stress behaves under a linear stress–slip relationship:

$$\tau = \kappa S \tag{8.3}$$

where κ is the slope of the shear strength diagram shown in Figure 8.3 and the slip S is defined as

$$S = (\delta_f - \delta_m) = \int_0^x \left[\varepsilon_f(x) - \varepsilon_m(x)\right] dx \tag{8.4}$$

where δ_f an δ_f are the elongation of the fiber and shortening of the matrix. Substituting Equation 8.4 in Equation 8.2 and taking the derivative with respect to x yields:

$$\frac{d^2 F}{dx^2} = \psi\kappa\left[\varepsilon_f(x) - \varepsilon_m(x)\right] \tag{8.5}$$

From Equation 8.2, if we substitute $\varepsilon_m = -F/A_m E_m$ into Equation 8.5 and rearrange the terms, the differential equation for the fiber pullout can be calculated as,

$$\frac{d^2 F}{dx^2} - \beta^2 F = 0 \tag{8.6}$$

where $\beta^2 = \psi\kappa Q$ and $Q = (1/A_f E_f) + (1/A_m E_m)$. The general solution of the second differential equation is in the form:

$$F(x) = C_1 e^{\beta x} + C_2 e^{-\beta x} \tag{8.7}$$

According to the test mechanism, the force boundary condition is equal to zero at the left end and the applied pullout load P at the right end is:

$$F(0) = 0$$
$$F(L) = P \tag{8.8}$$

When imposing two boundary conditions, the force distribution in embedded length L is obtained with

$$F(x) = P\frac{\sinh(\beta x)}{\sinh(\beta L)} \tag{8.9}$$

The corresponding shear stress can be derived from Equations 8.2 and 8.9:

$$\tau(x) = \frac{P\beta}{\psi}\frac{\cosh(\beta x)}{\sinh(\beta L)} \tag{8.10}$$

Pullout Response in Elastic Stage (Stage 1)

As long as the shear stress at the interface is less than the maximum shear strength, τ_{max}, and the yarn and matrix are fully bonded, the applied load is less than the maximum bonded load ($P_1 < P_{1b,max}$), and standard shear lag solutions apply as shown in Table 8.1. When the applied load is less than the maximum bonded load ($P_1 < P_{1b,max}$), as shown in Figure 8.3, the shear stress at interface is less than the maximum shear strength τ_{max}, and fiber and matrix are fully bonded as shown in Figure 8.4a. The slip at the end of the fiber is obtained by integrating Equation 8.3 up to $x = L$,

$$S_1 = \int_0^L \left[\frac{F(x)}{A_f E_f} + \frac{F(x)}{A_m E_m}\right] dx = \frac{P_1 Q}{\beta\sinh(\beta L)}\left[\cosh(\beta L) - 1\right] \tag{8.11}$$

TABLE 8.1
Parameters of the Fiber Pullout Model during Each Stage of Testing

Description	Pullout Force Distribution, $F_i(x)$	End force, $P(x)$	Slip	Boundary Conditions
1 Elastic case	$\tau(x) = \dfrac{P_1 \beta}{\psi} \dfrac{\cosh(\beta x)}{\sinh(\beta L)}$	$P_1 \dfrac{\sinh(\beta x)}{\sinh(\beta L)}$	$\dfrac{P_1 Q}{\beta \sinh(\beta L)} \left[\cosh(\beta L) - 1 \right]$	$\tau < \tau_{\max},\ P_1 < P_{1b,\max}$ $P_{1b,\max} = \dfrac{\tau_{\max} \psi}{\beta} \tanh(\beta L)$
2.a Nonlinear stage, bonded zone	$P_{2b,\max} \dfrac{\sinh(\beta x)}{\sinh(\beta(L-d))}$	$P_{2b,\max} = \dfrac{\tau_{\max} \psi}{\beta} \tanh(\beta(L-d))$	$\dfrac{P_{2b,\max} Q}{\beta} \left[\dfrac{\cosh(\beta(L-d)) - 1}{\sinh(\beta L)} \right]$	$0 \le x \le L - d$
2.b Nonlinear stage, debonded zone	$P_{2b,\max} + \tau_{FRC} \psi (x - L + d)$	$P_d = \tau_{FRC} \psi d$ $F = P_{2b,\max} + P_d$	$\dfrac{P_{2b,\max} Q}{\beta} \left[\dfrac{\cosh(\beta(L-d)) - 1}{\sinh(\beta L)} \right] + \dfrac{1}{2} Q d \left(\tau_{FRC} \psi d + 2 P_{2b,\max} \right)$	$L - d \le x \le L$
3. Dynamic stage, frictional shear	$\tau_{FRC} \psi x$	$P_{3,1st} = \tau_{FRC} \psi L$	$\dfrac{1}{2} Q \tau_{FRC} \psi L^2$	$\Delta_d = 0$
4 Rigid body motion	$P_{3,nth} + \tau_{dyn} \psi (x - L + \Delta_d)$	$P_{3,nth} = \tau_{dyn} \psi (L - \Delta_d)$	$\tau_d \psi \left(L\Delta_d - \dfrac{L^2}{2} - \dfrac{\Delta_d^2}{2} \right) + P_{3,nth}(L - \Delta_d)$	$0 \le x \le L - \Delta_d$

Note: $\beta^2 = \psi \kappa [(1/A_y E_y) + (1/A_m E_m)]$, where β is the equivalent circumference of the yarn and τ is the shear stress at yarn–matrix interface.

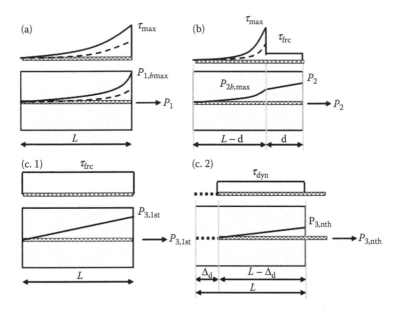

FIGURE 8.4 Shear stress and force distribution along the yarn: (a) elastic response, (b) debonding, (c) frictional pullout, and (d) sliding mode.

When the shear stress reaches the maximum strength (τ_{max}), the elastic response is terminated at the maximum bonded load, which is calculated by using the following:

$$P_{1b,max} = \frac{\tau_{max}\psi}{\beta}\tanh(\beta L)$$ (8.12)

PULLOUT RESPONSE IN THE NONLINEAR STAGE (STAGE 2)

As the load is increased beyond the elastic limit, debonding at the right end starts and extends by a distance d, with a frictional shear strength τ_{FRC}, whereas, on the left portion $(L-d)$, the two materials are still perfectly bonded. The criteria for the growth of debonding is characterized by a shear strength criterion with a constant frictional stress along the debonded zone, in addition to a shear lag model terminating with τ_{max} at the debonding junction as shown in Figure 8.4b. The resistant to pullout load, P_2, is calculated by the summation of two forces: the bonded $P_{2b,max}$ and debonded regions, P_d. Measurements of slip at the end of the yarn is obtained by summing the two bonded and debonded regions,

$$P_2 = P_{2b,max} + P_d$$ (8.13)

The maximum load in the bonded region, which now has a shorter embedded length $(L-d)$, can still be obtained by substituting $L-d$ for L, as shown in Equations 8.13 and 8.14.

$$P_{2b,max} = \frac{\tau_{max}\psi}{\beta}\tanh(\beta(L-d))$$ (8.14)

The resistant load from the debonded region is simply:

$$P_d = \tau_{FRC}\psi d$$ (8.15)

The force boundary condition is modified according to the bonded and debonded zones,

$$
\begin{aligned}
F(0) &= 0 \\
F(L - d) &= P_{2b,max} \\
F(L) &= P_2
\end{aligned}
\tag{8.16}
$$

When imposing boundary conditions (Equation 8.5 in Equation 8.7), the force distribution is obtained by:

$$
\begin{aligned}
F_b(x) &= P_{2b,max} \frac{\sinh(\beta x)}{\sinh(\beta(L - d))}, && 0 \le x \le L - d \\
F_d(x) &= P_{2b,max} + \tau_{FRC}\psi(x - L + d), && L - d \le x \le L
\end{aligned}
\tag{8.17}
$$

Slip at the end of the fiber is obtained in a same way as Equation 8.12, but is carried out for bonded and debonded regions as shown here:

$$
\begin{aligned}
S(L)_2 &= \int_0^{L-d}\left[\frac{F_b(x)}{A_f E_f} + \frac{F_b(x)}{A_m E_m}\right]dx + \int_{L-d}^{L}\left[\frac{F_d(x)}{A_f E_f} + \frac{F_d(x)}{A_m E_m}\right]dx \\
&= \frac{P_{2b,max}Q}{\beta}\frac{\left[\cosh(\beta(L - d)) - 1\right]}{\sinh(\beta L)} + \frac{1}{2}Qd\left(\tau_{FRC}\psi d + 2P_{2b,max}\right)
\end{aligned}
\tag{8.18}
$$

PULLOUT RESPONSE IN DYNAMIC STAGE (STAGE 3)

The dynamic response of the pullout in the third stage consists of two conditions: an initial stage up to complete debonding and a rigid body motion. It is assumed that until the yarn is completely debonded, the shear resistance still remains τ_{FRC}. Upon the completion of debonding, no sliding has occurred ($\Delta_d = 0$) until the yarn begins a rigid body motion. As the yarn starts to slide by a distance $\Delta_d > 0$, the resisting shear stress is assumed to drop to dynamic shear strength τ_{dyn}. During the initial stage ($\Delta_d = 0$), the shear resistance is a uniform τ_{FRC} throughout the yarn length, and the resulting slip at the end of the yarn in the initial stage is obtained in much the same way as the initial case this is shown in Figure 8.4c. During the rigid body motion stage ($\Delta_d > 0$), shear resistance drops to τ_{dyn}, and the embedded length reduces to $(L - \Delta_d)$. The measured slip is computed as an additive portion of end of Stage 2 (static) $S(L)_{2,last}$ and the slip in Stage 3 rigid body motion mode (Figure 8.4c).

ALGORITHM FOR PULLOUT SIMULATION

The pullout response obtained from experimentation can be simulated using the analytical model described in the preceding section. The simulation procedure can be summarized as follows:

1. Use a representative curve of the experiment pullout response.
2. In the elastic stage (Stage 1), calculate the pullout load is P_1 and calculate slip as $S(L)_1$ from Equation 8.11. Keep increasing P_1 until the it reaches $P_{1b,max}$ (Equation 8.12) then stop and go to the next stage.
3. In the nonlinear stage (Stage 2), keep imposing the debonded length d at the right end and calculate the corresponding load P_2 from Equation 8.13 and slip $S(L)_2$ from Equation 8.18. Because snap back is not observed in a fiber pullout experiment, the stage is terminated when slip $S(L)_2$ starts to decrease. Then the dynamic mode begins.

4. Finally, in the dynamic stage (Stage 3), the first set of data points $(P_{3,1st}, ST(L_{3,1st}))$, which corresponds to the last static stable ($\Delta_d = 0$), can be determined. When the fiber begins to move dynamically ($\Delta_d > 0$), the load and total slip $(P_{3,nth}, ST(L)_{3,nth})$ is calculated by equations provided in Table 8.1.

SINGLE-FIBER PULLOUT EXPERIMENTS

Li et al. [29] experimented with steel and glass fiber pullout from a cementitious matrix using slip-controlled closed-loop test. By comparing the experimental and the theoretical response, interface parameters such as stiffness, ω, adhesional and frictional shear strength, τ_y, and τ_f, and specific surface fracture energy, $p\Gamma$, were computed using a model similar to the previous section [29].

The experimental results of the pullout of AR glass fibers are shown in Figures 8.5a through 8.5d. The pullout response of glass fibers, which were subjected to a short aging period, are compared with as-is conditions, as shown in Figures 8.5a and 8.5b, for different curing periods and different fiber

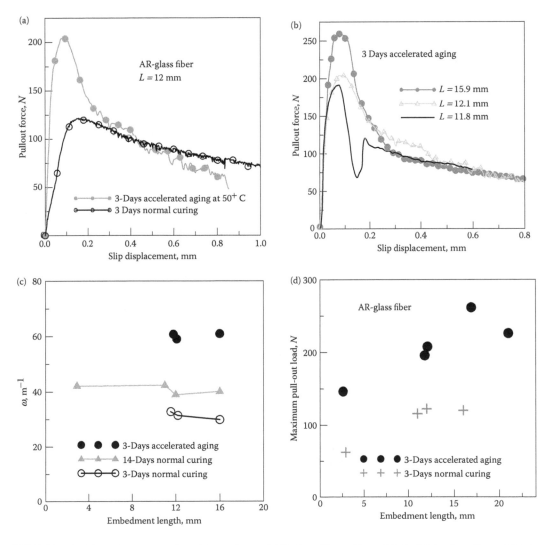

FIGURE 8.5 Experimental results of the pullout of AR glass fibers. (From Li, Z., Mobasher, B., and Shah, S. P., "Characterization of interfacial properties of fiber reinforced cementitious composites," *Journal of American Ceramic Society*, 74(9), 2156–2164, 1991. With permission.)

lengths. Figure 8.5a shows the results of CEMFIL 1 fibers (AR glass) tested before and after a short aging period. The drop in load beyond peak level for aged samples shows a substantial fiber breakage due to fiber embrittlement. Stiffness of the interface, ω, is measured from the slope of the ascending portion and increases with aging, as shown in Figure 8.5c. The maximum pullout load also increases significantly (roughly 75%), as shown in Figure 8.5d, where the maximum load versus the embedded length for various interface conditions can also be seen. Aging can be modeled by increasing the interface adhesional bond strength and the specific energy of interface fracture. Using this approach, one can predict the effect of composite age with the data obtained from the experimental tests [39].

The peak load is dependent on fiber length and interface conditions. The duration of curing from 14 to 28 days increases the adhesional and frictional bond strengths by as much as 300%. The post-peak region of the loading history is associated with significant energy dissipation, which are due to the frictional forces of pullout.

TEXTILE PULLOUT TESTS

Pullout experiments on fabric reinforced systems were conducted by Sueki et al. to address the effect of fabric type, mixture design, and processing methods such as pultrusion, vacuum casting, or rheology modification on the interface properties of textile composites [40]. These experimental variables and the material properties of yarn and cement pastes are provided in Tables 8.2 and 8.3. Using an assemblage of up to eight yarns, sections of a textile embedded in the cement matrix were tested at a crosshead rate of 0.25 mm/s. The testing mechanism is presented in Figures 8.6a and 8.6b. Because it may not be possible to hold the textile in the grip exactly at its exit point from the matrix, the compliance of the free sample length in between the edge of the sample and the grip must be taken into account [41]. Table 8.3 presents the summary of textiles used in this experiment.

A comparison of the effect of matrix type as shown in Figure 8.7a indicates that the glass fabric exhibits the best bond when high amounts of fly ash are used. The dual peak in the response of the fabric pullout is associated with the failure of the junction bonds and the transfer of the load down the length of the fiber. Effective stiffness of the interface transition zone is shown in Figure 8.7b and shows the bond stiffness parameter for various fabric systems. The pultrusion technique increases bond strength when noncoated multifilament was used to produce the fabric (polypropylene [PP] and polyvinyl acetate [PVA]), whereas the vacuum technique increases the AR glass bond but is ineffective with PP.

The stiffness of the interface transition zone as calculated from the linear portion of load–slip experiments is shown in Figure 8.7b and shows that there is a difference in magnitude between the bonding characteristics of different systems. The most basic measure of interfacial failure responses is the calculation of the nominal shear strength, which is defined as the averaged shear strength along the embedded length. This measure assumes a constant frictional mechanism along the embedded length, but does not take into account the interaction due to the transverse yarns. The nominal shear strength that is defined by the strength per fiber surface area is evaluated by the following,

---

TABLE 8.2
Properties of Yarn Making Up the Fabrics

Yarn Type	Yarn Nature	Strength (MPa)	Modulus of Elasticity (MPa)	Strain at Peak (mm/mm)	Filament Size (mm)	No. Filaments in a Bundle	Bundle Diameter (mm)
AR glass		1276–2448	78600		0.0135	400	≈0.27
PP	Bundle	500	6900	0.27	0.04	100	0.40
PVA		920	137[a]		0.025	200	
PE	woven						

[a] Modulus of elasticity for PVA is in cN/dtex.

TABLE 8.3
Material Parameters Used in Experiments and Simulations

Fabric	Series	Mixture and Processing	Yarn n Filaments	φ (mm)	E_y (MPa)	L (mm)
G	G105	Cast				12.7
	GP105	Pultrusion				12.7
	GV105	Vacuum	8	0.27	58,605	12.7
	FG105	FA, cast				12.7
	FGP105	FA, pultrusion				12.7
PP	PP103	Cast				7.6
	PP105	Cast				12.7
	PPP103	Pultrusion	8	0.40	493	7.6
	PPP105	Pultrusion				12.7
	PPV103	Vacuum				7.6
PE	PE105	Cast	8	0.25	965	12.7
PVA	PVA103	Cast	8	0.80	603	7.6
	PVAP103	Pultrusion				7.6
	FPVAP103	FA, pultrusion				7.6
	FPVAP105	FA, pultrusion				12.7

Note: Samples with dimension of 8.1 × 25.4 mm, e_m = 10 GPa, v_m = 0.2. FA, fly ash.

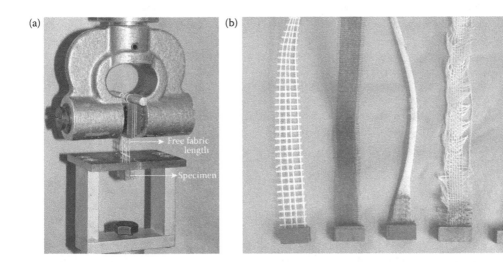

FIGURE 8.6 Schematics of pullout experiments on fabric reinforced systems.

$$\tau_{nom} = \frac{P_{ult}}{n\pi\phi L} \tag{8.19}$$

where P_{ult} is the ultimate load attained, n is the number of fibers resisting the load, ϕ is the nominal diameter of the fiber, and L is the embedded length. The average values and standard deviation of strength parameters are shown in Table 8.4 for sample results of different processes with 12.7 mm of embedded length. Note that shear strength values can reach as high as 4 MPa because of the inclusion of anchorage effects in an equivalent yarn model.

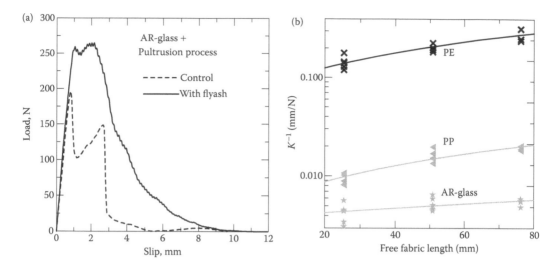

FIGURE 8.7   (a) The pullout response of an AR glass fabric using control and fly-ash-blended matrix; (b) the effective stiffness of the interface transition zone obtained from the pullout experiments.

TABLE 8.4
Basic Properties of Samples Tested Using a Uniform Shear Strength Approach

	Process	Slope of Deformation vs. Force (N·mm)	Maximum Load (N)	Toughness (N·mm)	Average τ_{max} (MPa)
AR glass fabric	Cast	246.51 (58.33)	183.59 (18.02)	589.86 (227.43)	2.28 (0.16)
	Pultrusion	218.45 (41.92)	145.50 (44.58)	368.78 (128.85)	1.76 (0.56)
	Vacuum	329.513 (47.24)	259.25 (31.31)	908.08 (167.13)	2.94 (0.33)
PVA fabric	Cast	47.94 (8.68)	145.92 (37.90)	477.52 (77.76)	1.54 (0.35)
	Pultrusion	46.19 (8.83)	272.74 (15.51)	1539.98 (349.56)	3.20 (0.26)
PP fabric	Cast	80.86 (38.73)	117.73 (49.87)	411.85 (236.14)	1.55 (0.62)
	Pultrusion	93.54 (20.63)	239.34 (70.09)	1042.29 (410.45)	2.91 (0.67)
	Vacuum	111.76 (14.01)	125.15 (24.90)	425.23 (178.11)	1.67 (0.38)
PP single yarn	Cast	15.52 (3.48)	24.66 (15.29)	124.06 (92.73)	2.52 (1.53)
	Pultrusion	18.80 (2.52)	27.44 (9.80)	136.33 (43.04)	2.85 (1.29)
	Vacuum	16.78 (3.31)	39.76 (10.96)	198.57 (42.34)	4.07 (1.09)

Note: The values in parentheses are the standard deviation.

The simplest pullout model parameters are defined by four variables: the initial stiffness of the pullout curve, the ultimate load, the slip at the ultimate load, and the energy absorbed in the process evaluated up to the ultimate load. The initial stiffness is obtained from the slope of the linear portion of the curve. The pullout response sometimes shows several peak loads in the case of textile materials. Toughness is defined by the area under the pullout slip. These parameters, however, are not material properties and depend on the fiber length, diameter, and testing conditions.

Some of the features of the experimental data are evident when the responses are plotted. Figure 8.8a shows the comparison of three fabric systems constructed of AR glass, PP, and polyethylene (PE). Note that the glass system is the stiffest and strongest system, whereas the PP woven fabrics result in the highest energy absorption. Figure 8.8b compares the results of three processing methods, including the cast, pultrusion, and vacuuming of the paste used the same AR glass. Results show that both the vacuum and pultrusion processing result in higher strength and energy absorption during pullout.

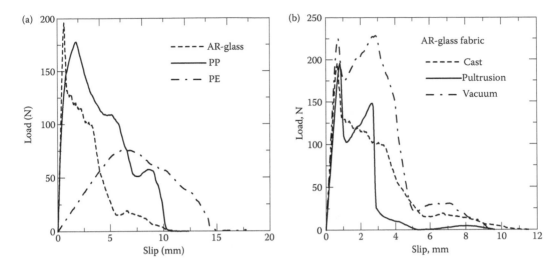

FIGURE 8.8 The response of the nonlinear textile debonding model.

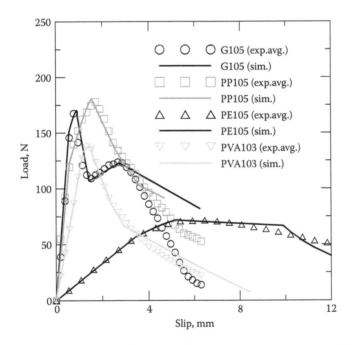

FIGURE 8.9 Simulation of representative pullout curve for control test series (G, PP, PE, and PVA).

As an alternative approach to parameter estimation, one can use models to back-calculate the material properties of interface. The response of each test series was simulated using the yarn and cement pastes provided in Tables 8.2 and 8.3 as input parameters [38]. The model fits various representative curves of different fabrics, matrices, embedded lengths, and processing methods. Internal interface parameters (κ, τ_{max}, τ_{FRC}, τ_{dyn}, and stiffness efficiency factor η) were measured by back-calculation.

The averaged responses and their simulations of the control test series (cast without fly ash G105, PP105, PE105, and PVA105) are presented in Figure 8.9. Material parameters, such as shear strength, were obtained by trial and error until the response of simulation matched with the experiment, as

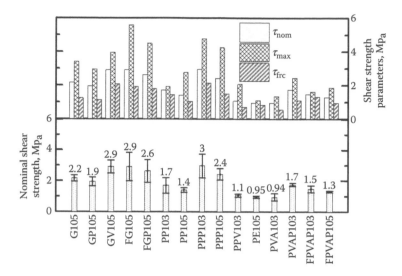

FIGURE 8.10 Shear strength parameters: (a) experimental nominal shear strength versus max and frictional shear strength from simulations; (b) nominal shear strength and standard deviation.

shown in Figure 8.10. The simulation clearly demonstrates the fit between the model and the experiments. The abrupt changes in the late postpeak responses are due to the sudden change of shear strength from τ_{FRC} to τ_{dyn}. The averaged response of the control test series (G105, PP105, PE105and PVA105) as shown in Figure 8.9 were selected to demonstrate the fit of the model and the experiments by the determination of the internal parameters (κ, τ_{max}, and τ_{FRC}). The values are listed in Table 8.5 alongside with experimental shear strength parameters.

The maximum shear strength (τ_{max}) and the frictional shear strength (τ_{FRC}) correlate well with the nominal shear strength, τ_{nom}. The magnitude of these stresses, however, are much higher than the tensile strength of the cement matrix phase, because mechanical bond mechanisms are included in the bond parameters. Nominal shear strength, defined as the average strength at maximum load, lies in between τ_{max} and τ_{FRC}, as shown in Figure 8.11. As the nominal shear strength increases, both the maximum and frictional shear strength values increase for a range of specimens regardless of conditioning or fabric types.

Because of the existence of the interface transition zone and the bundle effect in many filament-based fiber systems, an inefficient contact area exists and a parameter representing the efficiency of the yarn stiffness ($\eta < 1$) is introduced, which reduces the nominal contact stiffness to account for the inefficiencies in bonding. Typical values are shown in Table 8.5. An exact estimation of the stiffness of a yarn ($A_y E_y$) overestimates the experimental values significantly. Several contributing factors include the variations in bond of the interface zone, inefficiency due to the sleeve effect, lack of uniform strain in all the yarns during the test, the initial curvature in the yarns, and a lack of bonding of 100% of filaments due to imperfections and porosity. By defining a scalar parameter to represent the percentage of filaments actively contributing to the apparent axial stiffness ($\eta A_f E_f$) of the yarn, one can verify that the sleeve filaments are bonded to the matrix, which contributes to axial stiffness, whereas the core filaments only provide marginal stiffness when unbonded multifilament yarns are used [42].

The main factors that have a sizable effect on the bond strength and efficiency of the textile cement composite are: the fiber type, the matrix type, the embedded length, and processing methods. Results reveal that the bond strength of textiles ranked from the highest to the lowest are AR glass, PVA, PP, and PE, fly ash replacement of 40% by volume increases bond strength for AR glass but not in PVA textiles. The bond strength obtained from specimens with shorter embedded lengths yield slightly higher values with larger scattered results. The processing methods indicate that the

TABLE 8.5
Key Parameters Obtained from Experiments and Simulations

	Key Parameters									
	Experiments					**Simulations**				
Series	K (N·mm)	P_{max} (N)	Slip (at Max) (mm)	Toughness (at Max) (N·mm)	τ_{nom} (MPa)	κ (N·mm³)	τ_{max} (MPa)	τ_{FRC} (MPa)	τ_{dyn} (MPa)	η (%)
G105	210 (70)	186 (18)	0.74 (0.15)	67 (25)	2.16 (0.21)	4.24	3.41	1.24	1.59	5.0
GP105	281 (73)	168 (25)	0.69 (0.26)	64 (25)	1.94 (0.29)	4.24	2.96	1.14	1.24	5.7
GV105	380 (72)	250 (36)	0.63 (0.08)	85 (14)	2.91 (0.41)	4.24	3.96	2.07	2.62	9.5
FG105	258 (59)	251 (80)	1.77 (0.33)	316 (114)	2.91 (0.92)	6.28	5.58	1.93	1.52	4.0
FGP105	334 (77)	227 (63)	1.45 (0.43)	242 (137)	2.63 (0.73)	7.13	4.48	1.79	1.24	5.4
PP103	234 (137)	130 (39)	1.34 (0.75)	112 (61)	1.70 (0.51)	8.48	1.93	1.43	1.39	9.5
PP105	287 (70)	179 (20)	1.72 (0.07)	221 (28)	1.41 (0.16)	4.24	2.76	1.05	0.90	17.0
PPP103	190 (76)	227 (60)	2.00 (0.32)	294 (118)	2.96 (0.78)	4.24	4.76	2.14	1.96	10.0
PPP105	140 (27)	308 (47)	3.10 (0.17)	587 (106)	2.41 (0.36)	1.70	4.24	1.50	1.24	13.0
PPV103	237 (46)	135 (12)	1.85 (0.68)	178 (85)	1.06 (0.10)	5.09	2.07	0.72	0.55	13.5
PE105	17 (3)	76 (6)	6.25 (1.79)	299 (125)	0.95 (0.08)	0.17	1.10	0.83	0.83	24.0
PVA103	112 (36)	144 (35)	1.32 (0.16)	95 (20)	0.94 (0.23)	0.59	1.34	0.55	0.48	15.5
PVAP103	75 (15)	267 (16)	4.19 (1.28)	655 (273)	1.74 (0.10)	0.51	2.45	1.10	1.10	10.5
FPVAP103	208 (41)	225 (34)	1.24 (0.17)	166 (58)	1.47 (0.22)	0.34	1.62	1.31	1.21	32.0
FPVAP105	406 (193)	330 (14)	1.04 (0.41)	205 (76)	1.29 (0.05)	1.70	1.86	0.97	0.86	85.0

Source: Sueki, S., "An analytical and experimental study of fabric-reinforced, cement-based laminated composites," MS thesis, Arizona State University, 2003.

Note: Average and standard deviations of experiments on four replicate samples.

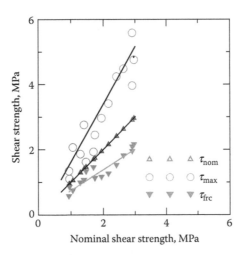

FIGURE 8.11 Correlation of shear strength parameters: nominal shear strength (τ_{nom}), maximum shear strength (τ_{max}), and frictional shear strength (τ_{FRC}).

pultrusion technique increases bond strength in PP and PVA textile but not in AR glass, whereas the vacuum technique increases bond strength in AR glass but not in PP.

In general, the highest bond strength is observed in glass textile systems, while PE, PP, and PVA show the lower strengths. Bonding of the glass textile is due to the high modulus of elasticity of fibers, the adhesion of glass to cementitious matrix, and the nature of the interlock in bonded textile.

The low bond of PVA may be attributed to the fact that the diameter of the bundle is relatively large. Similarly, PVA contains a high number of filaments (Table 8.1), which leads to a dense textile structure and poor paste penetration between the yarns and the filaments. An addition of fly ash to the glass system improves the durability of the glass textile as much as 35%. Variations in embedded length do not significantly influence the shear strength values for PP or PVA textiles. The use of the pultrusion increases bond strength in PP by roughly 72% and PVA by 85%, but not in AR glass (≈10%). The improvement in bonding by the pultrusion of the PP and PVA systems is due to the penetration of the cement matrix in between the filaments of the bundle [44]. In contrast, the vacuum technique increases bond strength in the AR glass, but decreases bond strength by 38% in the PP system. The higher viscosity and the poor penetrability of vacuumed paste within the fine grid and small filament spacing of PP textile are acknowledged. In comparison, the coarse grid AR glass textile, which is made from resin-coated bundles (Figure 8.8b), performs much better with the vacuum paste.

ENERGY DISSIPATION DURING PULLOUT

The load–slip response of the fiber can be used to compute parameter γ_p, which is defined as a measure of energy absorption during the debonding and pullout process:

$$\gamma_P = \int_0^{l_p} P*(a)\frac{du}{da}da \tag{8.20}$$

where $P*$ is the force at the fiber end, u represents the fiber slip as a function of debonding length and load, and l represents the length of the fiber pullout. The present model is compared with the work of Piggot [43], which combines the energy absorption due to pullout and fracture using a single-parameter, strength-based model for calculating the interface fracture energy:

$$G_i = \frac{V_f d\sigma_{fu}^3}{6\tau_i}\left[\frac{1}{8\tau_i s} + \frac{1}{E_f}\left(1 - \frac{\sigma_{fu}}{2\tau_i s}\right)\right] \tag{8.21}$$

where G_i represents toughness, σ_{fu} and τ_i represent fiber and interfacial shear strength, and E_f, d, and s represent the Young's modulus, the fiber diameter, and the aspect ratio, respectively. Toughness is defined as the summation of the strain energy stored in the fibers and the work done by the sliding of the fibers. A comparison of the current model and the model proposed by Piggot is shown in Figure 8.12. The dependence of toughness on the aspect ratio shows that optimum toughness coincides with the critical length of fibers. Fiber lengths longer than the critical length result in significant reduction of pullout work. Both models predict similar responses during the complete fiber pullout region. In the case of fiber fracture, Piggot's approach computes the average energy dissipated per unit length of pullout. The fracture-based model assumes that the entire debonded length pulls out, which is limited in nature since the fracture could take place at a point along the debonded length and followed by partial length pullout. As the fiber aspect ratio increases, the debonded length also increases, thus the energy of pullout is increased. The agreement between the critical length parameter computed for both models is due to using similar material parameters for both approaches.

Combinations of fiber length and interface properties as functions of strength and work of pullout (γ_p) maps show that the load increases proportional to the interface shear strength. As the fiber length and interfacial bond strength increase, both maximum load and pullout energy increase. A constant load region is an indication of fiber fracture. By increasing interface toughness and fiber length beyond this level, the work of pullout significantly decreases.

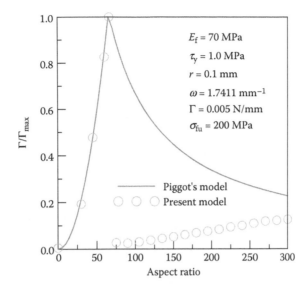

$E_f = 70$ MPa

$\tau_y = 1.0$ MPa

$r = 0.1$ mm

$\omega = 1.7411$ mm^{-1}

$\Gamma = 0.005$ N/mm

$\sigma_{fu} = 200$ MPa

——— Piggot's model

○ ○ ○ Present model

FIGURE 8.12 Effect of fiber aspect ratio on the energy dissipation during pullout.

FINITE ELEMENT SIMULATION

Several finite element models for the fiber pullout and pushout problem exist in two and three dimensions [21]. The form of the constitutive response of the interface is based on the work of Ballarini et al. [20] and Steif and Dollar [26], and consists of two linear elastic materials bonded by means of an interfacial layer with its unique elastic properties. In this test, a two-dimensional axisymmetric finite element formulation is used to formulate the sliding delamination of a two-layer interface system using a linear elastic single fiber in a cylindrical matrix [44]. This formulation allows the effect of residual stresses and Poisson's contraction on the sliding force–slip response to develop. Sliding contact surfaces are also imposed between the fiber and the substrate elements. These surfaces are initially bonded to one another, and debonding is based on a stress-based yield surface. A clamping pressure applied at the outside surface of the matrix results in a normal force on the interface. This simulates the effect of residual stresses and matrix shrinkage. The interface is characterized as an independent, linear, and elastic third phase with a stiffness significantly lower than that of the matrix.

Schematics of the model are shown in Figure 8.13. Clamping pressure is applied at the outer layer of the matrix elements in order to simulate the residual stresses. The interface is modeled as the third phase with elastic properties significantly lower than the matrix. Its failure mechanism is governed in a way similar to that in the yield surface of the Mohr-Coloumb method. As the yielding takes place at the interface, the constitutive response is replaced with the Coulomb friction model. This approach permits a study of the residual stresses at the interface and correlates the coefficient of friction at the interface to its resistance to interlaminar shear failure.

Sliding contact surfaces are defined at the space between the fiber and interface elements. These surfaces are initially bonded, but debond when a stress-based yield surface criterion is met. After the debonding, the stress continuity across the sliding contact elements is preserved. The slip of the interfacial region can be directly measured as a function of its position along the interface. Debonding criteria is defined in terms of parameters σ_n and τ_n, which represent the normal and the shear strength of the interface, respectively. Using these two parameters, an isotropic yield surface for debonding is expressed according to Equation 8.22:

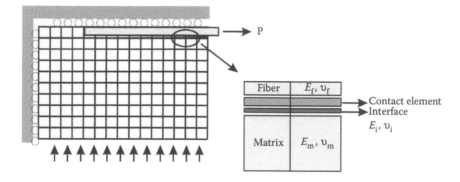

FIGURE 8.13 Schematics of the finite element model used for the simulation of debonding. (After Mobasher, B., and Li, C. Y., "Modeling of stiffness degradation of the interfacial zone during fiber debonding," *Journal of Composites Engineering*, 5, 10–11, 1349–1365, 1995.)

$$f = \sqrt{\left[\frac{\sigma_n}{\sigma}\right]^2 + \left[\frac{\tau_n}{\tau}\right]^2}\left(1 - f_{tol}\right) \le f \le \left(1 + f_{tol}\right) \tag{8.22}$$

The parameter f_{tol} represents the tolerance of convergence on the yield surface. Note that because of the normalization of stress components, the yield surface is elliptical. In comparison to with a Coulomb shear model, the present approach results in finite strength under high residual clamping stresses. Coulomb friction after debonding is expressed as:

$$\tau = \tau_0 - \mu\,\sigma_n < \tau_f \tag{8.23}$$

FRACTURE-BASED APPROACH

A fracture mechanics model for the debonding of a fiber from a cementitious matrix is used to calculate the stiffness degradation, which develops in response to the propagation of a stable debonding zone. Its formulation on the basis of the growth of a Mode 2 crack along an elastic perfectly plastic one-dimensional interface is used. The crack growth criterion is defined as or by using the strain energy release rate of a partially debonded interface. It is assumed that the fracture toughness of the interface increases as a function of the debonded length in the form of an R-curve. In addition, a frictional shear stress of constant magnitude acts over the debonded length. The R-curve parameters are obtained by calculating the numerical solution of the resulting differential equations. The load–slip response is obtained from the R-curves, and a parametric study of fiber type, length, and interface properties is conducted to show the sensitivity of the load-slip response to variations in the model parameters. Results are then compared with a finite element model on the basis of the Coulomb friction models (Figure 8.14).

STRAIN ENERGY RELEASE RATE

By assuming shear deformations to be localized, a formulation on the basis of a weak boundary layer in a rigid half plane developed by Stang et al. [8] has been used. A linear elastic circular fiber with radius r, length L, Young's modulus E_f, and ultimate tensile strength σ_{fu} (as shown in Figure 8.15) is considered in the Stang Model. A concentrated force, P^*, is applied at the end of the fiber. The displacement along the fiber length is $U(x)$, with $U(L)$ representing the displacement

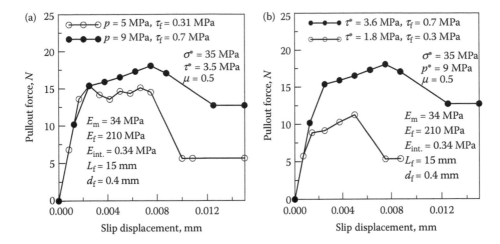

FIGURE 8.14 (a and b) Parametric study of the model indicates that, as the confining pressure on the fiber is increased, the resistance to pullout of the load increases. In addition, Panel B shows that, as the interfacial bond and the shear strength increase, the load-carrying capacity of the fiber also increases. (a) Effect of lateral clamping pressure on the fiber pullout response of steel fibers; (b) effect of bond shear strength and frictional shear strength of the interface on the fiber pullout. (After Mobasher, B., and Li, C. Y., "Modeling of stiffness degradation of the interfacial zone during fiber debonding," *Journal of Composites Engineering*, 5, 10–11, 1349–1365, 1995.)

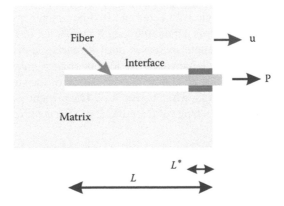

FIGURE 8.15 Schematics of the fiber pullout model by Stang and Shah. (From Stang, H., Li, Z., and Shah, S. P., "Pullout problem: Stress versus fracture mechanical approach," *Journal of Engineering Mechanics*, 116, 10, 2136–2149, 1990.)

at the fiber end. The constitutive response of the interface is shown in Figure 8.15 and expressed in Equation 8.24,

$$q = \begin{cases} K\,U(x) & 0 < x < L - a \\ q_f & L - a < x < L \end{cases} \tag{8.24}$$

The bonded interface is elastic-perfectly plastic with stiffness, K, and ultimate adhesional bond strength, $q_y\,(= 2\pi r \tau_y)$, which is the shear flow in force per unit length of fiber. The debonded length is denoted as a, with a constant frictional shear force per unit length, $q_f\,(= 2\pi r \tau_f)$, acting in this region. Equilibrium conditions are expressed as:

$$P^* = E_f\, A\, U,_x(L)$$
$$U,_x(0) = 0$$
$$U(L-a)^- = U(L-a)^+$$
$$U,_x(L-a)^- = U,_x(L-a)^+$$
(8.25)

where A represents the fiber cross-sectional area. The boundary conditions consist of the stress and displacement continuity across the bonded and debonded portion of the interface, the force at the fiber tip, and the zero force at the embedded fiber end (Figure 8.16). The subscript x represents the derivative with respect to length. The displacement at the fiber end is expressed by Stang et al. [8] as:

$$U(L) = \frac{(P^* - q_f\, a)\chi}{E_f\, A\, \omega} + \frac{P^* a - 0.5 q_f\, a^2}{E_f\, A}, \qquad \omega = \sqrt{\frac{k}{E_f\, A}}, \qquad \chi = \coth\left(\omega(L-a)\right)$$
(8.26)

The strain energy is the algebraic sum of the work done by external forces, W_{ex}, the strain energy in debonded and bonded system, W_e, and the inelastic work done by the friction, W_f. The energy release rate, G, is calculated by assuming an infinitesimal crack growth under constant displacement. Note that as the load increases, the debonding length increases until it reaches a critical level of debonding, beyond which as the debonding length increases, the load decreases. This suggests that the maximum load has been reached under partial debonding conditions.

Comparison of a two-parameter fracture-based approach by Mobasher and Li [38] with a single-parameter fracture-based approach by Stang et al. [8] is shown in Figure 8.16. The approach compares a energy release rate for a single-parameter and two-parameter fracture energy curves as a function of debonding; The load versus debonding length are shown in Figure 8.16b. Note that as the load increases, the debonding length increases until it reaches a critical level of debonding, beyond which as the debonding length increases, the load decreases pointing out to the maximum load reaching under partial debonding conditions. A simulation of the steel fiber pullout results of the Naaman and Shah experiments is shown in Figure 8.17. The present formulation is in agreement with the average values predicted for a range of fiber diameters. To achieve the prediction parameter

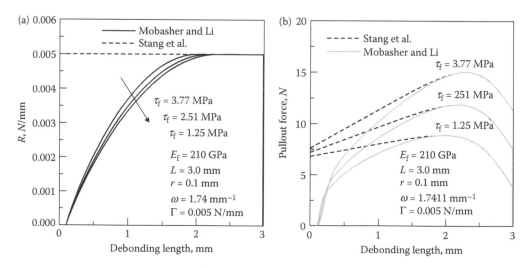

FIGURE 8.16 Comparison of a two-parameter fracture-based approach by Mobasher and Li [38] with a single-parameter fracture-based approach by Stang et al. [8]. (a) Comparison between the single-parameter and two-parameter fracture energy curves as a function of debonding; (b) load versus debonding length.

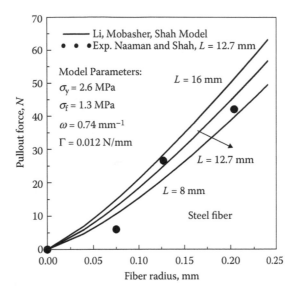

FIGURE 8.17 Effect of fiber diameter on the pullout response of steel fibers.

of τ_f in the range of 0.2 to 2 MPa, τ_y in the range of 0.4 to 4 MPa were used. Using a fracture-based approach described in the following sections as a failure criteria, a magnitude of $p\Gamma$ used as a measuring agent of the interfacial fracture toughness in the range of 0.0006 to 0.012 N is needed.

MODELING OF THE TRANSVERSE YARN ANCHORAGE MECHANISM

The load transfer across a matrix crack can be calculated using a closed form bond model [39, 45]. The effect of transverse yarns in woven and bonded textiles is handled using a periodic arrangement of linear springs providing anchorage at the warp–fill junctions (warp and fill are the yarns parallel and perpendicular to load direction, respectively). The point of intersection of the warp and fill yarns restrains the warp yarns, transfers, and redistributes the load to the fill yarns. The function of the fill yarns is in redistributing the load applied to the warp yarns. Anchorage may also be caused by the surface curvature of the yarn in a woven textile. Simulation results were compared with the experimental results of three textile types and three different sample-making procedures [46].

By using the superposition principal, one can add the contribution of the fiber junctions to the original solution of the bond. The steps in the solution are based on an incremental approach. Figure 8.18 shows two stages of intact and failed junction bonds. The resistance, which is due to the transverse yarns, contributes to the original solution. The shear force generated by the transverse yarns is obtained by multiplying the average slip at the location of the junction point with the stiffness of the junction. As the magnitude of the force at a junction exceeds the strength of the junction, it fails and the force reduces to zero.

The fabric load, as shown in Figure 8.19a, can be divided in three portions: load carried by the bonded region, P_y, load carried by the debonded region, P_d, and load transferred at the junction points to the transverse yarns, P_b, which is expressed as:

$$P = P_y + P_d + P_b \tag{8.27}$$

Two main parameters characterize the debonded region, the debonded length (L_d), and the constant frictional shear stress acting along this length, t_f. In the bonded region, shear lag is operating with a decaying shear stress because of parameter β, which models the stiffness of the interface. A portion

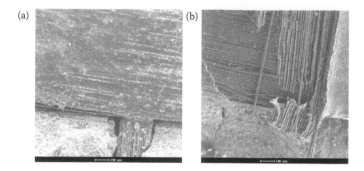

FIGURE 8.18 Two stages of yarn anchorage: (a) intact and (b) a failed junction bond.

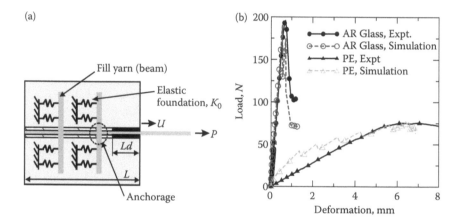

FIGURE 8.19 (a) Fabric pullout model across a matrix crack; (b) the response of the fabric debonding model for AR glass and PE fabrics.

of the load referred to as P_b ($P_b < P_N$) is transferred at the junction points to the transverse yarns by means of bending. The force P_b is obtained by using a beam on the elastic foundation analogy using the transverse stiffness of the fill yarns as they are resting on the supportive matrix. The load is, therefore, obtained by using the stiffness of the interface, which is treated as the elastic foundation multiplied by the nodal point displacement,

$$P_b = \sum_{i=1}^{n} K_b u(x_i) \tag{8.28}$$

A portion of the load referred to as P_b ($P_b < P_N$) is transferred at the junction points to the transverse yarns that react by means of shear. The general load–slip relationship for a debonding textile includes the sum of contributions of the number of active junctions, n, in redistributing the load:

$$P = P_d + P_y + P_b = \tau_f (2\pi r) L_d + \frac{-2\pi r \tau_{max}}{\beta_2 \coth(\beta_2(L - L_d))} + \sum_{i=1}^{n} K_b u(x_i) \tag{8.29}$$

$$U(L) = \frac{P - \tau_f L_d}{E_f \pi r^2 \beta_2} \coth\left(\beta_2 \left(L - L_d\right)\right) + \frac{P - (1/2)\tau_f L_d}{E_f \pi r^2} L_d \tag{8.30}$$

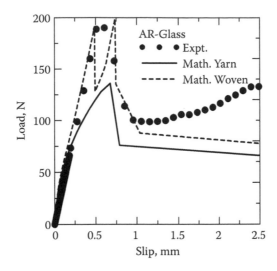

FIGURE 8.20 Various stages of contribution of the textile to the resisting the growth of the cracks.

where

$$K_b = \frac{2k}{\lambda} \frac{\sinh(\lambda l)+\sin(\lambda l)}{\cosh(\lambda l)+\cos(\lambda l)-2} \quad \beta_2 = \sqrt{\frac{G_i}{E_f \pi r^2}} \quad \lambda = \sqrt[4]{\frac{k}{4EI}} \quad (8.31)$$

In these equations, $k = bk_0$ k_0 is the modulus of foundation in N m³, b is the constant width of the beam in contact with the foundation, and EI is the flexural rigidity of the yarn when it is treated like a beam, as shown in Figure 8.19a. In the present approach, b is the thickness of yarn and I is calculated from the fill–yarn geometry [47]. k_0 and E are the values related to matrix and fiber interface. The simulation results were compared with the experimental results of different fabric types and matrix materials [39, 48] and verified the experimental results quite well [49, 45].

Figure 8.19b shows the effect of the inclusion of the junction points in the modeling of mechanical behavior using a simulation of the fabric pullout force–slip experiments for AR glass and PE fabrics. Compared with straight fiber, the inclusion of junction forces increases the load-carrying capacity with each failure of the junction bonds. The drop and the subsequent increase in the load are due to the failure of one junction and the subsequent transfering of the load to the following junctions. Figure 8.20 shows the comparison between straight fiber modeling with the textile-based approach that uses a periodic arrangement of spring elements along the length of the fiber. Note that the straight yarn is unable to capture the prepeak, nonlinear response of the yarn.

FINITE DIFFERENCE APPROACH FOR THE ANCHORAGE MODEL

An alternative formulation on the basis of the finite difference method was proposed by Soranakom and Mobasher [50]. The finite difference method, as shown in Figures 8.21a through 8.21c, is flexible in simulating material nonlinearity while demonstrating the main characteristics of fabric pullout response: slack in fabric due to the initial curvature, interface bond model, and nonlinear springs simulating the mechanical anchorage provided by transverse yarns. For typical fiber content, the axial stiffness of the matrix, $A_m E_m$, is much larger than that of the fiber, $A_f E_f$; hence, the effect of matrix strain can be ignored, whereas slip s is simplified as:

$$s = \int (\varepsilon_f - \varepsilon_m)\,dx = \int \varepsilon_f\,dx \quad \text{and} \quad s' = \varepsilon_f \quad (8.32)$$

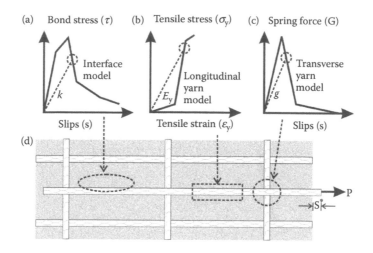

FIGURE 8.21 Fabric pullout mechanisms: (a) interface bond slip, (b) longitudinal yarn, (c) spring simulating anchorage strength at cross-yarn junction, and (d) fabric pullout specimen.

Figure 8.21d shows the load transfer mechanism between matrix and fiber with and without non-linear spring. The corresponding finite difference equations are:

$$F_{i+1} - F_{i-1} - 2ht_i - T_i = 0 \tag{8.33}$$

$$F_{i+1} - F_{i-1} - 2ht_i = 0 \tag{8.34}$$

By substituting the fiber force–strain relationship, $F = A_f E_f s'$, interface shear flow–slip relationship, $t = ks$, nonlinear spring force–slip relationship, $T = ms$, and forward finite difference, $s' = s_{i+1} - s_i$ or backward finite different, $s' = s_i - s_{i-1}$, in Equations 8.33 and 8.34, the finite difference equations can be obtained:

$$A_f E_{f_{i+1}} s_{i+1} - [A_f (E_{f_{i+1}} + E_{f_{i-1}}) + 2h^2 k_i + -m_i h] s_i + A_f E_{f_{i-1}} s_{i-1} = 0 \tag{8.35}$$

$$A_f E_{f_{i+1}} s_{i+1} - [A_f (E_{f_{i+1}} + E_{f_{i-1}}) + 2h^2 k_i] s_i + A_f E_{f_{i-1}} s_{i-1} = 0 \tag{8.36}$$

where h is a finite length between two nodes and i is the node number. For displacement control as shown in Figure 8.21c, two boundary conditions are imposed and the nodal slips are solved iteratively such that finite difference equations are satisfied. These boundary conditions as shown in Figures 8.22a and 8.22d are zero force at the left end and prescribed slip at the right end. Figures 8.22c and 8.22d show the load transfer mechanism between matrix and fiber with and without non-linear spring. The corresponding finite difference equations are derived based on the equilibrium of each node.

The pullout simulation of a fiber with an embedded length of 10 mm is shown in Figure 8.23. The transverse yarn that provides mechanical anchorage increases the pullout load and reduces slippage in the prepeak region. The simulations also show the drop in stress corresponding to the junction failure similar to what is measured in the experiments. Parametric studies of the effect of bond strength and presence of junctions have been studied elsewhere [40, 50]; In these publications materials constitutive laws and the parametric studies of pullout load–slip response addressing effects of bond strength, slack, and spring strength are addressed in these publications and recent work [49].

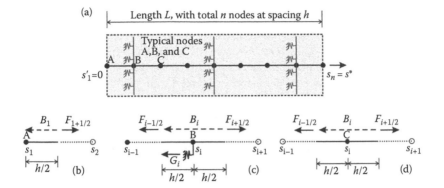

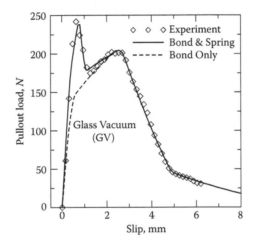

FIGURE 8.22 Finite difference fabric pullout model: (a) discretized fabric pullout model under displacement control; (b–d) free body diagram of three typical nodes "A," "B," and "C."

FIGURE 8.23 Pullout response of an AR glass fabric system using shear interface bond approach and the bond and junction model.

CHARACTERIZATION OF INTERFACIAL AGING

GFRC is made of AR glass fibers and cement paste or mortar [51], and shows resilient tensile strength, ductility, and fracture toughness at early ages of curing. However, it suffers a significant loss in strength and ductility as it ages with time. Several preventive measures have been experimentally investigated by many researchers [52,53,54]: using AR glass fiber to withstand chemical attack, adding pozzolanic materials to consume the calcium hydroxide built up in between glass fibers that causes stress concentration to the fibers, and adding a polymer emulsion to coat and protect fibers against chemical attack and the migration of cement hydration products.

Reducing the pore water solution's alkalinity provides a less aggressive environment for the fibers [35]. Test methods have been developed to characterize the pullout response of glass fibers from a portland cement matrix [32]. Methods are also being developed to characterize the strand effect in glass fiber composites. By determining strand perimeters using digital analysis of images captured from petrographical thin sections, one can obtain the effective perimeter parameter and use this to characterize and optimize the bond and fiber length parameters [36].

THEORETICAL MODELING OF INTERFACIAL AGING

Li et al. [29] studied the accelerated aging of the fiber–matrix interface in single-fiber pullout tests. Experimental data indicate that aging may be viewed as an increase in the adhesional and frictional bond strength of the interface coupled with an increase in the interface stiffness parameter. An empirical approach to model aging must consider increases in interfacial shear strength in the following way:

$$\tau = \tau^0(1 + e^{-\alpha t})$$ (8.37)

where τ represents the interfacial shear strength as a function of age, and τ^0 is the initial shear strength. The increase in shear strength due to aging may be incorporated through the two parameters "τ^0" and "a." The reduction in the tensile strength of fiber may be modeled as:

$$\sigma_{fu} = \sigma_{fu}^0(1 - e^{-bt})$$ (8.38)

Parameter σ_{fu}^0 represents the as-is strength of the glass fiber. Parameter b represents the aging coefficient and depends on the test temperature and the alkalinity of the medium. The experimental data will be used in conjunction with the analytical model for the fiber pullout in order to calibrate the effect of aging on the fiber pullout response. The fiber pullout response can then be used in the closing pressure to simulate the aging of the composite under uniaxial loading conditions [7] and compared with the theoretical and experimental study of durability in GFRC or textile reinforced concrete, as shown in the work by Shah et al. [52].

CONCLUSIONS

Models have been presented for the simulation of the pullout response of fibers and fabrics from a cement matrix. Various stages of modeling on the basis of solid mechanics shear lag, finite element, finite difference and fracture-based approaches have also been shown. In all these cases, it is clear that the bond adhesion and frictional shear parameters are needed to contrast the pullout force with the slip response. By modeling the additional restraint of the fill yarns as a beam on elastic foundation, which was subjected to a concentrated load and to a modified shear lag, or a finite difference, the pullout response of fabric reinforced composites have been simulated. Various models were used to characterize the changes due to the fabrication process, the fabric type, and the matrix formulations.

REFERENCES

1. Aveston, J., Cooper, G. A., and Kelly, A. (1971). "Single and multiple fracture," in *The Properties of Fiber Composites*. Surrey, UK: IPC Science and Technology Press, 15–26.
2. Kelly, A., and Tyson W. R. "Fiber-strengthened Materials," *Proceedings of the Second Berkeley International Materials Conference: High-Strength Materials—Present Status and Anticipated Developments*, Berkeley, CA, 1965, 578–602.
3. Bennison, S. J., and Lawn, B. R. (1989), "Role of interfacial grain-bridging sliding friction in the crack-resistance and strength properties of non-transforming ceramics," *Acta Metallurgica*, 37(10), 2659–2671.
4. Cox, B. N. (1993), "Scaling for bridged cracks," *Mechanics of Materials*, 15, 87–98.
5. McCartney, L. N. (1987), "Mechanics of matrix cracking in brittle-matrix fiber reinforced composites," *Proceedings of the Royal Society of London, Series A, Mathematical and Physical Sciences*, 409, 329–350.
6. Budiansky, B., Hutchinson, J. W., and Evans, A. G. (1986), "Matrix fracture in fiber reinforced ceramics," *Journal of the Mechanics and Physics of Solids*, 343(2), 167–189.
7. Marshall, D. B., Cox, B. N., and Evans, A. G. (1985) "The mechanics of matrix cracking in brittle matrix fiber composites," *Acta Metallurgica*, 33(11), 2013–2021.

8. Stang, H., Li, Z., and Shah, S. P. (1990), "Pullout problem: Stress versus fracture mechanical approach," *Journal of Engineering Mechanics*, 116(10), 2136–2149.
9. Naaman, A. E., Namur, G. G., Alwan, J. M., and Najm, H. S. (1991), "Fiber pull-out and bond slip. I: analytical study. II: experimental validation," *Journal of Structural Engineering*, 117(9), 2769–2800.
10. Sujivorakul, C., Waas, M., and Naaman A. E. (2000), "Pullout response of a smooth fiber with an end anchorage," *Journal of Engineering Mechanics*, 126(9), 986–993.
11. Abrishami, H. H., and Mitchell, D. (1996), "Analysis of bond stress distributions in pull-out specimens," *Journal of Structural Engineering*, 122, (3), 255–261.
12. Focacci, F., Nanni, A., and Bakis, C. E. (2000), "Local bond slip relationship for FRP reinforcement in concrete," *Journal of Composites for Construction*, 4, 24–32.
13. Marshall, D. B. (1992), "Analysis of fiber debonding and sliding experiments in brittle matrix composites," *Acta Metallurgica et Materialia*, 40(3), 427–441.
14. Marshall, D. B., and Oliver, W. C. (1987), "Measurement of interfacial mechanical properties in fiber reinforced ceramic composites," *Journal of the American Ceramic Society*, 70(8), 542–548.
15. Hsueh, C. H. (1990a), "Interfacial friction analysis for fiber-reinforced composites during fiber push-down (indentation)," *Journal of Materials Science*, 25, 818–828.
16. Hsueh, C. H. (1990b), "Interfacial debonding and fiber pullout stresses of fiber reinforced composites," *Materials Science and Engineering*, A123, 1–11.
17. Hutchinson, J. W., and Jensen, H. M. (1990) "Models of fiber debonding and pullout in brittle composites with friction," *Mechanics of Materials*, 9, 139–163.
18. Cox, B. N. (1990), *Acta Metallurgica et Materialia*, 38, 2411.
19. Dollar, A., and Steif, P. S. (1993), "Analysis of the fiber push-out test," *International Journal of Solids and Structures*, 30(10), 1313–1329.
20. Ballarini, R., Ahmed, S., and Mullen, R. L. "Finite Element Modeling of Frictionally Restrained Composite Interfaces," in *Interfaces in Metal–Ceramic Composites*, Proceedings of the International Conference on Interfaces in Metal Ceramics Composites, February 18–22, 1990, 349–388.
21. Mital, S. K., and Chamis, C. C. (1990), "Fiber pushout test: a three dimensional finite element computational simulation," *Journal of Composites Technology and Research*, 13(1), 14–21.
22. Gao, Y. C., Mai, Y. W., and Cotterell, B. (1988), "Fracture of fiber-reinforced materials," *Journal of Applied Mathematics and Physics*, 39, 550–572.
23. Leung, C. K. Y., and Li, V. C. (1991), "New strength-based model for the debonding of discontinuous fibers in an elastic matrix," *Journal of Materials Science*, 26, 5996–6010.
24. Zhou, L., M., Kim, J. K., and Mai, Y. W. (1992), "Interfacial debonding and fiber pullout stresses. Part II. A new model based on the fracture mechanics approach," *Journal of Materials Science*, 27, 3153–3166.
25. Liang, C., and Hutchinson, J. W. (1993), "Mechanics of the fiber pushout test," *Mechanics of Materials*, 14, 207–221.
26. Steif, P. S., and Dollar, A. (1992), "Models of fiber–matrix interfacial debonding," *Journal of the American Ceramic Society*, 75(6), 1694–1696.
27. Mobasher, B., and Li, C. Y. (1996), "Effect of interfacial properties on the crack propagation in cementitious composites," *Advanced Cement Based Materials*, 4(3), 93–106.
28. Li, C. Y., and Mobasher, B. (1998), "Finite element simulations of toughening in cement based composites," *Advanced Cement Based Materials*, 7, 123–132.
29. Li, Z., Mobasher, B., and Shah, S. P. (1991), "Characterization of interfacial properties of fiber reinforced cementitious composites," *Journal of the American Ceramic Society*, 74(9), 2156–2164.
30. Shao, Y., Li. Z., and Shah, S. P. (1993), "Matrix cracking and interface debonding in fiber-reinforced cement-matrix composites," *Advanced Cement Based Materials*, 1(2), 55–66.
31. Shao, Y, Ouyang, C., and Shah, S. P. (1998), "Interface behavior in steel fiber/cement composites under tension," *Journal of Engineering Mechanics*, 124, 1037–1044.
32. Mobasher, B., and Li, C. Y. (1996), "Effect of interfacial properties on the crack propagation in cementitious composites," *Advanced Cement Based Materials*, 4, 93–105.
33. Bentur, A., Diamond, S., and Mindess, S. (1985), "The microstructure of the steel fibre–cement interface," *Journal of Materials Science*, 20, 3610–3620.
34. Li, Z., Mobasher, B., and Shah, S. P. "Effect of Aging on the Interfacial Properties of GFRC," *Proceedings Materials Research Society*, MRS Fall Meeting, Boston, MA, 1990.
35. Purnell, P., Short, N. R., Page, C. L., Majumdar, A. J., and Walton, P. L. (1999), "Accelerated ageing characteristics of glass-fibre reinforced cement made with new cementitious matrices," *Composites Part A: Applied Science and Manufacturing*, 30(9), 1073–1080.

36. Purnell, P., Buchanan, A. J., Short, N. R., Page, C. L., and Majumdar, A. J. (2000), "Determination of bond strength in glass fibre reinforced cement using petrography and image analysis," *Journal of Materials Science*, 35(18), 4653–4659.
37. Naaman, E. A., Namur, G. G., Alwan, J. M., and Najm, H. S. (1991a), "Fiber pullout and bond slip. I: analytical study," *Journal of Structural Engineering*, 117, 2769–2790.
38. Naaman, E. A., Namur, G. G. Alwan, J. M., and Najm, H. S. (1991b), "Fiber pullout and bond slip. II: experimental validation," *Journal of Structural Engineering*, 117, 2791–2800.
39. Mobasher, B., "Modeling of Toughness Degradation and Embrittlement in Cement Based Composite Materials Due to Interfacial Aging," 13th ASCE Engineering Mechanics Division, Baltimore, June 13–16, 1999.
40. Sueki, S., Soranakom, C., Peled, A., and Mobasher, B. (2007), "Pullout–slip response of fabrics embedded in a cement paste matrix," *Journal of Materials in Civil Engineering*, 19, 9.
41. Sueki, S. (2003), "An analytical and experimental study of fabric-reinforced, cement-based laminated composites," MS thesis, Arizona State University.
42. Banholzer, B. (2004), "Bond behaviour of a multi-filament yarn embedded in a cementitious matrix," PhD dissertation, Aachen University, Germany.
43. Piggott. M. R. (2002), *Load Bearing Fibre Composites*, 2nd ed., New York: Elsevier, 475.
44. Mobasher, B., and Li, C. Y. (1995), "Modeling of stiffness degradation of the interfacial zone during fiber debonding," *Journal of Composites Engineering*, 5(10–11), 1349–1365.
45. Peled, A., Mobasher, B., and Sueki, S. (2004), "Technology methods in textile cement-based composites," in Kovelr, K., Marchand, J., Mindess, S., and Weiss, J. (eds.), *Concrete Science and Engineering: A Tribute to Arnon Bentur*, RILEM Proceedings PRO, Bagneux, France: RILEM Publications. 36, 187–202.
46. Peled, A., Sueki, S., and Mobasher, B. (2006), "Bonding in fabric-cement systems: Effects of fabrication methods," *Journal of Cement and Concrete Research*, 36(9), 1661–1671.
47. Hetényi, M. (1946), *Beams on Elastic Foundation*, Ann Arbor: University of Michigan Press, 255 pp.
48. Mobasher, B., Peled, A., and Pahilajani, J. (2006), "Distributed cracking and stiffness degradation in fabric-cement composites," *Materials and Structures*, 39, 317–331.
49. Soranakom, C., and Mobasher, B. (2010), "Modeling of tension stiffening in reinforced cement composites: Part I -Theoretical Modeling," *Materials and Structures*, 43, 1217–1230.
50. Soranakom, C., and Mobasher, B. (2009), "Geometrical and mechanical aspects of fabric bonding and pullout in cement composites," *Materials and Structures*, 42, 765–777. DOI 10.1617/S11527-008-9422-6.
51. Majumdar, A. J., and Nurse, R. W. (1974), Glass fiber reinforced cement. Building Research Establishment Current Paper, CP79/74, Building Research Establishment, England.
52. Shah, S. P., Ludirdja, D., Daniel, J. I., and Mobasher, B. (1998), "Toughness—Durability of glass fiber reinforced concrete systems," *ACI Materials Journal*, 85(5), 352–360.
53. Soroushian, P., Tlili, A., Yohena, M., and Tilsen, B. L. (1993), "Durability characteristics of polymer-modified glass fiber reinforced concrete," *ACI Materials Journal*, 90(1), 40–49.
54. Marikunte, S., Aldea, C., and Shah, S.P. (1997), "Durability of glass fiber reinforced cement composites: Effect of silica fume and metakaolin," *Advanced Cement Based Materials*, 5(3–4), 100–108.

9 Fracture Process in Quasi-Brittle Materials

INTRODUCTION

The main purpose of using fibers in concrete materials is to capitalize on their strength, bond, and, stiffness in reinforcing the brittle matrix. To demonstrate and to simulate the operating mechanisms, one has to address the problems raised from discrete fibers, the matrix, and interface characteristics. Volume fraction, stiffness, and nature of the bond between the fiber and the matrix also play a dominant role in the modeling of these systems. Once the model has been developed, it can be used in evaluating the effects of fiber type, size, length, volume fraction, anchorage, and interface. Scholars specializing in fracture mechanics discuss the underlying principles that govern initiation and propagation of cracks in materials.

Sharp internal or surface notches that exist in various materials intensify local stress distribution. If the energy stored at the vicinity of the notch is equal to the energy required for the formation of new surfaces, then crack growth can take place. The material at the vicinity of the crack relaxes, the strain energy is consumed as surface energy, and the crack grows by an infinitesimal amount. If the rate of release of strain energy is equal to the fracture toughness, then the crack growth takes place under steady-state conditions and failure is unavoidable.

The process of fracture can be visualized as a two-step process, the initiation of cracks begins the process and is followed by the propagation of those cracks. Both processes consume energy from internal or external forces in order to create new surfaces within a solid material. If the energy requirements for both mechanisms are similar and roughly in the same order of magnitude, then crack formation and crack propagation may occur simultaneously. If the amount of energy required by the growth phase can be supplied by the release of internal energy, and if there is an increasing amount of energy available for release as the crack grows, then the material is referred to as brittle, meaning that crack growth is instantaneous upon formation and may proceed at very fast speeds. However, if the energy required for the growth of the crack is greater than the initiation phase, then the material is ductile and consumes significantly more energy under stable crack growth conditions. These ductile to brittle transitions can occur because of environmental aging or thermal loading for the same material under reversible or irreversible conditions.

In an ideal elastic sample loaded by mechanical stress, deformation takes place, resulting in the storage of strain energy. This energy accumulation is higher at the vicinity of the flaws and notches because of stress concentration. The stored strain energy can be a source of energy needed to create the new surfaces. Energy conversion can be generally viewed as reversible or irreversible processes or a combination of both. For example, a perfectly elastic ball bouncing on a flat surface converts potential and kinetic energy back and forth in a reversible manner, whereas the rolling of that ball down a slope is a self-driving and irreversible process, which continuously converts the potential energy to kinetic energy during the rolling process and results in a speeding ball. In the same manner, the energy conversion from strain energy into surface energy is considered to be an irreversible process. It can be shown that if the rate of strain energy release is equal to the rate of extension of an existing crack, then the process will continue until the propagating crack traverses the entire available width of the specimen. The Griffith approach considers the largest flaw inside a material as the potential starter crack for failure that lays the foundation for the application of the energy release rate principal. All one has to do is apply a sufficient external load to a material with a given flaw size such that the strain energy is increased. Once the strain energy release rate is equal to the surface energy of the material, crack propagation is imminent, as is demonstrated in Figure 9.1 [1]. For every flaw size in the sample, there exists a critical load such that the load decreases for increasing flaw sizes.

In this chapter procedures to calculate the role of fibers on the toughening, tensile stress–strain response, and the fracture toughness of cement-based composites will be presented based on a fracture mechanics approach. Toughening and strengthening of FRC composites occur at different size scales. The bridging of the fibers alters the crack growth processes and contributes toughening mechanisms through changing the stiffness of the bridging zone. Fiber pullout can be characterized and incorporated in an R-curve based approach that is based on the growth of interfacial cracks. Parametric response of fiber length, interface properties, and strength using fiber pullout models can be then combined with the matrix crack growth problem to correlate the composite's performance to fiber, interface, and matrix properties. Dissipation of energy during fiber pullout can be characterized as a function of interface properties.

LINEAR ELASTIC FRACTURE MECHANICS

Linear elastic fracture mechanics (LEFM) first assumes that for an isotropic and linear elastic material, the stress field near the crack tip can be calculated using the theory of elasticity which results in singularity of the stress field near the crack tip.

The stress field near a crack tip is a function of its location, the loading conditions, and the geometry. Crack tips produce a singularity in the stress fields, which can be expressed as a product of $1/\sqrt{r}$ and a function of r and θ using the polar coordinate system as shown in Equation 9.1 and Figure 9.2. To avoid the singularity of the stress value in the vicinity of the crack, the intensity of the stress field is calculated and represented as the single parameter factor, K_I, called the "stress intensity factor,"

$$\lim_{x \to \infty} \sigma_{ij}^I = \frac{K_I}{\sqrt{2\pi r}} f_{ij}^I(\theta)$$
(9.1)

In Equation 9.1, the subscript I represents the opening mode of fracture. Similar derivations for other modes of failure such as shearing (II) and tearing modes (III) can also be expressed depending on the mode of loading applied to a crack. The detailed breakdown of stresses and the displacements for each mode are explained in standard handbooks on fracture mechanics [2]. For example, for a through crack of length "2a" in an infinite plate under uniform tension, σ, as shown in Figure 9.3, the stress intensity factor is defined in Equation 9.2. The dimension, K, is defined in terms of stress and crack length and has units of stress × (length)$^{1/2}$:

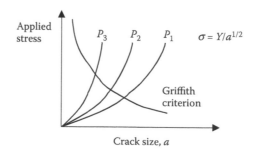

FIGURE 9.1   Crack propagation failure criterion. (Based on Zweben, C., and Rosen, B. W., "A statistical theory of material strength with application to composite materials," *Journal of Mechanics and Physics of Solids*, 18, 189–206, 1970.)

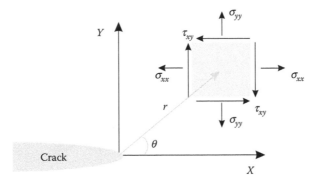

FIGURE 9.2   The stress field near a crack.

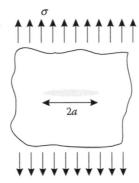

FIGURE 9.3   Single crack under tensile stress in infinite domain.

$$K_I = \sigma\sqrt{\pi a} \quad \frac{F}{L^2}\sqrt{L} = FL^{-3/2} \quad \text{psi}\sqrt{in} \times 1098.8 = \text{Nm}^{-3/2} = \text{MPa}\sqrt{m} \tag{9.2}$$

For a beam under three-point bending, the nondimensional parameters are defined as $\beta = S/D$ and $\alpha = a/D$, representing the span and the depth ratios:

$$\begin{aligned} K_I &= \frac{3P}{2B\sqrt{D}} \frac{\sqrt{\alpha}}{(1-\alpha)(1+3\alpha)}[\beta p_1(\alpha)+4p_2(\alpha)] \\ p_1(\alpha) &= 1.99+0.83\alpha-0.31\alpha^2+0.14\alpha^3 \\ p_2(\alpha) &= 0.99+0.42\alpha-0.82\alpha^2+0.31\alpha^3 \end{aligned} \tag{9.3}$$

These expressions are valid for any a and $b > 2.5$ with 0.5% error.

STRESS INTENSITY FACTOR AND FRACTURE TOUGHNESS

The stress intensity associated with the fracture toughness of the material is called the critical stress intensity factor K_{IC}. The fracture toughness of a material can be found through experimentation. An alternative approach to the representing of the stress intensity at the vicinity of the crack is to look for the strain energy available for release, which is defined as U. The energy release is expressed in rate form as a function of the crack extension and defined as G. These two factors, G, and stress intensity factor K, are, however, directly related by the following formulas:

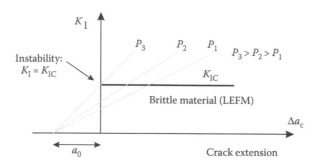

FIGURE 9.4 Process of fracture in brittle materials with little or no stable crack growth.

$$G_I = \frac{dU}{da} = \begin{cases} \dfrac{K_I^2}{E} & \text{plane stress} \\[2ex] \dfrac{K_I^2}{E}(1-v^2) & \text{plane strain} \end{cases} \qquad [G]=\frac{FL}{L^2}=FL^{-1}=\frac{\text{energy}}{\text{area}}=\text{N mm}^{-1} \qquad (9.4)$$

The definition and dimension of fracture toughness, G, based on plane stress failure criteria is:

$$G_{IC}=\frac{K_{IC}^2}{E} \qquad (9.5)$$

Figure 9.4 illustrates the process of fracture in brittle materials like glass, where there is little or no stable crack growth. The energy stored in the vicinity of the crack is a function of its length, applied load, and the geometry of loading. The stress intensity factor, K_I, relates the available rate of release of energy for the propagation of the crack by a unit length. As the load is gradually increased $(P_3 > P_2 > P_1)$, a point is reached when the strain energy release rate is equal to the fracture toughness of the material, K_{IC}, and an unstable propagation of the crack (horizontal line in Figure 9.4 indicating an unbounded increase in the crack length) becomes imminent so that the material fails. Such a failure can be modeled by using LEFM principles with a single parameter, namely, K_{IC}.

By assuming a linear elastic material response, one needs a single parameter to relate fracture toughness to failure load. The assumption of linearity during crack propagation, especially at the crack tip, is questionable. Additional insight into the laboratory tests and fracture process has helped with the evaluation of load necessary for crack initiation and the rate of crack propagation at various stress levels.

FRACTURE PROCESS ZONE

Crack propagation in concrete and in many quasi-brittle materials such as ceramics, rocks, and composites is characterized by a nonlinear (dissipative) zone at the crack tip. As defined in Figure 9.5, this nonlinear cracking zone is commonly referred to as the fracture process zone (FPZ), whose size is not negligible compared with usual structural sizes. Thus, the LEFM, according to which the FPZ is a point, does not apply so that the behavior of the FPZ needs to be simulated directly. The original attempt to apply LEFM to concrete materials was in 1961 by Kaplan [3], who used the Griffith Crack approach to explain the failure of concrete and mortar beams. Therefore, the next question is, which processes cause the formation of the FPZ and the subsequent propagation of cracks? One way of addressing this question is by examining the dissipation energy, where the fracture energy is obtained by measuring the work-of-fracture concept, which calculates the energy dissipated in the formation of a single crack from a notched specimen loaded in a tension.

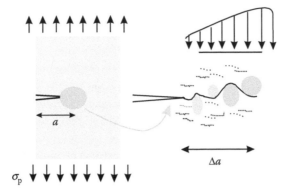

FIGURE 9.5   The process of fracture and toughening in plain concrete.

The Cohesive Crack Model (CCM) was introduced after the development of models for the fracture of metals. In this approach, the effect of plastic zone formation at the tip of a crack within a metal was incorporated in the original crack, and the energy dissipation due to the plastic work was also included in the energy balance criteria.

EQUIVALENT ELASTIC CRACKS

Dugdale's [4] model applies to the ductile fracture of elastoplastic materials and assumes the existence of a Mode I crack in an infinite sheet under uniform tensile stress. The material is ductile, and the plastic deformations are localized in a thin zone coplanar with the crack. The plastic zone is modeled through a fictitious crack (of unknown length), as shown in Figure 9.6, with a uniform distribution of cohesive tractions (σ_P = yield stress).

The length of the cohesive zone, a_p, is calculated by imposing the condition of smooth closure of the crack faces:

$$K_I = K_{I\sigma} - K_{IP} \tag{9.6}$$

$$K_{I\sigma} = \sigma\sqrt{\pi(a+a_P)} \tag{9.7}$$

The stress intensity factor due to the plastic stresses is calculated using the stress intensity, which is the result of a pair of concentrated forces, F, acting at x zx defined in Figure 9.7 as:

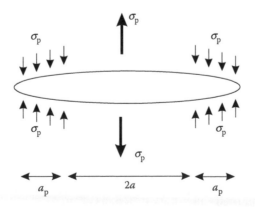

FIGURE 9.6   Schematics of the Dugdale model.

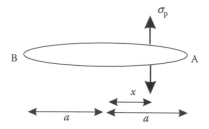

FIGURE 9.7 Parameter definition for Green's function calculation of pressure distribution.

$$K_I(A) = \frac{F}{\sqrt{\pi a}} \left(\frac{a+x}{a-x}\right)^{1/2}$$

$$K_I(B) = \frac{F}{\sqrt{\pi a}} \left(\frac{a-x}{a+x}\right)^{1/2} \tag{9.8}$$

$$K_{IP} = \frac{\sigma_P}{\sqrt{\pi(a+a_P)}} \int_a^{a_P} \left[\sqrt{\frac{(a+a_P)+x}{(a+a_P)-x}} + \sqrt{\frac{(a+a_P)-x}{(a+a_P)+x}} \right] dx \tag{9.9}$$

$$K_{IP} = \sigma\sqrt{\pi(a+a_P)} - 2\sigma_P\sqrt{\frac{a+a_P}{\pi}} \arccos \frac{a}{a+a_P} = 0 \tag{9.10}$$

$$\frac{a}{a+a_P} = \cos\frac{\pi\sigma}{2\sigma_P} \tag{9.11}$$

This equation predicts that in the two limits that as $\sigma = 0$, $a_p = 0$, and as $\sigma = \sigma_P$, $a_p \to \infty$. The Taylor series expansion of the cosine term and solving in terms of plastic zone length results in,

$$a_P = \frac{\pi}{8}\frac{K_{I\sigma}^2}{\sigma_P^2} \tag{9.12}$$

The energy dissipated is due to the strain energy release rate and is equal to

$$G_{IC} = \int_0^{\delta_a} \sigma_P d\delta = \sigma_P \delta_a \tag{9.13}$$

The opening of the crack in accordance to Dugdale's model is:

$$\delta_a = \frac{8\sigma_P\, a}{\pi E} \ln\left[\sec(\frac{\pi\sigma}{2\sigma_P}) \right] \tag{9.14}$$

Barenblatt [5] relaxed the constant stress approach and showed a similar approach for the closing pressure distribution as a material property that can explain brittle fracture. By visualizing a Mode I crack in a homogeneous, isotropic, linear-elastic infinite medium under uniform loading, the atomic bonds hold together the two halves of the body separated by the crack as cohesive forces acting along the edge regions of the crack and attract one side of the crack to the other. Barenblatt's [6] theory based on atomic forces is equivalent to Griffith's energy approach, provided that the integral of the cohesive forces is equal to the fracture energy or:

$$G_{\text{IC}} = \int_a^{a_P} \sigma_{\text{P}} \frac{\text{d}\delta}{\text{d}x} \text{d}x = \int_0^{\delta_a} \sigma(\delta)\text{d}\delta \tag{9.15}$$

Note that this relationship is similar to that shown in Equation 9.10, except that the condition for constant yield stress has been replaced by the stress–crack width relationship.

COHESIVE CRACK MODELS

The Dugdale–Barenblatt plastic crack tip model was used as a basis for CCM by Hillerborg et al. [7] in the development of a fictitious crack model. Using fracture mechanics in the finite element analysis, they assumed the existence of a crack longer than those measured in experiments. Preceding the crack tip, a process zone propagates where microcracking takes place. Temperature change, shrinkage, loading, aggregate–paste interface mismatch, and other factors serve to initiate microcracking. The essential element in CCM is the inclusion of this process zone in the total crack length as a fictitious crack and the assumption that a closing pressure acts over the process zone length. Shah and McGarry [8] noted that a relatively large FPZ results in the stable crack propagation, so that LEFM cannot be directly applied to cement-based composites.

The closing pressure in elastoplastic metals is assumed to be constant, and it equals the yielding stress. In concrete, the closing pressure is governed by the softening curve, which is measured in a stable displacement-controlled uniaxial tension test. The shape of this curve has been the subject of much scientific discussion over the past 30 years. Hillerborg et al. [7, 9] introduced closing pressure forces in the finite element formulation by measuring the stress–crack width. The softening function relates the stress transferred between the crack faces to the crack opening at each point, as shown in Figure 9.8. In theory, it is assumed to be a material property independent of geometry and size.

The differences observed between bending and tensile strengths and of the variation in bending strength with beam depth can be understood using the fracture method. The broader applicability of the fracture models is in applications where tensile properties contribute to the overall response of the structure. Examples include the determination of cracking moment of reinforced and unreinforced beams, the crack spacing and crack width in bending, the determination of deflection in

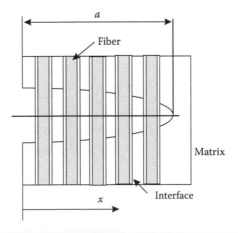

FIGURE 9.8 Propagation of a matrix crack resisted by the debonding of fibers, which results in crack closure.

beams, the formation of shear cracks and their effect on shear capacity, anchor pullout, microcracks in compression, and compression failure. The effects of multiaxiality, size effect, crack path trajectory, equivalent elastic crack approaches, and computation of the softening function by inverse analysis are some of the formulations that use the fictitious crack approach.

A CCM is characterized by the properties of the bulk material, the crack initiation, and the crack propagation. The simplest assumptions are:

1. The bulk material behavior follows a linear elastic and isotropic stress–strain relationship, with elastic modulus E and Poisson's ratio ν. This assumption is not essential, but does make computations easy.
2. The crack initiates at a point where the maximum principal stress reaches the tensile strength f'_t, and is normal to its direction.
3. After its formation, the crack opens and transfers stress from one face to another. The cohesive stress being transferred is a function of the crack opening displacement history. For Monotonic Mode I opening, the stress transferred is normal to the crack faces and is a unique function of the crack opening like the softening function or softening curve. This is the simplest form of this function and is affected by triaxiality, the process zone size, the specimen size, and other parameters. An experimental measurement of this curve may be done from a stable tensile test if a single crack is formed and opened while keeping its faces parallel. The development of flexural closed-loop tests for the measurement of fracture energy in bending is also another approach that consists of measuring the external work supplied to break a notched beam specimen.

Many models take into account the physical and natural effects of softening. In the past 30 years, the axiomatic assumptions and results of experimentation have been discussed in concrete fracture books and articles written by many authors, including Bazant et al. [10], Bazant and Planas [11], Bazant [12], Jenq and Shah [13], Planas et al. [14, 15], Planas and Elices [16], Karihaloo and Nallathambi [17], Nallathambi and Karihaloo [18], van Mier and van-Vliet [19], Reinhardt [20], Carpinteri and Chiaia [21], and many others. These efforts have led to a much better understanding of theoretical and experimental techniques and have aided the development of methods to measure the fracture energy, its properties, and its influences on the structural response.

According to the research of Petersson [22], an estimation of the fracture energy from a stable three-point bend test on a notched beam can be obtained from the area under the continuous load–deflection curve. The CCM uses two independent characteristic lengths:

$$l_0 = \frac{EG_F}{\sigma_0^2} \text{ and } l_1 = \frac{EG_f}{\sigma_0^2} \tag{9.16}$$

where G_F is the area under the entire softening stress–displacement curve, $\sigma = f(w)$, and G_f is the area under the initial tangent to this curve, which is equal to G_F only if the curve is simplified as linear elastic. This approach is by no means without criticism, because the softening response is a general interaction between the material and the test specimen size and geometry and, thus, must be viewed as a structural response. The overly simplified approach of the model in treating only the tension case has also been criticized. One must, however, appreciate the elegance and simplicity of this method as one of the most important contributions to the concrete materials research community in the past several decades.

CLOSING PRESSURE FORMULATIONS

The principle of superposition has been commonly used to add the effect of the process zone to the calculation of toughness. In fiber reinforced concrete (FRC), techniques based on fracture mechanics and closing pressure formulations relate the fiber, the interface, and the matrix properties to the strength, the toughness, and the fracture response. This is the case in the modeling of the critical volume fraction of fibers for distributed cracking by Li and Wu [23], Li and Leung [24], Soranakom and Mobasher [25, 26], Visalvanich and Naaman [27], and Tjiptobroto and Hansen [28]. The strengthening of the matrix phase and the critical volume fraction of fibers have been studied using micromechanics by Yang et al. [29] and Li et al. [30].

Short whiskers with a high aspect ratio strengthen the composite at the microlevel [31]. The frictional sliding of fibers is the main contribution of this reinforcement. The sliding shear resistance is used as a closing pressure that is inversely proportional to the crack width as shown in Figure 9.8. Using pullout experiments under closed-loop conditions, the interfacial region can be characterized as modeled by Li et al. [32]. As the fiber debonds, stiffness of the pullout force against the slip decreases and results in the nonlinear response of the ascending part of the pullout curve. To consider the evolution of the process zone before the steady-state conditions, an algorithm allowing for gradual fiber debonding is used to allow for the crack opening. To simulate the changes in compliance and of the pullout–slip response, nonlinear fracture models based on resistance curves (R-curves) have been recently used [33, 34].

R-CURVE APPROACH

The existence of stable crack growth before reaching a critical length results in nonlinear effects, because the actual crack length at the onset of instability is unknown. For granular-based materials, heterogeneous and tortuous crack growth is accompanied by aggregate interlock, microcracking, and inelastic deformation, which then results in the toughening of the material. This response is depicted as the increasing curve in Figure 9.9, representing the R-curve.

Lenain and Bunsell [35] used R-curves for asbestos cement mixtures. Mai obtained R-curves by using experiments on three-point bend and grooved double-cantilever beams. Foote et al. [36] studied R-curves in strain softening materials. Suzuki and Sakai [37], Hsueh and Becher [38], and Munz [39] used R-curves in the study of fracture in ceramics. R-curves for materials with a large process zone depend on the specimen geometry. Ouyang and Shah [40] studied the influence of geometry on the R-curves and the fracture response.

Figure 9.9 represents the fiber toughening process. As a material containing a small flaw is loaded, the flaw begins to grow (under an increasing applied stress intensity factor), and, as each fiber is engaged in resisting the growth of the crack, more energy is required. The fracture energy at the base level of R_m increases with incremental crack growth Δa because of the presence of bridging. The

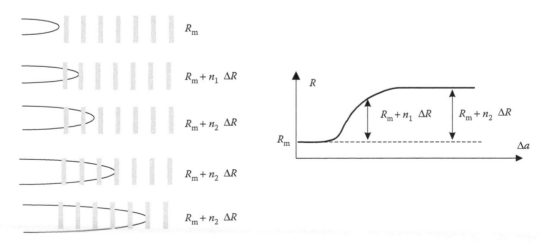

FIGURE 9.9 Schematics of R-curve definition.

toughening due to each intersected fiber may be accounted as $n_1\Delta R$. Once the zone has developed fully, then the whole crack may move forward with the process zone, remaining at a constant size, at an energy level of $R_m + n_2\Delta R$, resulting in a flat portion of the R-curve, which indicates steady-state conditions over certain limits of flaw size. This mechanism explains how the reduction of interfiber spacing results in the formation and growth of significant cracking without catastrophic fracture.

No standardized procedures to determine the R-curves for ceramics or cement-based materials exist. Because the energy dissipation in the process zone has to be related to an effective elastic crack length, approaches based on energy principles and unloading–reloading (compliance calibration) methods have been used. These procedures integrate the toughening mechanisms into the energy required for the propagation of a traction-free equivalent crack.

DERIVATION OF R-CURVES

On the basis of experimental observations, Krafft et al. [41] proposed that R-curve for metals (LEFM) is only a function of crack extension, Δa. Broek derived a semiempirical solution for the R-curve of a specimen containing a short crack [42]. A similar approach that is not restricted to short cracks is presented here. The first step is the definition of R-curves on the basis of G-curves, as defined in Equation 9.4 and Figure 9.10—the crack incrementally grows to be equal to R as the energy release rate, G, increases as a function of applied load. However, because the resistance of the material increases with growth of the crack, $dG/da < dR/da$, the steady-state conditions cannot be satisfied with such incremental growth, and the crack is arrested.

R is defined as an envelope of the energy release rates with different specimen sizes but the same initial notch length. On the basis of this definition, the R-curve can be derived by solving a differential equation. The energy release rate $G = G(a, P, \text{geometry})$, as shown in Figure 9.10, can be expanded as a Taylor's series at a critical point c,

$$G = G_c + \sum_{n=1}^{\infty} \frac{1}{n!} \left(\frac{d^n G}{d\Delta a^n} \right)_c (\Delta a - \Delta a_c)^n \tag{9.17}$$

Using $G = 0$ at $\Delta a = a_0$ results in:

$$G_c = -\sum_{n=1}^{\infty} \frac{1}{n!} \frac{d^n G}{d\Delta a^n} (a_0 - \Delta a_c)^n \tag{9.18}$$

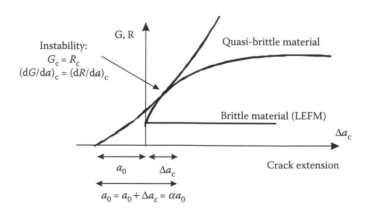

FIGURE 9.10 (a) The schematic model of the fiber toughening resulting in R-curve behavior, and (b) the model utilized for use of R-curves in crack growth and instability criterion.

Define the critical crack length at

$$a_c = \Delta a_c - a_0 = \alpha\, a_0 \tag{9.19}$$

$$G_c = -\sum_{n=1}^{\infty} \frac{1}{n!}\left(\frac{d^n G}{d\Delta a^n}\right)_c \left(\frac{\alpha}{\alpha-1}\right)^n (-\Delta a_c)^n \tag{9.20}$$

Define $\Delta a_c = e^x$ or $x = \ln(\Delta a_c)$. The previous equation is then transformed into a linear differential equation with constant coefficients. The auxiliary equation is represented as:

$$\sum_{n=1}^{\infty} \frac{1}{n!} d(d-1)(d-2)\ldots(d-n+1)\left(\frac{\alpha}{\alpha-1}\right)^n (-1)^n +1=0 \tag{9.21}$$

Parameter, d, represents the order of approximation of G, that is, $d = 1$, G is linear, and $d = 2$, G is represented by the solution of the quadratic characteristic equation:

$$G_c = \sum_{n=1}^{\infty} \beta_n (\Delta a_c)^{d_n} \tag{9.22}$$

Because at any given stable crack growth point, $R = G$, then $R = \sum_{n=1}^{\infty} \beta_n (\Delta a)^{d_n}$ It can be shown from Equation 9.20 that for $n = 1$,

$$-d\left(\frac{\alpha}{\alpha-1}\right)+1=0 \tag{9.23}$$

$$d_1 = \frac{\alpha-1}{\alpha} \tag{9.24}$$

$$R = \beta_1 (\Delta a)^{d_1} = \beta_1 (a-a_0)^{d_1} \tag{9.25}$$

which is the same as the equation by Krafft et al. [41]. For $n = 2$, G is represented by a quadratic expression:

$$-d\left(\frac{\alpha}{\alpha-1}\right)+d(d-1)\left(\frac{\alpha}{\alpha-1}\right)^2 +1=0 \tag{9.26}$$

$$R = \beta\left[1 - \frac{d_2}{d_1}\left[\frac{\alpha a_0 - a_0}{a-a_0}\right]^{(d_2-d_1)} (a-a_0)^{d_2}\right] \tag{9.27}$$

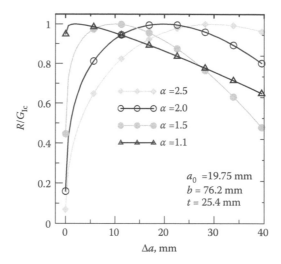

FIGURE 9.11　The effect of stable crack width on the rise of the R-curve.

$$d_i = \frac{1}{2} + \frac{\alpha-1}{\alpha} \pm \sqrt{\frac{1}{4} + \left(\frac{\alpha-1}{\alpha}\right)^2} \qquad i = 1,2 \qquad (9.28)$$

This equation is the basis of many calculations because it captures the nonlinear nature of the G-curve and the corresponding R-curves. Figure 9.11 shows the R-curve as the function of crack growth length for various critical crack lengths in a three-point bend specimen. Various critical crack length parameters, α_c, are specified; note that the larger the critical length, the more stable crack extension it takes for the R-curve to rise to the limit. As the crack extension takes place, the slope of the ascending portion of the R-curve decreases. For all the values of stable crack growth, the curve asymptotically reaches the maximum level specified at the point of $R = G_{IC}$. This assumption is valid for cases where the material is linear elastic with a fracture toughness of $R = G_{max}$ and $a_c = \alpha a_0$.

ALTERNATIVE FORMS OF R-CURVES

The effect of precritical crack growth on fracture toughness can be expressed in the form of R-curve in several potential ways, allowing for failure predictions on the basis of the loading conditions and initial flaw size [42, 43]. R-curves for bridging ceramics can be predicted on the basis of measured bridging stress functions in steady-state bridging zones [44, 45]. Summary expressions for various forms of R-curves are presented in Table 9.1. Parameter d in cases C, D, and E of Table 9.1 is the distance from the crack tip to the first bridging point when assuming isolated bridges.

STRESS–CRACK WIDTH RELATIONSHIP

Stress–crack width relationships are obtained from the descending branch of the stress–displacement curves in a uniaxial tension test and characterize the softening branch of the stress-displacement curve [56, 57]. Representation of the closing pressure requires a functional definition of the relationships between bridging stress profile, crack opening profile, and crack ligament length profile. The stress across the crack ligament is a function of both the crack opening and the crack ligament length. By assuming various functional relationships, both models of increasing and decreasing stress as a function of crack opening can be represented using similar parameters. The responses

TABLE 9.1
Various Forms of R-Curve Formulation

Case	Expression	Reference
A	$K_R = K_{max} - (K_{max} - K_0)\exp[-\Delta a/\lambda]$	[46, 47]
B	$K_R = K_0 + 2(K_{max} - K_0)\tan^{-1}[\Delta a/\lambda]/\pi$	[48, 49]
C	$K_R = K_{max} - (K_{max} - K_{min})\left[1 - \left[\dfrac{\Delta a - d}{a^* - d}\right]^{1/2}\right]^{m+1}, d < \Delta a < a^*$ $\quad K_R = K_0 \qquad \Delta a < d$ $\quad K_R = K_{max} \qquad \Delta a > a_s$	[50]
D	$K_R = K_{max} - (K_{max} - K_0)\left[1 - \left[\dfrac{a^*(\Delta a^2 - d^2)}{\Delta a(a^{*2} - d^2)}\right]^{1/2}\right]^{m+1}$	[51]
E	$K_R = K_0 + (K_{max} - K_0)\left[\dfrac{\Delta a - d}{a^* - d}\right]^{1/2}$	[52]
F	$K_R = Ad^n$	[47, 53–55]
G	$K_R = K_0\left[\dfrac{\Delta a + a_s}{a_s}\right]^n$	[47, 51–53]

Source: Munz, D., "What can we learn from R-curve measurements?" *Journal of the American Ceramic Society*, 90, 1, 1–15, 2007.

for bridging stress, σ_{br}, versus crack opening, δ, are expressed by various equations, as shown in Table 9.2. In Case 10, parameter $C(a)$ is the traction-free compliance (i.e., that of the same length crack, determined optically, but neglecting bridges), $C_u(x)$ is the observed compliance after cutting to the position x measured from the load line, and $C'(z) = dC(z)/dz$ (similar in nature to the model of Foote et al. [36]).

Methods have been developed for predicting R-curves on the basis of measuring steady-state bridging stress functions in ceramics [58, 59]. A general-purpose closing pressure profile can be used using a model proposed by Suzuki and Sakai [37] and implemented in the computation of the R-curves. Similar to items 1 and 2 in Table 9.1 and also comparable in nature with Mai [60], the stress distribution can be calculated as a function of both the crack opening and the crack ligament length using exponentially decaying parameters as shown in Equation 9.29. For example, the responses for both stress–crack opening and crack opening versus position (as shown in Figure 9.12) can be expressed by parameters σ_0 and δ_0, which represent characteristic stress and crack opening, whereas parameter l_b is equivalent to the stable crack growth length Δa_c.

$$u_b(x) = u_b^0\left(\frac{x}{l_b}\right)^n \qquad \sigma_b = \sigma_b^0\left(\frac{x}{l_b}\right)^q \tag{9.29}$$

By expressing the effects of crack extension on fracture toughness in terms of R-curves, failure predictions on the basis of the magnitude of loading and initial flaw size can be developed [61]. Depending on the form of the relationship, one can compute the toughening terms [62]. The criteria

TABLE 9.2
Various Forms of Closing Pressure Formulation

Model	Representation	Reference
1	$\sigma_{\mathrm{br}} = \sigma_0(1 - \delta/\delta_0)^m$	[60, 62, 63]
2	$\sigma_{\mathrm{br}} = \sigma_0(\delta/\delta_0)^\alpha$	[64]
3	$\sigma_{br} = \sigma_0\exp(-\delta/\delta_0)$	[65]
4	$\sigma_{br} = \sigma_0\delta/\delta_0\exp(-\delta/\delta_0)$	[66]
5	$\sigma_{br} = \sigma_0\delta/\delta_0(1 + \delta/\delta_0)\exp(-\delta/\delta_0)$	[23]
6	$\sigma_{br} = \sigma_0(1 + \delta/\delta_0)\exp(-\delta/\delta_0)$	[23]
7	$\sigma_{br} = \sigma_0\dfrac{\lambda}{\lambda - 1}[\exp(-\delta/\delta_0) - \exp(-\lambda\delta/\delta_0)]$	[23, 67]
8	$\sigma_{br} = 2\sigma_0[\exp(-\delta/\delta_0) - \exp(-2\delta/\delta_0)]$	[23, 67]
9	$\sigma_{br} = \sigma_0\exp(-\delta/\delta_0)\tanh(\lambda\delta/\delta_0)]$	
10*	$\sigma_{br}(x) = \sigma_{\max}\dfrac{C^2(a)C'(x)}{C'(a)C_u^2(x)}$	[68, 69]
11	$\sigma(\delta) = \sigma_c(1 - \delta/\delta_c)$	[70]
12	$\sigma(\delta) = \sigma_c[1 - (\delta/\delta_c)^k]$	[39, 71]

Source: Munz, D., "What can we learn from R-curve measurements?" *Journal of the American Ceramic Society,* 90, 1, 1–15, 2007.

for the cracking can be defined in terms of energy balance using the Potential Energy Approach, and expressing R based on the crack opening profile, $\delta(x)$,

$$\Delta R_b = 2\int_0^{l_b} \sigma_b(\delta)\left(\frac{\mathrm{d}\delta}{\mathrm{d}x}\right)\mathrm{d}x \tag{9.30}$$

Using the Green's function, the contribution of a closing pressure profile is integrated over the crack length shown in Figure 9.13 and is expressed as:

$$\Delta K_b(l_b) = \int_0^{l_b} \sigma_b(\delta)G(a, x)\mathrm{d}x \tag{9.31}$$

where $G(a,x)$ represents Green's function as described in the next sections, a is the crack length, l_b is the bridging zone length, and σ_b is the bridging stress.

Inverse approaches are also available that allow for estimating bridging tractions from the R-curve behavior. The first step is to use a stress–crack width relationship model and the tension σ–w curve as the material property, because bridging tractions are integrated and added as the ascending portions of the R-curves. The magnitude of toughening in the form of stress–crack opening integrals in the process zone is then calculated and converted to elastically equivalent fracture parameters.

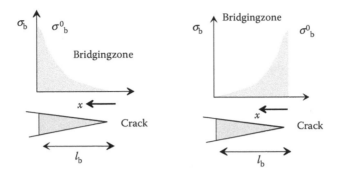

FIGURE 9.12 Closing pressure versus crack opening distribution according to the Sakai–Suzuki model.

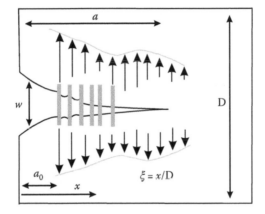

FIGURE 9.13 Cohesive crack and distribution of crack-bridging stresses.

STRESS INTENSITY APPROACH USING FIBER PULLOUT OR STRESS–CRACK WIDTH

Toughening due to crack bridging can be expressed using a stress intensity based algorithm. The bridging force of fibers, expressed in terms of the stress intensity factor, works to reduce the applied stress intensity factor. The available stress intensity at the crack tip is the net difference between the applied forces and the shielding offered by the fibers.

The steady-state conditions for crack growth involves the contribution of bridging, K_{IF}, in reducing the applied stress intensity as shown in Equation 9.32 [72]; superposition is then used to subtract this stress intensity from the corresponding value obtained from the applied far-field loads, K_{IP}, to obtain an intrinsic threshold for the tip of an unbridged crack, K_I^e, expressed as [59]:

$$K_I^e = K_{IP} - K_{IF} = K_{IC} \tag{9.32}$$

The strain energy release rate for a three-point bending specimen is obtained from the stress intensity factor handbook [2]. K_{IP} is given as:

$$K_{IP} = \frac{3s\,P\sqrt{\pi\alpha}}{d}f(\alpha) \quad \alpha = \frac{a}{d} \tag{9.33}$$

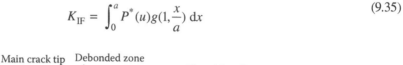

$$f(\alpha) = \begin{cases} \dfrac{1.99 - \alpha(1-\alpha)(2.15 - 3.93\,\alpha + 2.7\,\alpha^2)}{2b\sqrt{\pi d}\,(1+2\alpha)(1-\alpha)^{3/2}} & s/d = 4 \\[4mm] \dfrac{\left(1.0 - 2.5\alpha + 4.49\alpha^2 - 3.98\alpha^3 + 1.33\alpha^4\right)}{2bd(1-\alpha)^{3/2}} & s/d = 2.5 \end{cases} \tag{9.34}$$

where s represents the span, P is the far-field load, d is the beam depth, a is the matrix crack length, and b is the specimen width as shown in Figure 9.14. Figure 9.15 shows the stress intensity factor due to a unit load as a function of crack length for a three-point bend specimen.

The fiber pullout mechanism and the closing pressure are the primary parameters considered in this model. The stress intensity factors are directly obtained from the stresses that are required to pull the fiber out of the matrix and are expressed as:

$$K_{IF} = \int_0^a P^*(u)g(1,\frac{x}{a})\,dx \tag{9.35}$$

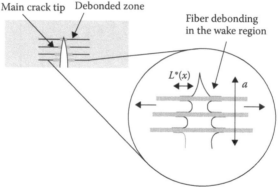

FIGURE 9.14 The bridged crack in a three-point bend flexural specimen.

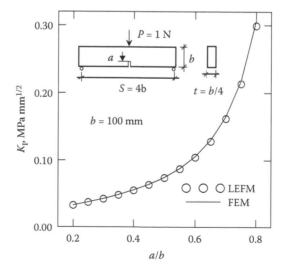

FIGURE 9.15 Computation of stress intensity factor as a function of crack length for a three-point bend specimen.

$$\mathrm{COD_f} = \frac{2}{E'} \int_{a_0}^{a} \int_{a_f}^{a} P^*(u) K_{\mathrm{IP}} \frac{\partial K_{\mathrm{IF}}}{\partial F} \mathrm{d}\xi \, \mathrm{d}\eta \tag{9.36}$$

where $P(u)$ represents the force carried by bridging fibers as a function of the crack opening. The fiber is located at distance x from the tip of a crack length, a. Parameter $g(1,x/a)$ represents Green's function, which in turn represents the stress intensity due to a unit load. Note that the stress–crack width relationships are a function of three variables: crack length, crack opening, and stress. Therefore, one needs to be aware of the interrelationship between these three parameters to properly model the toughening.

TERMINATION OF STABLE CRACK GROWTH RANGE

The stable crack growth process takes place as the load is increased; however, at some stage, the increase in applied load is no longer necessary for the growth of the crack. This occurs when either the crack is sufficiently long such that it interacts with the boundary and fails to grow anymore, or when the size of the process zone reaches a steady-state size such that the energy release is sufficient for the stable growth of the crack. In fracture terms, we can look at the strain energy release rate, $G(a)$, as the source of total energy available for crack extension. Once it reaches the critical value, G_{IC}, the instability condition is reached and crack propagation occurs. This instability is shown as the horizontal line in Figure 9.1 for brittle materials. Using Irwin's approximation, the strain energy release rate of the composite can be expressed as:

$$G_{\mathrm{IC}} = \frac{(K_{\mathrm{IC}} + K_{\mathrm{IF}})^2}{E'} \tag{9.37}$$

where $E' = E$ for plane stress and $E' = E / (1 - v)$ for plane strain conditions. K_{IF} represents the closing pressure of the fibers.

To characterize fracture toughness using a single parameter, G_{IC}, energy balance between the crack driving force and the crack resistance is used. The inequality condition ensures that during incremental crack growth, the resistance of the material exceeds the energy available, because of the crack growth. The stable crack growth is defined as:

$$R(a) = G(a)$$
$$\frac{\partial R}{\partial a} > \frac{\partial G}{\partial a} > 0 \quad \text{for } a < a_c \tag{9.38}$$

Steady-state crack growth condition determines the magnitude and occurrence of the peak load. When the crack length reaches the critical value, the R-curve is tangent to the strain energy release rate, the condition of which is defined as:

$$a_c = a_0 + \Delta a_c = \alpha a_0 \tag{9.39}$$

$$R(a_c) = G(a_c) = G_{\mathrm{IC}}(a_c)$$
$$\frac{\partial R}{\partial a} = \frac{\partial G}{\partial a} \quad \text{for } a = a_c \tag{9.40}$$

The parameters of the R-curve, which are referred to as α and β, depend on the strain energy release rate of the geometry and the failure criteria being used. This approach is applied to both the three-point bend and the fiber pullout [31].

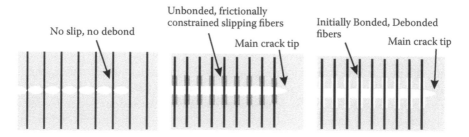

FIGURE 9.16   Steady-state matrix cracking under different interface conditions.

TOUGHENING UNDER STEADY-STATE CONDITION

Under the assumption that there is a steady-state crack growth, one can calculate the contribution of fibers to toughening and relate it to the additional gain in strength of the matrix phase. Various cases of steady-state cracking under different interface conditions have been discussed by Budiansky et al. [73], and are shown in Figure 9.16. The three cases include: no slip–no debond, unbounded but frictionally constrained slipping fibers, and initially bonded–debonded fiber.

The study of Case A, which generalizes the well-known Aveston–Cooper–Kelly theory [74–76] is based on the analysis of steady-state crack growth in the matrix. The concept is, in turn, based on the assumption that a "first" planar crack will propagate across the composite under an applied stress that becomes constant during the propagation as soon as the crack engulfs more than a few fibers. The strain capacity of the matrix is increased under these conditions. The parameter, ε_{mu}, is obtained using the Aveston–Cooper–Kelly approach [74], which predicts the strength of matrix phase in the presence of fibers. In this approach, γ is the fracture toughness and r is the fiber radius. This approach has been verified to be applicable for the cement-based materials, as the strength of the matrix is increased in the presence of fibers [77].

$$\varepsilon_{mu} = \left[\frac{12\tau\gamma_m E_f V_f^2}{E_c E_m^2 r V_m} \right]^{\frac{1}{3}} \tag{9.41}$$

DISCRETE FIBER APPROACH USING FIBER PULLOUT FOR TOUGHENING

A transient solution to modeling tougheing requires procedures for the incremental crack growth. An approach based on R-curves has been proposed by Ouyang et al. [78] and Mobasher et al. [79] that uses two parameters, which correspond to the load-deformation history of the specimen. Stress–crack width parameters represent the effect of fiber bridging and are incorporated in the R-curve, R versus crack length Δa. The effect of stress–crack width on closure of the crack faces and the reduction of the stress intensity factor is expressed as:

$$K_I^e = K_{IP} - \int_0^a \phi P^* \left(\frac{COD(x)}{2} \right) g\left(1, \frac{x}{a} \right) dx \tag{9.42}$$

$$COD(x) = COD_P(x) - \frac{2}{E'} \int_0^a \int_x^a \phi P^* (COD(\xi)/2) K_{IP}^* \left(1, \frac{\xi}{\alpha} \right) \frac{\partial K_{IF}^*(1,(\eta/a))}{\partial F} \, d\eta \, d\xi \tag{9.43}$$

where K_{IP} and COD_P are the stress intensity factor and the crack opening profile for an effective traction-free crack in the matrix due to far-field stress.

The first terms in these equations represent the stress intensity factor due to applied load. The second terms are responses to the closing pressure of fibers along the crack surface. Parameter ϕ represents the averaging factor needed to convert a single fiber pullout force to a distributed pressure acting on the crack face, which is defined as a fiber volume fraction normalized to the effective area of the specimen,

$$\phi = \frac{\lambda V_f}{\pi r^2} \tag{9.44}$$

where V_f represents the volume fraction of fibers. Parameter λ represents the effect of the fiber orientation and length distribution (according to Aveston and Kelly [75]) and was set equal to 1 for aligned fibers.

The contribution of the closing pressure can be obtained by integrating the pullout force across the crack face using the solution of a unit load over the crack length. The expression for the function $g(1,x/a)$ is obtained by visualizing a semi-infinite strip specimen as shown in Figure 9.17a. The stress intensity factor, which is due to a mutually opposite pair of unit forces acting on the crack faces at a distance x ($\lambda = x/a$) from the crack edge, can be generally expressed as a function of crack length, a ($\eta = a/d$), and distance x, as:

$$K = g(1,\lambda) = \frac{2}{\sqrt{a}} \frac{G(\lambda,\eta)}{(1-\eta)^{3/2}\sqrt{1-\lambda^2}} \tag{9.45}$$

$$G(\lambda, \eta) = g_1(\eta) + \lambda g_2(\eta) + \lambda^2 g_3(\eta) + \lambda^3 g_4(\eta)$$

$$g_1(\eta) = 0.46 + 3.06\eta + 0.84(1-\eta)^5 + 0.66\eta^2(1-\eta)^2$$

$$g_2(\eta) = -3.52\eta^2$$

$$g_3(\eta) = 6.17 - 28.22\eta + 34.54\eta^2 - 14.39\eta^3 - (1-\eta)^{3/2} - 5.88(1-\eta)^5 - 2.64\ \eta^2(1-\eta)^5$$

$$g_4(\eta) = -6.63 + 25.16\eta - 31.04\eta^2 + 14.41\eta^3 + 2.0\ (1-\eta)^{3/2} + 5.04(1-\eta)^{5/2} + 1.98\ \eta^2(1-\eta)$$

For the distributed bridging forces $P^*(u)$ over the crack faces, assuming a unit thickness where the stress intensity factor due to fibers, K_{IF}, can be obtained by integrating the distribution of the force over the crack face. The parameters are shown in Figure 9.17a:

$$K_I^f = \int_0^a P^*(U)g(1,\lambda)dx \tag{9.46}$$

where $P^*(u)$ is given by the fiber pullout force or the stress–crack width relationship as described in previous sections. The crack opening displacement for an edge crack subjected to a unit force was used as the Green's function [61], as is shown in Figure 9.17b. To calculate crack opening displacement, a single fiber located at point ξ along the crack causes displacements along the entire crack surface. Castigliano's theorem is used in the calculation of the crack opening due to a unit load applied at an arbitrary position along the crack length. Note that x is the distance from the specimen edge to the point of the applied unit force on the crack faces. The inner integral in the second term of Equation 9.42 computes the contribution of the fiber located at ξ to the closing of

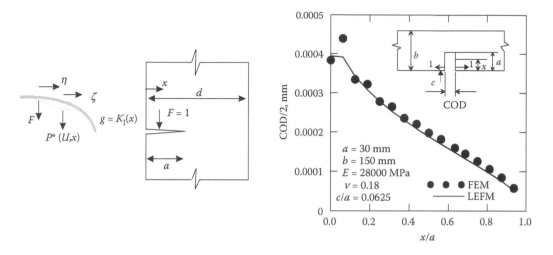

FIGURE 9.17 (a) Edge crack in an infinite strip subjected to point force, and (b) Green's function approach for crack closure due to a unit load applied at an arbitrary position. Comparison of FEM approach with LEFM-based equations.

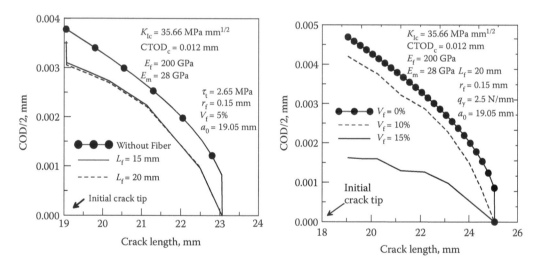

FIGURE 9.18 Effect of fiber length and fiber volume fraction on the closing of a propagating crack.

the crack. The independent variable, η, represents various points along the crack surface at which a displacement due to the force applied at point ξ is obtained.

For a random distribution of fiber lengths, this approach would require a development of the density function of the pullout responses and the use of $P^*(u)$ as a random variable. To simplify the solution algorithm, the assumption of a constant length of fibers across the crack was used to justify the use of a single fiber length pullout model.

The bridging force develops as the matrix crack propagates and opens in the process. Fiber interfacial parameters can be directly related to the toughening mechanisms. A three-point bend specimen with dimensions of $25.4 \times 76.2 \times 330.2$ mm and an initial notch length of $a_0 = 19.05$ mm was used for the parametric study [80]. No fibers are present in the initial notched region. An FRC composite with 5% steel fibers that are 15 and 20 mm long is compared with a plain mortar matrix with a traction-free crack as shown in Figure 9.18a. It is assumed that the matrix crack intersects the fibers at their midpoint. A variable 15- to 20-mm length fiber was used to generate the pullout slip response. Note that the fiber length (from 15 to 20 mm) does not show any effect during the initial crack growth period because the fiber closing pressure is mostly obtained from the ascending portion of the stress–crack width or the initial stiffness of the response. The effect of the fiber volume fraction is clearly evident as the crack tends to close much more with the addition of more fibers. This is shown in Figure 9.18b, which compares three different fiber loading levels.

The effects of the fiber content on the crack opening profile and simulated R-curves are shown in Figures 9.19a and 9.19b. As the stress intensity from the applied loads increases, the stress intensity due to the contribution of the fibers also increases, resulting in a differential value that can be equated to the constant steady-state matrix toughness. Because the contribution of the fiber phase increases with fiber loading, this approach can be used to simulate the effect of fibers in toughening. This is also shown in a parametric study of Figure 9.20. Note that by subtracting the contribution of fibers from

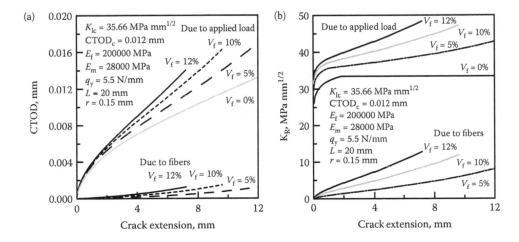

FIGURE 9.19 Effect of fiber on the crack opening profile and simulated R-curves of cement composites.

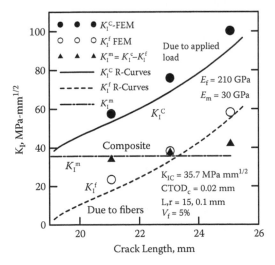

FIGURE 9.20 Contribution of fibers to the overall composite response is added to the transient or steady-state components of the matrix crack as shown by the FEM and also R-curves. (From Mobasher, B., and, Li, C. Y., "Effect of interfacial properties on the crack propagation in cementitious composites," *Journal of Advanced Cement Based Materials*, 4(3), 93–106 (November–December 1996). With permission.)

the overall composite response, the steady-state components of the matrix crack growth are extracted. This simulation was verified by both finite element analysis and the R-curve models [80].

COMPARISON WITH EXPERIMENTAL RESULTS

The R-curve approach can be used to simulate the contribution of fibers to the flexural response of concrete reinforced with steel and or alkali-resistant (AR) glass fibers. One of the ways to determine R-curve parameters is to define the instability conditions for the composite using two criteria: the critical stress intensity factor and the crack opening displacement at the tip of a fictitious crack. These equations are also stated in terms of critical crack length and critical far-field load and are expressed as:

$$K_I^e = K_{IC}^S \tag{9.47}$$

$$COD(a_0) = CTOD_c \tag{9.48}$$

Using these as failure criteria, the parameters of the R-curve are obtained and used in the simulation of the load–CMOD response by an incremental solving Equations 9.47 and 9.48 in conjunction with Equations 9.42 and 9.43. As the crack length is increased, toughness is obtained from the R-curve. The toughness is then used to compute the force necessary for the condition $G(a_i) = R(a_i)$, where a_i represents the increment of crack. Because the fiber closing pressures are already incorporated in the R-curve, the specimen is treated as an elastically equivalent material.

A closing pressure formulation is obtained from the fiber pullout slip over the stable crack growth. As the matrix crack propagates and opens, the fiber pullout leads to the growth of the debonding zone in the wake of the main crack. For a given crack opening, interface properties are used to compute the pullout force. The fiber pullout force distribution is computed iteratively on the basis of the equilibrium crack opening as a function of the specimen geometry, far-field applied load, debonding length of fibers, and the interface properties. This process is repeated for every increment of the matrix crack growth [81].

A composite specimen is idealized as having a series of straight, aligned, and short fibers of a constant length in the path of the matrix crack. Alternatively, one can use the results of an inclined fiber pullout test by Bažant et al. [10] as the constitutive response. Fiber parameters include stiffness, length, diameter, and volume. Interface properties are, on the other hand, defined using frictional and adhesional strength, fracture toughness, and stiffness.

Using standard nonlinear LEFM approaches, the equivalent material parameters in terms of G_f and u, or K_{IC} or $CTOD_c$, can be defined. These fracture parameters are used to obtain the parameters of the R-curve. The solution algorithm is defined by assuming the criteria for failure in terms of two parameters, namely, the stable crack growth length, a_c, and the scaling parameter in the R-curve defined earlier as β. Using these two parameters, the energy release required for growth, $R(a)$, is constructed. At this point, the crack is incrementally extended, and $R(a)$ and G are calculated and used in the equilibrium equation to solve for the parameters of the R-curve. The Newton–Raphson algorithm for nonlinear equation solution is used to solve the two equation and the two unknown problems. Once the parameters of the R-curve are calculated, the R-curve can be constructed, and the load-deformation response can be obtained by incrementing crack length a, setting $R(a) = G(a)$ to solve for P as a function of crack growth.

The load-deformation response is computed from the R-curve formulation using a compliance approach [77]. This procedure is then subjected to parameter optimization through an inverse solution to fit the experimental load-deformation response in terms of the stress–crack width relationship. As an added extra step, one can calculate and correlate closing pressure–crack length to energy in the process zone. Parameter optimization through an inverse solution can also be accomplished by fitting the experimental data with the model estimation.

Figure 9.21 present the simulation of the R-curve for steel FRC specimens tested in flexure [81]. Results of the experimental R-curves [83] are compared with the finite element simulations of the toughening and also the analytical solution for the R-curves on the basis of the proposed methodology. The stress intensity factor required for the growth of the crack increases with crack extension, and, although the matrix energy demand may remain constant, toughening due to fiber pullout determines the ascending shape of the R-curve. Figure 9.22 shows how to present the increase in the bend over point of carbon fiber cement composites [84]. Model simulations are also compared with the experimental results of Park et al. [85].

Composites with straight steel fibers at a volume fraction of 0.5% are simulated in Figure 9.23. Three different sizes of geometrically similar specimens with a constant thickness of $b = 50$ mm,

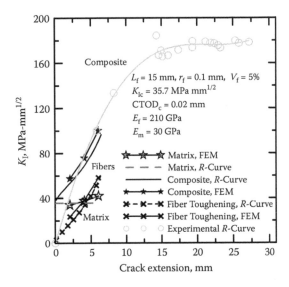

FIGURE 9.21 Comparison of experimental, simulation, and FEM predictions of R-curves for steel FRC specimens in flexure.

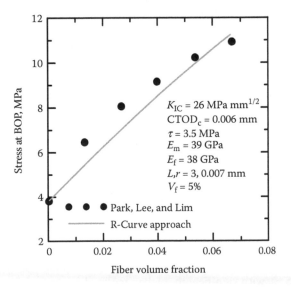

FIGURE 9.22 Prediction of stress at bend over point for experimental results of Park, Lee, and Lim.

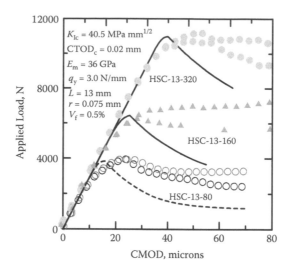

FIGURE 9.23 Comparison of experimental results with simulation of size effect of Steel FRC concrete specimens with 0.5% volume fraction of straight fibers ($L = 13$ mm).

F'_t (MPa)	β_1	α_c	R (N·mm)	G_f (N·mm)	ΔK (MPa mm$^{1/2}$)
3	0.013	3.67	0.067	0.067	31.8
4	0.020	3.427	0.089	0.089	37.2
5	0.028	3.221	0.111	0.111	41.8
6	0.037	3.057	0.133	0.133	46

depths of $d = 80, 160$, and 320 mm, span of $2.5d$, and a notch depths of $0.275d$ were studied by Bryars et al. [85]. Results for fiber lengths of 13 mm are shown [80]. The present approach underestimates the nonlinearity in the CMOD response by 15%, partly because of the simplifying assumptions of straight cracks and aligned fibers. The model also underestimates the postpeak load carrying capacity, partly because dissipative mechanisms such as crack deflection and tortuosity are not taken into account. The predictability improves with larger size specimens to within 0.3% and 3.2% of the experimental values for large and small samples, respectively. As the size increases, the ability of the fibers to bridge a longer process zone increases, resulting in more closing pressure toughening.

Figures 9.24a through 9.24d show the parametric study of the effect of tensile strength response on the R-curve and the resulting load-deformation response. A prismatic specimen consisting of $101.6 \times 101.6 \times 304.2$ mm in dimensions and an initial notch length of $a_0 = 12.75$ mm is used. The material parameters for the stress–strain response are $E = 25$ GPa, and the maximum width of a crack opening with traction, $\text{CTOD}_c = u = 0.06$ mm. Parameters $n = 0.16$, $n_i = 1.5$, and $q = 0.5$ were the power coefficients of the stress–softening and stress–crack length ligament response as defined by Equation 9.27 of the Sakai and Suzuki model [37]. The energy dissipation was modeled from the integration of Green's function over the crack length at the maximum load level,

$$\Delta K_b(l_b) = \int_0^{l_b} g(a, x)\sigma_b(x)dx \qquad (9.49)$$

Note that as the tensile strength and, thus, the stress–strain response of the specimen in the postpeak region is increased, the plateau value of the R-curve and the flexural load-deformation response as shown in Figure 9.24 is also increased. According to this simulation, the strength of a beam in flexure is as much as 57% with significantly higher energy dissipation in the postpeak response of the flexural curve.

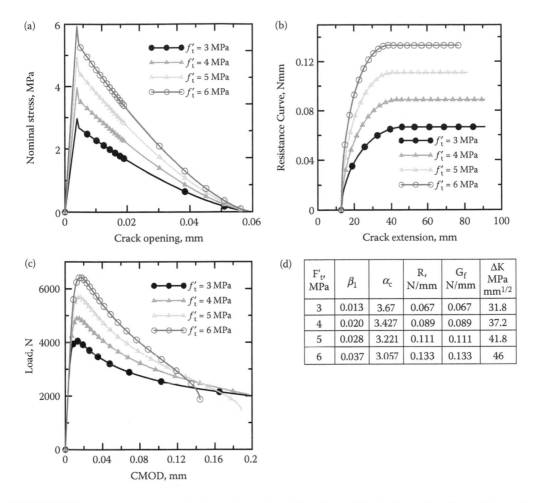

FIGURE 9.24 Parametric study of the stress–crack width relationship for increasing flexural strength. (a) The stress–crack width model. (b) Effect of stress strain response on the R-curve. (c) The load–CMOD response. (d) Components of stress intensity due to the closing pressure term.

Figures 9.25a and 9.25b show that, as the size of the stress–strain response of the specimen in the postpeak region is increased, it results in an increase in the flexural load-deformation response. Note that, according to this simulation, the maximum strength of the flexural response is as much as 70% with a significantly higher energy dissipation in the postpeak response of the flexural curve as the ultimate width is increased from 0.02 to 0.08 mm, a factor of four times.

SIMULATION OF GLASS FIBER CONCRETE

The flexural load-deformation of concrete reinforced with various levels of AR glass fibers was studied. Two types of AR glass fibers with different length combinations (referred to as high dispersion and high performance) were obtained from Vetrotex Cemfil and were considered and compared with control specimens [87].

By developing a nonlinear curve fit of the experimental data, one can back-calculate the stress–strain response of the composite by fitting the experimentally obtained load–CMOD response. Figure 9.26 presents the model fit for the effect of fiber volume fraction on the flexural load–CMOD response with Figures 9.26a and 9.26b showing the control and 10 kg/m³ of AR glass fibers of 24 mm in length,

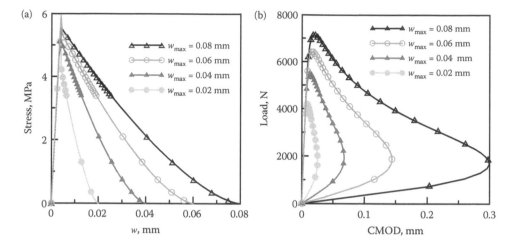

FIGURE 9.25 (a) Parametric study of the effect of postpeak range in the stress–strain response. (b) Parametric study of the effect of postpeak range on the resulting flexural load–CMOD response.

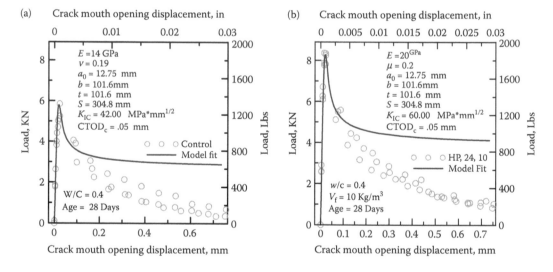

FIGURE 9.26 Model fit for the effect of fiber volume fraction on the flexural load–CMOD and the R-curve response.

respectively [87]. A nonlinear fit to the experimental load–CMOD responses yields two parameters: critical crack length Δa_c and also parameter β, representing the R-curve. In this case, the ranges of R values obtained are from 0.04 to 0.15 N.mm, and the range of critical crack extensions run from 20 to 35 mm. The effect of the fiber volume fraction in increasing load carrying capacity is clearly evident.

According to Figure 9.27, it is possible to model the effect of the duration of curing on the mechanical response by applying the nonlinear curve fit model to the experimental data for the flexural load–CMOD response based on R-curves.

Figure 9.28 represents the model fit parameters for the effect of fiber volume fraction on the flexural response, load-CMOD plot and b) the R-curve response. By conducting a nonlinear fit to the experimental load-CMOD responses, the two parameters, critical crack length Δa_c, and also the parameter β representing the R-curve are obtained. In this case the ranges of R values obtained are from 0.12 to 0.25 N/m and the range of critical crack extensions are in the range of 20-25 mm. Note

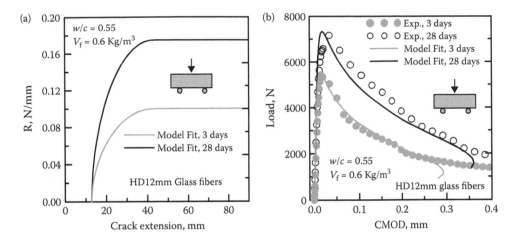

FIGURE 9.27   Modeling the effect of age on the flexural and R-curves of FRC. (a) The R-curve response and (b) the load-deformation response compared with experimental data.

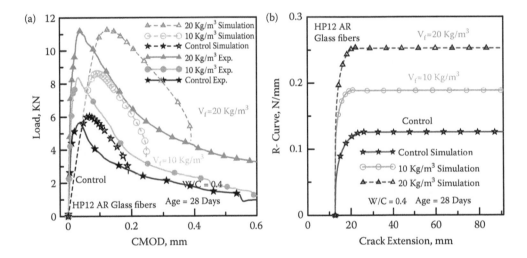

FIGURE 9.28   The model fit parameters for the effect of fiber volume fraction on the flexural response: (a) the load–CMOD plot and (b) the R-curve response.

that the predictability of the effect of fiber volume fraction in the increased load carrying capacity is significantly improved.

COMPLIANCE-BASED APPROACH

The compliance approach is based on the assumption that stable crack propagation leads to an increase in the compliance, which, if measured, can be used in a back-calculation procedure for obtaining an elastically equivalent, traction-free crack length. Using the crack mouth opening as a displacement measure, one can define the compliance of the specimen. The changing compliance can be expanded to calculate the increase in the fracture toughness as a function of crack length. The compliance, C, is defined by the inverse of the slope of the load–CMOD curve (Figure 9.28).

To obtain material properties, a notched beam is subjected to loading–unloading cycles under the three-point bending action as is illustrated in Figure 9.14. The unloading compliance of the specimen at various loading points can be used to calculate the effective crack length. The increase in the opening of the crack correlates with the fracture energy of the beam and, hence, can be used as a measure of toughness of the material. Results of this analysis can be used to compute properties such as the effective modulus of elasticity, E, the effective stress intensity factor at crack lengths corresponding to a given load, K_I, and the associated crack opening at the initial notch tip, which, because crack growth has taken place, is defined as the critical crack tip opening displacement, $CTOD_c$, and fracture toughness, G_f.

In this approach, the effect of fiber reinforcement closing pressure (because of the stress–crack width) is smeared in the overall compliance of the composite, which is treated as a homogeneous material. Using the LEFM formulas, the CMOD can be calculated as a function of the applied load, the specimen geometry, and the elastic properties of the material [89].

Figure 9.29a presents the cyclic loading and unloading curves used for obtaining the R-curves for composites with 1% alumina fibers and mortar. The loading and unloading compliance of polypropylene (PP) cement composites compared with plain mortar are shown in Figure 9.29b. Note that the rate of the increase of compliance in the plain matrix is much higher than the PP composites that contain the bridging zone. The change in compliance as a function of crack growth for the variety of cement composite systems discussed in Chapter 7 is shown in Figure 7.16, and points to the loading–unloading response of composites with 1% alumina fibers to 4% and 8% PP. Figure 9.29d

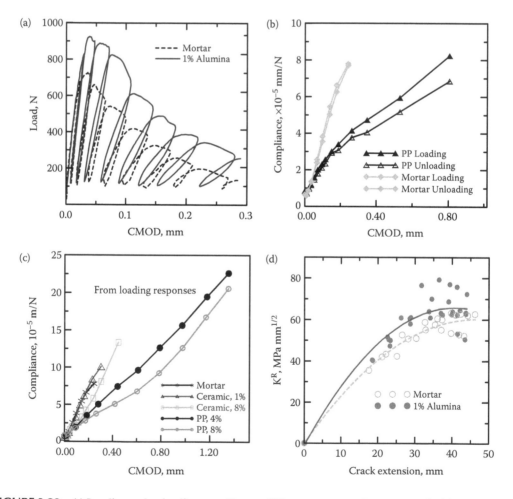

FIGURE 9.29 (a) Loading and unloading compliance of PP cement composites compared with plain mortar. (b) Change in compliance for a variety of cement composites. (c) Loading–unloading response of composites with 1% alumina fibers and mortar. (d) Experimental R-curves compared with plain mortar.

shows the R-curve response of specimens shown in Figure 9.29a; note that the rise in the R-curve is higher for the alumina composite samples.

REFERENCES

1. Zweben, C., and Rosen, B. W. (1970), "A statistical theory of material strength with application to composite materials," *Journal of the Mechanics and Physics of Solids*, 18, 189–206.
2. Tada, H., Paris, P. C., and Irwin, G. R. (1985), *The Stress Analysis of Cracks Handbook*, 2nd ed., St. Louis: Paris Productions, Inc.
3. Kaplan, M. E. (1961), "Crack propagation and fracture of concrete," *Journal of the American Concrete Institute*, 58(5), 591–610.
4. Dugdale, D. S. (1960), "Yielding of steel sheets containing slits," *Journal of the Mechanics and Physics of Solids*, 8, 100–104.
5. Barenblatt, G. I. (1959), "The formation of equilibrium cracks during brittle fracture: General ideas and hypotheses. Axially-symmetric cracks," *Journal of Applied Mathematics and Mechanics*, 23, 622–636.
6. Barenblatt, G. I. (1962), "The mathematical theory of equilibrium cracks in brittle fracture," in Dryden, H. L., and von Karman, T. (eds.), *Advanced in Applied Mechanics*, New York: Academic Press, 55–129.
7. Hillerborg, A., Modeer, M., and Petersson, P. E. (1976), "Analysis of crack formation and crack growth in concrete by means of fracture mechanics and finite elements," *Cement and Concrete Research*, 6, 773–782.
8. Shah, S. P., and McGarry, F. J. (1971), "Griffith fracture criterion and concrete," *Journal of the Engineering Mechanics Division*, 97(6), 1663–1676.
9. Hillerborg, A. (1985), "The theoretical basis of a method to determine the fracture energy GF of concrete," *Materials and Structures*, 18, 291–296.
10. Bažant, Z. P., Yu, Q., and Zi, G. (2002), "Choice of standard fracture test for concrete and its statistical evaluation," *International Journal of Fracture*, 118, 303–337.
11. Bazant, Z. P., and Planas, J. (1998), *Fracture and Size Effect in Concrete and Other Quasibrittle Materials*, Boca Raton, FL: CRC Press, 616 pp.
12. Bazant, Z. P. (1984), "Size effect in blunt fracture: concrete, rock and metals," *Journal of Engineering Mechanics*, 110(4), 518–535.
13. Jenq, Y. S., and Shah S. P. (1986), "Crack propagation in fiber-reinforced concrete," *Journal of Structural Engineering*, 112(1), 19–34.
14. Planas, J., Guinea, G. V., and Elices, M. (1993). "Softening curves for concrete and structural response," in Rossmanith, H. P. (ed.), *Fracture and Damage of Concrete and Rock*, London: E & FN Spon, 66–75.
15. Planas, J., Guinea, G. V., and Elices, M. (1999), "Size effect and inverse analysis in concrete fracture," *International Journal of Fracture*, 95, 367–378.
16. Planas, J., and Elices, M. (1990), "Fracture criteria for concrete: mathematical approximations and experimental validation," *Engineering Fracture Mechanics*, 35, (1/2/3), 87–94.
17. Karihaloo, B. L., and Nallathambi, P. (1987), "Notched Beam Test: Mode I Fracture Toughness," Draft report to RILEM Committee 89-FMT, Fracture Mechanics of Concrete: Test Method.
18. Nallathambi, P., and Karihaloo, B. L. (1986), "Determination of specimen—size independent fracture toughness of plain concrete," *Magazine of Concrete Research*, 38(135), 67–76.
19. van Mier J. G. M., and van-Vliet, M. R. A. (1998), "Experimental investigation of size effect in concrete under uniaxial tension," in Mihashi, H., and Rokugo, K. (eds.), *Proceedings of FRAMCOS-9*, Aedificatio, 1923–1936.
20. Reinhardt, H. W. (1984), "Fracture mechanics of an elastic softening material like concrete," *Heron*, 29, 2.
21. Carpinteri, A., and Chiaia, B. (1996), "Size effects on concrete fracture energy: Dimensional transition from order to disorder," *Materials and Structures*, 29, 259–266.
22. Petersson, P. E. (1980), "Fracture energy of concrete: Practical performance and experimental results," *Cement and Concrete Research*, 10, 91–101.
23. Li, V. C., and Wu, H. C. (1992), "Conditions for pseudo strain hardening in fiber reinforced brittle matrix composites," *Journal of Applied Mechanics Review*, 45(8), 390–398.
24. Li, V. C., and Leung, C. K. Y. (1992), "Theory of steady state and multiple cracking of random discontinuous fiber reinforced brittle matrix composites," *Journal of Engineering Mechanics*, 118(11), 2246–2264.
25. Soranakom, C., and Mobasher, B. (2007), "Closed-form moment-curvature expressions for homogenized fiber reinforced concrete," *ACI Materials Journal*, 104(4), July–August, 351–359.

26. Soranakom, C., and Mobasher, B. (2007), "Closed form solutions for flexural response of fiber reinforced concrete beams," *Journal of Engineering Mechanics*, 133(8), August, 933–941.

27. Visalvanich, K., and Naaman, A. E. (1983), "A fracture model for fiber reinforced concrete," *Journal of the American Concrete Institute*, 80(2), 128–138.

28. Tjiptobroto, P., and Hansen, W. (1993), "Tensile strain hardening and multiple cracking in high performance cement based composites," *ACI Materials Journal*, 90(1), 16–25.

29. Yang, C. C., Mura, T., and Shah, S. P. (1991), "Micromechanical theory and uniaxial tests of fiber reinforced cement composites," *Journal of Materials Research*, 6(11), 2463–2473.

30. Li, S. H., Shah, S. P., Li, Z., and Mura, T. (1993), "Micromechanical analysis of multiple fracture and evaluation of debonding behavior for fiber reinforced composites," *International Journal of Solids and Structures*, 30(11), 1429–1459.

31. Mobasher, B., and Li, C. Y. (1996), "Mechanical properties of hybrid cement based composites," *ACI Materials Journal*, 93(3), 284–293.

32. Li, Z., Mobasher, B., and Shah, S. P. (1991), "Characterization of interfacial properties of fiber reinforced cementitious composites," *Journal of the American Ceramic Society*, 74(9) 2156–2164.

33. Ouyang, C. S., Pacios, A. A., and Shah, S. P. (1994), "Pullout of aligned and inclined fibers from cement based matrices," *Journal of Engineering Mechanics*, 120(12), 2641–2659.

34. Mobasher, B., and, Li, C. Y. (1995), "Modeling of stiffness degradation of the interfacial zone during fiber debonding," *Journal of Composites Engineering*, 5(10–11), 1349–1365.

35. Lenain, J. C., and Bunsell, A. R. (1979), "The resistance to crack growth of asbestos cement," *Journal of Materials Science*, 14, 321.

36. Foote, R. M. L., Mai, Y.W, and Cotterell, B. (1986), "Crack–growth resistance curves in strain-softening materials," *Journal of the Mechanics and Physics of Solids*, 34(6), 593–607.

37. Suzuki, T., and Sakai, M. (1994), "A model for crack-face bridging," *International Journal of Fracture*, 65(4), 329–344.

38. Hsueh, C. H., and Becher, P. (1988), "Evaluation of bridging stress from R-curve behavior for non-transforming ceramics," *Journal of the American Ceramic Society*, 71(5), C234–C237.

39. Munz, D. (2007), "What can we learn from R-curve measurements?" *Journal of the American Ceramic Society*, 90(1), 1–15.

40. Ouyang, C., and Shah, S. P. (1991), "Geometry-dependent for quasi-brittle materials," *Journal of the American Ceramic Society*, 74(11), 2831–2836.

41. Krafft, J. M., Sullivan, A. M., and Boyle, R. W. (1961), "Effect of dimensions on fast fracture instability of notched sheets," Proceedings of Crack Propagation Symposium, College of Aeronautics, Cranfied, England, Vol. I, 8–28.

42. Broek, D. (1987), *Elementary Engineering Fracture Mechanics*, 4th ed., Hingham, MA: Marinus Nijhoff, 185–200.

43. Anderson, T. L. (1995), *Fracture Mechanics Fundamentals and Applications*, 2nd ed., Boca Raton, FL: CRC Press, p. 688.

44. Mai, Y.-W., and Lawn, B. R. (1987), "Crack-interface grain bridging as a fracture resistance mechanism in ceramics: II, theoretical fracture mechanics model," *Journal of the American Ceramic Society*, 70, 289.

45. Lawn, B. (1993). *Fracture of Brittle Solids*. Cambridge: Cambridge University Press, 378.

46. Ramachandran, N., and Shetty, D. K. (1991), "Rising crack-growth-resistance (R-curve) behavior of toughened alumina and silicon nitride," *Journal of the American Ceramic Society*, 74, 2634–2641.

47. Ramachandran, N., Chao, L.-Y., and Shetty, D. K. (1993), "R-curve behavior and flaw insensitivity of Ce-TZP/Al2O3 composite," *Journal of the American Ceramic Society*, 76, 961–969.

48. Evans, A. G. (1984). "Toughening mechanisms in zirconia alloys," in Claussen, N., Ruehle, M., and Heuer, A. H. (eds.), *Advances in Ceramics, Vol. 12, Science and Technology of Zirconia II*, Columbus, OH: American Ceramic Society, 193–212.

49. Shetty, D. K., and Wang, J.-S. (1989), "Crack stability and strength distribution of ceramics that exhibit rising crack-growth-resistance (R-curve) behavior," *Journal of the American Ceramic Society*, 72, 1158–1162.

50. Mai, Y.-W., and Lawn, B. R. (1987), "Crack-interface bridging as a fracture resistance mechanism in ceramics: II. Theoretical fracture mechanics model," *Journal of the American Ceramic Society*, 70, 289–294.

51. Cook, R. F., Fairbanks, C. J., Lawn, B. R., and Mai, Y.-W. (1987), "Crack resistance by interfacial bridging: Its role in determining strength characteristics," *Journal of Materials Research*, 2, 345–356.

52. Tanaka, K., Suzuki, K., and Tanaka, H. (1992), "Evaluation of critical defect size of ceramics," *Fracture Mechanics of Ceramics*, 9, 289–303.

53. Kendall, K., Alford, N. M., Tan, S. R., and Birchall, J. D. (1986), "Influence of toughness on weibull modulus of ceramic bending strength," *Journal of Materials Research*, 1, 120–123.
54. Krause, R. F. (1988), "Rising fracture toughness from the bending strength of indented alumina beams," *Journal of the American Ceramic Society*, 71, 338–343.
55. Krause, R. F., Fuller, E. R., and Rhodes, J. F. (1990), "Fracture resistance behavior of silicon carbide whisker-reinforced alumina composites with different porosities," *Journal of the American Ceramic Society*, 73, 559–566.
56. Gopalaratnam, V. S., and Shah, S. P. (1985), "Softening response of plain concrete in tension," *Journal of the American Concrete Institute*, 82(3), 310–323.
57. Stang, H., Li, V. C., and Krenchel, H. (1994), "Design and structural applications of stress–crack width in fiber reinforced concrete," *Materials and Structures,* 28, 210–219.
58. Mai, Y.-W., and Lawn, B.R. (June 1987), "Crack-interface grain bridging as a fracture resistance mechanism in ceramics: II, Theoretical fracture mechanics model," *Journal of the American Ceramic Society*, 70(4), 289–294.
59. Lawn, B. (1993), *Fracture of Brittle Solids*, Cambridge: Cambridge University Press, 378.
60. Foote, R. M. L., Mai, Y.W, and Cotterell, B. (1986), "Crack–growth resistance curves in strain-softening materials," *Journal of the Mechanics and Physics of Solids*, 34(6), 593–607.
61. Anderson, T. L. (1995), *Fracture Mechanics Fundamentals and Applications*, 2nd ed., Boca Raton, FL: CRC Press, 688.
62. Mai, Y. W. (2002), "Cohesive zone and crack-resistance (R)-curve of cementitious materials and their fibre-reinforced composites," *Engineering Fracture Mechanics*, 69(2), Jan, 219–234.
63. Ballarini, R., Shah, S. P., and Keer, L. M. (1984), "Crack growth in cement-based composites," *Engineering Fracture Mechanics*, 20, 433–445.
64. Cox, B. N. (1993), "Scaling for bridged cracks," *Mechanics of Materials,* 15, 87–98.
65. Fett, T., and Munz, D. (1993), "Evaluation of R-curve effects in ceramics," *Journal of Materials Science*, 28, 742–745.
66. Fett, T., and Munz, D.(1995), "Bridging stress relations for ceramic materials," *Journal of the European Ceramic Society*, 15, 377–383.
67. Fett, T., and Munz, D. (1990), "Influence of crack–surface interactions on stress intensity factor of ceramics," *Journal of Materials Science Letters*, 9, 1403–1406.
68. Kruzic, J. J., Cannon, R. M., Ager, J. W. III, and Ritchie, R. O. (2005), "Fatigue threshold R-curves for predicting reliability of ceramics under cyclic loading," *Acta Materialia*, 53, 2595–2605.
69. Wittmann, F. H., and Hu, X. (1991), "Fracture process zone in cementitious materials," *International Journal of Fracture*, 51, 3.
70. Fett, T., Munz, D., Dai, X., and White, K. W. (2000), "Bridging stress relation from a combined evaluation of the R-curve and post-fracture tensile tests," *International Journal of Fracture*, 104, 375–385.
71. Hu, X. Z., and Mai, Y. W. (1992), "A general method for determination of crack face bridging stresses," *Journal of Materials Science*, 27, 3502–3510.
72. Lathabai, S., Rodel, J., and Lawn, B. (1991), *Journal of the American Ceramic Society,* 74:1348.
73. Budiansky, B., Hutchinson, J. W., and Evans, A. G. (1986), "Matrix fracture in fiber-reinforced ceramics," *Journal of the Mechanics and Physics of Solids*, 34(2), 167–189.
74. Aveston, J., Cooper, G. A., and Kelly, A. "The properties of fiber composites," *Conference Proceedings, National Physical Laboratory* (IPC Science and Technology Press Ltd), Paper 1, 1971, 15.
75. Aveston, A. J., and Kelly, A. (1973), "Theory of multiple fracture of fibrous materials," *Journal of Materials Science*, 8, 352.
76. Hannant, D. J., Hughes, D. C., and Kelly, A. (1983), "Toughening of cement and of other brittle solids with fibres," *Philosophical Transactions of the Royal Society of London*, A310, 175–190.
77. Mobasher, B., and Shah, S. P. (1990), "Interaction between fibers and the cement matrix in glass fiber reinforced concrete," American Concrete Institute, ACI SP-124, 137–156.
78. Ouyang, C. S., Mobasher, B., and Shah, S. P. (1990), "An R-curve approach for fracture of quasi-brittle materials," *Engineering Fracture Mechanics*, 37, 4, 901–913.
79. Mobasher, B., Ouyang, C., and Shah, S. P. (1991), "Modeling of fiber toughening in cementitious composites using an R-curve approach," *International Journal of Fracture*, 50, 199–219.
80. Mobasher, B., and, Li, C. Y. (1996), "Effect of interfacial properties on the crack propagation in cementitious composites," *Journal of Advanced Cement Based Materials*, 4(3), 93–106.
81. Li, C. Y., and Mobasher, B. (1998), "Finite element simulations of toughening in cement based composites," *Journal of Advanced Cement Based Materials*, 7, 123–132.

82. Haupt, G. J. (1997). "Mechanical properties of cement based composite laminates," MS thesis, Arizona State University.

83. Mobasher, B., Li, C. Y., and Arino, A. (1995), "Experimental R-curves for assessment of toughening in micro-fiber reinforced hybrid composites," American Concrete Institute, ACI SP-155-5, 93–114.

84. Ouyang, C. S., and Shah, S. P. (1992), "Toughening of high strength cementitious matrix reinforced by discontinuous short fibers," *Cement and Concrete Research*, 22, 1201–1215.

85. Park, S. B., Lee, B. I., and Lim, Y. S. (1990), "Experimental study on the engineering properties of carbon fiber reinforced cement composites," *Cement and Concrete Research*, 21, 589.

86. Bryars, L., Gettu, R., Barr, B., and Arino, A. (1994), "Size effect in the fracture of fiber-reinforced high-strength concrete," in Bazant, Z. P., Bittnar, Z., Jirasek, M., and Mazars, J. (eds.), *Fracture and Damage in Quasibrittle Structures*. London: E & FN Spon, 319–326.

87. Desai, T., Shah, R., Peled, A., and Mobasher, B., "Mechanical properties of concrete reinforced with AR-glass fibers," Proceedings of the 7th International Symposium on Brittle Matrix Composites (BMC7), Warsaw, October 13–15, 2003.

88. Mobasher, B., and Peled, A., "Use of R-curves for characterization of toughening in fiber reinforced concrete," *Proceedings, International Conferences on Fracture Mechanics of Concrete and Concrete Structures* (FraMCoS V) Vail Colorado, 2004, 1137–1143.

89. RILEM 89-FMT. (1990), "Determination of fracture parameters (K_{IC}^S and $CTOD_c$) of plain concrete using three-point bend tests," *Materials and Structures*, 23(138), 457–460.

10 Tensile Response of Continuous and Cross-Ply Composites

INTRODUCTION

In most fiber reinforced cement (FRC) composites with short, randomly dispersed fibers at a low volume fraction, the matrix phase does not contribute to the overall strength of the composite until the first cracking takes place. Once the critical volume fraction of fibers is exceeded, the ultimate strength and ultimate strain increase considerably [1]. With a higher volume of closely spaced fibers, the microcracking can be controlled by crack arrest and bridging mechanisms [2]. This allows the stresses to be transferred back into the matrix and form additional microcracks while increasing the toughness considerably. The formation of parallel microcracks is a signature behavior in these systems, which is associated with a reduced stiffness in the composite and a tension stiffening effect. This chapter addresses basic tests on unidirectional glass and polypropylene (PP) unidirectional and cross-ply composites when tested under general tensile and flexural loading conditions. The mechanical response of continuous fiber laminate systems that exhibit distributed cracking is also discussed.

SPECIMEN PREPARATION

The samples presented in this section were manufactured using the pultrusion process in accordance to the procedures discussed in Chapter 3. The paste in these samples consisted of a type I/II portland cement with a silica fume content of 0% to 10% by weight of cement. A water to cement ratio of 0.35 was used. The silica fume was dispersed by premixing in a food processor. Mixing of the paste was completed in a Hobart mixer before the pultrusion process. The development of filament winding equipment is described in detail by Mobasher, Pivacek, and Haupt [3]. Properties of continuous, alkali-resistant (AR) glass and PP fibers are shown in Table 10.1. The volume fraction of the fibers was measured using the cross-sectional dimensions, the number of windings, and the cross-sectional area of a single-fiber strand [3]. A constant glass fiber volume fraction of 4.8% to 5% was used for all AR glass specimens studied [4, 5]. PP composites used continuous fibrillated PP fibers manufactured by Krenchel and Stang [6]. The properties are shown in Table 10.1. Rectangular specimens of 75×350 mm in dimension were machined into dog-bone samples specimens with a reduced cross section of 50×20 mm.

Direct tension tests were performed on an MTS 810 closed-loop controlled servohydraulic material system with a capacity of 220 kN (55 kips). Hydraulic grips were used with the pressure maintained between 5.0 and 7.5 MPa. The elongation of the specimen across a 90-mm gage length was measured using two linear variable differential transformers (LVDTs) of ± 1.27 mm (± 0.05 in.) range, as is shown in Figure 10.1. The LVDTs were mounted on both sides of the specimen, and their average response was used as the feedback control.

Several parameters characterize the tensile response. These include the modulus of elasticity (E), the first cracking strain, the bend over point (BOP), the slope in the quasi-linear region beyond the BOP (defined as E_2), which may be treated as a decaying function or constant, the pseudo-strain-hardening region, the ultimate strength, and the toughness. Test results of uniaxial, cross-ply, and angle-ply laminates are presented in Table 10.2.

Figure 10.2 represents the typical tensile stress–strain response of an AR glass fabric reinforced cement composite. Several features of both elastic and inelastic response are of interest. The initial elastic response follows the rule of mixtures quite closely. The average elastic moduli for composites is in the 18- to 25-GPa range, which is in accordance to the rule of mixtures. Summary results of some of the tensile tests conducted are presented in Table 10.3. A detailed set of data can be obtained from published work in this area [7, 8].

The tensile response shows a linear behavior up to about 3 to 8 MPa. Beyond this level, the stress–strain response becomes nonlinear, and a major change in the stiffness of the sample occurs at BOP around 3 to 4 MPa. The BOP in unidirectional fibers is characterized by a knee in the stress–strain curve. The specimen continues to carry load at a significantly lower stiffness up to an ultimate strain level of 0.6%. In the region between the BOP and the ultimate strength, individual cracks form sequentially until a crack saturation state is reached. Beyond the crack saturation level, no new cracks form, however widening of existing cracks takes place. The parallel cracking in tensile specimens is shown in Figure 10.3. The experimental data can be analyzed to obtain a range of parameters in terms of the initial stiffness, BOP stress (initial and final levels indicated by – and + signs), BOP strain, post-BOP stiffness, ultimate strength capacity, toughness, and pullout stiffness [9].

TABLE 10.1
Properties of AR glass and PP Fibers

Fiber	L_f (mm)	d_f (µm)	Ultimate Strength (MPa)	Elastic Modulus (GPa)	Density (g/cm³)
AR glass	Continuous	12	1725	70	2.70
PP	Continuous	35 × 250	340–500	8.5–12.5	0.91

FIGURE 10.1   Tension setup.

TABLE 10.2
Tensile Test Response of Unidirectional and Cross-Ply AR Glass and PP Composites

Layup	q	V_f (%)	E (GPa)	E_2 (GPa)	σ_{BOP} (MPa)	ε_{BOP} (mm/mm)	$\sigma_{ultimate}$ (MPa)	$\varepsilon_{ultimate}$ (%)
0 G	10	4.8	22.33	9.15	12.06	0.92E–3	45.22	0.82
0/90/0 G	4	4.8	8.46	2.82	6.91	1.41E–3	21.81	1.25
(0/90)$_s$ G	5	5.0	21.63	1.89	9.86	2.39E–3	38.27	1.84
90/0/90 G	4	5.0	11.31	2.47	11.58	1.27E–3	34.29	1.41
[45/–45]$_s$ G	3	9.6	1.354	–	0.77	0.96E–3	5.12	0.48
[0/45/–45]$_s$ G	3	5.6	14.23	–	5.35	0.40E–3	44.20	1.35
[0/–45/45/90]$_s$ G	3	8.8	17.14	–	9.71	0.56E–3	50.32	1.22
PP 0	5	7.0	22.50	0.22	7.80	0.90E–3	14.02	3.94

Note: G, AR glass.

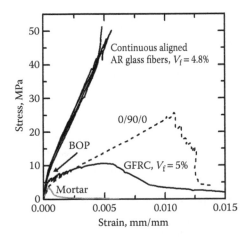

FIGURE 10.2 Stress–strain response of AR glass fibers composites as compared with mortar.

TABLE 10.3
Mechanical Properties of Glass Fiber Composite Laminates

Layup	No. Replicates	E (GPa)	Stress at BOP (MPa)	Strain at BOP	Ultimate Strength (MPa)	Ultimate Strain (mm/mm)
0	9	23.6	10.0	4.11E–04	45.4	7.62E–03
		(4.4)	(2.2)	(1.14E–04)	(6.2)	(4.90E–03)
0/90/0	5	10.2	5.2	6.43E–04	23.1	1.44E–02
		(0.8)	(1.0)	(2.30E–04)	(4.5)	(3.07E–03)
0/90/90/0	5	24.5	11.8	6.08E–4	38.3	2.26E–2
		(6.4)	(3.8)	(3.95E–4)	(9.2)	(1.29E–2)
90/0/90	4	11.6	11.6	1.27E–03	31.5	1.22E–02
		(7.6)	(5.8)	(3.46E–04)	(7.0)	(2.67E–03)

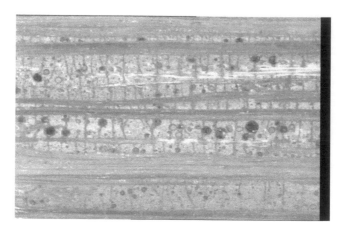

FIGURE 10.3 Mechanisms of parallel cracking in tensile specimens.

We, furthermore, compared the stress–strain curve of unidirectional composites with the $(0/90)_s$ specimens, plain mortar, and glass FRC (GFRC), as shown in Figure 10.2. GFRC uses approximately 5% chopped and randomly distributed glass fibers [10]. It should be noted that the tensile response is significantly improved with the use of continuous fibers. By comparing the unidirectional fiber composites with the GFRC samples, we observed that the efficiency of fiber length and alignment results in a threefold increase in the ultimate tensile strength. The initial response of the composites is linear up to the BOP.

The BOP refers to the point at which the ultimate strength of the brittle matrix phase in an FRC composite is reached. Although the characterization of BOP is quite evident in the composites with short fibers or intermediate fiber volumes, in the present data, we observed it as a knee in the stress–strain response. This is partly due to the significant stiffness of the fibers, which provide traction across the cracked matrix and make the BOP determination difficult at best. Another alternative is to address this range in two distinct lower and upper limits.

Beyond the BOP, the response remains linear up to the ultimate strength. The stiffness between the BOP level and the ultimate strength is referred to as E_2. This parameter represents the equivalent elastic modulus of a composite containing matrix cracks. Because of the fracture of fibers at the point of ultimate strength, very little postpeak ductility is observed; however, that may also be attributed to the mode of testing, which is stroke control, and the fact that obtaining strain softening using strain control is not accurate. The failure is dominated by a complex interaction of fiber failure, matrix cracking, shear cracks in the composite, debonding, delamination, and, occasionally, crushing under the clamping pressure at the grips.

After the transition of the stress into the pseudo-strain-hardening state, the stiffness of composites is directly related to the contribution of the fiber as $E_f V_f$, or 3.75 GPa (under assumed values, $E_f = 75$ GPa and $V_f = 5\%$). On average, uniaxial samples had a secondary slope (E_2) of 9.15 GPa, which is more than twice the expected value. This represents the contribution of the matrix phase in stiffening the composite as tension stiffening, which will be addressed under a separate section. The E_2 values signify that while the matrix continues to crack, it is carrying a portion of the applied load. In cross-ply composites, this secondary stiffness (E_2) was approximately 50% to 75% of the expected value computed by means of ply discount. This is expected because the 90° layers have no load carrying capabilities due to cracking of the matrix.

The essential feature in the response of unidirectional 0° laminates is an ultimate strength of 50 MPa at an ultimate strain of 0.6%. Because of the various modes of failure involved, this strength level may not yet represent the total strength of glass fibers. The comparison of the results with uniaxial response of GFRC [11] indicates that for the similar volume fraction of fibers, the strength of composites with continuous fibers is as much as five times higher than the short, randomly distributed fiber composites.

(0/90) COMPOSITE LAMINATES

The tensile response of a sample with 0/90/90/0-ply orientation is compared with a unidirectional sample in Figure 10.4. Note that the strength of the composite is reduced to an approximate level of 40 MPa. Because of averaging, the strength of the 90° plies in tension is closer to the tensile strength of the plain matrix. The ductility of the composite laminate, however, is significantly increased when compared to the unidirectional samples. Ultimate strain capacity is as high as 1.5% as compared with 0.6% for the unidirectional composites, an increase of 250%. The increase in ductility is attributed to two mechanisms: interlaminar cracking and parallel microcracking in the 90° plies. These mechanisms result in a loss of stiffness in the composite, but add a degree of ductility. The increased strain capacity (as high as 2%) is clearly indicated as the specimen maintains tensile stresses of 35 MPa. The extent of cracking in the 90° plies is indicative of the ability of the composite to distribute the microcracks throughout the matrix phase.

The cross-ply composites exhibited much larger ultimate strain capacities when compared with the uniaxial specimens despite a reduced ultimate strength level. Cross-ply composites showed ultimate strains in the range of 1% to 4%. The average ultimate strain values increased 124% for (0/90)$_s$ samples, whereas the 90/0/90 and (90/0)$_s$ composites increased 71% and 52% when compared with the uniaxial samples. BOP strains were also much larger with (0/90)$_s$ composites, since they are 260% larger than the uniaxial samples. With average values around 38 MPa, cross-ply composites experienced approximately 15% lower strength when compared with uniaxial specimens. These cross-ply composites, however, exhibited tensile strength in excess of 30 MPa with an ultimate strain more than 1%. This added ductility is attributed to the cracking of the 90° plies over the entire specimen length. The experimental modulus of elasticity of the composite (E) for the uniaxial and (0/90)$_s$ specimens closely matches the values from the rule of mixtures and other models [12]. The initial stiffness values for 0/90/0 and (90/0)$_s$ composites were lower than the theoretical values, perhaps because of the lack of bond and the rotation of the specimen in the grips.

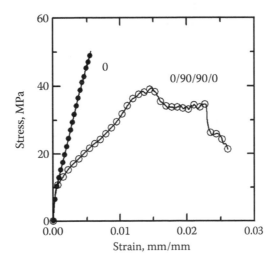

FIGURE 10.4   Tensile response of unidirectional composites and a stacking sequence of 0/90/90/0.

The failure mechanism in cross-ply composites is characterized by the formation of intralaminar cracks that form parallel to the loading direction along the specimen width. These cracks form at the 90° layers and grow toward the 0° to 90° interface, resulting in the debonding of the laminates. This mechanism was the major mode of failure in the cross-ply composites used in our study. The failure of the 90° layers resulted in an increased stress in the 0° layers. Ultimate strengths in four uniaxial and one (0/90)$_s$ specimen exceeded 50 MPa, two of these having ultimate strains over 1.2%. It was observed that variations in the failure modes were influenced by the gripping parameters; thus showing that testing conditions influence the ultimate strength of the composite.

(+45) COMPOSITE LAMINATES

The interlaminar shear between the highly orthotropic 0° and 90° layers can be reduced by incorporating 45° layers in between. The tensile response of a four-ply (±45)$_s$ composite laminate is shown in Figure 10.5. Although the initial portion of the response is linear, the linearity terminates quite rapidly as the low fiber–matrix bond strength dominates shear failure mechanisms at a stress of 1.2 MPa and 0.5% strain. This represents 5% to 10% of the average ultimate strength of uniaxial specimens. Beyond this level, one major inclined crack along the interfacial zone of the layers leads to composite failure.

As shown in Figure 10.6, the tensile stress–strain curve of a [0/–45/45/90]$_s$ composite is characterized by an ultimate tensile strength of 50 MPa and an ultimate strain capacity of more than 1%. The addition of a 45° lamina in between cross-ply laminates substantially decreases the shear stress, as shown in Figure 10.6. The +45° layer marginally contributes to the load carrying capacity by reducing the interlaminar shear stresses while improving the strain capacity by distributing the cracking in the composite. When placed between the 0° and the 90° layers, nominal strength improves, surpassing both unidirectional and [0/+90]$_s$ composites. Postpeak response is far superior as well. The extent of cracking is indicative of the 90° layers that distribute microcracks throughout, whereas the 45° layers reduce the shear stresses in between the plies. Even at a 2% axial strain, the load carrying capacity remains significantly high.

COMPRESSION RESPONSE

The compression response of composite laminates is shown in Figure 10.7. Note that the response is elastic and the modes of strain capacity that exist in tension field are not present. The failure strain is limited to approximately 0.003 mm/mm, which matches the plain matrix's failure strain quite well. The failure

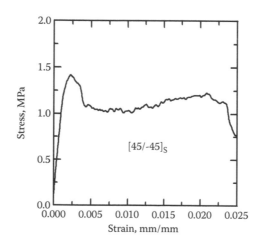

FIGURE 10.5   Tensile stress–strain response of a (+45)s composite.

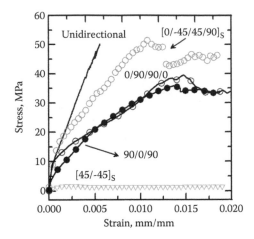

FIGURE 10.6   Ductile composites using cross-ply laminates.

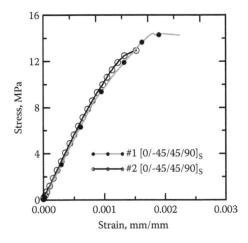

FIGURE 10.7   Compression response of composite laminates.

in compression is also dominated by the delamination of two adjacent layers in response to the interlaminar shear, which is followed by crack growth and the eventual buckling of delaminated regions.

PP FIBER LAMINATES

PP fiber composites were also filament wound into uniaxial specimens during the study. Figure 10.8 represents the tensile stress–strain response of composites containing 7% PP fibers. The stress–strain curves are in agreement with the results of the manually pultruded composites tested by Krenchel and Stang [6] and Mobasher et al. [14]. Note that, as shown in Figure 10.8, the straining was continued up to a 3% strain level and could be continued up to 8% in uniformly prepared specimens, as well [3]. The stress–strain response was semilinear up to approximately 8 MPa and ended with a very distinct BOP. As this demonstrates, the BOP stress value is dependent on volume fraction of fibers, water–cement ratio, and bonding efficiency of the fiber–matrix interface. The modulus of elasticity value was well within the predicted range of rule of mixtures and represents the matrix properties, especially since the fiber modulus was weaker.

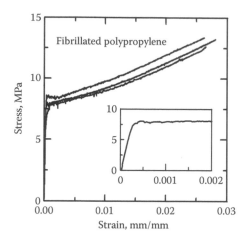

FIGURE 10.8 Tensile response of composites with unidirectional PP-based fibers, V_f = 7%. Inset of the figure shows the initial ascending part of the curve.

The inset of the figure above shows the initial ascending portion of the curve. This region is characterized by a quasi-linear increase in stress up to MPa, approximately. The BOP is quite apparent in these composites. This BOP level is a function of the fiber volume fraction, the interface properties, and the matrix formulation. The present experimental results are comparable with the uniaxial tension results of composites pultruded manually by Krenchel and Stang [13] and Mobasher et al. [14]. Theoretical analysis for the prediction of the BOP response has been provided through the use of a fracture mechanics model [15, 16]. The ultimate strain capacity of these composites is well beyond the 2% level shown in the figure. Formation of visible cracks parallel to and transverse to the fiber direction is confirmed through microscopic evaluations of thin sections. At this level of strain, significant distributed microcracking was observed in the specimens used in the study, and the load-carrying capacity was observed to be primarily due to the straining of the fibers.

The multiple cracking phase is signified by the increasing strain at a reduced stiffness follows the BOP. This phase ends as the cracking is saturated in the specimen, and the composite stiffness depends on the contribution of fibers. The theoretical $E_f V_f$ value for PP fibers (E_f = 7.0 "initial" GPa and V_f = 7%) is 0.49 GPa, even though the secondary slope in the experiment obtained only 0.22 GPa. This variation points out to the tension stiffening effect of the matrix.

FLEXURAL RESPONSE

Flexural testing of continuous AR glass or cement specimens under closed-loop control four-point bending also confirms these tension results. Linear elastic behavior was assumed in the calculation of flexural stiffness and strength measures throughout our testing. The proportional elastic limit (PEL) is considered to be the point at which the flexural stress–deflection curve deviates from linearity. The modulus of rupture is computed on the basis of the elastic section properties. Flexural stress–deflection curves for several unidirectional fiber composites are shown in Figure 10.8. It is notable that the majority of the composites fail at an equivalent flexural stress above 50 MPa, which is as much as ten times the plain matrix strength.

Having said that, the postpeak response, is varied since it depends on the nature of distributed cracking and the initiation of delamination failure and shear cracking in the composites. The flexural modulus of elasticity is observed to be in the range of values for glass/cement composites containing random fibers. Figures 10.9a and 10.9b represent a comparison between the unidirectional fiber composites and the 0/90/0 specimens containing 4.8% AR glass fibers with the portland cement paste (Figure 10.10). An addition of the fibers increases the strength of the composite by an

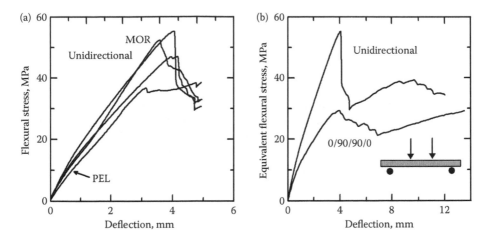

FIGURE 10.9 Flexural stress–deflection curve of uniaxial and 0/90/90/0 specimens containing 4.8% AR glass fibers compared with portland cement paste.

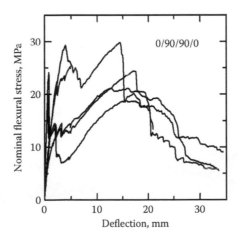

FIGURE 10.10 Flexural stress deflection of 0/90/90/0 composites with 4.8% AR glass fibers.

order of magnitude. Postpeak ductility of the unidirectional composites is lower than the 0/90/0 composites.

The PEL represents a loss of linearity as a result of the formation of the first crack. In the flexural test, the PEL may not be evident since there is only a minor change in the slope of stress deflection. The PEL values are in the range of 11 MPa for uniaxial samples and 13.8 MPa for $(0/90)_s$, which compares well with previously reported values of 7.2 to 9.0 MPa [17]. The deflection at PEL was nearly 1 mm for the majority of the uniaxial specimens and 0.85 mm for $(0/90)_s$. This is well above the deflection values of 0.18 mm that are reported for short fiber composites that use glass fibers [18]. The pseudo-linear response of the flexural stress–deflection curve exists up to the 40 MPa in most cases. The specimen in the stage from the PEL to the peak level exhibits a much stiffer response than short fiber composites. This illustrates the increased efficiency attained by fiber alignment and placement.

The PEL level is associated with the first major cracking, signifies the permanent material damage, and is synonymous to the BOP level in tension. The average deflection at the PEL for $(0/90)_s$ composites was less than one half the deflection of unidirectional composites. The modulus of rupture was significantly higher than the results reported for short fiber cement composites [15, 16, 19].

MICROSTRUCTURAL DAMAGE AND TOUGHNESS

The definition of toughness in brittle matrix composites varies widely. The tensile toughness is defined as the area under the stress stain curve. The ultimate toughness values may not represent the material property because the failure mechanisms are quite varied. Tests may be terminated because of a variety of reasons, such as formation of longitudinal cracks along the interface or shear cracking induced by the gripping pressure. A higher level of energy absorption is observed for the cross-ply laminates. The uniaxial PP specimens absorb a higher amount of energy than the glass specimens. This is due to the extended strain range of PP fibers. Results reported in the above-mentioned studies [2,3,5] indicate that at the 8.5% strain level, an approximate amount of 108 N/mm is absorbed. The formation of large transverse cracks in the 90° layers and their coalescence into major delamination accounted for this behavior. Figure 10.11 represents the type of ply delamination cracking observed in a flexural specimen. The delamination microcracking in a flexural 0/90/90/0 AR glass fiber composite is shown in Figure 10.12. Note that large deflections are observed due to shear cracking and the loss of stiffness of the composites while the 0 degree

FIGURE 10.11   Ply delamination cracking in a flexural specimen (scale markers = 1 mm).

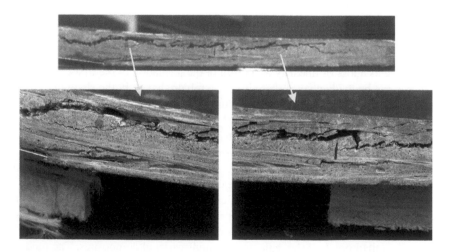

FIGURE 10.12   Delamination microcracking in a 0/90/90/0 AR glass fiber composite.

compression layers are still able to carry load. The $(\pm 45)_s$ specimens exhibited a low toughness of approximately 1/5 the uniaxial composites. This behavior was expected because fiber orientations were aligned in an off axis direction, which resulted in delamination.

The toughness of flexural samples was measured as the area under the load-deflection curve, which was divided by the cross-sectional area. The uniaxial specimens had the highest energy absorption capacity, whereas the $(0/90)_s$ specimens formed delamination cracks after the formation of transverse cracks in the 90° layers. When compared with short fiber composites, aligned-fiber laminates had a dramatically increased flexural toughness. Although the toughness is normally viewed as the work to fracture many specimens in our study failed to achieve a complete fracture.

REFERENCES

1. Balaguru, P. N., and Shah, S. P. (1992), *Fiber-Reinforced Cement Composites*, New York: McGraw-Hill Inc.
2. Li, C. Y. (May 1995), "Mechanical behavior of cementitious composites reinforced with high volume content of fibers," Doctoral dissertation, Arizona State University.
3. Mobasher, B., Pivacek, A., and Haupt, G. J. (1997), "Cement based cross-ply laminates," *Journal of Advanced Cement Based Materials*, 6, 144–152.
4. Pivacek A., and Mobasher, B., "A filament winding technique for manufacturing cement based cross-ply laminates," Innovations Forum, *ASCE Journal of Materials Engineering*, May 1997, 55–58.
5. Mobasher, B., and Pivacek, A. (1998), "A filament winding technique for manufacturing cement based cross-ply laminates," *Journal of Cement and Concrete Composites*, 20, 405–415.
6. Krenchel, H., and Stang, H. (1994), "Stable microcracking in cementitious materials," in Brandt, A. M., Li, V. C., and Marshall, I. H. (eds.), *Brittle Matrix Composites 1*, Cambridge, UK: Woodhead Publishing Limited, 20–33.
7. Haupt, G. J. (May 1997), "Mechanical properties of cement based composite laminates," MS thesis, Arizona State University.
8. Pivacek, A. (May 2001), "Development of a filament winding technique for manufacturing cement based materials," MS thesis, Arizona State University.
9. Mobasher, B., Peled, A., and Pahilajani, J. (2006), "Distributed cracking and stiffness degradation in fabric-cement composites," *Materials and Structures*, 39, 317–331.
10. Shah, S. P., Ludirdja, D., Daniel J. I., and Mobasher, B. (1988), "Toughness-durability of glass fiber reinforced concrete systems," *ACI Materials Journal*, Sept–Oct., 352–360.
11. Mobasher, B., and Shah, S. P. (1989), "Test parameters in toughness evaluation of glass fiber reinforced concrete panels," *ACI Materials Journal*, Sept–Oct., 86, 5, 448–458.
12. Brandt, A. M. (1995). *Cement Based Composites: Materials, Mechanical Properties, and Performance*. London: E & FN Spon.
13. Krenchel, H., and Stang, H. (1994), "Stable microcracking in cementitious materials," in Brandt, A M., Li, V. C., and Marshall, I. H. (eds.), *Brittle Matrix Composites 1*, Cambridge, UK: Woodhead Publishing Limited, 20–33.
14. Mobasher, B., Stang, H., and Shah, S. P. (1990), "Microcracking in fiber reinforced concrete," *Cement and Concrete Research*, 20, 665–676.
15. Mobasher, B., and Li, C. Y. (1996), "Effect of interfacial properties on the crack propagation in cementitious composites," *Journal of Advanced Cement Based Materials*, 4(3), 93–106.
16. Mobasher, B., and Li, C. Y. (1995), "Modeling of stiffness degradation of the interfacial zone during fiber debonding," *Journal of Composites Engineering*, 5, (10–11), 1349–1365.
17. Majumdar, A. J., and Laws, V. (1991), *Glass Fiber Reinforced Cement*, Oxford: BSP Professional Books.
18. Beaudoin, J. J. (1990), *Handbook of Fiber-Reinforced Concrete: Principals, Properties, Developments and Applications*, Park Ridge, NJ: Noyles Publications.
19. Bentur, A., and Mindness, S. (1990), *Fiber Reinforced Cementitious Composites*, London: Elsevier Science Publishers Ltd.

11 Inelastic Analysis of Cement Composites Using Laminate Theory

INTRODUCTION

The filament winding process allows for the development of cross-ply cement-based composite laminates using continuous alkali-resistant (AR) glass, polyethylene, and polypropylene fibers. Tensile stress–strain response, which are measured using closed-loop strain-controlled tests, have indicated that there is a significant amount of strengthening and ductility. The tensile strength of composites can exceed 50 MPa with a strain capacity of 1% to 2% using 5% AR glass fibers. The flexural strength exceeds 50 MPa as well. Experimental results have shown that varying the stacking sequences can increase the ultimate strain capacity and result in values as high as 2%. These results have been extended to sandwich composites containing a lightweight aggregate core composite. Proper modeling of these composites requires the step-by-step implementation of simplifying assumptions with an increasing level of complexity to address the various competing mechanisms.

This chapter applies the principles of composite laminate theory to unidirectional and cross-ply laminates. The modeling begins with the elastic analysis of a lamina, and continues with procedures for the calculation of stiffness parameters. The modeling here extends to the incorporation of a change of stiffness of the composite as a function of applied strain using a damage mechanics approach. As the failure of matrix phase is included, stiffness degradation is addressed, but only in the context of damage parameters. This procedure is followed by the demonstration of a failure of an entire lamina.

STIFFNESS OF A LAMINA

A theoretical model composed on the basis of the composite laminate theory can predict the response of composites subjected to axial and flexural loads. The formulation for the stiffness of a single ply consisting of matrix and fiber is used to calculate the initial linear elastic response of a lamina in tension and compare it with the experimentally obtained data to complete a parameter back-calculation. The procedure requires the foreknowledge of the elastic properties of each phase, in addition to the geometrical parameters and volume fraction of each phase. The elastic response of the ply to stresses parallel and normal to the fiber axis requires the axial and transverse Young's moduli (E_1 and E_2), the shear modulus (G_{12}), and Poisson's ratios. These parameters can be used in conjunction with failure theories to characterize the entire stress–strain response of the composite systems. The experimental results agree quite favorably with theoretical predictions on the basis of the ply discount method [1–3].

In this study, we used a general approach to the treatment of composites made with various fiber and matrix materials as continuous and cross-ply laminates. Each lamina is modeled as an orthotropic sheet in plane stress with Direction 1, representing the longitudinal direction of alignment of fibers, and Direction 2, representing the transverse direction, as shown in Figure 11.1. The parameters h_k and h_{k+1} represent the coordinates and the top and bottom of lamina number k in a stack of n laminates. Angle θ represents the orientation of the fiber with respect to the direction of

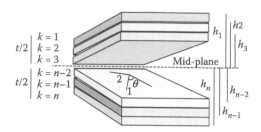

FIGURE 11.1 Definition of lamina and coordinates used in generating stiffness coefficients.

the application of load, so that a 0° lamina represents the load being applied in the direction of the fibers, and a 90° lamina represents the load being applied transverse to the direction of the fibers. The fiber is assumed to be linear elastic, and the effect of the fiber volume fraction is incorporated in the elastic properties of each lamina. On the basis of the layer model, the property of each layer is calculated using the material properties and volume fraction of components. Using the stacking sequence, the overall axial and bending stiffness matrices are obtained. The equivalent elastic stiffness of each lamina is obtained using the sum of the contributions from each phase to the overall value. Depending on the state of strain (normal and shear) and curvature distribution, strains at the top and bottom of the lamina are calculated. The strain distribution is then applied to the orthotropic model to calculate ply stress.

STIFFNESS OF A PLY ALONG MATERIAL DIRECTION

The stiffness of a single ply, in either axial or transverse directions, can be calculated using the elastic properties of individual phases. From these values, the stresses in a cross-ply laminate (when it is loaded parallel to or at an angle inclined to the fiber direction) can readily be calculated. By using the stiffness of the laminate, along with the applied strain in the loading direction, the average stress in each ply can be obtained. The stresses in individual fiber and matrix plies can also be derived from these average stresses and by imposing failure criteria using input values of fiber/matrix stiffness ratio and fiber content.

The constitutive relations for a general orthotropic material require that we calculate the compliance matrix, S, or its inverse, the stiffness matrix, Q, which relates the stress and the strain within a lamina loaded in its principal directions [4]. A uniaxial tension test of a 0° lamina is represented as:

$$\varepsilon_j = S_{ij}\sigma_i \quad \varepsilon_1 = S_{11}\sigma_1 \quad \varepsilon_2 = S_{12}\sigma_1 \; S_{11} = \frac{1}{E_1}; \quad S_{12} = -\frac{v_{12}}{E_1} \tag{11.1}$$

Knowing the material properties obtained from these elementary tests, we can generate the stress–strain relations for a general state of stress, as described in Equation 11.1. Alternatively, one can express the relationship in terms of the inverse of compliance, S, or the stiffness matrix, Q, which is defined as:

$$\begin{bmatrix} \sigma_1 \\ \sigma_2 \\ \tau_{12} \end{bmatrix} = S_{ij}^{-1}\varepsilon_j = Q_{ij}\varepsilon_j = \begin{bmatrix} Q_{11} & Q_{12} & 0 \\ Q_{21} & Q_{22} & 0 \\ 0 & 0 & Q_{66} \end{bmatrix} \begin{bmatrix} \varepsilon_1 \\ \varepsilon_2 \\ \gamma_{12} \end{bmatrix} \tag{11.2}$$

with its components:

$$Q_{11} = \frac{E_1}{1 - v_{12}v_{21}} \qquad Q_{22} = \frac{E_2}{1 - v_{12}v_{21}} \qquad Q_{12} = \frac{v_{12}E_2}{1 - v_{12}v_{21}} \qquad Q_{66} = G_{12} \tag{11.3}$$

For a composite laminate consisting of several lamina, each with a fiber orientation of θ^m, where m represents the first to the nth ply, classical lamination theory results in the derivation of laminate stiffness components as:

$$\bar{A}_{ij} = \sum_{m=1}^{n} \bar{Q}_{ij}^m (h_m - h_{m-1}), \quad \bar{B}_{ij} = \frac{1}{2} \sum_{m=1}^{n} \bar{Q}_{ij}^m (h_m^2 - h_{m-1}^2), \quad \bar{D}_{ij} = \frac{1}{3} \sum_{m=1}^{n} \bar{Q}_{ij}^m (h_m^3 - h_{m-1}^3) \quad (11.4)$$

The form of submatrices $\bar{A}, \bar{B},$ and $\bar{D}$ is discussed by Agarwal and Broutman [6], in which they indicate that $\bar{A}$ represents the extensional, $\bar{D}$ the bending, and $\bar{B}$ the coupling stiffness. With the knowledge of laminate strains and curvatures, the stress distribution per lamina can be computed for each loading step in an incremental fashion. M represents the moment per unit length, N represents the force per unit length of cross section, and ε^0 and κ represent the midplane strains and the curvature of the section, respectively the following examples demonstrate the use of Equations 11.1 through 11.4 in the context of unidirectional and cross-ply cement composite laminates.

PROBLEM 1

Compute the stiffness properties of a unidirectional cement composite with a thickness of $t = 24$ mm and a volume fraction of 4.8% of AR glass fibers. Assume all layers are homogeneous and discuss the variability of the test results and effect of various parameters. Use $v_m = 0.18$, $v_f = 0.2$, $E_m = 20,000$ MPa, and $E_m = 70,000$.

Solution: The rule of mixtures calculation on the basis of material properties and volume fractions are used initially:

$$E_c = E_f V_f + E_m (1 - V_f) = 70 \times 0.048 + 20 \times (1 - 0.048) = 22.4 \text{ GPa} \quad (11.5)$$

Sensitivity is analyzed by varying the fiber volume fractions. If the fiber is assumed in the range of 5% to 9%, then the elastic stiffness would be in the range of 22.5 to 25 GPa. Therefore, the measurement of the effective fiber content from the elastic modulus is not quite accurate because of the averaging effects. The analytical data are plotted against the experimental data obtained by Pivacek and Mobasher [2]. A stiffness of 23.5 GPa is obtained from experimental results that compare favorably with the rule of mixtures calculation, as is shown in Figure 11.2.

PROBLEM 2

Use the Halpin–Tsai equations to calculate the orthogonal properties and the laminate properties of Problem 1.

Solution: The stiffness of the lamina in the transverse direction is dominated by the matrix properties. Transverse modulus E_2 and v_{12} is achieved using the Halpin–Tsai equations by assuming a value of $\xi = 0.2$ [5]:

$$E_2 = \frac{E_m (1 + \xi \eta V_f)}{1 - \eta V_f} \quad \eta = \frac{E_f - E_m}{E_f + \xi E_m} \quad (11.6)$$

Using these parameters and the generalized forms of the Halpin–Tsai equations, one obtains a matrix dominated value of $E_2 = 20,804$ MPa, $v_{12} = 0.181$, and $G_{12} = 15,289$ MPa. Because the matrix is brittle, there is a range of applicability of these equations that must be evaluated in consideration to the strength of each layer. For a value of $V_f = 4.8\%$, one calculates the following elastic properties: $S_{11} = 44.64 \times 10^6$, $Q_{11} = 23,102$ as the dominant stiffness parameter relating applied strain to the axial force. The Q elastic matrix is defined and computed as:

$$\sigma_i = S_{ij}^{-1} \varepsilon_j = Q_{ij} \varepsilon_j \quad (11.7)$$

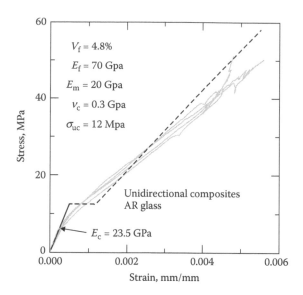

FIGURE 11.2 Comparison of the theoretical and mechanical response of a three ply 0° laminate with the ply discount method.

$$Q_{ij} = \begin{bmatrix} Q_{11} & Q_{12} & 0 \\ Q_{21} & Q_{22} & 0 \\ 0 & 0 & Q_{66} \end{bmatrix} = \begin{vmatrix} \dfrac{E_1}{1-v_{12}v_{21}} & \dfrac{v_{12}E_2}{1-v_{12}v_{21}} & 0 \\ \dfrac{v_{12}E_2}{1-v_{12}v_{21}} & \dfrac{E_2}{1-v_{12}v_{21}} & 0 \\ 0 & 0 & G_{12} \end{vmatrix} = \begin{vmatrix} 23102 & 3881 & 0 \\ 3881 & 21456 & 0 \\ 0 & 0 & 15289 \end{vmatrix}$$

One, furthermore, needs to specify the thickness and the number of layers: for $t = 24$ mm, $n = 3$, and for a unidirectional layer composite all three lamina are oriented in the same direction, so: $\theta_1 = \theta_2 = \theta_3 = 0°$. Parameters for A, B, and D matrices are obtained by:

$$A_{ij} = \sum_{m=1}^{n} \bar{Q}_{ij}^m (z_m - z_{m-1}) = \begin{vmatrix} 522480 & 86103 & 33469 \\ 86103 & 561010 & -29135 \\ 33469 & -29135 & 359899 \end{vmatrix}, \text{N} \cdot \text{mm}$$

$$B_{ij} = \sum_{m=1}^{n} \bar{Q}_{ij}^m (z_m^2 - z_{m-1}^2) = 0$$

$$D_{ij} = \frac{1}{3} \sum_{m=1}^{n} \bar{Q}_{ij}^m (z_m^3 - z_{m-1}^3) = \begin{vmatrix} 25.0790 & 4.1329 & 1.6065 \\ 4.1329 & 26.9285 & -1.3985 \\ 1.6065 & -1.3985 & 17.2751 \end{vmatrix} \times 10^6 \text{ N} \cdot \text{mm}$$

Note that if you assume 1, 2, or n layers, and assume that all the layers are oriented along 0, the parameters of the A, B, and D matrices are the same. These coefficients allow any stress or curvature value to be correlated with the resulting force or bending moment. For example, at a strain of 133 μstr (microstrains) and using the elastic stiffness of $A_{11} = 5.2248 \times 10^5$ N·mm, a force of 69.66 N·mm is obtained, which corresponds to a stress level of 69.66/24 mm = 2.9 MPa.

Assuming that the matrix may fail under a plain tensile strength between 4 and 6 MPa, one can limit the range of applicability depending on the approach used in modeling, whether it is using the plain matrix properties or the enhanced matrix properties due to the toughening effect of fibers. Beyond the matrix cracking stress, any load is transferred to the fibers, and, if the stiffness of the fiber phase is sufficiently high, then the composite continues to resist the load. This procedure can be implemented in the context of setting the stiffness of the matrix as zero and calculating the effective stiffness of the composite as the stiffness of the fiber phase, or $V_f E_f$.

PLY DISCOUNT METHOD

In the elastic range, one can apply the rule of mixtures to the longitudinal modulus and the Halpin–Tsai [1] estimates of the transverse modulus. The elastic zone is terminated by the cracking of the matrix phase using a stress-based criterion [6], which is designated as σ_{t1}.

The ply discount method updates the stiffness of the composite by accounting for the surviving laminates. This method is a conservative way to update the stiffness, and does not take into account the effect of matrix tension stiffening. Using this method, one can assume that some or all of the matrix exhibits cracking and, therefore, a total loss of stiffness is imposed. The case is demonstrated by following up with the Problem 2 discussed in the previous section. Parameter a_{11} is the primary stiffness coefficient of the composite; therefore, one can assume and calculate updated values of a by allowing for failure of 1, 2, or all of the layers. This would require the calculation of Equation 11.1 under the assumption that $E_m = 0$ for any layer that is marked as having failed. Under these assumptions, the stiffness drops from 5.2248×10^5 to 3.5395×10^5 N/mm for a one layer failure and 1.8543×10^5 N/mm and 16,917 N·mm for a failure mechanism of two and three layers. Note that in the case of all layers failing in a multiple ply composite, it is only the stiffness of the fiber phase that contributes to carrying the load. Figure 11.3 compares these three assumptions for the overall nominal stress vs. strain of the sections.

Figure 11.4 shows the application of the ply discount method to two different cross-ply laminates, the first being a 90/0/90 and a 0/90/90/0. Note that the mechanical responses of the two systems are similar in the postcracking region and that their stiffnesses are more or less similar. We should also note, however, that the application of a single 90/0/90 model on the basis of the ply discount method allows for the development of the two upper, and lower bounds of the response of these systems.

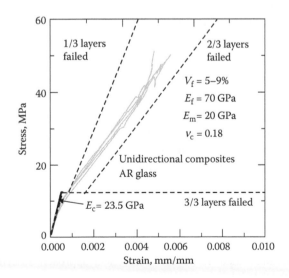

FIGURE 11.3 Failure of a three ply 0° laminate under the assumptions of a one, two, or three layer failure.

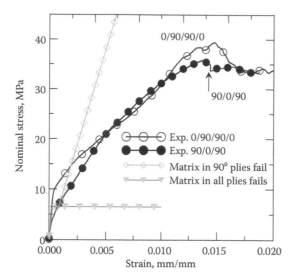

FIGURE 11.4　Comparison of the mechanical response of a composite with a 90-0-90 and a composite with 0-90-90-0 and the ply discount method.

DAMAGE-BASED MODELING USING A NONLINEAR-INCREMENTAL APPROACH

In this section, a damage-based formulation is used to compute the effective stiffness of a degrading matrix. The material degradation is formulated by means of a scalar damage parameter. Figure 11.5 shows three distinct modes of failure: delamination, unbridged matrix cracks, and bridged matrix cracks. In Figure 11.5a, the matrix cracking and ply delamination failure modes are shown in a cross-ply laminate. Figure 11.5b shows the cracking within a single unidirectional lamina. The cracking in case (a) is assumed to result in unbridged cracks in which the effect of the shear transfer of forces is quite important. In the case of Figure 11.5b, the cracks are bridged, and toughening mechanisms are dominant.

Two different modeling approaches can be used, as shown in Figures 11.6a and 11.6b. In Method 1, which is featured in Figure 11.6a, a shear-lag-based formulation is used to incorporate the load redistribution between fibers and matrix. This modeling has been effectively conducted using a finite difference approach [7]. Using a linear elastic matrix and fiber and a nonlinear interface law, one can develop algorithms to extract the tension stiffening effect of fibers. In the second approach, as shown in Figure 11.6b, the composite laminate theory is used, and the matrix is assumed to behave as an elastic-degrading-softening solid, while the fiber can be treated as an elastic. There is no need to incorporate an interface formulation in this approach. In this manner, a direct correlation can be developed so that instead of modeling for the tension stiffening, one uses an assumption on the basis of the strain softening response of the matrix.

In the procedures described earlier for the ply-discount method, the gradual stiffness degradation was not taken into account. Once a layer failed, the stiffness of the matrix phase was set equal to zero for that layer and updated stiffness values were calculated for the lamina. Alternatively, one can approach the problem by allowing for a gradual decrease in the stiffness in a method proposed by Mobasher et al. [8] where they present several case studies and theoretical results are compared with experimentally obtained data. In another theoretical approach described by Mobasher [9], the degradation of stiffness is taken into account by using a strain-based scalar damage-softening model. The load-carrying capacity of the matrix phase in each lamina decreases after cracking and

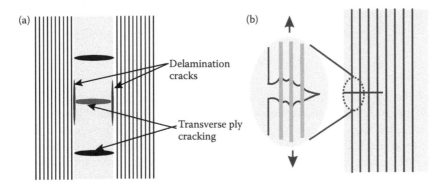

FIGURE 11.5 (a) Two modes of failure in laminate composites; (b) mode of matrix cracking and fiber debonding in a single ply.

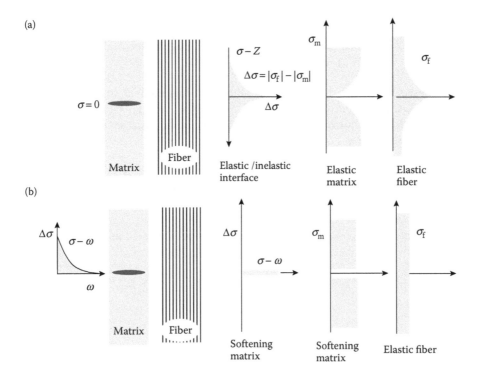

FIGURE 11.6 Two approaches for the modeling nonlinear effects of composite laminates: (a) a fiber, matrix, and interface approach based on shear lag; (b) a fiber–matrix approach base on matrix strain softening applicable to composite laminate theory.

the stiffness of the lamina degrades such that the composite response asymptotically approaches levels predicted by the ply discount method.

Three zones of behavior need to be considered for the matrix phase. These include the elastic range, the stiffness degradation due to initiation and generation of parallel crack formations, and the strain softening range that leads to tension stiffening. The tension softening must be used to address the effective stress transfer across the sample. Alternatively, one can assume that the matrix does not have any postpeak strength contribution and obtain the tension stiffening indirectly by taking into account the shear lag effects.

The stiffness degradation can be addressed using to a single scalar damage parameter ω. The form of the evolution of the damage parameter as a function of strain is expressed as

$$\omega_i = \omega_1 + \alpha(\varepsilon_1 - \varepsilon_{um})^\beta \quad \sigma_{t1} < \sigma_1 < \sigma_{um} \tag{11.8}$$

The form of the function in Equation 11.8 was based on a model proposed by Karihaloo and Bhushan [10] and relates the damage versus strain as shown in Figure 11.7. The damage parameter ω is calculated at various strain levels with constants α, β, H, and ω_1. The values of these constants are of $\alpha = 0.16$, $\beta = 2.3$, $\omega_1 = \varepsilon_{t1}$, and $H = 0.05$, where H is the gage length of the specimen used. σ_{t1} and $\varepsilon_{t1} = \sigma_{t1}/E_{m0}$ were used to represent the ultimate strength and strain at failure under uniaxial tension for the paste in an unreinforced condition. Within the cracked matrix range, as the strain is increased, the stiffness of the matrix decreases in terms of a damage evolution law of Horii et al. [11] and also of Nemat-Nasser and Hori [12] to estimate the degradation of stiffness as a function of strain. The stiffness defined as a function of damage is $E_m(\omega)$ and expressed in Equation 11.9 as a function of uncracked matrix elastic modulus E_{m0}:

$$E_m(\omega) = E_{m0}\left[1 + \frac{16}{3}\omega(1-v_m^2)\right]^{-1} \tag{11.9}$$

This value is used in the rule of mixtures to obtain the longitudinal stiffness $E_1(\omega)$, as defined in Equation 11.3. Calculation of the transverse modulus E_2 and v_{12} can be followed using the Halpin–Tsai equations as shown in Equation 11.6. The value of ξ was set equal to 2 in the present study [6],

$$E_1(\omega) = E_f V_f + (1-V_f) E_{m0}\left[1 + \frac{16}{3}\omega(1-v_m^2)\right]^{-1} \tag{11.10}$$

The stress in the matrix phase beyond the elastic range is calculated incrementally as:

$$\sigma_1^i(\omega) = \sigma_{t1} + \sum_{n=1}^{i} E_m(\omega)(\varepsilon_n - \varepsilon_{n-1}) \quad \varepsilon_i < \varepsilon_{mu} \tag{11.11}$$

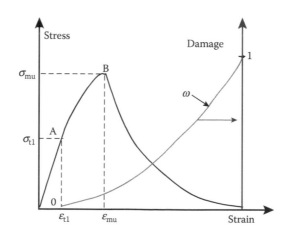

FIGURE 11.7 The stiffness degradation as a function of damage parameter ω.

Equation 11.11 computes the stress using an incremental approach that adds the contribution of strain increments at each level. In this manner, the degraded stiffness at each strain value is updated. The definition of strain in this region is gage length dependent, and the present approach uses the mean strain over the length of several cracks in the matrix.

FAILURE CRITERIA FOR LAMINA

The $0°$ plies may be subjected to significant cracking transverse to the loading direction. The matrix phase in the $90°$ plies loaded in tension may also be subjected to parallel cracking in response to the shear lag of adjacent layers. Because of the bridging effect of fibers, a cracked matrix in a ply may still carry a significant amount of stress, whereas a cracked matrix in a $90°$ layer may be stress free due to the lack of fiber bridging. Therefore, the initial cracking and the final cracking of the matrix must be differentiated. Additionally, the complete failure of lamina as a response to the failure of the fiber phase must also be considered. The failure criteria for the first cracking of the matrix and the final cracking of the matrix based on the state of stress and represented as the yield surface, F_1 and F_2, are as follows:

$$F_1(\sigma_1, \sigma_2, \tau_{12}) = 1 \qquad \sigma_1 \geq \sigma_{t1} \qquad \sigma_2 \geq \sigma_{t2} \qquad \tau_{12} \geq \tau_{12} \tag{11.12}$$

$$F_2(\sigma_1, \sigma_2, \tau_{12}) = 1 \qquad \sigma_1 \geq \sigma_{mu} \qquad \sigma_2 \geq \sigma_{t2} \qquad \tau_{12} \geq \tau_{12} \tag{11.13}$$

After each incremental loading, the stresses are checked against the failure surface to update the material properties for the subsequent iteration. The second form of the yield surface F_2 can be used to address the strength of the matrix in the presence of fibers or σ_{mu}. After the cracked matrix is updated in the stiffness equations, the ultimate tensile strength is set equal to the strength of the fiber phase, and represented as:

$$\sigma^{ut}(\theta) = \max (V_f \, \sigma^{fu} \cos^2 \theta, \sigma^{t2}) \tag{11.14}$$

In addition, for an off-axis lamina subjected to shear, the matrix phase may have a brittle failure because of shear cracking with little or no softening response.

GENERALIZED LOAD DISPLACEMENT FOR THE COMPOSITE RESPONSE

Because the incremental approach updates the elastic stiffness of the matrix due to cracking, an elastically equivalent compliance matrix $\bar{S}$ needs to be defined in which the bar indicates the use of updated elastic properties. In the term, S_{jk}^i, parameter i represents the load increment, j is the direction of applied the strain, and k is the stress component. The stress–strain relationship is represented in an incremental form for each loading increment i, as:

$$\Delta \varepsilon_j^i = \bar{S}_{jk}^i \, \Delta \sigma_k \qquad \sigma_k^i = \left(\bar{S}_{jk}^i \right)^{-1} \Delta \varepsilon_j^i + \sigma_k^{i-1} \tag{11.15}$$

The strains and forces are updated incrementally in accordance with the matrix form representation:

$$\begin{bmatrix} \Delta N \\ \Delta M \end{bmatrix} = \begin{bmatrix} \bar{A} & \bar{B} \\ \bar{B} & \bar{D} \end{bmatrix} \begin{bmatrix} \Delta \varepsilon^0 \\ \Delta \kappa \end{bmatrix} \tag{11.16}$$

For each iteration, the incremental loads and strains are determined and update the loads and strains of previous increments. The applied load in the x direction at the ith interval of the jth lamina is represented as $N_{x,i}^j$ according to:

$$N_{x,i}^j = N_{x,i-1}^j + \Delta N_{x,i}^j = N_{x,i-1}^j + \left[\bar{A}\right]_i \left[\Delta \varepsilon^0\right] \tag{11.17}$$

Similarly,

$$M_{x,i}^j = M_{x,i-1}^j + \Delta M_{x,i}^j = M_{x,i-1}^j + \left[\bar{D}\right]_i \left[\Delta \kappa\right] \tag{11.18}$$

The solution algorithm imposes the strain and curvature distributions incrementally. At each increment of the strain, the updated stiffness is used to calculate the stress. The stress is checked against the failure criteria for the plain matrix failure, the bridged matrix failure, and the composite failure. If the failure criteria are met, then the stress level and the stiffness of that layer are adjusted according to the constitutive response. Subsequent loading of a cracked layer results in a change in the magnitude of the damage parameter. This indicates that at any stress level, the degradation of elastic properties is primarily related to the magnitude of crack density and overall strain response. Using the updated damage parameter, the quasi-elastic stiffness parameters $\bar{A}$, $\bar{B}$, and $\bar{D}$ are obtained and used to calculate the load and moment for that increment. The procedure is repeated for the next strain increment. A complete description and the parametric evaluation of the model are provided elsewhere [9].

PERFORMANCE OF MODEL: SIMULATION OF TENSILE LOAD

Several case studies involving various systems are presented in order to evaluate the applicability of the model to composites under tension and bending. Figure 11.8 represents the effect of a fiber volume fraction on the tensile response of the composite system. Note that as the fiber volume fraction is increased, there is a significant increase in load-carrying capacity. The stiffness degradation

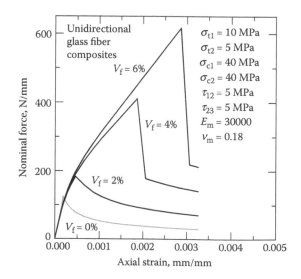

FIGURE 11.8 Parametric study of the effect of fiber content on the gradual degradation of the matrix phase. (Data from Mobasher, B., "Micromechanical modeling of filament wound cement-based composites," *ASCE, Journal of Engineering Mechanics*, 129(4), 373–382, 2003.)

is clearly observed since the stiffness of the matrix phase is allowed to degenerate as opposed to suddenly changing in accordance to the ply discount method. The shape of the postpeak response is determined by the assumptions in the modeling of the stress decay in accordance to the stress crack width law used.

Figures 11.9a and 11.9b present the simulated results for 0/90/0 stacked laminates subjected to a uniformly applied tensile strain level. A constant strain level is imposed across the depth of the cross section, as shown in Figure 11.9b, and this strain is associated with the stress distribution across the section at each depth level. As seen in Figures 11.9a and 11.9b, the cracking starts with the matrix cracks forming in the 0° and 90° layers. Damage is allowed to accumulate in the 0° layers because of multiple matrix cracking in accordance to the damage evolution law. The loading in the transverse direction (90° layers) is limited to the ultimate tensile strength, σ_{2t}. Note that as the fiber volume fraction is increased, the response of specimens in carrying the forces and distributing the cracks beyond the initial cracking phase are also enhanced. As the damage accumulation increases, it results in a reduction of the stiffness for the overall composite. The load-carrying capacity of the composite extends well beyond the matrix-cracking point, and, as damage accumulates, stiffness decays. The stress in the longitudinal layers increases to a maximum level determined by the fiber fracture strength, or an effective strength determined by $V_f\sigma_{fu}$. The Successive failures of the 0° and 90° layers are apparent in the angle-ply samples, which are shown in Figure 11.3b. Note that in the transverse direction, both stiffness and strength are significantly lower than the 0° layers. The stiffness degradation that is due to damage results in a nonlinear response, is also shown in the load versus deformation response. However, this is not clearly visible because of the high relative stiffness of glass as compared with the cement matrix. Other systems such as $[0/+45/-45]_s$ and $[0/45/-45/90/90]_s$ glass cement systems have been addressed elsewhere by Mobasher in detail [9].

Figure 11.10 represents a comparison of the theoretical predictions with the experimental results for continuous AR glass fiber systems for both unidirectional and cross-ply lamina. The experimental procedures are described in detail by Mobasher and Pivacek [1]. A uniform strain is imposed in the Principal Material Direction 1 across the 18-mm thickness of the cross section at several stages. As the ultimate strength of the matrix phase is reached, there is a shift in the slope of the stress–strain response, also known as the bend over point (BOP). The load-carrying capacity extends well beyond the matrix-cracking phase and continues as damage accumulates the stiffness decays. Results are also compared with the response of a $[0/90]_s$ stacked lamina ($V_f = 9\%$). The loading in the 90° layers is limited to the ultimate tensile strength σ_{2t}. This results in a lower stress in the

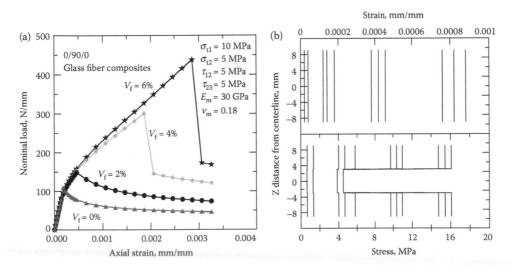

FIGURE 11.9 Comparison of model predictions with experiments for [0/90/0].

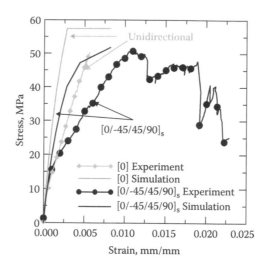

FIGURE 11.10 Comparison of model predictions with experiments for unidirectional and [0/90]$_s$ glass cement systems.

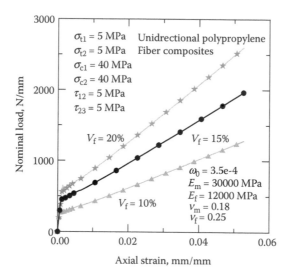

FIGURE 11.11 Parametric study of effect of lamina orientation on the mechanical response. Unidirectional, 0/90/0, and 90/0/90 glass cement systems are compared.

90° layers. The maximum load is attained when the stress in the remaining 0° longitudinal lamina reaches a stress equal to the effective strength of the fiber phase or $V_f\sigma_{fu}$.

The model was further extended to composites with fibrillated polypropylene fibers. The values $E_m = 30,000$ MPa, $E_f = 7000$ MPa, $\nu_m = 0.18$, $\nu_f = 0.25$, and the lamina strength $\sigma_{t1} = \sigma_{t2} = 6$ MPa are used. Figure 11.11 represents the model predictions for the response of unidirectional (0), 0/90/0, and 90/0/90 laminates with polypropylene fiber composites. There is a major drop in the stiffness of the composite as the strength of the matrix is reached at the BOP. This is attributed to the low stiffness of the polypropylene fibers. As a 0° lamina is replaced by 90° layers, it is observed that both the first crack strength and the post-BOP stiffness drop markedly. However, the benefit of this layup arrangement is found in improvements in the transverse properties of the layers. The response exhibited in the 0/90/0 and the 90/0/90 laminates demonstrates the behavior of an ideal composite for use under a biaxial loading condition, because both transverse and longitudinal directions are ductile and strong, whereas the 0° laminates show a very strong and ductile response in the longitudinal direction, even though the transverse response is brittle.

Figure 11.12 compares model predictions with experiments for unidirectional polypropylene fiber composite systems [7]. Similar to the case of glass fabrics, at the fiber volume fraction of 6% PP Fibers, a BOP strength level of 8 MPa is obtained. Because of the high ultimate strain capacity of the polypropylene fibers, the overall strain in the sample may be of the order of several percent. The choice of the crack spacing-stiffness degradation model in the matrix is quite important in the response of these composites (Figure 11.13). The results shown here are for a constant strain softening coefficient of $\omega = 50$, $\alpha = 5$, and $\beta = 0.8$, which is used in Equation 11.8.

Response simulation of the shear response is not very promising, since elastic properties and failure responses are not modeled very accurately. Figure 11.14 shows the simulation of the shear strength results of +45 layers. Note that the effective strength is quite low in the range of 2 MPa, the

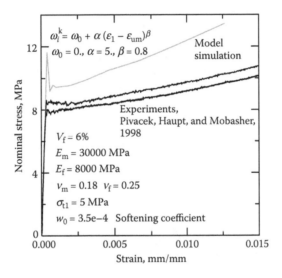

FIGURE 11.12 Comparison of model predictions with experiments for unidirectional polypropylene fiber composite systems. (From Pivacek, A., and Mobasher, B., "A filament winding technique for manufacturing cement based cross-ply laminates," Innovations Forum, *ASCE Journal of Materials Engineering*, 55–58, May 1997; Mobasher, B., and Pivacek, A., "A filament winding technique for manufacturing cement based cross-ply laminates," *Journal of Cement and Concrete Composites*, 20, 405–415, 1998. With permission.)

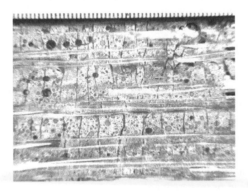

FIGURE 11.13 Parallel cracking in unidirectional PP fiber cement composites (marker = 1 mm).

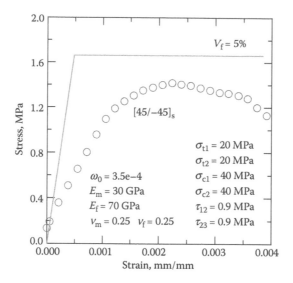

FIGURE 11.14 Simulation of the tensile response of +45 composite laminate. (From Soranakom, C., and Mobasher, B., "Geometrical and mechanical aspects of fabric bonding and pullout in cement composites," *Materials and Structures*, 42, 765–777, 2009. With permission.)

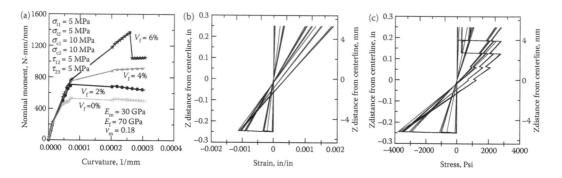

FIGURE 11.15 Model predictions of the moment curvature response for unidirectional composites containing a different volume fraction of fibers, (a) effect of fiber volume fraction, (b) stages of strain distribution, and (c) stages of stress distribution.

stiffness is overestimated by the composite laminate theory, and the postpeak response is not well modeled using a straight strength criterion.

SIMULATION OF FLEXURAL RESULTS

Figure 11.15 shows the model predictions of the moment–curvature response for unidirectional composites containing a different volume fraction of fibers. The flexural moment–curvature response of a unidirectional laminate with AR glass fibers shows distinct levels of cracking because of the failure of lamina in tension. The moment–curvature responses indicate the improved deformation capacity of composites with higher fiber fractions. Significantly, as the fiber volume fraction increases, the initial stiffness remains the same even though the point of the first cracking is increased. Above a certain critical level of fibers, it is possible for the composite to carry loads beyond the first cracking load or proportional elastic limit. Also, note that as cracking takes place, the nominal stiffness decreases. The effect of stages of strain and stress distribution as a function of sample depth are shown in Figures 11.15b and 11.15c. In the case of flexural loading, the neutral axis of the beam would move toward the compression region in the case of matrix cracking. The computation of the neutral axis would require an additional equation to be developed and solved for internal equilibrium. This nonlinear equation is expressed in terms of the normal stress distribution for an assumed neutral axis depth and can be calculated in the following way:

$$f(z) = \sum_{j=1}^{n} N_{x,i}^{j} = \sum_{j=1}^{n} N_{x,i-1}^{j} + \sum_{j=1}^{n} \Delta N_{x,i}^{j} = \sum_{j=1}^{n} N_{x,i-1}^{j} + \sum_{j=1}^{n} \left[\bar{A}\right]_i \left[\Delta \varepsilon^0\right]^j = 0 \qquad (11.19)$$

To solve Equation 11.19, the strain distribution is obtained based on the assumed neutral axis. Using this strain distribution and the values of effective stiffness, the parameters of axial stiffness, A_{ij}, are obtained and used to calculate the stress distribution. Integrating the stress distribution across the depth of the section, as shown by the summation sign from $j = 1$ to n, results in the axial force distribution. The location of the neutral axis is obtained by setting this axial force equal to imposed axial force, which, in a case of plain bending, is equal to zero.

The simulation of the flexural load-deflection response of a unidirectional laminate is shown in Figures 15. The various stages of loading are obtained by increasing the magnitude of strain that changes linearly across the thickness of the specimen. The longitudinal stress distribution results in cracking in the tension zone and is followed by distributed cracking and strain softening. The compression zone is assumed to behave elastically. In the present analysis, the neutral axis is obtained by solving for the equilibrium of internal forces. Using the location of the neutral axis and the strain at the extreme fiber, the resulting moment–curvature response of the cross section can be obtained by integrating the first moment of the stress distribution through the thickness. The curvature distribution is obtained from the strain magnitude.

The simulation response of a unidirectional specimen is compared with a [0/90/90/0] composite in Figure 11.16. The cross-ply laminated composite exhibits cracking and a loss of its load-carrying capacity. This leads to nonlinear behavior, which is shown by multiple cracking and stiffness reduction. The unidirectional sample is twice stronger in flexure, a property that is well captured in the modeling work.

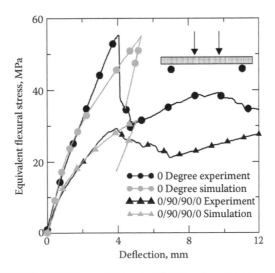

FIGURE 11.16 Comparison of model predictions with experiments for unidirectional and [0/90/90/0] glass cement systems.

REFERENCES

1. Mobasher, B., Pivacek A., and Haupt, G. J. (1997), "Cement based cross-ply laminates," *Journal of Advanced Cement Based Materials*, 6, 144–152.
2. Pivacek, A., and Mobasher, B. (1997), "A filament winding technique for manufacturing cement based cross-ply laminates," Innovations Forum, *ASCE Journal of Materials Engineering*, 55–58.
3. Mobasher, B., and Pivacek, A. (1998), "A filament winding technique for manufacturing cement based cross-ply laminates," *Cement and Concrete Composites*, 20, 405–415.
4. Jones, R. M. (1975), *Mechanics of Composites Materials*, New York: McGraw-Hill Book Co.
5. Halpin, J. C., and Tsai, S. W. (1967), "Environmental Factors in Composite Materials Design, Air Force Materials Research Lab," Technical Report, AFML-TR-67–423.
6. Agarwal, B. D., and Broutman, L. J.(1990). *Analysis and Performance of Fiber Composites*, 2nd ed., New York: Wiley.
7. Soranakom, C., and Mobasher, B. (2009), "Geometrical and mechanical aspects of fabric bonding and pullout in cement composites," *Materials and Structures*, 42, 765–777.
8. Mobasher, B., Pahilajani, J., and Peled, A. (2006), "Analytical simulation of tensile response of fabric reinforced cement based composites," *Cement and Concrete Composites*, 28(1), 77–89.
9. Mobasher, B. (2003), "Micromechanical modeling of filament wound cement-based composites," *Journal of Engineering Mechanics*, 129(4), 373–382.
10. Karihaloo, B. L. (1995), *Fracture Mechanics and Structural Concrete*, Harlow, Essex, England: Longman Scientific & Technical.
11. Horii, H., Hasegawa, A. and Nishino, F. (1987), "Fracture process and bridging zone model and influencing factors in fracture of concrete," in Shah, S. P. and Swartz, S. E. (eds.), Fracture of Concrete and Rock SEM-RILEM Int. Conf., Houston, 1989, 205–219.
12. Nemat-Nasser, S., and Hori, M. (1993), *Micromechanics: Overall Properties of Heterogeneous Materials*. North-Holland Publishing Co., 687.

12 Tensile and Flexural Properties of Hybrid Cement Composites

INTRODUCTION

The limited strain capacity of cementitious materials makes the tension weak, brittle, and considerably notch sensitive. Research on composites reinforced with steel and polymeric fibers like PP show that at the low-dosage levels, fibers do not increase the tensile or flexural strength [1]. Composite toughness, however, is directly related to the fiber pullout mechanism so that superior improvements in stiffness, strength, and ductility are reported as the amount of fibers increases and the specific fiber spacing decreases.

Because various fibers have inherently different properties, hybrid reinforcement provides an added dimension to the design of fiber reinforced concrete (FRC) composites. Significant strength increase is observed with the addition of short whiskers, including carbon, alumina, and steel. Because of the inability to bridge the macrocracks, the whisker's contribution is limited to increasing the ultimate strength. Since the postpeak toughening effects are limited, fracture toughness may be increased by the addition of polymeric fibers like polypropylene (PP). The pullout of PP fibers from the matrix is the main toughening mechanism and results in significant energy dissipation. Hybrid composites like carbon, alumina, and PP fibers are shown to have superior properties when compared with either reinforcement. The Tensile and flexural properties of cementitious composites, which are reinforced with carbon and alumina whiskers in addition to PP fibers, are presented to address the following research objectives:

1. To examine the strength enhancement of microfibers in FRC materials.
2. To examine the feasibility of manufacturing a high V_f short fiber composite using various methods, including high-energy shear mixing, extrusion, and compression molding (CM).
3. To examine hybrid fiber composites using whiskers and ductile fibers to optimize both strength and toughness.

Because of the heterogeneous microstructure, the fracture process is associated with discontinuous microcracking around the strong phases that bridge the unbroken ligaments. The inelastic energy dissipation results in a marginal toughening of the matrix phase in a region referred to as the Fracture Process Zone. If the specific fiber spacing and the stiffness of fibers bridging the microcracks are high enough to result in crack bridging mechanisms in the process zone, then fiber reinforcement results in a marked increase in the tensile strength and ductility [2]. As the sample undergoes stable crack growth, energy dissipation in the process zone continues to increase the load-carrying capacity, even in the prepeak region.

Random fibers intersect the cracks, debond, and pullout, resulting in energy absorption and toughening [3]. This effect is normally measured by evaluating the postpeak region of the load–deformation response in the tension or bending mode. As fibers inhibit crack localization, a homogeneous state of microcracking is formed within the stressed material. By controlling the microcracks at a microscopic level, the macroscopic constitutive response is enhanced and dissipates energy over the entire volume [4].

High volume of fibers in the process zone can dramatically influence the failure mode of the matrix, increase the ultimate tensile strength up to fourfold, and enhance the strain capacity by

211

orders of magnitude. The toughness is increased by orders of magnitude as well [5–7]. These toughening mechanisms have been observed for composites with continuous steel, glass, and fibrillated
PP fibers. This chapter documents various mechanisms that exist and uses a hybrid approach to
incorporate both strengthening and toughening modes.

Results indicate that the addition of short whiskers increases the flexural and tensile strengths
in of cement-based composites in addition to enhancing its ductility. The most profound of this
strengthening effects can be observed in the increased load-carrying capacity. The toughening
effects of alumina and carbon fibers are not significant. The addition of PP fibers results in hybrid
composites with both high strength and toughness.

MANUFACTURING TECHNIQUES AND MATERIALS

The effects of various manufacturing techniques for short fiber composite systems are best addressed
by studying the mechanical properties of composites reinforced with high volumes of different brittle and ductile fibers. Because various fibers are inherently different in their mechanical response,
hybrid reinforcement can provide an added dimension to the design of FRC composites.

The use of two types of brittle, short fibers, including carbon whiskers and alumina-based fibers,
in conjunction with PP fibers is presented first. The choices of alumina fibers are few. Use of these
fibers in cement-based materials has been limited. The main reinforcing potential is in providing early age strength for concrete when the marginally high tensile stresses can lead to plastic
cracking. High purity, kaolin-based alumina-silicate fibers were obtained from Carborundum Corp.
(Fiberfrax® fibers; Niagra Falls, New York). Physical and mechanical properties of the alumina
fibers are shown in Table 12.1. Carbon whiskers are also evaluated for their ability to increase the
strength of cement composites. By combining carbon or alumina and PP fibers, the is a potential for
producing composites with both improved strength and toughness.

EXPERIMENTAL PROGRAM

Flexural and tensile test coupons were prepared with a mortar matrix containing different types
and volume fractions of fibers. Specimen dimensions for both three-point bending and direct tension were $25.4 \times 76.2 \times 330.2$ mm ($1 \times 3 \times 13$ in.) [8]. Two types of brittle microfibers, namely,
alumina and carbon, in addition to ductile PP fibers, were used. High purity kaolin-based alumina-
silicate fibers were obtained from Carborundum Corp. (Fiberfrax®). Carbon fibers were obtained
from Ashland Chemical (Lexington, KY). PP fibers were Krenit fibers obtained from Danaklon,
Denmark [9]. Physical properties of the fibers are shown in Table 12.1 [10].

Mixtures were tested for three-point bending, flexural cyclic, double-edge tension, and unnotched
uniaxial tension. Morphology of the fracture surface, in addition to the fiber pullout and fracture
mechanisms, was studied by scanning electron microscopy.

TABLE 12.1
Physical Properties of Fibers Used in Hybrid Reinforcement Study

Fiber		E (GPa)	$\sigma_{ultimate}$ (MPa)	Density (g/cm^3)	Length (mm)	Diameter (μm)
Alumina	Carborandum	105	1725	2.70	0.762	2.5
PP	Krenit Fibers, Denmark	8.5–12.5	340–500	1.70	12.0	35×250
Carbon	Ashland Chemical	230	1800	1.90	1.0	25
Alkali-resistant glass		70	500	2.0	cont.	12
Steel	Novocon Fibers	230	345	7.8	3.0	500

SPECIMEN PREPARATION

The control mortar matrix was composed of type I/II cement, sand, and silica fume with a weight proportion of 1:0.8:0.15. The nominal maximum particle size of sand was 4.75 mm. Water-to-cementitious solids ratio for all mixes was 0.3. Details of the mixture design are provided by Mobasher and Li [11].

Conventional drum-type concrete mixers are unable to accommodate high-volume fractions of fibers. Thus, a high-energy omni mixer that uses a flexible rubber casing was used, because the flexibility of the mixing chamber allows for the breakup of the clumps and contributes to the mixing action [12]. The vigorous shearing action also enabled homogeneous distribution of fibers. Using this mixer, up to 16% volume fraction of carbon fibers and 8% volume fraction PP fibers could be easily mixed.

FLEXURAL THREE-POINT BENDING TESTS

Monotonic flexural and cyclic loading–unloading tests were carried out on three-point bending specimens using the Crack Mouth Opening Displacement (CMOD) as the control variable as discussed in Chapter 7. The stiffness was defined as the slope of the load–deformation curve in the initial linear range. The first crack load was defined as the point at which the load–deformation response deviated from linearity. Toughness, G_f, was defined as the area under the load–deflection curve normalized with respect to the cracked ligament.

DIRECT TENSION TESTS

Uniaxial tension tests were conducted for both double-edge notched and unnotched specimens under strain-controlled conditions as discussed in Chapter 7. The nominal stress at the point of deviation of the stress–strain response from linearity was used as the first cracking point. The average strain was obtained using the elongation measurement over a fixed gage length. Note that after the first cracking takes place, the strain measurement defined above is gage length-dependent, with the majority of the deformation localized as the crack opening. The toughness of the composite was defined as the area under stress–strain curve.

BRITTLE FIBERS

The load versus CMOD plot of a specimen with 8% carbon fibers is compared with plain mortar in Figure 12.1. The shapes of these curves are similar and show the increase in load-carrying capacity. An addition of 8% carbon fibers increases the peak load by 89%, as compared with the mortar specimens. The response is quasi-linear up to the peak load, and the specimens retain a major portion of their tangential stiffness. The first crack load increases up to approximately 85% of the peak load for fiber composites as well. Despite of the high-volume fraction of fibers, the load-carrying capacity decreases rapidly in the postpeak region. The brittle response is attributed to the short length of chopped fibers that affect the prepeak microcracking. Whiskers stabilize the microcracks that occur before the peak load and, thus, increase the ultimate strength. Mechanisms of fiber debonding and pullout seem to be marginal in these composites, because the short length of fibers cannot provide significant bridging across the crack faces.

A similar response in is shown in Figure 12.2, where the load–deflection plots of flexural cyclic specimens with 2% and 6% alumina fibers are compared with plain mortar. The response is quasi-linear up to the peak load, and the specimens retain a major portion of their tangential stiffness. Whiskers increase the ultimate strength by stabilizing the microcracks such that addition of 6% alumina fibers increases the peak load by 100%. Short length of chopped fibers can only affect the prepeak microcracking zone. Because of the localization of failure in the plain matrix, the lowest

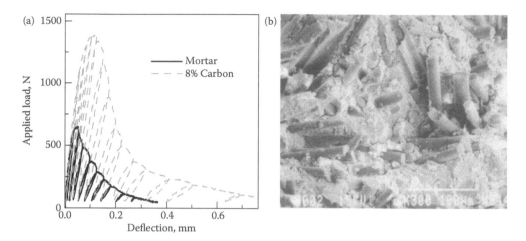

FIGURE 12.1   (a) Cyclic flexural load–deflection response of composites with 8% carbon compared with mortar. (b) Fracture surface of a carbon fiber cement composite containing 16% volume fraction. (Adapted from Mobasher, B., and Li, C. Y., "Mechanical properties of hybrid cement based composites," *ACI Materials Journal*, 93(3), 284–293, 1996.)

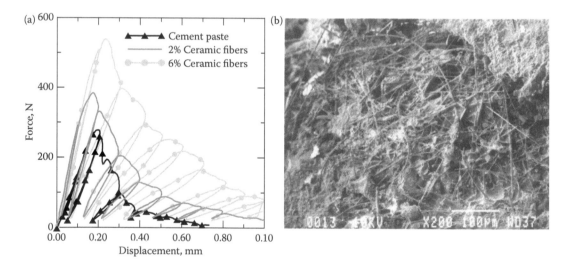

FIGURE 12.2   (a) Cyclic flexural load–deflection response of composites with 2% and 6% alumina fibers compared with mortar. (b) Fracture surface of an alumina fiber reinforced cement-based composite containing 6% volume fraction. (Adapted from Mobasher, B., and Li, C. Y., "Mechanical properties of hybrid cement based composites," *ACI Materials Journal*, 93(3), 284–293, 1996.)

postpeak response is obtained in the unreinforced sample. The similarity in the shapes of the fiber reinforced curves follow the increase in load-carrying capacity with the fiber addition. Despite the high-volume fraction of fibers, the load-carrying capacity decreases rapidly in the postpeak region. The cyclic loading response indicates that before the peak load, the irreversible deformation is relatively small. After the peak load, the irreversible deformation continues to increase, indicating material damage and a large plastic deformation. This may be due to the irreversible fiber pullout leading to a mismatch of the crack surfaces.

DUCTILE FIBERS

At low or nominal volume fractions, PP fibers increase ductility by bridging the matrix cracks. At high-volume fractions, composite strength may be significantly increased if a sufficiently high bridging force is provided before microcrack growth becomes unstable.

Figure 12.3 shows the responses of specimens reinforced with 4% and 8% PP fibers when compared with the mortar specimens. At the 4% level, the peak load increases appreciably and the toughness increase is significant. The material exhibits major nonlinearity before the peak load, whcih tells us what? With 8% PP fibers, both the peak load and the toughness increase significantly. Stiffness degradation of the composite is not significant before peak load, as the path to the ultimate load is quasi-linear. Also, compared with mortar, there is more than a 700% increase in the ultimate strength of the composite.

A large residual displacement during the loading–unloading cycles is observed, which is attributed to cracks that do not close upon unloading, because the fiber pullout mechanism makes the crack faces unable to close.

In comparison with the plain matrix, the toughness of PP composites is shown to change from 9.8 to 1150 N·mm, an increase of more than 100 times. The highly nonlinear response beyond the maximum load is due to significant cracking. Composites exhibit a ductile behavior, because the fiber pullout generates a dominating closing pressure on the crack surfaces. The material behavior may be modeled as an elastic, perfectly plastic solid.

HYBRID COMPOSITES

Composites with both the added strength from brittle fibers and improved toughness, are brought about by the ductile fibers. Two types of hybrid composites are addressed. The first series used 4% alumina and 4% PP fibers, and the second series used 4% carbon fibers and 4% PP fibers. Because the two fiber lengths are of various orders of magnitude, their mixing in the matrix does not pose a problem. The load–CMOD responses are shown in Figure 12.4. Note that the ultimate load of both series if hybrid composites is as much as 75% higher in comparison to the PP composites. The increased strength is attributed to the contribution of short alumina and carbon fibers and indicates their ability in transferring load across the faces of the precritical crack. Composites with alumina show a drop in load after the first major crack. This drop is recovered as the PP fibers begin to pull out from the matrix. The toughness of both composites is quite comparable and reflects an

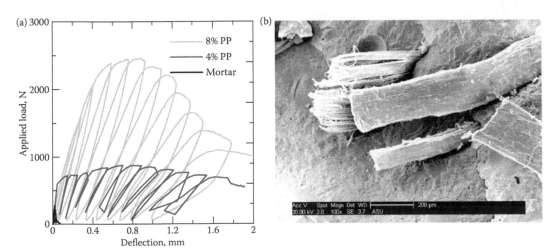

FIGURE 12.3 Flexural cyclic load–deflection response of composites with 4% and 8% PP fibers compared with mortar. (From Mobasher, B., and Li, C. Y., "Mechanical properties of hybrid cement based composites," *ACI Materials Journal*, 93, 3, 284–293, 1996.)

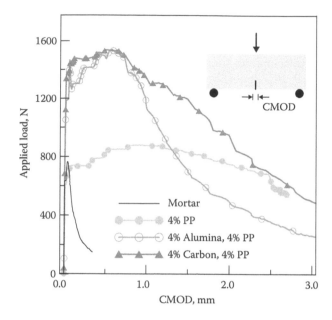

FIGURE 12.4   Flexural load–CMOD response of hybrid cement composites.

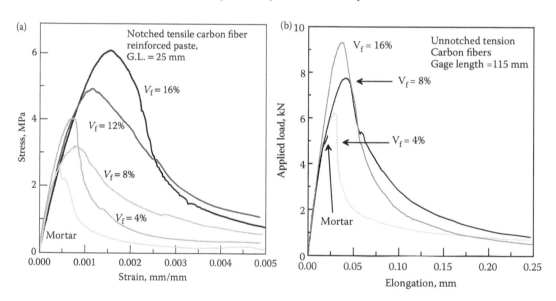

FIGURE 12.5   Tensile response of 2% to 16% of notched carbon reinforced specimens; (a) notched tensile sample and (b) unnotched tensile samples. (After Mobasher, B., and Li, C. Y., "Mechanical properties of hybrid cement based composites," *ACI Materials Journal*, 93, 3, 284–293, 1996.)

increase of approximately 20% compared with the PP composites. This is mainly due to the strength increase. Composites with carbon and PP fibers have a better postpeak performance than those with alumina and PP fibers. The overall effects of carbon and alumina fibers are similar.

TENSION RESULTS

Figure 12.5a shows the typical tensile response of the notched specimens reinforced with up to a 16% volume fraction of carbon fibers. As the fiber volume increases, the nominal tensile strength increases by a factor of 2.4 for a 16% volume fraction of carbon fibers. A small degree of nonlinearity can be observed, and, after the peak load, the strain-softening region drops rapidly, indicating that the weak fiber toughening is in effect, as mentioned in the flexural section. Figure 12.5b represents the response of similar samples tested without a notch. It is not possible to capture the postpeak response of the unreinforced samples using the closed loop testing system. The load elongation response, especially for higher volume fraction fibers, is uniform and indicates a stable softening portion.

The load carried by the unnotched specimens is much smaller than that of notched specimens. This points out to the fact that forcing the failure zone become relatively localized would increase the tensile strength of the material. The difference between the fibers diminishes as the fiber volume fractions increases. For instance, as the fiber volume fractions increase, the material toughens and the notch sensitivity is removed considerably. Therefore, the average strength obtained is the same with or without the notch.

Figures 12.6a and 12.6b compare the response of the various levels of alumina fibers on the tensile response for both notched and unnotched specimens. Similar to the carbon fibers, alumina fibers increase the strength of the composite. The first crack stress is approximately 85% peak stress. The enhancement in strength due to the fiber addition increases proportionately with the fiber volume fractions. The load–deformation response is almost linear up to peak value, and the specimen retains its tangential stiffness much better than mortar. Beyond the peak load, the load-carrying capacity decreases rapidly for all specimens, indicating a brittle response. Minor postpeak ductility is due to the very short fiber length and results in the sudden drop in load after the peak load. The diminishing improvement in strength due to the fiber volume increases is attributed to the improper consolidation and potential fiber damage as a result of vigorous mixing. It is therefore clear that the main role of whiskers is in the toughening of the prepeak microcracking process and, thus, increasing the ultimate strength.

The typical tensile response of hybrid composites reinforced with PP fibers and carbon fibers are shown in Figure 12.7. The composite strength increases from 2.58 MPa for mortar to 3.34 MPa for hybrid composites. The load response does not show much nonlinearity before peak stress, as can be observed in three-point bending test. After the peak stress, the crack opening increases and the PP fibers pull out from the matrix. The use of the first cracking strength as a normalizing parameter for the toughness index would cause a great degree of scatter. This value is practically useless in normalizing the data due to the significant variations observed. Because a significant improvement

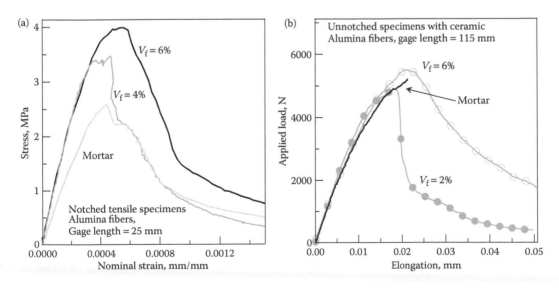

FIGURE 12.6   Tensile response of alumina fiber composites using (a) notched and (b) unnotched samples.

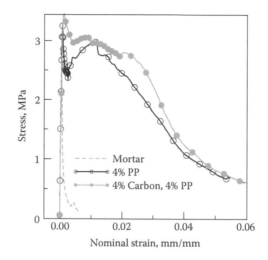

FIGURE 12.7 Tensile response of a hybrid composite with PP and carbon fibers.

in the material property is observed in the postcracking response, the entire area under the stress–strain curve is used as a measure of toughness. The first crack toughness is approximately 0.002 to 0.0028 N/mm², which is significantly smaller then the overall toughness. The relative toughness of these materials increases from 0.0028 to 2.623 N/mm², which represents a three orders of magnitude change in energy absorption. The maximum load of composite with 4% PP and 4% alumina fibers is more than twice the mortar specimens, and the toughness increases a tenfold more.

Average material properties for both notched and unnotched specimens from the direct tension tests are summarized in Table 12.2. The reported values include the Young's modulus, the first cracking stress and strain, the peak stress and strain, the first crack toughness, and the total toughness. Because linear variable differential transformer (LVDT's) measurement after the first cracking is actually the crack opening, strain parameters cannot be defined beyond this point, and only average strains calculated by using the measured elongation over the gage length can be used. Young's modulus obtained from the unnotched tests is shown to increase as the volume fraction of the fibers increase.

COMPARISON OF INJECTION MOLDING AND COMPRESSION MOLDING

Figure 12.8 shows the flexural load–crack mouth opening (CMOD) response of a specimen reinforced with 2% alumina fibers manufactured using the injection molding (IM) process compared with the compression molding (CM) process. A detailed set of data can be obtained from published work in this area [13]. Note that the response of this material for both manufacturing methods is quite brittle, as shown by the marginal stable crack growth before the peak and the significant drop in load-carrying capacity after the peak load. The IM process yields samples that are slightly stronger than the CM process. This may be attributed to the high shear stresses observed in the IM process as opposed to the hydrostatic pressure observed in the CM method. The high shear stresses tend to distribute the fibers and reduce the porosity in the interface of the high surface area fibers and the mortar matrix.

An addition of PP fibers increases the toughness significantly, as shown in Figure 12.9. The non-linear postcracking range is enhanced because of the use of PP fibers, and the composites exhibit postpeak ductility. In the presence of PP fibers, matrix cracking does not lead to a sudden failure and drop in the load level, since the fiber pullout from matrix becomes the main reinforcing mechanism. The response is quasi-ductile with a flat or rising plateau, which indicates that the forces that are released after matrix cracks transfer to the fibers that bridge the crack. The load drop is recovered

TABLE 12.2
Three-Point Bend Test Data

Specimen	Fiber Type	W/C	V_f (%)	Notch Size (mm)	E (GPa)	Peak Load (N)	Peak CMOD (mm)	Peak Deflection (mm)
EPA1	Mortar	0.40	N/A	25.4	8.97	311.5	0.022	0.075
EPB1	Mortar	0.40	N/A	25.4	9.00	277.8	0.019	0.074
CC2A[a]	Alumina	0.30	2	12.7	15.43	382.5	0.017	0.042
CC2B[a]	Alumina	0.30	2	12.7	12.10	319.3	0.019	0.045
CC2C[a]	Alumina	0.26	2	12.7	18.65	405.8	0.015	N/A
CC2D[a]	Alumina	0.35	2	12.7	16.19	452.5	0.019	N/A
CC2P2B[a]	Alumina/PP	0.23	2/2	12.7	7.35	568.6	0.508	N/A
CC4A[a]	Alumina	0.4	4	12.7	10.86	342.6	0.021	0.051
CC6A[a]	Alumina	0.35	6	12.7	16.39	541.5	0.023	0.087
CP2A[a]	PP	0.30	2	12.7	11.44	757.9	0.428	0.688
EC2A1	Alumina	0.40	2	25.4	7.27	219.1	0.030	0.077
EC2B1	Alumina	0.40	2	25.4	12.46	442.6	0.026	0.022
EC2G	Alumina	0.40	2	19.05	11.18	345.5	0.016	0.066
EC2H	Alumina	0.40	2	19.05	11.84	666.3	0.025	0.061
EC2P2A	Alumina/PP	0.40	2/2	19.05	7.31	338.7	0.037	0.115
EC2P2A1	Alumina/PP	0.40	2/2	25.4	9.26	317.4	0.037	0.108
EC2P2B	Alumina/PP	0.40	2/2	19.05	6.03	286.1	0.043	0.102
EC2P2B1	Alumina/PP	0.40	2/2	25.4	8.04	258.1	0.037	0.101
EC4B	Alumina	0.40	4	19.05	10.89	566.9	0.020	N/A
EC4C	Alumina	0.40	4	19.05	14.30	711.4	0.021	0.131
EP2A1	PP	0.40	2	25.4	5.56	466.5	1.005	1.858
EP2B1	PP	0.40	2	25.4	5.78	253.6	0.846	1.595
EP2C	PP	0.40	2	19.05	4.58	309.1	1.024	1.882
EP2D	PP	0.40	2	19.05	2.60	407.0	0.527	0.982
ES5A1	Steel	0.40	5	25.4	14.18	1419.2	0.285	0.564
ES5A2	Steel	0.40	5	25.4	22.28	2816.2	0.474	1.271
ES5B	Steel	0.40	5	19.05	10.78	2387.3	0.605	1.311

[a] CM specimen.

as the PP fibers pull out from the matrix. The slope of loading–unloading curve is also indicative of the damage that takes place. For the same volume fractions, the IM process results in a much higher increase in the response of the sample compared with the CM process.

The response of a hybrid composite is shown in Figure 12.10. The composite reinforced with 2% PP fibers and 2% alumina exhibits an increase in the load-carrying capacity, which is linear up to the peak value and then drops gradually. Using a hybrid response, the two fiber types become complimentary to each other, and the ceramic fibers contribute to an increase in the strength while the PP fibers contribute to the additional toughness of the sample.

A comparison of the behavior of this sample with those manufactured using the CM indicates that the effect of hydrostatic pressure on our samples is significant in reducing the initial porosity of the specimen and in decreasing the available water content. Preliminary studies show that the level of water was reduced by as much as 30% to 40% because of the hydrostatic pressure and the seepage of the water through the filter paper. Similar techniques are commonly used in the Hatscheck

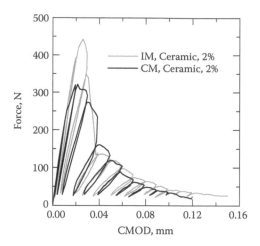

FIGURE 12.8 Load versus CMOD graph for test specimen for CM and IM.

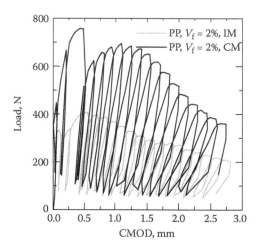

FIGURE 12.9 Load versus CMOD graph for test specimen EP2D.

process. Figure 12.11 shows the response of a composite with 5% steel fibers of 3 mm in length manufactured using the IM process. Note that this composite has an exceptionally high bending resistance and a pseudo-plastic behavior. The load-carrying capacity of these fibers is as much as three times the flexural response of alumina and carbon fibers. The ductility is also significant, and the level of energy absorption in terms of flexural toughness is in order of 2500 N·mm, which is more than twice the composites presented earlier.

Fracture Resistance Curves

An alternative means of measuring the toughening due to fibers is to characterize the resistance growth of the cracks by means of R-curves. The stress intensity factor and the K_R curves are computed on the basis of an effective crack approach using principles of linear elastic fracture mechanics [14]. The loading–unloading curves are used to calculate the compliance of the specimen for various CMOD levels. At each level, the compliance is used to calculate the effective crack length, and the compliance versus crack length response is plotted. Using numerical differentiation, the derivative

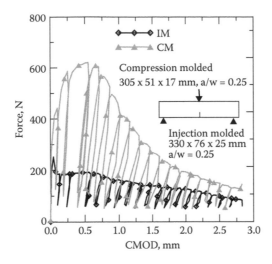

FIGURE 12.10   Comparison of load versus CMOD for CM and injection-molded composites.

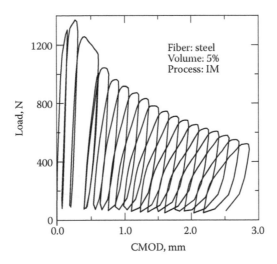

FIGURE 12.11   Load versus CMOD for injection-molded steel fiber composites.

of the rate of change in compliance is calculated and used to compute the strain energy release rate. The inelastic displacement is further used to compute the R-curve response of the specimen on the basis of the approach developed by Mobasher et al. [15]. Both the fracture toughness, which is based on the two parameter fracture model and the R-curve approach have been developed and discussed in detail elsewhere. Only the R-curve results are presented here.

This curve has units of toughness, MPa mm$^{0.5}$, and represents the relationship between the crack length and the energy required for the propagation of the crack. The plateau values of K_R exceeds the elastic equivalent fracture toughness parameter, $K_{IC}{}^S$, which is measured according to the RILEM approach [16, 17]. The R-curve is material, geometry, and size dependent; therefore, it can conveniently be used to study stable crack growth and toughening for similar specimens. Using R-curves, it is possible to study the fracture process and stable growth of the single crack. The toughening in response to various fiber combinations is effectively measured and characterized using the present approach.

Figure 12.12 illustrates the R-curves for alumina fiber composites. The addition of 4% alumina fibers slightly increased the toughness but did not affect the crack extension, whereas the addition of

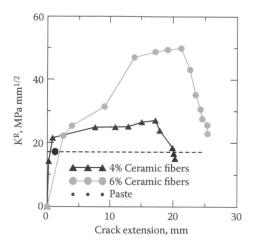

FIGURE 12.12   R-curves for 4% and 6% alumina specimens.

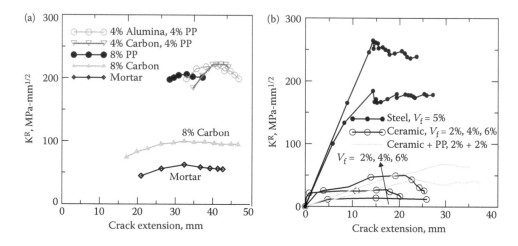

FIGURE 12.13   (a) R-curves for brittle and hybrid ductile samples compared with mortar. (b) Comparison of R-curves for Ductile composite systems.

6% alumina fibers greatly increased the toughness during crack extension. The plain mortar specimen has a toughness of $K_{IC} = 17.1$ MPa mm$^{1/2}$ and a crack extension of 1.2 mm, which is shown by a single data point. By adding 4% alumina fibers, the toughness increases to $K_{IC} = 27.8$ MPa mm$^{1/2}$ and the crack extension remains close at 1.7 mm. With the addition of 6% alumina fibers, the toughness increases to $K_{IC} = 50$ MPa mm$^{1/2}$ and the crack extension increases by 40 mm. This illustrates the effect of different fiber types. The alumina fibers increase the toughness by bridging the microcracks, but once cracking reaches beyond this stage they are unable to resist further crack growth.

The K_R curves for various composites are shown in Figures 12.13a and 12.13b. Toughening due to various fiber combinations is effectively represented by the R-curve response. Results indicate that addition of short whiskers increases the flexural and tensile strength in addition to improving the ductility of the composites. The most profound effects can be observed in the increased load-carrying capacity. The toughening effects of alumina and carbon fibers from total energy dissipation are not significant even at high fiber loading levels. This weak toughening effect may be attributed to the short length of the fibers and their inability to effectively bridge the macrocracks. The alumina fibers'

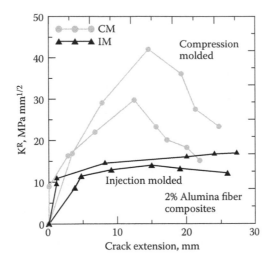

FIGURE 12.14 R-curves for 2% alumina specimens manufactured by CM and IM.

main contribution remains in increasing the ultimate strength of composites. The addition of PP fibers results in hybrid composites with both high strength and toughness. Also, the stable crack growth region is significantly increased for composites with ductile PP fibers.

Figure 12.13b represents a comparison between a specimen containing 2% PP fibers and a specimen containing both 2% PP 5% steel and 2% alumina fibers. The specimen with only the PP fibers exhibited greater toughness along the order 70 to 100 MPa mm$^{1/2}$, whereas the other specimen was on the order of 40 to 50 MPa mm$^{1/2}$. The specimen containing both fibers exhibited an increased crack extension. The PP fibers are able to bridge both the microcracks and the larger cracks and, therefore, increase the crack extension and specimen ductility.

Figure 12.14 represents the IM and the CM specimen containing 2% alumina fibers. All four extrusion specimens displayed a toughness ranging between 12 and 16 MPa mm$^{1/2}$. However, the specimen with an initial notch length of 25 mm displayed a more stable crack growth. The CM specimens displayed toughness values twice that of the extrusion specimen and a crack extension similar to the 25-mm notched extrusion specimen.

CONCLUSION

Composites with high-volume fraction of chopped hybrid fibers were developed using high-energy mixing that resulted in a uniform fiber distribution. Short alumina and carbon fibers are able to stabilize the microcracks and increase the composite strength. The toughening effect after peak load is not significant because of the fiber's short length. PP fibers increase the toughness of the composites significantly. There exists a critical volume fraction of the fibers, above which both load-carrying capacity and toughness increase significantly. Combining short alumina and carbon fibers with PP fibers can lead to composites that embody enhanced strength and ductility.

REFERENCES

1. Krenchel, H., and Jensen, H. W., "Organic reinforcement fibres for cement and concrete," *Proceedings of the Symposium on Fibrous Concrete (CI80)*, The Construction Press Ltd., Lancaster, 1980, 87–94.
2. Hannant, D. J. (1978), *Fibre Cements and Fibre Concretes*, Chicester, UK: John Wiley & Sons, 219 pp.
3. Li, Z., Mobasher, B., and Shah, S. P. (1991), "Characterization of interfacial properties in fiber-reinforced cement based composites," *Journal of the American Ceramic Society*, 74(9), 2156–2164.

4. Mobasher, B., Castro-Montero, A., and Shah, S. P. (1990), "A study of fracture in fiber reinforced cement-based composites using laser holographic interferometry," *Experimental Mechanics*, 30, 286–294.

5. Mobasher, B., and Shah, S. P., "Interaction between fibers and the cement matrix in glass fiber reinforced concrete," Thin-Section FRC and Ferrocement, ACI Special Publication, SP-124, 1990, 137–148.

6. Stang, H., Mobasher, B., and Shah, S. P. (1990), "Quantitative damage characterization in polypropylene fiber reinforced concrete," *Cement and Concrete Research*, 20, 540–558.

7. Mobasher, B., Stang, H., and Shah, S. P. (1990), "Microcracking in fiber reinforced concrete," *Cement and Concrete Research*, 20, 665–676.

8. Mobasher, B., and Li, C. Y. (1996), "Mechanical properties of hybrid cement based composites," *ACI Materials Journal*, 93(3), 284–293.

9. Krenchel, H., and Stang, H., "Stable microcracking in cementitious materials," *Proceedings of the International Symposium on Brittle Matrix Composites 2, (BMC2)* Cedzyna, Poland, September 1988, 20–33.

10. "Ceramic Source®," Annual Company Directory and Buyer's Guide Compiled by the American Ceramic Society, Inc., Vol. 7, 1991–1992, 338.

11. Mobasher, B., and Li, C. Y. (1996), "Mechanical properties of hybrid cement based composites," *ACI Materials Journal*, 93(3), 284–293.

12. Garlinghouse, L. H., and Garlinghouse, R. E. (1972), "The omni mixer—A new approach to mixing concrete," *Journal of the American Concrete Institute*, 69, 220–223.

13. Haupt, G. J. (May 1997), "Mechanical properties of cement based composite laminates," MS thesis, Arizona State University.

14. Shah, S. P. (1990), "Determination of fracture parameters (K^s_{IC} and $CTOD_c$) of plain concrete using three-point bend tests," *Materials and Structures*, 23, 457–460.

15. Mobasher, B., Li, C. Y., and Arino, A. "Experimental R-curves for assessment of toughening in fiber reinforced cementitious composites," 1995 American Concrete Institute (ACI) Spring Conference, Testing of Fiber Reinforced Concrete—SP-155, ed. D. J. Stevens, N. Banthia, V. S. Gopalratnam, and P. C. Tatnall, Salt Lake City, Utah, 1995, 93–114.

16. Jenq, Y. S., and Shah, S. P. (1985), "A fracture toughness criterion for concrete," *Engineering Fracture Mechanics*, 21(5), 105.

17. Jenq Y. S., and Shah S. P. (1985), "Two parameter fracture model for concrete," *Journal of Engineering Mechanics*, 111, 1227–1241.

13 Correlation of Distributed Damage with Stiffness Degradation Mechanisms

INTRODUCTION

There is a great potential for the use of textiles as reinforcement for cement components. Textiles provide benefits such as excellent anchorage and bond development [1–5]. The mechanical anchoring is provided by the nonlinear geometry of individual yarns within the textile and directly relates to its structure [4, 6]. The textile structure induces tortuosity in the path of the propagation of a crack, which promotes crack deflection processes that lead to the enhanced strength and ductility of the composite [7]. Another important parameter is the effectiveness of the matrix impregnation into the textile microstructure. The matrix penetration in the interstitial spaces between individual fibers, in between two parallel yarns, and furthermore into the vicinity of the junction point of the orthogonal yarns of the textile contributes to the anchorage and increases the ability of the yarns to carry loads. The geometry of the textile is, thus, responsible for the transfer of the load between the matrix and fibers, significantly reducing the dependence on the interface transition zone and mitigates the effect of processes such as debonding and pullout, which normally serve as the main load transfer mechanism. The net result of these mechanisms is the development of a continuous system of microcracks, which serve to toughen the composite. Characterizing the underlying toughening mechanisms for the use of textiles in cement products leads to better processing and an optimization of the textile structure and geometry for these composite materials.

ROLE OF MICROCRACKING CEMENT COMPOSITES IN TENSION

The behavior of textile–cement composites exposed to tensile loads is characterized by multiple cracking. Transverse cracks form gradually as the composite is loaded beyond the initial cracking stages. The nature of multiple cracking and the resulting stress–strain curve, toughness, and strength are dependent on the properties of the reinforcing textiles, the cement matrix, as well as the interface bond and the anchorage of the textiles developed. Therefore, the parameters that lead to crack formation ultimately control the textile–cement composite properties, and, as the strain increases, the interaction of crack spacing and stiffness can be correlated with damage development. Microstructural features such as crack spacing, width, and density are related to the evolution of damage as a function of applied displacement. The composite stiffness in the postcrack range is an important design property. By monitoring the crack spacing, one can correlate textile type and mixture formulations with increases in the modulus of rupture, ductility, and an increased strain capacity.

TENSILE RESPONSE OF TEXTILE REINFORCED CEMENT COMPOSITES

The general stress–strain behavior of a textile reinforced composite involves interactions between the matrix, the textile, and the interface. Compared with GRC, these materials are as much as five times stronger and six times more ductile. Compared with plain, cement-based materials, these proportions are 8 and 400 more times [8]. Figure 13.1 shows the transverse and orthogonal directions of a glass textile consisting of bonded layers of glass yarns.

Schematics of the general tensile behavior of glass fabric are shown in Figure 13.2 [9]. Four distinct zones are identified using roman numerals with two zones coming before the Bend Over Point (BOP) and two zones being after the BOP range. Zone 1 corresponds to the elastic-linear range where both the matrix and the textile behave linearly. In this zone, one can assume a perfect bond between the two components, such that the strain distribution is uniform in both longitudinal and transverse directions without any debonding, cracking, slip, or matrix cracks. One can use composite laminate theory as shown in Chapter 11 with an isostrain model to relate the properties of textile and matrix to the composite response, that is, the rule of mixtures for longitudinal modulus. The effect of fiber stiffness and volume fraction can be determined as the characteristic linear response.

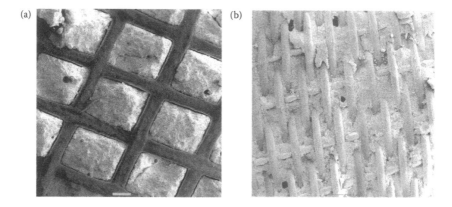

FIGURE 13.1   Scanning electron microscope micrograph of matrix embedment in two types of textile (scale bar = 500 μm). (a) AR glass textile. (b) PE textile.

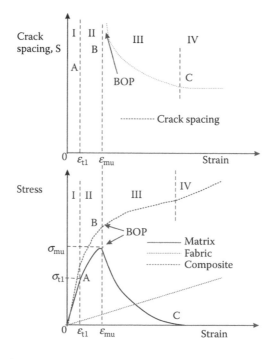

FIGURE 13.2   The stages in the tensile response of textile composites.

Because of the low volume fraction of textile (normally less than 10%), the stiffness of the lamina is dominated by matrix properties; hence, one reaches stresses of 2 to 3 MPa with as low as 100 to 150 μstr. This represents a stress level sufficient to initiate cracking in the matrix. The elastic response in Zone 1 is terminated by the initial crack formation in the matrix phase at a point labeled "A" and designated as σ_{t1} at the strain level ε_{t1} [10]. This is designated with stress levels of σ_{BOP}^-. After the initiation of cracks in the matrix, its load-carrying capacity does not vanish as the cracks are bridged by the longitudinal yarns. The second range, identified as Zone 2, is defined as being between the two stress levels of σ_{BOP}^- and σ_{BOP}^+ and is further defined as a stable cracking-nonlinear range. In this zone matrix cracks begin to form and propagate across the width of the specimen; Having said that, no single crack has traversed the entire length of the specimen. This range (Zone 2) defines the formation of first complete crack across the cross section between points "A" and "B." The stiffness degradation in this section is due to the formation of a dilute concentration of cracks in the medium. Parameter σ_{t1} ($= \sigma_{BOP}^-$) represents the ultimate strength of matrix paste, and $\varepsilon_{t1} = \sigma_{t1}/E_{m0}$ ($= \varepsilon_{BOP}^-$) is defined as the strain at failure under uniaxial tension for the paste in an unreinforced condition, as shown in Figure 13.2. The gradual decrease in the stiffness of the matrix starts at the plain matrix strength of σ_{BOP}^-. The maximum stress in the matrix phase is achieved at a stress and strain level of (σ_{mu}, ε_{mu}). This zone terminates when the first crack traverses the entire width of the specimen. The degradation of the overall sample stiffness begins at ε_{t1} continues up to the BOP strain level ($\varepsilon_{mu} = \varepsilon_{BOP}^+$). Zone 2 terminates at a damage level corresponding to the stress at the BOP (σ_{BOP}^+) level, which is also ultimate strength of the matrix in the presence of fibers σ_{mu} [11].

The parameter ε_{mu} may theoretically be obtained using the Aveston–Cooper–Kelly (ACK) approach [12] or other methods [13], which predict the strain capacity of the matrix phase in the presence of fibers, as shown in Equation 13.6. The ACK approach is applicable to the cement-based materials because it has been verified that the strength of the matrix increases in the presence of fibers [14]. In the ACK-based approach, γ is the fracture toughness and r is the fiber radius. The experimentally obtained values for ε_{mu} correspond to the values reported as ε_{BOP}, as reported in experimental programs. Depending on the textile–cement system used, γ_m/r in the range 0.5 to 5.0 N·mm/mm results in a good correlation between experimental (σ_{BOP}^+) and theoretical simulations of the stress and strain at BOP level,

$$\varepsilon_{mu} = \left[\frac{12\tau\gamma_m E_f V_f^2}{E_c E_m^2 r V_m} \right]^{1/3}$$

(13.1)

The post-BOP stage is characterized by the formation of distributed cracking in Zone 3. The stiffness of the reinforcement system is sufficiently high in this case to keep the newly formed cracks from widening and thus, promoting additional cracking. This stiffness affects the rate of reduction of crack spacing as the sample is strained. Zone 3 is dominated by formation of parallel microcracking. Once the formation of new cracking seizes, a condition referred to as crack saturation is defined. Additional straining is unable to transfer sufficient load to the matrix to cause further cracking; hence, crack formation stabilizes and crack spacing remains constant at this point. The gradual reduction of matrix stress levels across the cracked matrix is referred to as the softening zone. In this zone, the matrix cracks widen, and although there may be no localization in the strain softening zone, the response is modeled by contributions from a softening matrix and the textile pullout force.

The definition of strain in this region is gage length dependent and represents mean strain over the length of several cracks in the matrix. As the matrix phase in this zone undergoes strain softening, it can be viewed as a material with a decaying stiffness.

The transition point between Zone 3 and Zone 4 is asymptotically defined using the saturation crack spacing. Debonding of the textile becomes the dominant mode of failure at this stage. Zone 4 is dominated by a crack widening stage, ultimately leading to failure by textile pullout or tensile failure of fibers. The behavior of both the matrix and textile, in addition to their interaction with each other is studied in each of these four ranges.

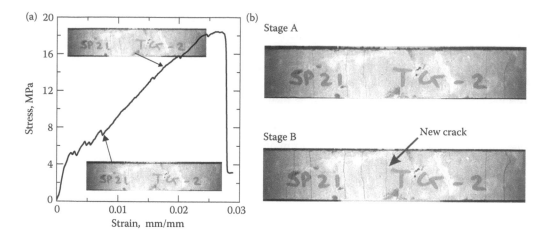

FIGURE 13.3   (a) Various cracking stages within a strain range of 1%. (b) New cracks form between existing cracks while the existing cracks are widening.

CRACK SPACING MEASUREMENT

The significance of the crack measurements is related to several factors such as the efficiency of the fiber matrix bond, the stiffness of the cracked composite, the durability aspects, and also the tension stiffening effect. The crack spacing can also be used as a design parameter.

Parallel microcracks form in brittle matrix composites containing a high-volume fraction of fibers [8]. Crack density is a measure of parallel crack spacing and can be used to ascertain the degree of damage, as well as the bond strength between the fibers and the matrix. The first step is to document the evolution and the sequential nature of parallel crack formation, their spacing, and width as a function of applied strain.

A method based on digital image processing was developed to monitor the various stages of crack formation under load [15]. By comparing high-resolution images of each state of strain, the role of textiles in reinforcing the microstructure can be correlated with crack initiation and parallel crack formation. Photographs of the specimen at regular time intervals also document the crack development throughout the loading cycle of a tensile test. A monochromatic light source was used to illuminate the specimen while a digital frame grabber captured images. Figures 13.3a and 13.3b present the crack development at two different stages (recorded at $\varepsilon = 2.72\%$, and $\varepsilon = 4\%$). A close examination of these pictures reveals the formation of a new crack in between two existing cracks.

IMAGING PROCEDURES FOR MEASUREMENT OF CRACK SPACING

Formation of the cracking pattern throughout the loading cycle can be quantitatively measured using the superposition of images at two different stages. The procedure is based on the assumption that the intensity of the pixels determines the existence of a crack. Each image is sharpened using standard routines such as Laplacian filters and were subjected to segmentation. This is a process completed to separate the crack from the rest of the image by specifying a threshold intensity for the selection of pixels [16]. All the pixels below certain intensity are designated as a crack. A starting image is selected, which is the picture of the specimen before any crack has started to form. Two programs are used. The first program uses a manual approach of identification and tracing of the newly formed cracks in each image and adds this information to data from previous loading increment. The second program measures the crack spacing from the traced cracks.

The crack spacing, measured in pixels, is calibrated using conventional techniques. The photograph shown in Figure 13.4a is the specimen before any cracking, whereas Figure 13.4b represents

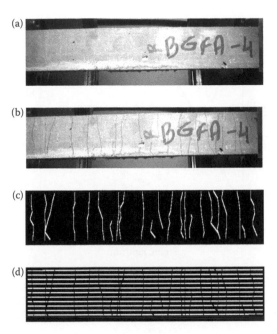

FIGURE 13.4 Various steps in the measurement of crack spacing: (a) the beginning of the loading, (b) the end of the loading, (c) the binary image at any intermediate stage, and (d) the crack spacing measurements. (After Mobasher, B., Peled, A., and Pahilajani, J., "Distributed cracking and stiffness degradation in fabric–cement composites," *Materials and Structures*, 39, 317–331, 2006.)

the same sample at the end of the test. Figure 13.4c represents the profile of the trace of cracks represented as a binary image. The measurement of the crack spacing (i.e., density) on the basis of the information in Figure 13.4c is as follows: a binary image consisting of a series of parallel lines is generated traversing the entire length of the specimen. By using a sequence of "AND" and an "OR" operators between this set of parallel lines and the Figure 13.4c, the original parallel lines are broken into segments, each representing a measurement of crack spacing as shown in Figure 13.4d. The next step is to count the distribution of the length segments and statistical parameters of crack spacing. By relating the image number with the corresponding strain value, a profile of the crack spacing with respect to the applied average strain is obtained.

Figure 13.5 shows the crack spacing as a function of the strain. Note that, as the crack spacing reduces with an increasing strain, the standard deviation of the spacing parameter also decreases. A constant crack spacing is obtained at the late stages of loading, which is referred to as saturation crack spacing and indicates that beyond this level, only existing cracks are opening as opposed to new cracks forming.

Parameters of crack spacing as a function of applied strain are correlated with the stress–strain plot, as is shown in Figure 13.5. As the strain is increased, the magnitude of average crack spacing and its standard deviation decrease [17, 15]. Initially, the crack spacing is large, but as straining of the sample proceeds, the crack spacing becomes more uniform and the standard deviation decreases. The next step is to document the crack spacing and the deformation fields in the vicinity of the bridged cracks as measurements of the internal damage evolution.

Crack spacing distributions within distinct regions can be evaluated by their range and distribution. Figure 13.6 represents the distribution of the crack spacing for three different ranges in the strain history of samples with alkali-resistant (AR) glass textile. During the initial loading stage ($\varepsilon < 0.015$), the distribution varies in two distinct ranges. This is the region where the spacing between any two cracks is sequentially reduced by the formation of a new crack. As the strain is increased, the crack distribution homogenizes so that at strain levels of 3.87%, as many as 80% of the cracks have a spacing of 10 mm or less.

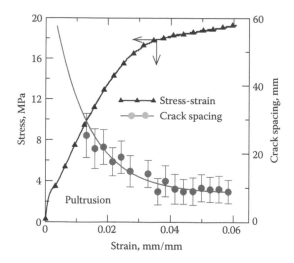

FIGURE 13.5 Correlation of damage evolution using crack spacing measures and the stress–strain response of AR glass fiber composites. (Adapted from Mobasher, B., Peled, A., and Pahilajani, J., "Distributed cracking and stiffness degradation in fabric-cement composites," *Materials and Structures*, 39, 317–331, 2006.)

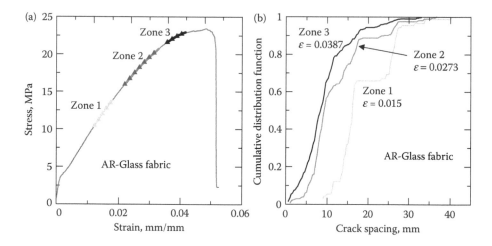

FIGURE 13.6 Crack spacing distribution in AR glass composites.

Figure 13.7 represents a similar analysis conducted for the polyethylene (PE) specimens. The same homogenization effect is taking place here while the final crack spacing numbers are much lower than the samples with glass fabrics. Figure 13.8 represents the cumulative probability distribution plot parameters on the basis of a Weibull distribution response reported as a function of the strain level. Maximum likelihood estimates of the parameters of the Weibull distribution, a and b, are defined as:

$$P_f = F(x|a,b) = \int_0^x abt^{b-1}\, e^{-at^b}\, dt$$

where P_f is the Weibull cumulative distribution function.

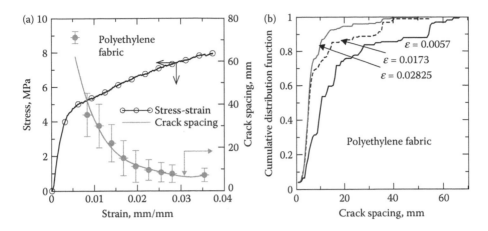

FIGURE 13.7 Crack spacing distribution in PE Fiber composites.

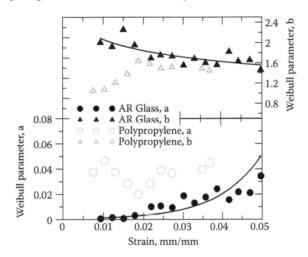

FIGURE 13.8 Weibull parameters *a* and *b* for glass fiber and PE composites.

EFFECT OF FABRIC TYPE

Figure 13.9a compares the effect of AR glass and PE fabric types on the tensile stress–strain response [8]. In both cases, a similar matrix is used. The tensile response of the glass composite shows a linear behavior up to approximately 2–3 MPa. The BOP$^-$ ranges from the initiation of nonlinearity (approximately 2–3 MPa) to a point where a complete and pronounced change in the slope of the stress–strain response, which occurs at approximately 4 to 5 MPa (BOP$^+$). Beyond this level, there is a knee in the curve, and the stress measure increases with a reduced stiffness to levels as high as 20 MPa. The post-BOP range is referred to as the strain hardening behavior in the tensile response and is observed for both AR glass and PE fabrics, although the PE fabrics are made from low modulus fibers. The ultimate strain of both composites reaches as high as 4%. Similar to the glass fabric case, the curve of the PE fabric composite is linear up to approximately 4.5 MPa, and beyond this level, the formation of the knee is quite evident since the postcrack stiffness is significantly lower than the glass textile case. The ultimate strength of the sample is in the range of 7 MPa at a strain level of 4%.

The crack spacing measurements are also shown in Figure 13.9a for both glass and PE textile composites. The volume fraction of the PE textile is 9.5% and that of the glass textile is 4.4%. The figures show a general decrease in the crack spacing until a steady state is reached. This constant

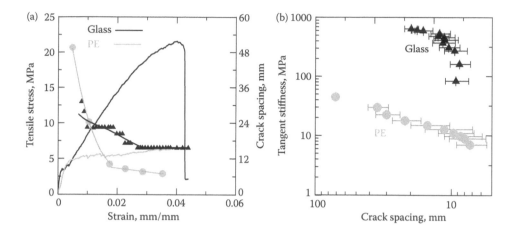

FIGURE 13.9 Comparison of AR glass and PE fabrics composites: (a) tensile stress and crack spacing versus strain and (b) tangent stiffness versus crack spacing. (After Mobasher, B., Peled, A., and Pahilajani, J., "Distributed cracking and stiffness degradation in fabric–cement composites," *Materials and Structures*, 39, 317–331, 2006.)

level of crack spacing is designated as saturation crack spacing. Beyond this point, reduction in crack spacing is not observed. Additional imposed strain results in the widening of the existing cracks. Measurements of crack width as a function of applied stress have shown that the primary mode of displacement in this region is mainly opening of existing cracks [18]. Widening of the matrix cracks results in textile debonding [19]. The PE composites show much smaller crack spacing compared with glass textile composites, which can be related to the smaller size of the textile opening in the PE system and a better bond. Figure 13.9a indicates that crack widening in the glass system starts at much later stage (80% final strain) than that of the PE system (50% final strain). Crack widening is the dominant mechanism in the PE strain capacity because of the lower overall stiffness, whereas in the glass system, the distributed cracking and composite stiffness are maintained over a longer range of response.

Comparative evaluation indicates that AR glass textiles show a stiffer response in the postcrack stress–strain relationship than PE textiles (Figure 13.9a). The tangent stiffness and the crack spacing decrease as a function of applied strain as shown in Figure 13.9b, where the correlation of crack spacing with the stiffness degradation is shown. A major reduction in the tangent stiffness takes place at very low strain values, which corresponds to matrix strength. Beyond this point, as loading continues and successive cracks develop, there is a steady reduction in the composite stiffness. The stiffness degradation with AR glass samples is very high, up to the saturation crack spacing of 18 mm, which correlates with crack widening mechanism. Although the stiffness reduction in glass is very high, its tangent stiffness is still much higher than the PE at the postcracked stage (Figure 13.9b), but there is not a significant reduction in this value as a function of strain. An exponential decaying model was used to represent the crack spacing–strain relationship. The function is expressed as a four-parameter model as

$$S(\varepsilon_i) = S_1 + S_0 e^{-\alpha(\varepsilon_i - \varepsilon_{mu})} \quad \varepsilon_i > \varepsilon_{mu} \tag{13.2}$$

where $S(\varepsilon_i)$ is the crack spacing as a function of strain; ε_{mu} is the strain at the BOP level or where the first set of measurements were obtained; ε_i is the independent parameter, input strain, S_0; α is the constant representing the initial length of the specimen and rate of crack formation as a function of strain; and S_1 is the saturation crack spacing. Typical values of S_1, S_0, ε_{mu}, and α for different matrix and fabric combinations are presented in several publications [11, 15, 18].

On the basis of the previous discussion, one can differentiate three regions of response. The first portion of the curve represents the behavior of the uncracked composite as a linear-elastic material. The primary response in the second portion is transverse crack formation and its distribution in the matrix. In the third region, crack widening is the dominant mechanism, mainly because of the degradation of interface. Comparison of the mechanical behavior of the composites with the PE and AR glass textiles shows that the primary mechanism of the PE composite is crack widening by textile debonding, whereas with the glass textile composites, crack widening is not the governing mechanism up to approximately 80% of ultimate strength has reached (Figure 13.3a). The glass textile is well bonded to the matrix and the tensile behavior is representative of the entire composite.

EFFECT OF MINERAL ADMIXTURES

Use of fly ash and silica fume (SF) in matrix is beneficial because the pozzolanic reactions reduce both the porosity and the amount of calcium hydroxide at the interface transition zone. Furthermore, mineral admixtures improve the long-term durability of composites with low durability fiber systems [20]. Moreover, the particle morphology of the fly ash reduces the yield strength and viscosity of the mixture. The penetration of cement paste in between the opening of the textiles is a controlling factor in the development of bond. As such penetration depends on the size of the textile opening, rheology, and viscosity of fresh paste. Figure 13.10 indicates an improvement in the mechanical behavior of the cement composites with high levels of fly ash as compared with similar specimens without fly ash. In this case, 60% by volume of cement was replaced by fly ash. A tensile strength of approximately 25 MPa at a strain capacity of approximately 5% is observed. Detailed studies of the effect of fly ash on the properties of composite laminates are discussed by Mobasher, Peled, and Pahilajani [21].

Superior results obtained with matrices containing high levels of fly ash are due to enhanced paste impregnation in between fabric layers and changes in rheology of the mix. The viscosity of fresh mixtures characterized by shear rheometry indicated that shear stresses reduced from 3672 to 586 dyn/cm² with the use of fly ash [21].

Composites with various SF contents as replacement for cement are shown in Figure 13.11 [8]. In this case, composites with 5% by volume of SF perform significantly better than those with 10%. This improvement was as much as 50% for both AR glass and PE fabrics and is related to the rheological properties of mixtures with low SF content. Viscosity measurement of mixtures with various SF contents indicated a reduction in shear stress from 697 to 367 Pa when the SF content decreased

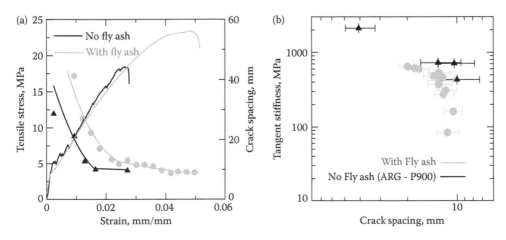

FIGURE 13.10 Effect of fly ash on response of AR glass fabric composites: (a) tensile stress and crack spacing versus strain and (b) tangent stiffness versus crack spacing. (After Mobasher, B., Peled, A., and Pahilajani, J., "Distributed cracking and stiffness degradation in fabric–cement composites," *Materials and Structures*, 39, 317–331, 2006.)

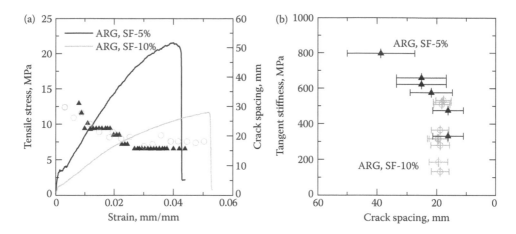

FIGURE 13.11   Effect of silica fume on response of AR glass fabric composites: (a) tensile stress and crack spacing versus strain and (b) tangent stiffness versus crack spacing. (After Mobasher, B., Peled, A., and Pahilajani, J., "Distributed cracking and stiffness degradation in fabric–cement composites," *Materials and Structures*, 39, 317–331, 2006.)

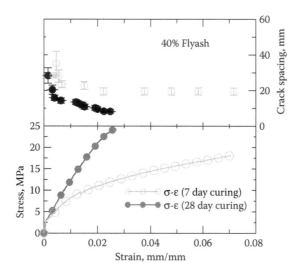

FIGURE 13.12   Effect of curing on the stress–strain and crack spacing formation for samples at 7 and 28 days curing. (After Mobasher, B., Peled, A., and Pahilajani, J., "Distributed cracking and stiffness degradation in fabric–cement composites," *Materials and Structures*, 39, 317–331.)

from 10% to 5% by volume [14]. Difference in crack spacing observed in Figure 13.11 correlates with the effects of SF. Composites with lower content of SF showed smaller crack spacing, improvement in bond, leading to better tensile response.

The influence of fly ash on crack spacing is observed in Figures 13.12a and 13.12b. In general, when comparing different composites, lower crack spacing suggests higher interfacial bond strength for systems having similar fabric content. Crack widening is observed as a primary mechanism in composites without fly ash and starts at approximately 50% of composite strength. In composites with fly ash, only the late stage of loading is governed by crack widening because of fabric debonding (Figure 13.10a). The correlation between crack spacing and bond strength has been well established using models such as the ACK approach [22].

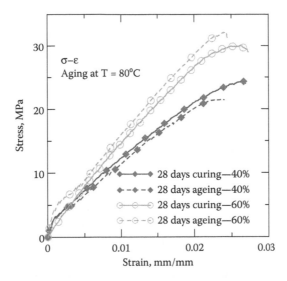

FIGURE 13.13 Stress–strain response of continuous fabric composites after 28 days accelerated aging compared with control samples. (After Mobasher, B. , Peled, A., and Pahilajani, J., "Pultrusion of fabric reinforced high flyash blended cement composites," *Proceedings, RILEM Technical Meeting, BEFIB*, 1473–1482, 2004.)

A comparison of the stiffness of the composites with and without fly ash shows no significant benefit for the fly ash addition in the low strain ranges. However, the strain capacity in composite with fly ash is much higher (from 3% up to 5%), although it is accompanied by a significant reduction in the tangent stiffness of the composite (Figure 13.10b). Although the glass fabric in the composite without fly ash is fractured, the fabric may pullout in the presence of fly ash. This behavior might be attributed to differences in fabric–matrix interface. The effect of curing was examined for various composites as shown in Figure 13.12. Note that extension of curing from 7 to 28 days increases the postcracking stiffness but reduces the ultimate strain capacity of the composite.

EFFECT OF ACCELERATED AGING

Accelerated aging was conducted by storing specimens in a saturated calcium hydroxide solution for 28 days [22] at 80°C. The influences of the aging process on the mechanical response of glass fabrics composites are presented in Figure 13.13. It is observed that the accelerated aging has a limited effect on the response of the AR glass fiber fabrics because the samples are loaded in the direction of yarns, and the mode of fiber pullout is not a dominant contributor to toughness in these directions. For the two different levels of fly ash studied, there is no significant difference in the mechanical performance.

RHEOLOGY AND MICROSTRUCTURE

The difference in the mechanical behavior of the samples at different levels of fly ash could also be attributed to the rheology of the matrix. Rheological properties of fresh paste were measured using a Brookefield rheometer. Figure 13.14 shows the change of viscosity and yield strength of paste as a function of fly ash content. The viscosity of the control samples is much higher than the three different levels of fly ash, indicating that the presence of fly ash causes a slower rate of stiffness gain. The yield strength of the sample decreases with the addition of fly ash as well. This may correlate the effect of high fly ash content mixtures with developing a better bond strength since the paste is able to better infiltrate the textile openings. It is also observed that the shear rate increases with time but with an increasing level of fly ash, this strength gain is slow.

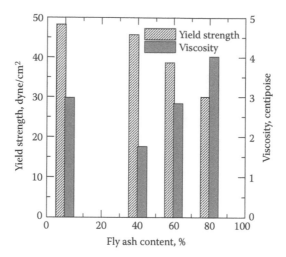

FIGURE 13.14 Viscosity yield strength for various levels of fly ash in the matrix. (After Mobasher, B., Peled, A., and Pahilajani, J., "Pultrusion of fabric reinforced high flyash blended cement composites," *Proceedings, RILEM Technical Meeting, BEFIB*, 1473–1482, 2004.)

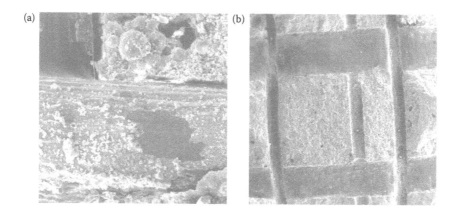

FIGURE 13.15 (a) Fly ash (60% mixture) in the vicinity of the yarn junction. (b) The matrix penetration in between the two adjacent fabric layers (60% mixture). (After Mobasher, B., Peled, A., and Pahilajani, J., "Pultrusion of fabric reinforced high flyash blended cement composites," *Proceedings, RILEM Technical Meeting, BEFIB*, 1473–1482, 2004.)

The microstructure of the specimens was studied under scanning electron microscope to observe impregnation aspects of the matrix materials. Fly ash addition improves the microstructure of the fabric paste bond as shown in Figure 13.15a, where the cenospheres of fly ash occupy the capillary voids and the transition zone in the vicinity of the inter yarn spacing. This packing effect explains why the bonding is so strong in mixtures with high levels of fly ash. The fly ash particles lower the viscosity and cause better flow of the matrix around the surface of the textiles. This is clearly reflected from Figure 13.15b.

EFFECT OF CURING

Two curing processes were examined: 3 days of accelerated curing at 80°C in +90% RH followed by 4 days of stabilizing in room temperature at 20% RH and 28 days of curing at 23°C at 100% RH. The influence of the curing process on the mechanical response of glass fabrics composites is presented in Figure 13.16a. As observed by the strain capacity, accelerated cured composites perform better than those cured for 28 days at room environment. PE textile composites showed a similar

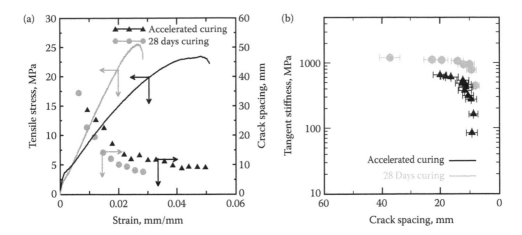

FIGURE 13.16 Effect of curing of AR glass fabric composites: (a) tensile stress and crack spacing versus strain and (b) tangent stiffness versus crack spacing. (After Mobasher, B., Peled, A., and Pahilajani, J., "Distributed cracking and stiffness degradation in fabric–cement composites," *Materials and Structures*, 39, 317–331, 2006.)

response. As shown in Figure 13.16b, extended curing increases the stiffness in the postcracking stage. Although the strength remains relatively the same, the strain capacity decreases. However, when high content of fly ash was added to the mix, the trend was reversed, and composite cured for 28 days at 23°C performed significantly better than the accelerated cured composite.

The brittleness observed at 28 days curing suggests improvement in the textile–matrix interfacial properties, leading to improvement in bond strength [23–25]. Such increase in bonding may cause textiles to fracture before pullout. Crack widening is seen in the large strain capacity of composites with accelerated curing as shown in Figure 13.16a. Crack spacing profiles indicate that in composites cured for accelerated curing, the crack spacing remains approximately constant throughout most of the strain range from 2.2% up to 5%, indicating that no new cracks are forming. The stiffness in the postcracking stage of the 28 days cured specimens is greater than that of the accelerated cured specimen (Figures 13.16a and 13.6b), thus implying an improved fabric–matrix bond strength.

EFFECTS OF PRESSURE

After forming the laminates with the pultrusion process, pressure is applied to facilitate matrix penetration in between the textile openings. For the pressure effects, two levels were examined for glass textile composites at 1.7 and 15.3 kPa (1:9 ratio). Figure 13.17a presents the effect of the applied pressure on the stress–strain and crack density–stiffness response. Many parameters such as initial cracking stress, postcrack stiffness, ultimate strength, and also mean crack density are dependent on the level of applied pressure. By increasing the processing pressure, the tensile strength is increased by approximately 40%; however, the ductility is reduced as much. The tangent stiffness is also higher for composites with higher pressure (Figures 13.17a and 13.17b). The crack density increases as a function of applied hydrostatic pressure suggesting better bonding, whereas the composite with the low pressure developed fewer cracks suggesting a poorer bond (Figure 13.17a). Moreover, Figure 13.17a also shows that crack widening is the operating mechanism in the low-pressure composites, which do not show a significant cracking activity beyond a strain of 2%, a value less than half of the ultimate strain. In contrast, the entire tensile response in high-pressure systems is controlled by multiple cracking throughout. This is attributed to the stress transfer between fabric and matrix. Interaction between the textile and the matrix is affected by the poor bond of the low-pressure system and improved by an enhanced high-pressure processing.

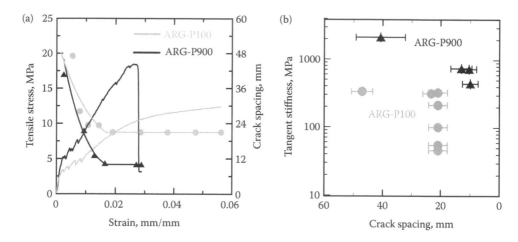

FIGURE 13.17 Effect of the pressure applied after the pultrusion process: (a) tensile stress and crack spacing vs. strain and (b) tangent stiffness versus crack spacing. (After Mobasher, B., Peled, A., and Pahilajani, J., "Distributed cracking and stiffness degradation in fabric–cement composites, *Materials and Structures*," 39, 317–331, 2006.)

Crack spacing is directly correlated with tangent modulus reduction as two independent measures of damage. Both parameters decrease as a function of applied strain at early stage of loading; however, beyond a certain saturation crack spacing, further increase in loading only reduces the stiffness. At an early range of loading, reduction of stiffness can be attributed to the formation of first transverse crack, and during the later loading stages, this reduction is due to the widening of the cracks under increased strains, resulting in textile delamination. At the final stages, the slope of the stress–strain curve is controlled mainly by the properties of the textile–matrix interface and textile performance.

MICROCRACK–TEXTILE INTERACTION MECHANISMS

Figure 13.18 shows a side view of glass textile composite at the end of the tensile test. The crack propagates through the thickness of the specimen from one textile layer to the next. The reinforcing textile's role in crack arrest and bridging is clear as it leads to high mechanical performance and energy absorption in tensile responses. As cracks propagate from one layer to the next, delamination cracks form between the textiles and matrix (Figure 13.18c).

Crack arresting and bridging by textile layers is observed for the PE system in Figure 13.19. The textile yarns have a woven geometry and matrix cracks in the vicinity of the fabric junctions. As the reinforcing yarn is straightened during tensile loading, tensile stresses are developed in the matrix

FIGURE 13.18 Crack propagation in pultruded glass textile composite: (a and b) crack propagation across the loading direction shown by arrows and (c) cracking across and along the loading direction at failure.

FIGURE 13.19 (a and b) Crack propagation across the loading direction of woven PE textile–cement composite and (c) schematic representation of stresses developed at the fabric junction.

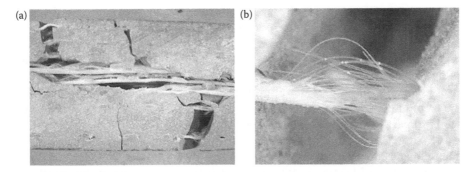

FIGURE 13.20 (a) Failure of the fabric and distributed cracking at the end of testing in glass fabric composite. (b) Fiber buckling due to crack closure during unloading.

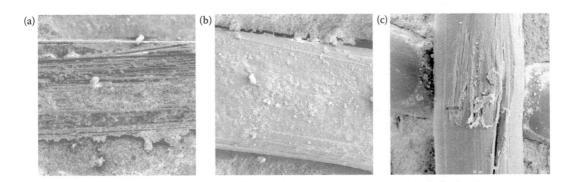

FIGURE 13.21 Scanning electron micrographs of the textile in pultruded cement composite with high content of silica fume (10%): (a) glass yarn along the pultrusion direction, (b) glass yarn perpendicular to the pultrusion direction, and (c) PE textile at the junction of the yarns (scale bar = 100 μm).

around the pressed perpendicular yarn (Figure 13.19c), leading to further damage at this area, as clearly observed in Figure 13.19b.

The failure of the glass textile due to total delamination at the end of test is shown in Figure 13.20. Figure 13.20b shows the fiber buckling due to crack closure during unloading. The fiber pullout during crack opening is irreversible and as cracks unload, closing results in compression, leading to buckling of the yarn.

Figure 13.21 shows the micrographs of the textile in pultruded cement composite with high content of SF (10%). Figure 13.21a shows a glass yarn along the pultrusion direction with a significant damage to the sizing material due to sliding and straining of the yarns. A glass yarn perpendicular to the pultrusion direction as shown in Figure 13.21b is however is not affected as severely. Figure 13.21c shows the damage of a PE textile at the junction of the yarns.

It is clear that during manufacturing, the high viscosity of the SF mix (shear strength of 697.2 Pa) resulted in surface damage to the yarns in both textile systems, glass (Figure 13.21a) and PE (Figure 13.21c). Such damage was observed mainly at the surface of the yarn located along the pultrusion process. The sizing is damaged at the surface of the glass yarn as it passed through the matrix bath during the pultrusion process, whereas with the PE textile system, damage is observed at the junction points of the textile. Such damage was not observed for the mix with lower content of SF or for the yarns perpendicular to the pultrusion process (Figure 13.21c).

CONCLUSIONS

Because the stiffness of the bond in textile reinforced cement composites is quite well developed and leads to distributed crack formation as the primary and dominant mode during a large strain range. The transition from distributed cracking to textile debonding is the secondary mode of response and takes place at the later loading stages. The transition from the primary to secondary mode depends of the textile type.

A variety of processing and material properties were evaluated for continuous fabric systems. The technique of pultrusion can be used to manufacture textile reinforced cement-based system with properties superior to conventional cement-based composite systems. A comparative study of AR glass and PE textiles was used to highlight the potential mechanisms. The use of pozzolanic materials such as SF and fly ash were shown to affect the mechanical properties. The main mechanism of loading in PE composite systems was crack widening due to textile debonding and pullout, whereas in the glass textile composites, crack widening was not the governing mechanism until a majority of the strain capacity was reached. Although the stiffness reduction in glass is very high, it still has a higher tangent stiffness than the PE textile systems at the postcracked stage. The intensity of the static pressure applied after casting affects the mechanical behavior of the pultruded composites. Increasing the pressure improves the tensile strength.

REFERENCES

1. Swamy, R. N., and Hussin, M. W. (1990), "Continuous woven polypropylene mat reinforced cement composites for applications in building construction," in Hamelin, P., and Verchery, G. (eds.), *Textile Composites in Building Construction, Part 1*, Pluralis, Paris, 57–67.
2. Perez-Pena, M., Mobasher, B., and Alfrejd, M. A. "Influence of pozzolans on the tensile behavior of reinforced lightweight concrete," Materials Research Society, Innovations in the Development and Characterization of Materials for Infrastructure, December 1991, Boston, MA.
3. Kruger, M., Ozbolt, J., and Reinhardt, H. W. "A new 3D discrete bond model to study the influence of bond on structural performance of thin reinforced and prestressed concrete plates," *Proceeding of the Fourth International RILEM Workshop on High Performance Fiber Reinforced Cement Composites (HPFRCC4)*, eds. A. E. Naaman and H. W., Reindhart, Ann Arbor, 2003, 49–63.
4. Meyer, C., and Vilkner, G. "Glass concrete thin sheets prestressed with aramid fiber mesh," *Proceeding of the Fourth International RILEM Workshop on High Performance Fiber Reinforced Cement Composites (HPFRCC4)*, ed. A. E. Naaman and H. W. Reindhart, Ann Arbor, 2003, 325–336.
5. Häußler-Combe, U., Jesse, F., and Curbach, M. "Textile reinforced concrete—overview, experimental and theoretical investigations," in Li, V. C. et al. (eds.), *Fracture Mechanics of Concrete Structures*. Proceedings of the Fifth International Conference on Fracture Mechanics of Concrete and Concrete Structures/Vail, Colorado, USA, April 12–16, 2004, Ia-FraMCos, 204, 749–756.
6. Bentur, A., Peled, A., and Yankelevsky, D. (1997), "Enhanced bonding of low modulus polymer fibers–cement matrix by means of crimped geometry," *Cement and Concrete Research*, 27(7), 1099–1111.
7. Peled, A, Bentur, A, and Yankelevsky, D. (1999), "Flexural performance of cementitious composites reinforced by woven fabrics," *Journal of Materials in Civil Engineering*, 11(4), 325–330.

8. Peled, A., and Mobasher, B. (2006), "Properties of fabric–cement composites made by pultrusion," *Materials and Structures*, 39(8), 787–797.

9. Peled, A., and Mobasher, B. "Cement based pultruded composites with fabrics," *Proceedings of the 7th International Symposium on Brittle Matrix Composites (BMC7)*, Warsaw, Poland, 2003, 505–514.

10. Agarwal, B. D., and Broutman, L. J. (1990), *Analysis and Performance of Fiber Composites*, 2nd ed., New York: Wiley.

11. Mobasher, B., Pahilajani, J., and Peled, A. (2006), "Analytical simulation of tensile response of fabric reinforced cement based composites," *Cement and Concrete Composites*, 28(1), 77–89.

12. Aveston, J., Cooper, G. A., and Kelly, A., "The properties of fiber composites," *Conference Proceedings, National Physical Laboratory* (IPC Science and Technology Press Ltd), Paper 1, 1971, 15.

13. Mobasher, B., and Li, C. Y. (1996), "Effect of interfacial properties on the crack propagation in cementitious composites," *Advanced Cement Based Materials*, 4(3), 93–106.

14. Mobasher, B., and Shah, S. P., "Interaction between fibers and the cement matrix in glass fiber reinforced concrete," American Concrete Institute, ACI SP-124, 1990, 137–156.

15. Mobasher, B., Peled, A., and Pahilajani, J. (2006), "Distributed cracking and stiffness degradation in fabric–cement composites," *Materials and Structures*, 39, 317–331.

16. Stang, H., Mobasher, B., and Shah, S. P. (1990), "Quantitative damage characterization in polypropylene fiber reinforced concrete," *Cement and Concrete Research*, 20(4), 540–558.

17. Mobasher, B. (2003), "Micromechanical modeling of angle ply filament wound cement based composites," *Journal of Engineering Mechanics*, 129(4), 373–382.

18. Pahilajani, J. (2004), "Fabric-reinforced, cement-based laminated composites: An experimental and theoretical study," MS thesis, Arizona State University.

19. Sueki, S. (2003), "An analytical and experimental study of fabric-reinforced, cement-based laminated composites," MS thesis, Arizona State University.

20. Marikunte, S., Aldea, C., and Shah, S. P. (1997), "Durability of glass fiber reinforced cement composites," *Advanced Cement Based Materials*, 5, 100–108.

21. Mobasher, B., Peled, A., and Pahilajani, J., "Pultrusion of fabric reinforced high flyash blended cement composites," *Proceedings, RILEM Technical Meeting, BEFIB*, 2004, 1473–1482.

22. Aveston, J., Cooper, G. A., and Kelly, A., "The properties of fiber composites," Paper 1, *Conference Proceedings, National Physical Laboratory*, IPC Science and Technology Press Ltd., Surrey, England, 1971, 15–26.

23. Mobasher, B., and Shah, S. P. (1989), "Test parameters in toughness evaluation of glass fiber reinforced concrete panels," *ACI Materials Journal*, 86(5), 448–458.

24. Zhu, W., and Bartos, P. J. M. (1997), "Assessment of interfacial microstructure and bond properties in aged GRC using a novel microindentation method," *Cement and Concrete Research*, 27(11), 1701–1711.

25. Trtik, P., Reeves, C. M., and Bartos, P. J. M. (2000), "Use of focused ion beam (FIB) techniques for production of diamond probe for nanotechnology-based single filament push-out tests," *Journal of Materials Science Letters*, 19, 903–905.

14 Flexural Model for Strain-Softening and Strain-Hardening Composites

INTRODUCTION

The nonlinear behavior of fiber reinforced concrete (FRC) is best characterized by tension or flexure tests. Analysis of the results however points out to different behaviors and conclusions. Figure 14.1a presents tensile stress–strain response as compared with equivalently elastic flexural stress versus deflection of alkali-resistant (AR) glass textile reinforced composite material [1]. It is imperative to note that the general shapes of these curves are similar as they represent initial linear portions followed by a range with a reduced stiffness that is due to distributed cracking. There is however a fundamental difference in the magnitudes of nominal elastically equivalent stress from the tests and the associated deformations. The two main parameters characterizing the tensile response are the first cracking tensile strength or the bend over point (BOP) and the ultimate tensile strength (UTS). In the flexural loading case, the first cracking is referred to as the limit of proportionality (LOP) and ultimate strength as the modulus of rupture (MOR). The fact that the MOR value may be several times higher than the UTS can be attributed to several parameters, including the nature of equations used, the size effect, and also the nature of the loading. This discrepancy has been well known in the field [2] and is best shown by comparing the cumulative probability distribution functions for the four strength parameters. The distribution of BOP and UTS in tension followed by LOP and MOR in flexure is shown in Figure 14.1b using the results from thin section textile cement composites [1]. Fundamental overpredictions of first cracking and ultimate strength in flexure as compared with the tension are by as much as 300%. Therefore, use of flexural data as fundamental material properties in the design of cement composites may be misleading and nonconservative.

Closed-form relationships that explain such differences between the tensile and the flexural strengths have been recently proposed for both strain-hardening [3,4] and strain-softening-type composites [5]. The present formulation combines and extends the available modeling techniques using a unified approach.

To correlate tensile and flexural data for various materials, an approach for closed-form solution of moment–curvature response of homogenized materials is presented. Direct explicit derivations make iterative procedures for handling material nonlinearities unnecessary. This method is attractive for use in inverse analysis algorithms to back-calculate material parameters from convenient flexural tests. Furthermore, the closed-form solution of moment–curvature response can be used as an input section property for a beam element in nonlinear finite element analysis to predict flexural behavior of more complex structures. Finally, the closed-form solutions of the proposed equations can be subsequently used in a simplified design procedure for cement composites.

243

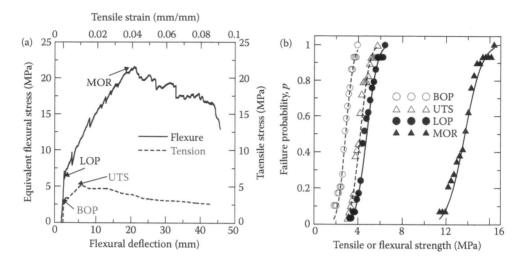

FIGURE 14.1 Correlation of uniaxial tension and flexural test of textile cement composites; (a) experimental responses and (b) cumulative probability distributions of LOP, MOR, BOP, and UTS.

CORRELATION OF TENSILE AND FLEXURAL STRENGTH FROM WEIBULL STATISTICS PERSPECTIVE

The strength of brittle material can be analyzed by the Weibull distribution function. The fracture probability can be given as

$$F(\sigma) = 1 - \exp\left[-V\left(\frac{\sigma}{\sigma_0}\right)^m\right] \tag{14.1}$$

where $F(\sigma)$ is the fracture probability to a stress σ, σ_0 is the scale parameter, V is the effective volume, and m is the shape parameter known as the Weibull modulus. The mean $\bar{\sigma}$ and the coefficient of variation (CV) of the strength are given by

$$\bar{\sigma} = \sigma_0 V^{-1/m} \Gamma\left(1 + \frac{1}{m}\right) \tag{14.2}$$

$$CV = \left[\frac{\Gamma(1+(2/m))}{\Gamma^2(1+(1/m))} - 1\right]^{1/2} \times 100 \tag{14.3}$$

where, for a specimen under uniform tension, the effective volume V is the tested specimen volume. In the case of the bend test, a stress gradient exists in the specimen. The effective volume, V, for the three-point bend strength is given by [6]

$$V = \frac{1}{2(m+1)^2 V_0} \tag{14.4}$$

where V_0 is the volume of the specimen between two supports. The effective volume of the four-point bend strength tested with a load span of one-third of a support span is given by

$$V = \frac{m+3}{6(m+1)} V_0 \qquad (14.5)$$

Assuming that the strength data follow weakest-link scaling and that the stress state is elastic in the flexure specimens, the σ_b/σ_t ratio can be estimated using the following equation derived from the comparison of the failure probabilities of the tensile and the three-point bending test specimens:

$$\frac{\sigma_b}{\sigma_t} = \left[2(m+1)2\left(\frac{V_t}{V_b}\right) \right]^m \qquad (14.6)$$

where V_t is the volume of the tensile and V_b is the volume of the three-point bending specimens [7].

DERIVATION OF CLOSED-FORM SOLUTIONS FOR MOMENT–CURVATURE DIAGRAM

Figure 14.2 presents a constitutive model for homogenized strain-softening and strain-hardening FRC. As shown in Figure 14.2a, the linear portion of an elastic–perfectly plastic compressive stress–strain response terminates at yield point (ε_{cy}, σ_{cy}) and remains constant at compressive yield stress σ_{cy} until the ultimate compressive strain ε_{cu}. The tension model in Figure 14.2b is described by a trilinear response with an elastic range defined by E and then postcracking modulus E_{cr}. By setting E_{cr} to either a negative or a positive value, the same model can be used to simulate strain-softening or strain-hardening materials. The third region in the tensile response is a constant stress range defined with stress σ_{cst} in the postcrack region. The constant stress level μ can be set to any value at the transition strain, resulting in a continuous or discontinuous stress response. Two strain measures are used to define the first cracking and transition strains (ε_{cr}, ε_{trn}). The tensile response terminates at the ultimate tensile strain level of ε_{tu}. The stress–strain relationship for compression and tension can be expressed as

$$\sigma_c(\varepsilon_c) = \begin{cases} E_c \varepsilon_c & 0 \le \varepsilon_c \le \varepsilon_{cy} \\ E_c \varepsilon_{cy} & \varepsilon_{cy} \le \varepsilon_c \le \varepsilon_{cu} \\ 0 & \varepsilon_c > \varepsilon_{cu} \end{cases} \qquad (14.7)$$

$$\sigma_t(\varepsilon_t) = \begin{cases} E\varepsilon_t & 0 \le \varepsilon_t \le \varepsilon_{cr} \\ E\varepsilon_{cr} + E_{cr}(\varepsilon_t - \varepsilon_{cr}) & \varepsilon_{cr} \le \varepsilon_t \le \varepsilon_{trn} \\ \mu E\varepsilon_{cr} & \varepsilon_{trn} \le \varepsilon_t \le \varepsilon_{tu} \\ 0 & \varepsilon_t > \varepsilon_{tu} \end{cases} \qquad (14.8)$$

where, σ_c, σ_t, ε_c, and ε_t are compressive and tensile stresses and strains, respectively. To derive the closed-form solutions for moment–curvature response in nondimensional forms, the material parameters shown in Figures 14.2a and 14.2b are defined as a combination of two intrinsic material

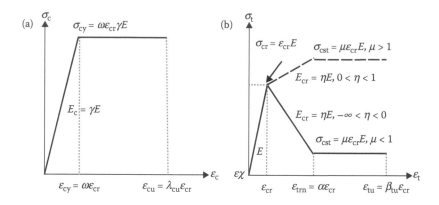

FIGURE 14.2 Material models for homogenized FRC: (a) compression model and (b) tension model.

parameters: the first cracking tensile strain ε_{cr} and tensile modulus E in addition to seven normalized parameters with respect to E and ε_{cr} as shown in Equations 14.9 through 14.11:

$$\omega = \frac{\varepsilon_{cy}}{\varepsilon_{cr}}; \quad \alpha = \frac{\varepsilon_{trn}}{\varepsilon_{cr}}; \quad \beta_{tu} = \frac{\varepsilon_{tu}}{\varepsilon_{cr}}; \quad \lambda_{cu} = \frac{\varepsilon_{cu}}{\varepsilon_{cr}} \tag{14.9}$$

$$\gamma = \frac{E_c}{E}; \quad \eta = \frac{E_{cr}}{E} \tag{14.10}$$

$$\mu = \frac{\sigma_{cst}}{E\varepsilon_{cr}} \tag{14.11}$$

The normalized tensile strain at the bottom fiber β and compressive strain at the top fiber λ are defined as

$$\beta = \frac{\varepsilon_{tbot}}{\varepsilon_{cr}}; \quad \lambda = \frac{\varepsilon_{ctop}}{\varepsilon_{cr}} \tag{14.12}$$

They are linearly related through the normalized neutral axis parameter, k,

$$\frac{\lambda\varepsilon_{cr}}{kd} = \frac{\beta\varepsilon_{cr}}{d-kd} \quad \text{or} \quad \lambda = \frac{k}{1-k}\beta \tag{14.13}$$

Substitution of all normalized parameters defined in Equations 14.3 through 14.6 into Equations 14.1 and 14.2 results in the following normalized stress–strain models:

$$\frac{\sigma_c(\lambda)}{E\varepsilon_{cr}} = \begin{cases} \gamma\lambda & 0 \le \lambda \le \omega \\ \gamma\omega & \omega < \lambda \le \lambda_{cu} \\ 0 & \lambda_{cu} < \lambda \end{cases} \qquad \frac{\sigma_t(\beta)}{E\varepsilon_{cr}} = \begin{cases} \beta & 0 \le \beta \le 1 \\ 1 + \eta(\beta-1) & 1 < \beta \le \alpha \\ \mu & \alpha < \beta \le \beta_{tu} \\ 0 & \beta_{tu} \le \beta \end{cases} \tag{14.14}$$

In the derivation of moment–curvature diagram for a rectangular cross section with a width b and a depth d, the Kirchhoff hypothesis of plane section remaining plane for flexural loading is applied. By assuming linear strain distribution across the depth and by ignoring shear deformation, the stress–strain relationships in Figures 14.2a and 14.2b are used to obtain the stress distribution across the cross section as shown in Figure 14.3 at three stages of imposed tensile strain: $0 < \beta < 1$, $1 < \beta < \alpha$, and $\alpha < \beta < \beta_{tu}$. For stages 2 and 3, there are two possible scenarios: the compressive strain at top fiber is either elastic $(0 < \lambda < \omega)$ or plastic $(\omega < \lambda < \lambda_{cu})$. These cases will be treated in subsequent sections. Normalized heights of compression and tension zones with respect to beam depth d and the normalized magnitudes of stress at the vertices with respect to the first cracking stress $E\varepsilon_{cr}$ are presented in Tables 14.1 and 14.2, respectively. The area and the centroid of stress in each zone represent the force components and lines of action. Their normalized values with respect to cracking tensile force $bdE\varepsilon_{cr}$ and beam depth d are presented in Tables 14.3 and 14.4, respectively.

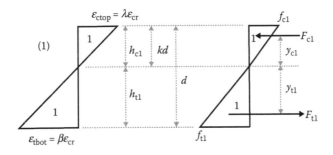

FIGURE 14.3 Stress–strain diagram at stage 1 of normalized tensile strain at the bottom fiber (β): (a) $0 < \beta < 1$ and $\lambda < \omega$.

TABLE 14.1
Normalized Height of Compression $(h' = h/d)$ and Tension Zones for Each Stage of Normalized Tensile Strain at Bottom Fiber (β)

Stage	1	2.1	2.2	3.1	3.2
Tension	$0 < \beta < 1$	$1 < \beta < \alpha$	$1 < \beta < \alpha$	$\beta > \alpha$	$\beta > \alpha$
Compression	$\lambda < \omega$	$\lambda < \omega$	$\omega < \lambda < \lambda_{cu}$	$\lambda < \omega$	$\omega < \lambda < \lambda_{cu}$
h'_{c2}	–	–	$\dfrac{k\beta - \omega(1-k)}{\beta}$	–	$\dfrac{k\beta - \omega(1-k)}{\beta}$
h'_{c1}	k	k	$\dfrac{\omega(1-k)}{\beta}$	k	$\dfrac{\omega(1-k)}{\beta}$
h'_{t1}	$1-k$	$\dfrac{1-k}{\beta}$		$\dfrac{1-k}{\beta}$	
h'_{t2}	–	$\dfrac{(1-k)(\beta-1)}{\beta}$		$\dfrac{(1-k)(\alpha-1)}{\beta}$	
h'_{t3}	–	–	–	$\dfrac{(1-k)(\beta-\alpha)}{\beta}$	

TABLE 14.2

Normalized Stress ($f' = f/Ee_{cr}$) at Vertices in the Stress Diagram for Each Stage of Normalized Tensile Strain at Bottom Fiber (β)

Stage	1	2.1	2.2	3.1	3.2
Tension	$0 < \beta < 1$	$1 < \beta < \alpha$	$1 < \beta < \alpha$	$\beta > \alpha$	$\beta > \alpha$
Compression	$\lambda < \omega$	$\lambda < \omega$	$\omega < \lambda < \lambda_{cu}$	$\lambda < \omega$	$\omega < \lambda < \lambda_{cu}$
f'_{c2}	–	–	$\omega\gamma$	–	$\omega\gamma$
f'_{c1}	$\dfrac{\gamma\beta k}{1-k}$	$\dfrac{\gamma\beta k}{1-k}$	$\omega\gamma$	$\dfrac{\gamma\beta k}{1-k}$	$\omega\gamma$
f'_{t1}	β	1		1	
f'_{t2}	–	$1+\eta(\beta-1)$		$1+\eta(\alpha-1)$	
f'_{t3}	–	–	–	μ	

TABLE 14.3

Normalized Force Component $F' = F/bdEe_{cr}$ as a Function of Tensile Strain (β)

Stage	1	2.1	2.2	3.1	3.2
Tension	$0 < \beta < 1$	$1 < \beta < \alpha$	$1 < \beta < \alpha$	$\beta > \alpha$	$\beta > \alpha$
Compression	$\lambda < \omega$	$\lambda < \omega$	$\omega < \lambda < \lambda_{cu}$	$\lambda < \omega$	$\omega < \lambda < \lambda_{cu}$
F'_{c2}	–	–	$\dfrac{\omega\gamma}{\beta}(\beta k + \omega k - \omega)$	–	$\dfrac{\omega\gamma}{\beta}(\beta k + \omega k - \omega)$
F'_{c1}	$\dfrac{\beta\gamma k^2}{2(1-k)}$	$\dfrac{\beta\gamma k^2}{2(1-k)}$	$\dfrac{\omega^2\gamma}{2\beta}(1-k)$	$\dfrac{\beta\gamma k^2}{2(1-k)}$	$\dfrac{\omega^2\gamma}{2\beta}(1-k)$
F'_{t1}	$\dfrac{\beta}{2}(1-k)$	$\dfrac{(1-k)}{2\beta}$		$\dfrac{(1-k)}{2\beta}$	
F'_{t2}	–	$\dfrac{(1-k)(\beta-1)(\eta\beta-\eta+2)}{2\beta}$		$\dfrac{(1-k)(\alpha-1)(\eta\alpha-\eta+2)}{2\beta}$	
F'_{t3}	–	–	–	$\dfrac{(1-k)(\beta-\alpha)\mu}{\beta}$	

Using the conditions of plane sections remaining plane and the equilibrium of internal forces, the strain distribution, the stress distribution, and the internal forces can be obtained using standard analytical approaches

$$F = C + T = \sum_{i=1}^{n} F_i = \int_{z=0}^{d} \sigma(z)bdz = 0 \tag{14.15}$$

$$M = \sum_{i=1}^{n} M_i = \int_{z=0}^{d} \sigma(z)zbdz \tag{14.16}$$

The location of neutral axis, defined as k, and the normalized moment and curvature ($m-\varphi$) relationships can be obtained in closed form as parametric representations of β, κ, γ, and η and represented in terms of two independent materials parameters E_1 and ε_{cr}.

TABLE 14.4
Normalized Internal Moment Arm ($y' = y/d$) as a Function of Tensile Strain β

Stage	1	2.1	2.2	3.1	3.2
Tension	$0 < \beta < 1$	$1 < \beta < \alpha$	$1 < \beta < \alpha$	$\beta > \alpha$	$\beta > \alpha$
Compression	$\lambda < \omega$	$\lambda < \omega$	$\omega < \lambda < \lambda_{cu}$	$\lambda < \omega$	$\omega < \lambda < \lambda_{cu}$
y'_{c2}	–	–	$\dfrac{\beta k + \omega(1-k)}{2\beta}$	–	$\dfrac{\beta k + \omega(1-k)}{2\beta}$
y'_{c1}	$\dfrac{2}{3}k$	$\dfrac{2}{3}k$	$\dfrac{2}{3}\dfrac{\omega(1-k)}{\beta}$	$\dfrac{2}{3}k$	$\dfrac{2}{3}\dfrac{\omega(1-k)}{\beta}$
y'_{t1}	$\dfrac{2}{3}(1-k)$	$\dfrac{2}{3}\dfrac{(1-k)}{\beta}$		$\dfrac{2}{3}\dfrac{(1-k)}{\beta}$	
y'_{t2}	–	$\dfrac{2\eta\beta^2 - \eta\beta - \eta + 3\beta + 3}{3\beta(\eta\beta - \eta + 2)}(1-k)$		$\dfrac{2\eta\alpha^2 - \eta\alpha - \eta + 3\alpha + 3}{3\beta(\eta\alpha - \eta + 2)}(1-k)$	
y'_{t3}	–	–	–	$\dfrac{(\alpha + \beta)}{2\beta}(1-k)$	

STAGE 1: $(0 < \beta < 1)$ AND $(\lambda < \omega)$

During stage 1, the tensile and the compressive zones are both elastic. This derivation is shown to familiarize the average person based on elastic mechanics of materials approach. Note that because of differences in the elastic modulus the neutral axis may not be at the center of the rectangular section, that is, $k = 0.5$ only in the case where $\gamma = 1$. The general case of $\gamma > 1$ is presented here. The depth of neutral axis is defined as:

$$h_{c1} = kd, \quad h_{t1} = (1-k)d \tag{14.17}$$

The stresses are obtained on the basis of the stress at the tension fiber or $\beta\varepsilon_{cr} < \varepsilon_{cr}$:

$$f_{c1} = \frac{\gamma\beta k}{1-k}E\varepsilon_{cr}, \quad f_{t1} = E\varepsilon_{cr}\beta \tag{14.18}$$

Tensile and compressive forces and their line of action are obtained by integration of the stresses across the following depth:

$$F_{c1} = \int_{z=0}^{h_{c1}} \sigma(z)bdz = \int_{z=0}^{h_{c1}} \frac{\gamma\beta k}{1-k}E\varepsilon_{cr}bdz = \frac{\beta\gamma k^2}{2(1-k)}bdE\varepsilon_{cr}$$

$$F_{t1} = \int_{z=0}^{h_{t1}} \sigma(z)bdz = \int_{z=0}^{h_{t1}} E\varepsilon_{cr}\beta bdz = \frac{\beta}{2}(1-k)bdE\varepsilon_{cr} \tag{14.19}$$

$$y_{c1} = \frac{2}{3}kd$$

$$y_{t1} = \frac{2}{3}(1-k)d \tag{14.20}$$

The force and the moment equilibrium require that

$$\sum F = -F_{c1} + F_{t1} = 0$$
$$\sum M = F_{c1}y_{c1} + F_{t1}y_{t1}$$

(14.21)

The solution to the first equation in terms of value of neutral axis k is defined as k_1 and obtained as

$$k_1 = \begin{cases} \dfrac{1}{2} & \text{for } \gamma = 1 \\[2mm] \dfrac{-1+\sqrt{\gamma}}{-1+\gamma} & \text{for } \gamma < 1 \text{ or } \gamma > 1 \end{cases}$$

(14.22)

The magnitude of the internal moment is obtained using the values of k_1 in Equation 14.16. The answer for $\gamma = 1$ results in the elementary solution for linear elastic materials, whereas the second equation yields

$$M'_1 = \frac{2\beta\left[(\gamma-1)k_1^3 + 3k_1^2 - 3k_1 + 1\right]}{1 - k_1}$$

(14.23)

According to the geometry of Figure 14.3, the curvature as a function of neutral axis depth is

$$\phi'_1 = \frac{\beta}{2(1-k_1)}$$

(14.24)

STAGE 2: $1 < \beta < \alpha$

There are two potential regions when the elastic stage 1 ends and the tensile cracking takes place. This zone is defined as the tensile cracking region and identified in terms of parameter ($1 < \beta < \alpha$). The compression zone is initially elastic, and it may or may not enter the plastic range. The first of these two cases when we have an elastic compression response is defined as zone 2.1 ($1 < \beta < \alpha, \lambda < \omega$). If the compression enters the plastic range, then the zone is defined as cracking tension-plastic compression defined by zone 2.2 ($1 < \beta < \alpha, \lambda > \omega$) (Figure 14.4). The first of these zones (zone 2.1) has an operating range in the domain $1 < \beta < \alpha, \lambda < \omega$, which is derived first:

$$h_{c1} = kd$$
$$h_{t1} = \frac{1-k}{\beta}d$$
$$h_{t12} = \frac{(1-k)(\beta-1)}{\beta}d$$

(14.25)

$$f_{c1} = \frac{\gamma\beta k}{1-k}E\varepsilon_{cr}$$
$$f_{t1} = E\varepsilon_{cr}$$
$$f_{t2} = \left[1+\eta(\beta-1)\right]E\varepsilon_{cr}$$

(14.26)

The stresses and the line of action of forces are shown in Figure 14.5 and listed in Tables 14.2 through 14.4. The equilibrium of internal forces results in a solution to location of neutral axis defined by

$$k_{21} = \frac{\beta^2\gamma + D_{21} - \sqrt{\gamma^2\beta^4 + D_{21}\gamma\beta^2}}{D_{21}}, \quad D_{21} = \eta(\beta^2 - 2\beta + 1) + 2\beta - \beta^2\gamma - 1 \quad (14.27)$$

The moment in this range and the curvature defined as $\varepsilon_c/(kd)$ are obtained as

$$M'_{21} = \frac{(2\beta\gamma + C_{21})k_{21}^3 - 3C_{21}k_{21}^2 + 3C_{21}k_{21} - C_{21}}{1 - k_{21}}, \quad C_{21} = \frac{-2\eta\beta^3 + 3\eta\beta^2 - 3\beta^2 - \eta + 1}{\beta^2} \quad (14.28)$$

$$\phi'_{21} = \frac{\beta}{2(1 - k_{21})} \quad (14.29)$$

As the sample is loaded beyond the ultimate tensile strain cracking ($\beta > 1$), the compression zone may potentially enter the yielding stage. Stage 2.2 is defined as the compression yielding ($\omega < \lambda < \lambda_{cu}$). Its governing equilibrium equations are defined by two zones in tension and two zones in compression. Its derivation follows similar steps with results presented in Tables 14.1 through 14.5.

$$k_{22} = \frac{D_{22}}{D_{22} + 2\omega\gamma\beta}, \quad D_{22} = \eta(\beta^2 - 2\beta + 1) + 2\beta + \omega^2\gamma - 1 \quad (14.30)$$

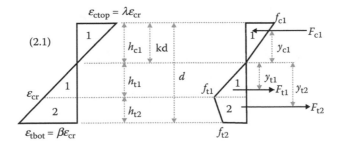

FIGURE 14.4 Stress–strain diagram at stage 1 of normalized tensile strain at the bottom fiber (β): (2.1) $1 < \beta < \alpha$ and $\lambda < \omega$.

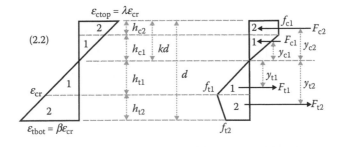

FIGURE 14.5 Stress–strain diagram at stage 1 of normalized tensile strain at the bottom fiber (β): (2.2) $1 < \beta < \alpha$ and $\omega < \lambda < \lambda_{cu}$.

$$M'_{22} = (3\omega\gamma + C_{22})k_{22}^{2} - 2C_{22}k_{22} + C_{22}, \quad C_{22} = \frac{2\eta\beta^{3} - 3\eta\beta^{2} + 3\beta^{2} - \omega^{3}\gamma + \eta - 1}{\beta^{2}} \quad (14.31)$$

$$\phi'_{22} = \frac{\beta}{2(1 - k_{22})} \quad (14.32)$$

STAGE 3: $\beta > \alpha$

There are two potential regions when the stage 2 ends and the tensile softening initiates depending on whether the transition takes place form region 2.1f or 2.2. The tensile-softening region is identified in terms of a parameter ($\beta > \alpha$). The compression zone may be in the elastic or the plastic range. The first of these two cases is defined as zone 3.1 ($\beta > \alpha$, $\lambda < \omega$). If the compression enters the plastic range as well, then the zone is defined as tension-softening plastic compression defined by zone 3.2 ($\beta > \alpha$, $\lambda > \omega$). The first of these zones (zone 3.1) with an operating range in the domain $\beta > \alpha$, $\lambda < \omega$ is derived first.

Stage 3.1: $\beta > \alpha$ and $\lambda < \omega$

As shown in Figure 14.6, the depths of each region and the stresses are denoted as

$$\frac{h_{c1}}{d} = k, \quad h_{t1} = \frac{1-k}{\beta}d, \quad h_{t2} = \frac{(1-k)(\alpha-1)}{\beta}d, \quad h_{t3} = \frac{(1-k)(\beta-\alpha)}{\beta}d \quad (14.33)$$

$$f_{c1} = \frac{\gamma\beta k}{1-k}E\varepsilon_{cr}, \quad f_{t1} = E\varepsilon_{cr}, \quad f_{t2} = [1 + \eta(\alpha - 1)]E\varepsilon_{cr}, \quad f_{t3} = \mu E\varepsilon_{cr} \quad (14.34)$$

The moment in this range and the curvature are obtained as

$$k_{31} = \frac{D_{31} - \sqrt{\gamma\beta^{2}D_{31}}}{D_{31} - \beta^{2}\gamma}, \quad D_{31} = \eta(\alpha^{2} - 2\alpha + 1) + 2\mu(\beta - \alpha) + 2\alpha - 1 \quad (14.35)$$

$$M'_{31} = \frac{(C_{31} - 2\beta\gamma)k_{31}^{3} - 3C_{31}k_{31}^{2} + 3C_{31}k_{31} - C_{31}}{k_{31} - 1}$$

$$C_{31} = \frac{3(\mu\beta^{2} - \mu\alpha^{2} - \eta\alpha^{2} + \alpha^{2}) + 2\eta\alpha^{3} + \eta - 1}{\beta^{2}} \quad (14.36)$$

$$\phi'_{31} = \frac{\beta}{2(1 - k_{31})} \quad (14.37)$$

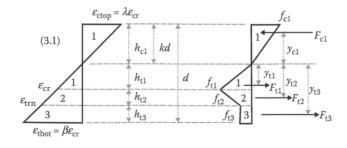

FIGURE 14.6 Stress–strain diagram at stage 3.1 of normalized tensile strain at the bottom fiber (β): (3.1) $\alpha < \beta < \beta_{tu}$ and $\lambda < \omega$.

Stage 3.2: $\beta > \alpha$ and $\omega < \lambda < \lambda_{cu}$

Stage 3.2 corresponds to elastic softening tension and elastic–plastic compression and is computed as (Figure 14.7)

$$k_{32} = \frac{D_{32}}{D_{32} + 2\omega\gamma\beta}, \quad D_{32} = \omega^2\gamma + \eta\alpha^2 + 2(\mu\beta - \eta\alpha - \mu\alpha + \alpha) + \eta - 1 \tag{14.38}$$

$$M'_{32} = (C_{32} + 3\omega\gamma)k_{32}^2 - 2C_{32}k_{32} + C_{32}, \quad C_{32} = \frac{3\mu\beta^2 - 3\alpha^2(\mu + \eta - 1) + 2\eta\alpha^3 - \omega^3\gamma + \eta - 1}{\beta^2} \tag{14.39}$$

$$\phi'_{32} = \frac{\beta}{2(1 - k_{32})} \tag{14.40}$$

Table 14.5 shows the steps in determination of net section force, moment, and curvature at each stage of applied tensile strain, β. The net force is obtained as the difference between the tension and the compression forces, equated to zero for internal equilibrium, and solved for the neutral axis depth ratio k. The expressions for net force in stages 2 and 3 are in the quadratic forms and result in two solutions for k. With a large scale of numerical tests covering a practical range of material parameters, only one solution of k yields the valid value in the range $0 < k < 1$, and it is presented in Table 14.6. The internal moment is obtained by operating on the force components, and their distance from the neutral axis and the curvature is determined as the ratio of compressive strain at top fiber ($\varepsilon_{ctop} = \lambda\varepsilon_{cr}$) to the depth of neutral axis kd. The moment M_i and curvature ϕ_i at each stage i

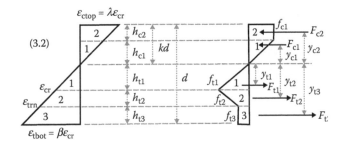

FIGURE 14.7 Stress–strain diagram at stage 1 of normalized tensile strain at the bottom fiber (β): 3.2) $\alpha < \beta < \beta_{tu}$ and $\omega < \lambda < \lambda_{cu}$.

TABLE 14.5
Equilibrium of Force, Moment, and Curvature for Each Stage of Normalized Tensile Strain at the Bottom Fiber (β)

Stage	Tension	Compression	Force Equilibrium	Internal Moment
1	$0 < \beta < 1$	$\lambda < \omega$	$-F_{c1} + F_{t1}$	$F_{c1}y_{c1} + F_{t1}y_{t1}$
2.1	$1 < \beta < \alpha$	$\lambda < \omega$	$-F_{c1} + F_{t1} + F_{t2}$	$F_{c1}y_{c1} + F_{t1}y_{t1} + F_{t2}y_{t2}$
2.2	$1 < \beta < \alpha$	$\omega < \lambda < \lambda_{cu}$	$-F_{c1} - F_{c2} + F_{t1} + F_{t2}$	$F_{c1}y_{c1} + F_{c2}y_{c2} + F_{t1}y_{t1} + F_{t2}y_{t2}$
3.1	$\beta > \alpha$	$\lambda < \omega$	$-F_{c1} + F_{t1} + F_{t2} + F_{t3}$	$F_{c1}y_{c1} + F_{t1}y_{t1} + F_{t2}y_{t2} + F_{t3}y_{t3}$
3.2	$\beta > \alpha$	$\omega < \lambda < \lambda_{cu}$	$-F_{c1} - F_{c2} + F_{t1} + F_{t2} + F_{t3}$	$F_{c1}y_{c1} + F_{c2}y_{c2} + F_{t1}y_{t1} + F_{t2}y_{t2} + F_{t3}y_{t3}$

Note: Curvature $= \varepsilon_c / (kd)$.

TABLE 14.6

Neutral Axis Parameter k, Normalized Moment m, and Normalized Curvature ϕ for Each Stage of Normalized Tensile Strain at the Bottom Fiber (β)

Stage	Parameters	k	$m = M/M_{cr}$	$\phi = \Phi/\Phi_{cr}$
1	$0 < \beta < 1$	$k_1 = \begin{cases} \dfrac{1}{2} & \text{for } \gamma = 1 \\ \dfrac{-1+\sqrt{\gamma}}{-1+\gamma} & \text{for } \gamma \neq 1 \end{cases}$	$m_1 = \dfrac{2\beta\left[(\gamma-1)k_1^{\,3} + 3k_1^{\,2} - 3k_1 + 1\right]}{1-k_1}$	$\phi'_1 = \dfrac{\beta}{2(1-k_1)}$
2.1	$1 < \beta < \alpha$ $0 < \lambda < \omega$	$k_{21} = \dfrac{D_{21} - \sqrt{D_{21}\gamma\beta^2}}{D_{21} - \gamma\beta^2}$ $D_{21} = \eta\left(\beta^2 - 2\beta + 1\right) + 2\beta - 1$	$M'_{21} = \dfrac{\left(2\gamma\beta^3 - C_{21}\right)k_{21}^{\,3} + 3C_{21}k_{21}^{\,2} - 3C_{21}k_{21} + C_{21}}{1-k_{21}}$ $C_{21} = \dfrac{(2\beta^3 - 3\beta^2 + 1)\eta + 3\beta^2 - 1}{\beta^2}$	$\phi'_{21} = \dfrac{\beta}{2(1-k_{21})}$
2.2	$1 < \beta < \alpha$ $\omega < \lambda < \lambda_{cu}$	$k_{22} = \dfrac{D_{22}}{D_{22} + 2\omega\gamma\beta}$ $D_{22} = D_{21} + \gamma\omega^2$	$M'_{22} = \left(3\gamma\omega\beta^2 + C_{22}\right)k_{22}^{\,2} - 2C_{22}k_{22} + C_{22}$ $C_{22} = C_{21} - \dfrac{\gamma\omega^3}{\beta^2}$	$\phi'_{22} = \dfrac{\beta}{2(1-k_{22})}$
3.1	$\alpha < \beta < \beta_{tu}$ $0 < \lambda < \omega$	$k_{31} = \dfrac{D_{31} - \sqrt{D_{31}\gamma\beta^2}}{D_{31} - \gamma\beta^2}$ $D_{31} = \eta\left(\alpha^2 - 2\alpha + 1\right) + 2\mu(\beta-\alpha) + 2\alpha - 1$	$M'_{31} = \dfrac{\left(2\gamma\beta^3 - C_{31}\right)k_{31}^{\,3} + 3C_{31}k_{31}^{\,2} - 3C_{31}k_{31} + C_{31}}{1-k_{31}}$ $C_{31} = \dfrac{(2\alpha^3 - 3\alpha^2 + 1)\eta - 3\mu\left(\alpha^2 - \beta^2\right) + 3\alpha^2 - 1}{\beta^2}$	$\phi'_{31} = \dfrac{\beta}{2(1-k_{31})}$
3.2	$\alpha < \beta < \beta_{tu}$ $\omega < \lambda < \lambda_{cu}$	$k_{32} = \dfrac{D_{32}}{D_{32} + 2\omega\gamma\beta}$ $D_{32} = D_{31} + \gamma\omega^2$	$M'_{32} = \left(3\gamma\omega\beta^2 + C_{32}\right)k_{32}^{\,2} - 2C_{32}k_{32} + C_{32}$ $C_{32} = C_{31} - \dfrac{\gamma\omega^3}{\beta^2}$	$\phi'_{32} = \dfrac{\beta}{2(1-k_{32})}$

are then normalized with respect to the values at cracking M_{cr} and ϕ_{cr}, respectively, and their closed-form solutions are presented in Table 14.6:

$$M_i = M_i' M_{cr}; \quad M_{cr} = \frac{1}{6}bd^2 E\varepsilon_{cr} \tag{14.41}$$

$$\phi_i = \phi_i' \phi_{cr}; \quad \phi_{cr} = \frac{2\varepsilon_{cr}}{d} \tag{14.42}$$

As mentioned earlier, the compressive strain at the top fiber λ in stage 2 or 3 could be either in elastic ($0 < \lambda < \omega$) or in plastic ($\omega < \lambda < \lambda_{cu}$) range, depending on the applied tensile strain β and neutral axis parameter k. The range can be identified by assuming $\lambda < \omega$ (Figure 14.4, case 2.1, or Figure 14.6, case 3.1) and using the expression k_{21} or k_{31} in Table 14.6 to determine λ from Equation 14.7. If $\lambda < \omega$ holds true, the assumption is correct; otherwise, $\lambda > \omega$ and the expression k_{22} or k_{32} is used instead. Once, the neutral axis parameter k and the applicable case are determined, the appropriate expressions for moment and curvature in Table 14.6 and Equations 14.9 and 14.10 are used.

The nominal moment capacity M_n is obtained by taking the first moment of force about the neutral axis, $M_n = F_{c1}y_{c1} + F_{t1}y_{t1} + F_{t2}y_{t2}$, and it is expressed as a product of the normalized nominal moment m_n and the cracking moment M_{cr} as follows:

$$M_n = m_n M_{cr}, \quad M_{cr} = \frac{\sigma_{cr}bh^2}{6} \tag{14.43}$$

$$k = \frac{C_1 - \sqrt{\beta^2 C_1}}{C_1 - \beta^2}, \quad C_1 = \eta\left(\beta^2 - 2\beta + 1\right) + 2\beta - 1 \tag{14.44}$$

If the full stress–strain response is desired, then the location of neutral axis and moment capacity are obtained under the definitions provided in Table 14.1. In Table 14.1, the derivations of all potential combinations for the interaction of tensile and compressive response are presented. Note that depending on the relationship among material parameters, any of the zones 2.1 and 2.2 or 3.1 and 3.2 are potentially possible.

The contribution of fibers is mostly apparent in the postcracking tensile region (Figure 14.1a). The postcrack modulus E_{cr} is relatively flat with values of $\eta = 0.00$–0.4 for a majority of cement composites. The tensile strain at peak strength ε_{peak} is relatively large compared with the cracking tensile strain ε_{cr} and may be as high as $\alpha = 100$ for polymeric-based fiber systems. These unique characteristics cause the flexural strength to continue to increase after cracking. Because typical strain-hardening FRC do not have significant postpeak tensile strength, the flexural strength drops after passing the tensile strain at peak strength. Furthermore, the effect of postcrack tensile response parameter μ can be ignored for a simplified analysis. In the most simplistic way, one needs to determine two parameters in terms of postcrack stiffness η and postcrack ultimate strain capacity α to estimate the maximum moment capacity for the design purposes.

The bilinear tension and the elastic compression model shown in Figures 14.1a and 14.1b indicate the maximum moment capacity at a point when the normalized tensile strain at the bottom fiber ($\beta = \varepsilon_t/\varepsilon_{cr}$) reaches the tensile strain at peak strength ($\alpha = \varepsilon_{peak}/\varepsilon_{cr}$). However, the simplified equations (Equations 14.8 through 14.10) for moment capacity are applicable for the compressive stress in elastic region only. The elastic condition must be checked by computing the normalized compressive strain developed at the top fiber λ and compare it with the normalized yield compressive strain ω. The general solutions for all the cases are presented in Table 14.1. Using the strain diagram in Figure 14.2a, one can obtain the relationship between the top compressive strain and the bottom tensile strain as follows:

$$\frac{\varepsilon_c}{kh} = \frac{\varepsilon_t}{(1-k)h} \tag{14.45}$$

By substituting $\varepsilon_c = \lambda\varepsilon_{cr}$ and $\varepsilon_t = \beta\varepsilon_{cr}$ in Equation 14.13 and limit of maximum compressive strain to yield compressive strain $\varepsilon_{cy} = \omega\varepsilon_{cr}$, the condition can be expressed in a normalized form as

$$\lambda = \frac{k}{1-k} \quad \beta \leq \omega \tag{14.46}$$

SIMPLIFIED EXPRESSIONS FOR MOMENT–CURVATURE RELATIONS

CASE 2.1: $1 < \beta < \rho$ AND $0 < \lambda < \omega$

The case represented by case 2.1 of Table 14.6, where the tensile behavior is elastic–plastic while the compressive behavior is still elastic, is studied first. Equations for other cases can also be developed.

The general solution presented in Table 14.6 can be simplified by representing the location of neutral axis represented as a function of applied tensile strain β:

$$k = \frac{\sqrt{A}}{\sqrt{A} + \beta\sqrt{\gamma}} \quad A = \eta(\beta^2 + 1 - 2\beta) + 2\beta - 1 \tag{14.47}$$

This equation can be simplified by assuming equal tension and compression stiffness ($\lambda = 1$). Furthermore, for an elastic–perfectly plastic tension material ($\eta = 0$), the equation reduces to

$$k = \frac{\sqrt{2\beta - 1}}{\sqrt{2\beta - 1} + \beta} \tag{14.48}$$

Table 14.7 presents the case of ($\lambda = 1$) for different values of postcrack stiffness $\eta = 0.5, 0.2, 0.1$, 0.05, 0.01, and 0.001. Note that the neutral axis is a function β and can be used in calculation of the moment or the moment–curvature relationship. These general responses are shown in Figures 14.8a and 14.8b and show that with an increase in applied tensile strain, the neutral axis compression zone decreases; however, this decrease is a function of postcrack tensile stiffness factor. The moment–curvature relationship in this range is ascending; however, its rate is a function of the postcrack tensile stiffness. The parameter-based fit equations in the third and fourth column are obtained by curve fitting the simulated response from the closed-form derivations and are applicable within 1% accuracy of the closed-form results. Using these equations, one can generate the moment capacity and moment–curvature response for any cross section using basic tensile material parameters in the 2.1 range as defined in Table 14.7.

Figures 14.8a and 14.8b show the depth of neutral axis and the moment as a function of curvature for various stiffness values in the postcrack region η. As η increases, the depth of the neutral axis remains stable and stays closer to the midpoint of the section. The moment response however increases significantly as a function of postcrack stiffness. As η reaches 0.2, the moment capacity almost doubles at $\varphi = 10$ compared with an elastically perfect plastic material.

TABLE 14.7
Location of Neutral Axis, Moment, and Moment–Curvature Response of a Strain-Hardening Composite Material (Case 2.1) with $\gamma = 1$ and $\eta = 0.0001–0.5$

η	$A\left(k = \dfrac{\sqrt{A}}{\sqrt{A} + \beta}\right)$	$M'(k)$	$M'(\varphi)$
0.5	$0.5(\beta^2 + 1 - 2\beta) + 2\beta - 1$	$-0.773 + 0.108 \times 10^{-1} k^{-6}$	$0.507 + 0.686\varphi$
0.2	$0.2(\beta^2 + 1 - 2\beta) + 2\beta - 1$	$0.654 + 0.516 \times 10^{-2} k^{-6}$	$1.105 + 0.383\varphi$
0.1	$0.1(\beta^2 + 1 - 2\beta) + 2\beta - 1$	$1.276 + 0.289 \times 10^{-2} k^{-6}$	$1.461 + .234\varphi$
0.05	$0.05(\beta^2 + 1 - 2\beta) + 2\beta - 1$	$1.645 + .1632 \times 10^{-2} k^{-6}$	$1.720 + .1401\varphi$
0.01	$0.01(\beta^2 + 1 - 2\beta) + 2\beta - 1$	$0.852 + 0.456\, k^{-1}$	$1.342 + 0.371\sqrt{\varphi}$
0.0001	$0.0001(\beta^2 + 1 - 2\beta) + 2\beta - 1$	$3.177 - 3.068\, k$	$3.021 - 2.047/\sqrt{\varphi}$

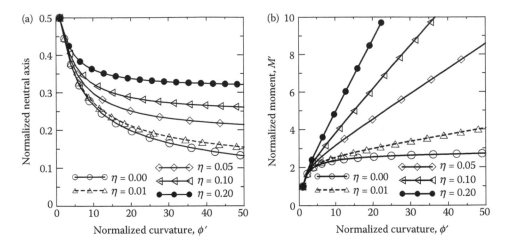

FIGURE 14.8 Effect of (a) the depth of neutral axis on the moment capacity of a section and (b) the moment–curvature response in the range 2.1.

TABLE 14.8
Location of Neutral Axis, Moment–Neutral Axis Depth, and Moment–Curvature Response of a Strain-Hardening Composite Material with $\lambda = 1$, $\eta = 0.1$, and $\alpha = 5$

μ	$k = \dfrac{\sqrt{A}}{\sqrt{A} + \beta}$	$m(k)$	$m(\varphi)$
0.05	$A = 10.1 + 0.1\beta$	$0.113 + 14.358\,k^2$	$0.148 + 37.635\varphi^{-2}$
0.10	$A = 9.6 + 0.2\beta$	$0.228 + 13.556\,k^2$	$0.292 + 35.173\,\varphi^{-2}$
0.20	$A = 8.6 + 0.4\beta$	$0.463 + 11.927\,k^2$	$0.572 + 30.381\,\varphi^{-2}$
0.40	$A = 6.6 + 0.8\beta$	$1.099 + 20.343\,k^3$	$1.105 + 21.185\,\varphi^{-2}$
0.60	$A = 4.6 + 1.2\beta$	$1.651 + 71.169\,k^5$	$1.655 + 47.915\,\varphi^{-3}$

Case 3.1: $\alpha < \beta < \beta_{TU}$ and $0 < \lambda < \omega$

In zone 3.1, the conditions imposed are $(\beta > \alpha)$ and $(0 < \lambda < \omega)$. Simplification of the values presented in Table 14.8 points out to the closed-form equations for parameter k, the neutral axis depth as a function of applied strain in range 3.1. If the compressive strain exceeds the yield limit, the tensile strain must be decreased less than the tensile strain at peak strength until the corresponding compressive stress developed the top fiber is within the elastic limit.

In the case of a strain-hardening material with equal tension and compressive stiffness, $\gamma = 1$, a postcrack stiffness that is 10% of the elastic stiffness, $\eta = 0.1$, and ultimate strain capacity of $\alpha = 5$, one can simplify the equations of case 3.1 to calculate the strength in terms of moment capacity, and expressed as a function of applied tensile strain or β. The columns in Table 14.8 show the location of neutral axis, moment capacity versus neutral axis depth, and moment–curvature response. The proper way to use the table is to choose a parameter μ representing postpeak tensile strength capacity, assume a tensile strain β, and use the first column to calculate A and then k, the neutral axis depth. With these values, moment can be obtained from the third column, or moment–curvature from the last column. Parameters of Table 14.8 assume that the compression response in is elastic range. Similar relationships can be developed for case 3.2.

Figures 14.9a and 14.9b show the response of the neutral axis depth and also the moment as a function of the curvature for various values of α. The parameters used for this simulation are $\eta = 0.1$, $\gamma = 1$, and $\mu = 0.2$. As the strain capacity parameter, α, increases, the stiffness in the postcrack region remains unchanged, but the stress increases, thus reducing the rate of change of the neutral axis as it stays closer to the midpoint of the section. The moment response however increases significantly as a function of postcrack stiffness. As η reaches 0.2, the moment capacity increases by as much as twice at $\varphi = 10$ compared with an elastically perfect plastic material.

In Zone 3.1, the effect of postpeak tensile response is also observed. This response is modeled in the context of the parameter μ. Figures 14.10a and 14.10b show the response of the neutral axis depth and also the moment as a function of the curvature for various values of μ.

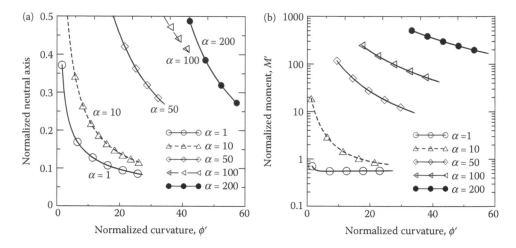

FIGURE 14.9 The location of (a) the neutral axis depth and (b) the moment capacity as a function of the curvature for various values of ultimate tensile capacity α within zone 3.1.

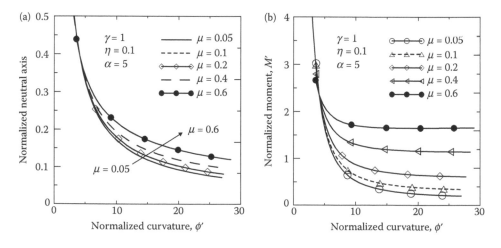

FIGURE 14.10 Location of (a) the neutral axis depth and (b) the moment capacity as a function of the curvature for various values of postpeak tensile strength ratio μ within zone 3.1.

The parameters of this simulation are $\alpha = 5$, $\gamma = 1$, and $\eta = 0.1$. Note that as the postpeak tensile stress capacity parameter μ increases, the stiffness in the postcrack prepeak slightly increases and the moment capacity also increases. The range of moment as a function of applied strain is ascending because the effect of parameter μ is mostly in the ascending part of the moment–curvature response.

In zone 3.1, it is clearly shown that changing the precrack stiffness has a marginal effect on the location of neutral axis or the moment capacity as shown in Figure 14.11. The simulations shown represent the results of $\alpha = 10$, $\gamma = 1$, $\mu = 0.2$, and $\eta = 0.1–0.2$.

The interaction between zone 2.1 and zone 3.1 is shown in Figure 14.12. Note that the transition from one mode of failure to another is at the intersection of the two curves, allowing for the minimum load to govern the transition. Three different values of cracking stiffness, namely, $\eta = 0.0$,

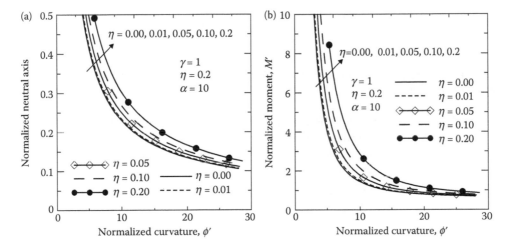

FIGURE 14.11 The location of (a) the neutral axis depth and (b) the moment capacity as a function of the curvature for various values of postpeak tensile stiffness ratio η within zone 3.1.

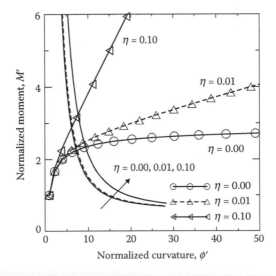

FIGURE 14.12 Interaction diagram between zones 2.1 and 3.1, allowing for transition from one mode of failure to another at the intersection of the two curves.

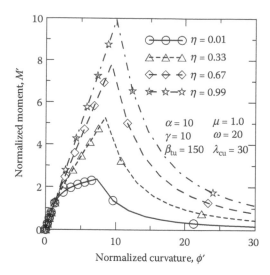

FIGURE 14.13 Superposition of three modes of failure for a range of cracking stiffness ratios as a function of cracked stiffness parameter.

0.01, and 0.1, are used. Parameter η is dominant in zone 2.1 but has little effect in the zone 3.1, as the descending portion is associated with the postpeak response.

Figure 14.13 shows the superposition of two modes of failure for a range of postcracking stiffness ratios. Note that as parameter η increases, the composite strength in the zone 2.1 increases. As discussed earlier, parameter η has a marginal effect in zone 3.1. The overall response is dominated by the three mechanisms of linear elastic, cracking, and postpeak tension. The compression response in all these three modes is linear elastic.

CRACK LOCALIZATION RULES

When a flexural specimen is loaded beyond the peak strength, the load decreases and two distinct zones develop as the deformation localizes in the cracking region while the remainder of the specimen undergoes general unloading. By the same assumptions that allow for correlating the stress–crack width relationship as a stress–strain approach, localization of major cracks is simulated as an average response over the crack spacing region. Results are used as a smeared crack in conjunction with the moment–curvature diagram to obtain load–deformation behavior.

Figure 14.14a presents the schematic moment–curvature diagram with crack localization rules, and Figure 14.14b shows a four-point bending test with localization of smeared crack occurring in the mid-zone, while the zones outside the cracking region undergo unloading during softening [8, 9]. The length of the localized zone is defined as cS, representing the product of a normalized parameter c and loading point spacing $S = L/3$, where L is the clear span in the four-point bending and $S = L/2$ for three-point bending. For simulation purposes, cracks are assumed to be uniformly distributed throughout the mid-zone and a value of $c = 0.5$ was used.

Moment distribution along the length of a beam is obtained by static equilibrium, and the corresponding curvature is obtained from a moment–curvature relationship. As shown by a solid curve in Figure 14.14a, a typical moment–curvature diagram is divided into two portions: an ascending curve from 0 to M_{max} and a descending curve from M_{max} to M_{fail}. For a special case of low-fiber volume fraction where an ascending curve from 0 to M_{max1}, representing the tensile cracking strength is followed by a sharp drop in the postpeak response, the postpeak moment–curvature response exhibits two portions: a descending curve from M_{max1} to M_{low} and ascending again from M_{low} to M_{max2}. In this case, there are two local maxima, which either point could be the global maximum.

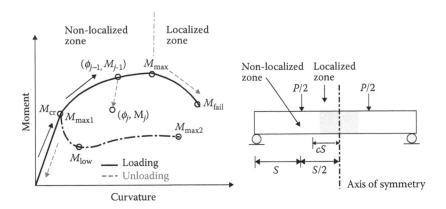

FIGURE 14.14 (a) Moment–curvature diagram and crack localization rules, (b) four-point bending test.

To predict load–deflection response, static equilibrium is used and an array of load steps is derived from a series of discrete data points along a moment–curvature diagram. For each load step, the moment and corresponding curvature distribution along the beam are calculated. Although the specimen is loaded from 0 to M_{max} (or M_{max1}), the ascending portion of the diagram is used. Beyond the maximum load, as the specimen undergoes softening, the curvature distribution depends on the localized or nonlocalized zones and its prior strain history (uncracked or cracked). For an uncracked section, the curvature unloads elastically. If the section has been loaded beyond M_{cr}, the unloading curvature of cracked sections follows a quasi-linear recovery path expressed as

$$\phi_j = \phi_{j-1} - \xi \frac{(M_{j-1} - M_j)}{EI} \tag{14.49}$$

where ϕ_{j-1} and M_{j-1} represent the previous moment–curvature state and ϕ_j and M_j are the current state. E and I represent the elastic modulus and the moment of inertia of uncracked section. The unloading factor ξ is between 0 and 1; $\xi = 0$ indicates no curvature recovery, whereas $\xi = 1$ is unloading elastically with initial stiffness EI. An unloading factor $\xi = 0$ used in the present study under the assumption that cracks do not close when material softens in displacement control. For a section in the localized zone, the unloading curvature is determined from the descending portion of the moment–curvature diagram (M_{max} to M_{fail}) or (M_{max1} to M_{low}). For a special case of low fiber content, the moment–curvature diagram is divided into three portions; the curvature corresponding to the load step beyond the M_{low} is determined by the third portion (M_{low} to M_{max2}).

ALGORITHM TO PREDICT LOAD–DEFLECTION RESPONSE OF THE FOUR-POINT BENDING TEST

The load deflection response of a beam can be obtained by using the moment–curvature response, crack localization rules, and moment–area method as follows:

1) For a given cross section and material properties, the normalized tensile strain at the bottom fiber β is incrementally imposed to generate the moment–curvature response using Equations 14.37 and 14.38 and the expressions given in Table 14.6. For each value of β in stages 2 and 3, the condition for compressive stress $\lambda < \omega$ or $\lambda > \omega$ is verified in advance of moment–curvature calculation.
2) The moment–curvature diagram determines the maximum load; hence, using a discrete number of moment magnitudes along the diagram, a load vector $P = 2M/S$ is generated.

3) The beam is segmented into finite sections. For each load step, static equilibrium is used to calculate moment distribution along the beam and used with the moment–curvature relationship (with crack localization rule) to obtain the curvature along the beam.

4) The deflection at midspan is calculated by numerical moment area method of discrete curvature between the support and the midspan. This procedure is applied at each load step until a complete load deflection response is obtained. A simplified procedure for direct calculation of the deflection is presented in an earlier work [5].

PARAMETRIC STUDY OF MATERIAL PARAMETERS

Parametric studies address the behavior of strain-softening and strain-hardening materials. The flexural strength and the ductility are expressed using the normalized moment–curvature response, which is independent of section size and first cracking tensile strength. Figure 14.14 presents the parametric study of a typical strain-softening material with material parameters specified as a compressive-to-tensile strength ratio $\gamma\omega = 10$, normalized ultimate compressive strain $\lambda_{cu} = 30$, and ultimate tensile strain $\beta_{tu} = 150$. For each case study, all parameters were held constant at the typical values while a range of variables were studied. To avoid a discontinuous tensile response at the transition strain α for strain-softening materials, the postpeak modulus η is determined by Equation 14.44,

$$\eta = -\frac{(1-\mu)}{(\alpha-1)} \tag{14.50}$$

Figure 14.14a shows the compression and the tension model with the transition strain α varied from 1.01 to 15. According to Figure 14.14b, an increase in the transition strain α increases both flexural strength and ductility. Figure 14.14c shows that the material model with residual tensile strength μ varied from 0.01 to 0.99 simulates a range of brittle to elastic–perfectly plastic response of high volume fraction FRC. Figure 14.14d shows that the moment–curvature diagram is quite sensitive to the variations in parameter μ as it affects both the pre- and postpeak response. The flexural response changes from a brittle to ductile material as μ changes from 0.01 to 0.99. To study the effect of compressive stiffness, the range of parameters γ and ω were used together to represent the increase in relative compressive to tensile stiffness from $0.1 < \gamma/\omega < 16$ (1/10 to 4/2.5) at a fixed compressive-to-tensile strength ratio ($\gamma\omega = 10$) as shown in Figure 14.14e. According to Figure 14.14f, changes in the relative stiffness slightly affect the peak moment from 1.7 to 2.1 and marginally increase the stiffness of the moment–curvature response. It is also concluded that the normalized compressive modulus γ and compressive yield strain ω have a marginal effect on the predicted moment–curvature response as long as the compressive strength is about one order of magnitude higher than the first cracking tensile strength.

Figure 14.15 presents the parametric study of a typical strain-hardening material, with a compressive-to-tensile cracking strength ratio of $\gamma\omega = 10$, normalized ultimate compressive strain $\lambda_{cu} = 30$, and ultimate tensile strain $\beta_{tu} = 150$ similar to the previous case study. The postpeak response was ignored for this case by setting μ to a very low value of 0.01. Figure 14.15a shows the compression and tension model of a typical strain-hardening material with varying $\alpha = 1.01$–15. Figure 14.15b shows that the increase in α directly increases the normalized moment and curvature. Figures 14.15c and 14.15d show that increasing the postcrack modulus η also increases the moment–curvature diagram. Similar to the strain-softening materials, the increase in relative compressive to tensile stiffness at constant compressive-to-tensile strength ratio $\gamma\omega$ (as shown in Figure 14.15e) has a subtle effect to the moment–curvature response as shown in Figure 14.15f. These results indicate that the most significant parameters affecting the moment capacity are the transition strain α and the stiffness in the postcracking tensile range η.

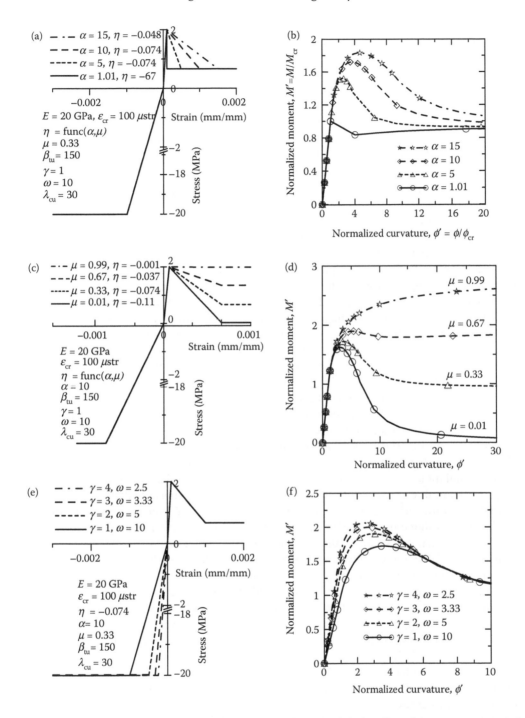

FIGURE 14.15 Parametric study of a typical strain-softening material: the effect of parameters α, μ, and γ and ω to normalized moment–curvature diagram.

PREDICTION OF LOAD–DEFORMATION RESPONSE

The algorithm to simulate load–deflection response of a beam under four-point bending test was used as a predictive tool to study three classes of materials: steel FRC (SFRC) with 0.5% and 1.0% volume fraction representing strain-softening–deflection-softening and deflection-hardening materials, respectively, and engineered cementitious composite (ECC) with 2.0% volume fraction representing strain-hardening material.

Steel FRC

Two mixes of SFRC (H1 and H21) that used hook-end fibers at volume fraction levels of 0.5% and 1.0% were selected [10, 11] to demonstrate the algorithm to predict load–deflection response. Tensile "dog-bone" specimens of $100 \times 70 \times 200$ mm in net dimensions with an enlarged width of 100 mm at the ends were used. The flexural four-point bending specimens were $100 \times 100 \times 1000$ mm with a clear span of 750 mm. The average material properties were compressive strength $f'_c = 34$ MPa, initial compressive modulus $E_{ci} = 28.5$ GPa, and initial tensile modulus $E = 25.4$ GPa. First cracking tensile strain ε_{cr} for mix H1 and H21 are 111 and 116 µstr (microstrain), respectively.

As shown in the parametric studies, compressive properties have a marginal effect on the predicted flexural response as long as the compressive strength is as much as nine times greater than the tensile strength [5]. Therefore, typical values for compression parameters can be estimated without severely affecting the results. The compressive yield stress f_{cy} was assumed to be $0.85f'_c$, and compressive modulus E_c was estimated to be $0.85 E_{ci}$. The normalized compressive modulus λ was then obtained by E_c/E and the normalized compressive yield strain by $\omega = f_{cy}/(E_c\varepsilon_{cr})$. The range of ultimate compressive strain ε_{cu} between 0.0035 and 0.004 was suggested by several researchers [12,13]. Using a value of 0.004, the corresponding normalized value was calculated by $\lambda_{cu} = \varepsilon_{cu}/\varepsilon_{cr}$. The material parameters for tension model were determined by fitting the model to the uniaxial tension test result as shown by the solid line in Figures 14.16a and 14.16c. All other parameters used in the simulation of flexural bending of mix H1 and H21 are provided in the figures. The constant postpeak stress level $\mu = 0.24$, postcrack stiffness $\eta = -0.76$, postcrack stiffness ultimate strain capacity $\alpha = 2.0$, total tensile strain level $\beta_{tu} = 136$, compression-to-tension stiffness ratio $\gamma = 0.95$, and compression-to-tension strength ratio of $\omega = 10.8$ were used.

Figure 14.16b shows the predicted flexural response of deflection softening material (mix H1) from the simulation process. If one chooses to directly use the uniaxial tension data in the curve fitting, the flexural response will be underpredicted. This is attributed to differences in the stress distribution profiles discussed in the following section using Weibull statistics. In the tension test, the entire volume of the specimen is a potential zone for crack initiation. Comparatively, in the flexural test, a fraction of the tension region is subjected to the highest tensile stress. To quantify the differences between the equivalent tensile strengths of tension and flexure, a single scaling parameter was used [14]. This multiplier applies to the first cracking tensile strain ε_{cr}, resulting in scaling of associated strains and stresses and a uniform increase in material strength. Inverse analysis by trial and error indicates a scaling parameter of 40%. The inverse analysis procedure applied to tensile and flexural responses of several samples can help establish the statistical relationship between the tensile and the flexural responses as have been done for other brittle materials [15].

For deflection-hardening materials represented by mix H21, the solid curve of tension model that fits to the uniaxial tension test data is shown in Figure 14.16c. The constant postpeak stress level $\mu = 0.8$, postcrack stiffness $\eta = -0.1$, transition strain $\alpha = 3.0$, ultimate tensile strain level $\beta_{tu} = 129$, compression-to-tension stiffness ratio $\lambda = 0.95$, and compression-to-tension strength ratio of $\omega = 10.3$ were used. The predicted flexural response is shown as a solid line in Figure 14.17d. A direct algorithm slightly underestimates the flexural response. The inverse analysis shows that the strength of the uniaxial models should be increased by 8% for a reasonable prediction of flexural results as shown by the dashed lines in Figures 14.16c and 14.16d. The correlation of experimental and simulated responses in the deflection-hardening range is quite reasonable; the discrepancy between the fitted and the modified tension models is much lower than that of strain-softening materials.

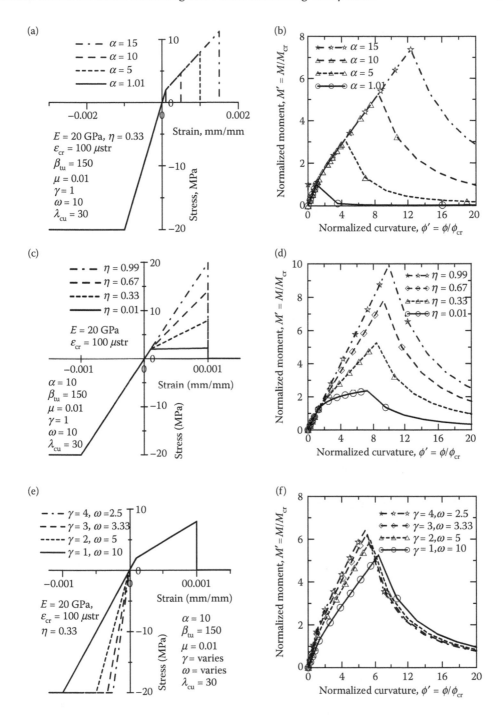

FIGURE 14.16 Parametric study of a typical strain-hardening materials: the effect of parameters α, η, and γ and ω to normalized moment–curvature diagram.

ENGINEERED CEMENTITIOUS COMPOSITES (ECC)

ECC composites with $V_f = 2\%$ polyethylene (PE) fibers was used for simulation of load–deflection response as a strain-hardening material [14, 15]. The flexural specimens for the four-point bending test were 76.2 × 101.6 × 355.6 mm with a clear span of 304.8 mm. The average material properties were as follows: compressive strength $f'_c = 60$ MPa. The initial compressive modulus $E_{ci} = 4.35$ GPa was obtained by back calculation of the initial flexural load deflection response. Initial tensile modulus $E = 4.75$ GPa and first cracking tensile strain $\varepsilon_{cr} = 600$ microstrains were obtained directly from the uniaxial tensile test results. The constant postpeak stress level $\mu = 1.49$, postcrack stiffness $\eta = 0.009$, postcrack strain capacity $\alpha = 95.0$, ultimate tensile strain level $\beta_{tu} = 117$, compression-to-tension stiffness ratio $\gamma = 0.92$, and compression-to-tension strength ratio of $\omega = 18.4$ were also used.

The compressive yield stress f_{cy} was assumed to be $0.80\,f'_c$, and the compressive modulus E_c was estimated equal to E_{ci}. The ultimate compressive strain ε_{cu} was assumed to be 0.012. The material parameters for tension model were determined by fitting the model to the uniaxial tension test result as shown by the solid line in Figure 14.17a. Parameters used in the simulation are provided in the same figure. The solid curve in Figure 14.17b shows the predicted flexural response for the strain-hardening material during the pre- and postcrack stages that agreed well with the experimental results. Formation of the distributed crack system and ductility can be adequately described by the smeared pseudo-strain model (Figure 14.18).

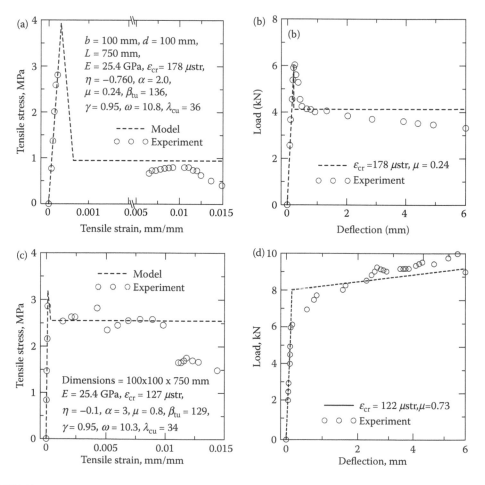

FIGURE 14.17 Simulation of an SFRC: (a and b) tension model and flexural response of mix H1 ($V_f = 0.5\%$); (c and d) tension model and flexural response of mix H21 ($V_f = 1.0\%$). (After Soranakom, C., and Mobasher, B., "Closed form solutions for flexural response of fiber reinforced concrete beams," *Journal of Engineering Mechanics*, 133, 8, 933–941, 2007.)

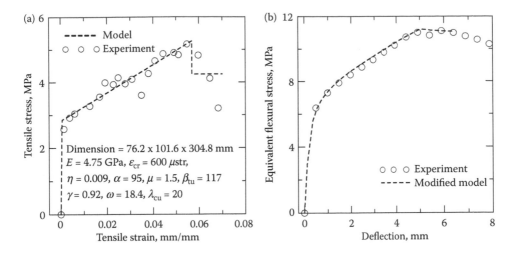

FIGURE 14.18 Simulation of ECC: (a) tension model and (b) flexural response.

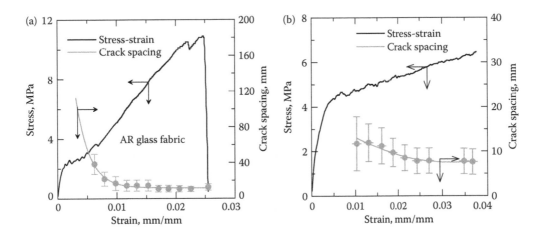

FIGURE 14.19 Uniaxial tension stress–strain response of (a) AR glass and (b) PE textile composites.

AR GLASS AND PE TEXTILE REINFORCED CEMENT COMPOSITES

Two types of textile systems, including bonded AR glass and woven PE mesh, are addressed. AR glass textile manufactured by Saint-Gobain Technical Fabrics Inc., Saint Andrews Canada were used with a cement paste using a water-to-cement ratio of 0.45 [14]. Two layers of textile were used at the top and bottom of the specimens to provide reinforcement in each direction: $V_L = V_T = 0.70\%$. Figure 14.19a shows the tensile stress–stain response of the AR glass sample with an initial linear response of up to 2 MPa, followed by reduced stiffness due to subsequent cracking and formation of parallel cracks [14, 16]. As multiple cracking proceeds to approximately 20 mm and reaches saturation crack spacing, the stiffness decreases. The saturation of crack spacing correlates to a tensile strain of 1.5%. Similar behavior is also observed from the PE textile composites with the stress–strain and crack spacing shown in Figure 14.19b. Note that the slope in the postcrack phase is significantly lower, the strain capacity is higher, and the crack spacing is smaller than AR glass composites.

Correlation procedure for tension and bending responses used experimental data from tension specimens of $10 \times 25 \times 200$ mm and flexural four-point bending specimens of $10 \times 25 \times 200$ mm in dimension with a clear span of 152 mm. Figure 14.20a shows the tension test results with the fitted tension model. Material properties for simulation were $\alpha = 100$, $\mu = 0.1$, $\eta = 0.05$, $\gamma = 1.0$, and $\omega = 20.4$, whereas the limits were $\beta_{tu} = 250$ and $\lambda_{cu} = 150$. The Young modulus of tension and compression response (E) was obtained by assuming a value of 18 GPa, and a first crack strain capacity of $\varepsilon_{cr} = 200$ μstr was used. The ultimate compressive strength of cement pastes was assumed to be the typical value of $f'_{cr} = E\omega\varepsilon_{cr} = 20.4(18)\,(0.0002) = 73.4$ MPa.

Figure 14.20b shows the predicted flexural load deflection response of cement composites. The steps in calculation of load–deflection response from the moment–curvature have been discussed in detail in recent publications dealing with strain-hardening and strain-softening-type composites [4, 17, 18]. Note that in these systems, the high tensile stiffness and strength of the composite leads to high values for the load and distributed flexural cracking. Analysis of the samples indicates formation of diagonal tensions cracks in the samples because of the shear failure mechanism. No provisions for shear cracking were accounted in for the present approach. No attempt was made to simulate the response beyond the first major flexural crack.

The material properties for the simulation of PE textile composites were $\alpha = 150$, $\mu = 0.1$–0.4, $\eta = 0.008$, $\lambda = 1.0$, and $\omega = 20.4$. The constants were $\varepsilon_{cr} = 200$ μstr and $E = 18$ GPa, whereas the limits of the modeling were $\beta_{tu} = 250$ and $\lambda_{cu} = 150$. The overall tensile response for the PE composites is shown in Figure 14.21a.

The flexural specimens for the four-point bending test were $25 \times 10 \times 200$ mm with a clear span of 152 mm. The average material properties were compressive strength $f'_c = 73.4$ MPa. The initial tensile modulus $E = 18$ GPa and the first cracking tensile strain $\varepsilon_{cr} = 200$ μstr were used as they correlated with the uniaxial tensile test results. The ultimate compressive strain ε_{cu} was assumed to be $\alpha\varepsilon_{cr} = 150(0.0002) = 3\%$. All parameters used in the simulation are provided in the figure. The prediction for the strain-hardening material during the pre- and postcrack stages agreed well with the experimental results.

Figures 14.20b and 14.21b show the predictions of the equivalent load–deflection response for AR glass and PE textiles, respectively. Simulation using direct tension data underestimates the equivalent flexural stress. This may be due to several factors including the size effect and the uniformity in loading in tension versus the linear strain distribution in flexure. The underestimation of the flexural capacity of the beam can be eliminated by increasing the tensile capacity by scaling parameters as discussed in earlier publications. Alternatively, one can use the flexural response and develop a back-calculation procedure to directly fit experimental data.

The three simulations of FRC materials, ranging from low to high fiber contents and material response changing from softening to hardening, indicate that as the postpeak tensile strength of

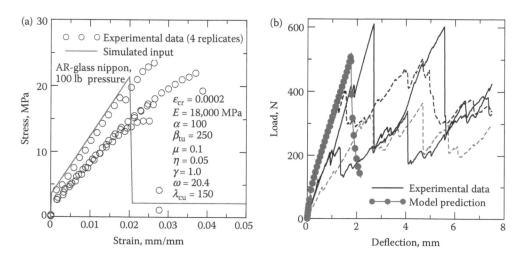

FIGURE 14.20 Tensile stress–strain response input model and predicted load deflection response of AR glass textile composites.

material increases, the use of uniaxial response to predict flexural response becomes more accurate. It also implies that the increase of fiber contents in the experiments can effectively suppress the flaws from initiating crack that leads to premature failure, especially under uniaxial tensile conditions. Therefore, the flaw size distribution in material is less sensitive to the constant versus linear distribution of tensile stress patterns in uniaxial and flexural specimens, respectively.

CLOSED-FORM MOMENT–CURVATURE SOLUTIONS FOR FRC BEAMS WITH REINFORCEMENT

The solutions provided for an FRC beam can be extended to a reinforced concrete section containing reinforcement. Geometrical parameters are defined as a combination of normalized parameters and beam dimensions: width b and full depth h. Figure 14.22a shows a beam cross section that contains an area of steel $A_s = \rho_g bh$ at the reinforced depth $d = \alpha h$. Note that the reinforcement ratio ρ_g is defined per gross sectional area bh, as opposed to effective area bd normally used for reinforced concrete. Figure 14.22b presents the elastic–perfectly plastic steel model, which is similar to the compression model using yield strain $\varepsilon_{sy} = \kappa\varepsilon_{cr}$ and yield stress $f_{sy} = \kappa n\varepsilon_{cr}E$ as defined by normalized parameters: κ and n. There is no termination level specified for steel strain as it is assumed infinite plastic material. The material models for tension and compression of FRC are as before for the

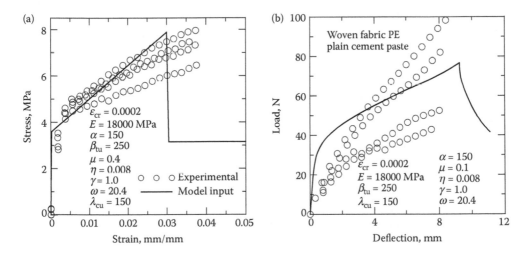

FIGURE 14.21 Tension stress–strain response input model and predicted load deflection response of PE textile composites.

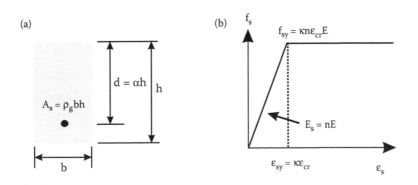

FIGURE 14.22 Material model for single reinforced concrete design: (a) steel model and (b) beam cross section.

special case of elastic softening response ($\eta = 0$ and $\alpha = 1$) discussed in Equations 14.7 through 14.10; the models for FRC and steel rebar are presented as

$$\frac{\sigma_t(\beta)}{E\varepsilon_{cr}} = \begin{cases} \beta & 0 \le \beta \le 1 \\ \mu & 1 < \beta \le \beta_{tu} \\ 0 & \beta > \beta_{tu} \end{cases}; \quad \frac{\sigma_c(\lambda)}{E\varepsilon_{cr}} = \begin{cases} \gamma\lambda & 0 \le \lambda \le \omega \\ \gamma\omega & \omega < \lambda \le \lambda_{cu} \\ 0 & \lambda > \lambda_{cu} \end{cases} \tag{14.51}$$

$$f_s(\varepsilon_s) = \begin{cases} E_s\varepsilon_s & 0 \le \varepsilon_s \le \varepsilon_{sy} \\ E_s\varepsilon_{sy} & \varepsilon_s > \varepsilon_{sy} \end{cases}; \quad \frac{f_s(\chi)}{E\varepsilon_{cr}} = \begin{cases} n\chi & 0 \le \chi \le \kappa \\ n\kappa & \chi > \kappa \end{cases} \tag{14.52}$$

where normalized strains are defined as $\beta = \varepsilon_{tt}/\varepsilon_{cr}$, $\lambda = \varepsilon_c/\varepsilon_{cr}$, and $\chi = \varepsilon_s/\varepsilon_{cr}$.

In derivation of moment–curvature equations, the Kirchhoff hypothesis of plane section remaining plane is assumed. The full derivation is presented in detail by Soranakom [18]. A normalized compressive strain at the top concrete fiber λ is used as the independent variable to incrementally impose flexural deformation, which is defined in three stages. The first stage or the elastic range ($0 < \lambda < \lambda_{R1}$) corresponds to the compressive strain ranging from zero to the point where the tensile strain at the bottom fiber reaches the first cracking tensile strain. Stage 2 ($\lambda_{R1} < \lambda < \omega$) corresponds to the compressive strain in the elastic range, and the tensile strain is in the postcrack region. Finally, stage 3 ($\omega < \lambda < \lambda_{cu}$) corresponds to the compressive strain in plastic range while the tensile strain is in postcrack range. For stages 2 and 3, two possible scenarios exist: the steel is either elastic ($\varepsilon_s < \varepsilon_{sy}$) or yielding ($\varepsilon_s > \varepsilon_{sy}$).

The steps in the determination of net section force, moment, and curvature at each stage of normalized compressive strain, λ, are similar to the previous cases, and complete derivation can be found in [17, 18]. When steel is elastic in stages 1, 2.1, and 3.1, the expressions for net force are in the quadratic forms and result in two possible solutions for k. With a large scale of numerical tests covering a practical range of material parameters, only one solution yields the valid value in the range $0 < k < 1$. During stage 1, the singularity of k_1 is found when $\gamma = 1$; thus, additional expression for k_1 is derived by taking the limit as $\gamma \to 1$. On the other hand, when steel yields in stage 2.2 or 2.3, there is only one valid solution for k. The moment M_i, curvature ϕ_i, and effective flexural stiffness K_i for each stage i are normalized with respect to their values at cracking of plain FRC M_{cr}, ϕ_{cr}, and K_{cr}, respectively, and their closed-form solutions M_i', ϕ_i', and K_i' are presented in Table 14.9:

$$M_i = M_i' M_{cr}; \quad M_{cr} = \frac{1}{6}bh^2 E\varepsilon_{cr} \tag{14.53}$$

$$\phi_i = \phi_i' \phi_{cr}; \quad \phi_{cr} = \frac{2\varepsilon_{cr}}{h} \tag{14.54}$$

$$K_i = K_i' K_{cr}; \quad K_{cr} = \frac{1}{12}bh^3 \tag{14.55}$$

Because the compressive modulus E_c may not be equal to the tensile modulus E, the normalized compressive strain corresponding to end of elastic region 1 (λ_{R1}) must be determined from the strain gradient diagram,

$$\frac{\lambda_{R1}\varepsilon_{cr}}{kh} = \frac{\varepsilon_{cr}}{(1-k)h} \tag{14.56}$$

By substituting k_1 from Table 14.9 for k in Equation 14.56 and by solving for λ_{R1}, one obtains

TABLE 14.9
Normalized Neutral Axis, Moment, Curvature, and Stiffness for Each Stage of Normalized Compressive Strain at Top Fiber (λ)

Stage	Neutral axis, k	Normalized moment, M'
1	$0 < \lambda < \lambda_{R1}\ \ k_1 = k_{1\gamma}$	$M_1' = \dfrac{2\lambda}{k_1}\left[C_1 k_1^3 + C_2 k_1^2 + C_3 k_1 + C_4\right]$
2.1	$\lambda_{R1} < \lambda \le \omega \quad \varepsilon_s \le \varepsilon_{sy}$ $k_{21} = \dfrac{\lambda}{B_1}\left(B_2 + \sqrt{B_3 + 2\alpha\rho_g n B_1}\right)$	$M_{21}' = \dfrac{1}{\lambda^2 k_{21}}\left[C_5 k_{21}^3 + C_6 k_{21}^2 + C_7 k_{21} + C_8\right]$
2.2	$\lambda_{R1} < \lambda \le \omega \quad \varepsilon_s > \varepsilon_{sy}$ $k_{22} = \dfrac{B_4}{B_1}$	$M_{22}' = \dfrac{1}{\lambda^2}\left[C_5 k_{22}^2 + C_9 k_{22} + C_{10}\right]$
3.1	$\omega < \lambda \le \lambda_{cu}\ \varepsilon_s \le \varepsilon_{sy}$ $k_{31} = \dfrac{\lambda}{B_5}\left(B_2 + \sqrt{B_3 + 2\alpha\rho_g n B_5}\right)$	$M_{31}' = \dfrac{1}{\lambda^2 k_{31}}\left[C_{11} k_{31}^3 + C_6 k_{31}^2 + C_7 k_{31} + C_8\right]$
3.2	$\omega < \lambda \le \lambda_{cu}\ \varepsilon_s > \varepsilon_{sy}$ $k_{32} = \dfrac{B_4}{B_5}$	$M_{32}' = \dfrac{1}{\lambda^2}\left[C_{11} k_{32}^2 + C_9 k_{32} + C_{10}\right]$

Note: Where, the coefficients are:

$B_1 = \gamma\lambda^2 + 2\mu(\lambda+1)-1;\quad B_2 = \mu - \rho_g n\lambda;\quad B_3 = \rho_g n(\rho_g n\lambda^2 - 2\mu\lambda)+\mu^2;\quad B_4 = 2\lambda(\rho_g n\kappa + \mu);$

$B_5 = 2\gamma\omega\lambda - \gamma\omega^2 + 2\mu(\lambda+1)-1;\quad C_1 = \gamma-1;\quad C_2 = 3(\rho_g n+1);\quad C_3 = -3(2\rho_g n\alpha+1);\quad C_4 = 3\rho_g n\alpha^2 +1;$

$C_5 = 2\gamma\lambda^3 + 3\mu(\lambda^2-1)+2;\quad C_6 = 6\lambda^2(\rho_g n\lambda - \mu);\quad C_7 = 3\lambda^2(\mu - 4\rho_g n\alpha\lambda);\quad C_8 = 6\rho_g n\alpha^2\lambda^3;$

$C_9 = -6\lambda^2(\rho_g n\kappa + \mu);\quad C_{10} = 3\lambda^2(2\rho_g n\alpha\kappa + \mu);\quad C_{11} = 3\gamma\omega\lambda^2 - \gamma\omega^3 + 3\mu(\lambda^2-1)+2$

$$k_{1\gamma} = \begin{cases} \dfrac{1 + \rho_g n - \sqrt{\rho_g^2 n^2 + 2\rho_g n(1-\alpha+\alpha\gamma)+\gamma}}{1-\gamma} & \gamma \ne 1 \\[2ex] \dfrac{2\rho_g n\alpha + 1}{2(\rho_g n+1)} & \gamma = 1 \end{cases}$$

The curvature and stiffness for each stage are defined using normalized parameters ϕ', and K' and obtained using normalized strain λ, neutral axis, k, and moment expressions M', as:

$$\phi' = \frac{\lambda}{2k}\ K' = \frac{M'}{\phi'}, i.e.\ \left[\phi_{21}' = \frac{\lambda}{2k_{21}},\ K_{21}' = \frac{M_{21}'}{\phi_{21}'}\right]$$

$$\lambda_{R1} = \begin{cases} -\dfrac{1+\rho_g n - \sqrt{\rho_g^2 n^2 + 2\rho_g n(1-\alpha+\alpha\gamma)+\gamma}}{\gamma+\rho_g n - \sqrt{\rho_g^2 n^2 + 2\rho_g n(1-\alpha+\alpha\gamma)+\gamma}} & \text{when } \gamma \ne 1 \\[2ex] -\dfrac{2\rho_g n\alpha + 1}{2\rho_g n(\alpha-1)-1} & \text{when } \gamma = 1 \end{cases} \qquad (14.57)$$

The yield condition for tensile steel can be checked by first assuming that it yields and then using k_{22} or k_{32} in Table 14.6 for k in Equation 14.58 to calculate the steel strain ε_s:

$$\varepsilon_s = \frac{\alpha - k}{k} \lambda \, \varepsilon_{cr} \tag{14.58}$$

If ε_s obtained by Equation 14.58 is greater than ε_{sy}, the assumption is correct; otherwise, steel has not yielded and one has to use k_{21} or k_{31}. Once, the neutral axis parameter k and the applicable case are determined, the appropriate expressions for moment, curvature, and stiffness in Table 14.9 and Equations 14.53 through 14.55 are then used to generate moment–curvature response and its flexural stiffness.

To avoid compression failure occurring in the ultimate stage, the steel used in flexural members must be less than the balanced reinforcement ratio $\rho_{g,bal}$. Parameter $\rho_{g,bal}$ represents the simultaneous achievement of concrete compressive strain at failure ($\varepsilon_c = \varepsilon_{cu}$) with the steel reaching its yield limit ($\varepsilon_s = \varepsilon_{sy}$). The strain gradient in stage 3.2 represents the compressive strain in plastic range, and the tensile strain in the postcrack region is used to derive the balance reinforcement ratio:

$$\frac{\lambda_{cu} \varepsilon_{cr}}{kh} = \frac{\kappa \varepsilon_{cr}}{(\alpha - k)h} \tag{14.59}$$

By substituting λ_{cu} in the expression for k_{32} in Table 14.9 and then by using it for k in Equation 14.59, one can solve for the balance reinforcement ratio as:

$$\rho_{g,bal} = \frac{2\mu\left(\lambda_{cu}(\alpha - 1) + \alpha - \kappa\right) + \alpha \, \gamma \, \omega(2\lambda_{cu} - \omega) - \alpha}{2n\kappa(\lambda_{cu} + \kappa)} \tag{14.60}$$

PARAMETRIC STUDIES

Parametric studies of postcrack tensile strength and reinforcement ratio as two main reinforcing parameters were conducted. Variations in the location of neutral axis, moment–curvature response, and stiffness degradation of a beam section as flexural deformation increases are presented as normalized quantities with respect to first cracking parameters of plain FRC. Figure 14.23 shows typical material models for SFRC and steel rebar used in the parametric studies. Two material parameters, the tensile modulus E of 24 GPa and the first cracking tensile strain ε_{cr} of 125 µstr, were used. Other normalized parameters for tension and compression models of SFRC were $\beta_{tu} = 160$, $\gamma = 1$, $\omega = 8.5$, and $\lambda_{cu} = 28$. The normalized parameters for steel rebar were $n = 8.33$, $\kappa = 16$, and $\alpha = d/h = 0.8$. Postcrack tensile strength parameter μ varied from 0.00 to 1.00, and reinforcement ratio ρ_g varied from 0.0 to 0.03.

Figure 14.24 shows the change of neutral axis depth ratio k as the compressive strain at top fiber λ increases. For plain FRC system ($\mu > 0$, $\rho_g = 0$), Figure 14.24a shows that the neutral axis starts at 0.5 as expected for a material with equal compressive and tensile modulus ($\gamma = 1$) and then drops, at different rates, depending on the level of postcrack tensile strength parameter μ. For brittle material represented by $\mu = 0.00$, k instantaneously drops to zero after initiation of cracking. As μ increases from 0.00 to 1.00, the rate of decrease in neutral axis k becomes slower. A very ductile FRC with elastic–plastic tensile behavior defined as $\mu = 1.00$ yields the maximum value of $k = 0.12$ at the ultimate compressive strain $\lambda_{cu} = 28$. For conventional reinforced concrete system ($\mu = 0$, $\rho_g > 0$), Figure 14.24b shows the effect of reinforcement ratio to the change of neutral axis. With reinforcement present in plain concrete ($\mu = 0$), the initial value of k is slightly higher than 0.5 because of the equivalent section criteria; however as ρ_g increases, the descending rate is at a much slower rate. With a

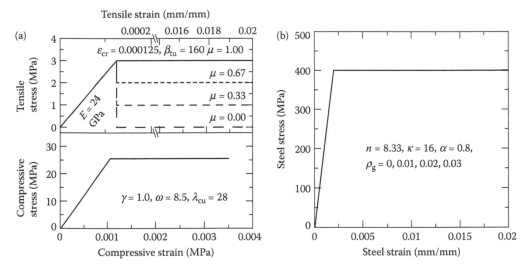

FIGURE 14.23 Material model for typical FRC and rebar used in parametric studies: (a) concrete model and (b) steel model.

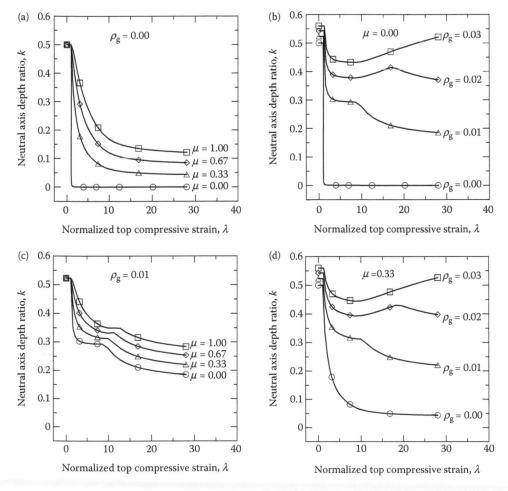

FIGURE 14.24 Parametric studies of neutral axis depth ratio for different levels of postcrack tensile strength parameter μ and reinforcement ratio ρ_g.

small amount of $\rho_g = 0.01$, the k reduces to 0.19 at $\lambda_{cu} = 28$, which is higher than $k = 0.12$ for ductile FRC ($\mu = 1.00$). At higher levels of ρ_g between 0.02 and 0.03, k initially decreases then increases to a relatively high value compared with the starting value. Figure 14.24c shows that when the effect of increasing postcrack tensile strength for a fixed reinforcement ratio $\rho_g = 0.01$, the response is closer to those of plain FRC. Figure 14.24d shows the effect of increasing reinforcement ratio for a fixed level of postcrack tensile strength $\mu = 0.33$.

Figure 14.25 shows the effect of parameters μ and ρ_g to the normalized moment–curvature response. Figure 14.25a shows that the moment–curvature response of plain FRC system when postcrack tensile strength increases from brittle ($\mu = 0$) to ductile ($\mu = 1$). At a value of $\mu = 0.33$, close to $\mu_{crit} = 0.35$ as defined in earlier sections [5], the flexural response is almost perfectly plastic, in which deflection softening starts to shift to deflection hardening. The elastic–plastic tensile response of FRC ($\mu = 1$) yields an upper bound normalized moment capacity of 2.7. In a more efficient reinforced concrete system that uses steel reinforcing bars as the main flexural reinforcement (Figure 14.25b), the maximum normalized moment capacity of 5.8 can be achieved by using ρ_g of only 0.01. Note that as ρ_g increases, the response changes from ductile underreinforced to brittle overreinforced section. Figure 14.25c reveals the response when varying postcrack tensile strength for a fixed amount of reinforcement ratio ($\rho_g = 0.01$). The responses are similar to the curve using

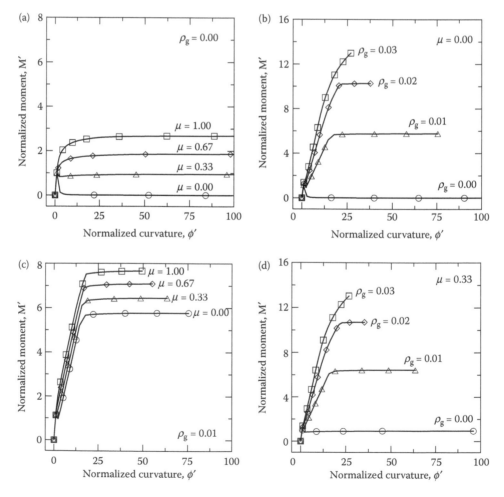

FIGURE 14.25 Parametric studies of normalized moment–curvature diagram for different levels of postcrack tensile strength parameter μ and reinforcement ratio ρ_g.

$\rho_g = 0.01$ in Figure 14.25b plus additional strength from postcrack tensile strength ($\mu = 0.00 - 1.00$) as demonstrated in Figure 14.25a. This system provides intermediate strength between the weaker plain FRC and the stronger conventional reinforced concrete. Figure 14.25d reveals that there is little benefit to use postcrack tensile strength μ of 0.33 or less to the reinforced concrete system as the moment capacity slightly increases from the reinforced concrete without any fibers (Figure 14.25b). It is noted that the ductility of each curve shown in Figure 14.25d is less than that of Figure 14.25b. This is due to the fact that the tension capacity increases for an ultimate compressive strain capacity of 0.0035. In actual concrete mixtures, discrete fibers increase both the postcrack tensile strength and the ultimate compressive strain and may thus yield more ductile responses.

The ultimate moment capacity as a function of postcrack tensile strength and reinforcement ratio can be presented as a convenient design chart for any combination of concrete and steel properties used in a beam section. The yielding condition of steel reinforcement can be identified by comparing ρ_g with the reinforcement ratio at balance failure as defined by Equation 14.60. Once steel condition is determined, appropriate expressions for neutral axis k_{31} or k_{32} are then used to calculate the ultimate moment capacity M'_{31} or M'_{32}.

Figure 14.26 shows a design chart for the concrete and steel previously used in the parametric studies. The material parameters are also provided in the chart. The normalized moment capacity strongly depends on the reinforcement ratio, whereas extra capacity is provided by the postcrack tensile strength. Underreinforced sections are below the balance failure points ($\rho_g < \rho_{g,bal}$), and the moment capacity increases proportional to the reinforcement ratio. As ρ_g exceeds $\rho_{g,bal}$, the strength of all curves marginally increases as the added steel fails to yield. To design flexural members with this design chart, the ultimate moment M_u due to factored load is determined initially and then normalized with the M_{cr} cracking moment of the plain FRC to obtain the required ultimate moment capacity M'_u. The chart is then used to select any combination of normalized postcrack tensile strength μ and reinforcement ratio ρ_g that provides sufficient strength with reasonable safety factor for M'_u. Additional guides for design using this approach and comparison with experimentally conducted test results are provided by Soranakom [18].

CONCLUSIONS

This chapter used closed-form solutions to generate moment–curvature diagrams of a rectangular beam made of homogenized FRC. The algorithm for moment–curvature diagram can be used

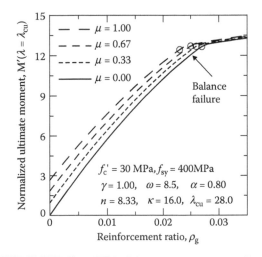

FIGURE 14.26 Design chart of normalized ultimate moment capacity for different levels of postcrack tensile strength μ and reinforcement ratio ρ_g.

with crack localization rules and moment area method to predict load deformation of a beam under three- and four-point bending tests. The normalized moment–curvature relationships were used in parametric studies of strain-softening and strain-hardening materials. For a typical strain-softening material, the most important factors to flexural strength and ductility are the constant residual tensile strength and the strain at UTS. For a typical strain-hardening material, both tensile strain at transition and postcrack modulus are almost equally important to the flexural behavior of a beam. Strain-softening and strain-hardening materials share the similarity that the change of relative compressive to tensile stiffness slightly affects the flexural response as long as the compressive-to-tensile strength ratio is greater than 9.

Load deflection responses of four-point bending tests revealed that the use of uniaxial tensile data tends to underpredict the flexural response for the material that has a relatively low postcrack tensile strength. This was in part due to a difference in stress distribution between the uniaxial tension and the bending tests. The underpredicted load–deformation response can be corrected by increasing strength of the uniaxial material model using a scaling parameter to the first cracking strain. With this approach, other associated strains and stresses will subsequently be increased by the same factor. With proper scaling parameters, the predicted responses agree well with experimental observations. It is also observed that as fiber content or bond is increased such that the postcrack tensile strength is improved, the size effect observed in predicting the response of flexural samples tends to be reduced, and the use of uniaxial response to predict flexural response becomes more accurate.

NOMENCLATURE

B	beam width
C	a parameter for normalized moment in Table 14.6
c	a parameter to define localized crack length
D	a parameter for neutral axis depth ratio in Table 14.6
d	beam depth
E	tensile modulus
E_c	compressive modulus
E_{ci}	initial compressive modulus
E_{cr}	postcrack tensile modulus
F	force component
f	stress at vertex in stress diagram
h	height of each zone in stress diagram
I	moment of inertia
k	neutral axis depth ratio
L	clear span
M_{cr}	moment at first cracking
M_{fail}	moment at failure
M_i'	normalized moment M/M_{cr}
M_{low}	moment at the lowest point before continue the second ascending curve
M_{max}	maximum moment in moment–curvature diagram
M_{max1}, M_{max2}	local maximum moment 1 and moment 2 in moment–curvature diagram
P	total load applied to four-point bending beam specimen
S	spacing in four-point bending test
y	moment arm from neutral axis to center of each force component
α	normalized transition strain
β	normalized tensile strain at bottom fiber
β_{tu}	normalized ultimate tensile strain
ε_c	compressive strain
ε_{cr}	first cracking tensile strain
ε_{ctop}	compressive strain at the top fiber
ε_{cu}	ultimate compressive strain
ε_{cy}	compressive yield strain
ε_t	tensile strain
ε_{tbot}	tensile strain at bottom fiber
ε_{trn}	strain at transition point

$\varepsilon_{\mathrm{tu}}$	ultimate tensile strain
γ	normalized compressive strain
η	normalized postcrack modulus
λ	normalized compressive strain
λ_{cu}	normalized ultimate compressive strain
σ_{c}	compressive stress
σ_{cr}	cracking tensile strength
σ_{cst}	constant tensile stress at the end of tension model
σ_{cy}	compressive yield stress
σ_{t}	tensile stress
ω	normalized compressive yield strain
ξ	unloading factor

SUBSCRIPTS

t1, t2, and t3 tension zones 1, 2, and 3
c1, c2 compression zones 1 and 2
1, 21, 22, 31, 32 = stages 1, 2.1, 2.2, 3.1, and 3.2 according to the value of β

REFERENCES

1. Aldea, C. M., Mobasher, B., and Jain, N., Cement-based matrix-grid system for masonry rehabilitation, SP-244-9, ACI Special Publications, 2007, 141–155.
2. Majumdar, A. J., and Laws, V. (1983), "Composite materials based on cement matrices," *Philosophical Transactions of the Royal Society of London. Series A, Mathematical and Physical Sciences*, 310, 191–202.
3. Soranakom, C., and Mobasher, B. (2007), "Closed-form moment–curvature expressions for homogenized fiber-reinforced concrete," *ACI Materials Journal*, 104(4), 351–359.
4. Soranakom, C., Mobasher, B., and Bansal, S., "Effect of material non-linearity on the flexural response of fiber reinforced concrete," *Proceedings of the Eighth International Symposium on Brittle Matrix Composites BMC8,* Warsaw, Poland, 2006, 85–98.
5. Soranakom, C., and Mobasher, B. (2007), "Closed form solutions for flexural response of fiber reinforced concrete beams," *Journal of Engineering Mechanics*, 133(8), 933–941.
6. Weil, N. A., and Daniel, I. M. D. (1964), "Analysis of fracture probabilities in nonuniformly stressed brittle materials," *Journal of the American Ceramic Society*, 47(6), 268–274.
7. Calard, V., and Lamon, J. (2002), "A probabilistic-statistical approach to the ultimate failure of ceramic-matrix composites—Part I: Experimental investigation of 2D woven SiC/SiC composites," *Composites Science and Technology*, 62, 385–393.
8. Ulfkjaer, J., Krenk, S., and Brincker, R. (1995), "Analytical model for fictitious crack propagation in concrete beams," *Journal of Engineering Mechanics*, 121(1), 7–15.
9. Olesen, JF. (2001), "Fictitious crack propagation in fiber-reinforced concrete beams," *Journal of Engineering Mechanics*, 127(3), 272–280.
10. Lim, T. Y., Paramasivam, P., and Lee, S. L. (1987), "Analytical model for tensile behavior of steel-fiber concrete," *ACI Materials Journal*, 84(4), 286–298.
11. Lim, T. Y., Paramasivam, P., and Lee, S. L. (1987), "Bending behavior of steel-fiber concrete beams," *ACI Structural Journal*, 84(6), 524–536.
12. Swamy, R. N., and Al-Ta'an, S. A. (1981), "Deformation and ultimate strength in flexural of reinforced concrete beams made with steel fiber concrete," *ACI Structural Journal*, 78(5), 395–405.
13. Hassoun, M. N., and Sahebjam, K. "Plastic hinge in two-span reinforced concrete beams containing steel fibers," Proceedings of the Canadian Society for Civil Engineering, 1985, 119–139.
14. Peled, A., and Mobasher, B. (2005), "Pultruded fabric-cement composites," *ACI Materials Journal*, 102(1), 15–23.
15. Noguchi, K., Matsuda, Y., and Oishi, M. (1990), "Strength analysis of Yttria-stabilized tetragonal zirconia polycrystals," *Journal of the American Ceramic Society*, 73(9), 2667–2676.
16. Mobasher, B., Peled, A., and Pahilajani, J. (2006), "Distributed cracking and stiffness degradation in fabric-cement composites," *Materials and Structures*, 39(287), 317–331.

17. Soranakom, C., and Mobasher, B. (2008), "Moment–curvature response of strain softening and strain hardening cement based composites," *Cement and Concrete Composites*, 30(6), 465–477.
18. Soranakom, C. (2008), "Multi-scale modeling of fiber and fabric reinforced cement based composites," PhD dissertation, Arizona State University, Tempe, AZ.

15 Back-Calculation Procedures of Material Properties from Flexural Tests

INTRODUCTION

Several test methods have been proposed to characterize the postcrack tensile strength of fiber reinforced concrete (FRC). Their merits and drawbacks were discussed in details by Gopalaratnam [1], who addressed both experimental and analytical limitations. The material properties obtained in testing may be slightly different from those used in the design calculations. Because of the lack of a comprehensive testing program that can be put to use before design and construction, many material properties used in current design procedures are estimates based on the relationship of tensile and flexural responses with uniaxial compressive strength f_c'. In the proposed approach, the material properties can be estimated on the basis of prior experience and the material properties measured using testing during the construction quality control phase. As the sample must meet the specified strength in design calculations, the strength of a flexural member is often associated with its compressive strength. Other test methods that have been developed for the characterization of ductility and toughness may be great for the differentiation between various comparative samples, but their utility for design purposes is questionable. It would be ideal to obtain fundamental tensile properties from flexural tests; however, it is imperative that the discrepancy between the two must be addressed in design calculations, and the actual values obtained from testing should be subjected to appropriate safety factors.

To verify the theoretical derivations for flexural loading, the model applied to FRC beams must be subjected to several cases of deflection softening, deflection hardening, and strain hardening under three- and four-point bending tests under conditions where tensile data may or may not be available. The closed-form solutions for moment–curvature diagrams can be used to predict load deformation response and followed through a back-calculation procedure in order to measure the material properties from flexural tests. The material parameters for tension models are found by fitting the model to simulate the flexural response. If tension results are available, one can use a forward calculation to predict the flexural response. However, if tensile results are not available, one can use the flexural data to obtain effective tensile properties. Because the material parameters for compression models are less dominant in the prediction of the load deflection, they, too, can be estimated from the uniaxial compressive strength or tensile data.

The model presented in Figure 15.1 represents the set of derivations for the moment–curvature responses of both a stress-softening and a strain-hardening material with different tensile and compressive responses. Note that by setting the parameter $\alpha = 1$ and η as an extremely small number, one can obtain the strain-softening parameters from this model. The behavior of a specimen under flexure is dominated by cracking, which initiates at the notch and grows along the depth of the specimen. As such a test progresses, the deformation localizes at the notch and leads to crack propagation. Because the critical deformations are present at the opening of the crack tip, which may be measured at the base of the notch, the best-controlled variable in flexure tests is the Crack Mouth Opening Displacement (CMOD), also referred to as Crack Mouth Opening Deformation.

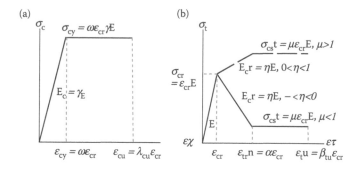

FIGURE 15.1 Model for tensile strain-softening, strain-hardening response.

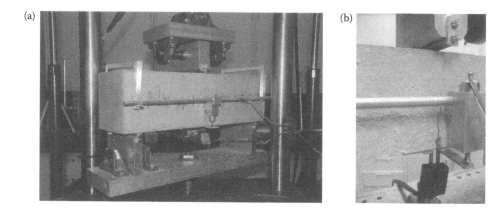

FIGURE 15.2 (a) Closed-loop flexural test setup. (b) Crack formation in a fiber reinforced specimen developed from the notch through the concrete mass.

To capture the postpeak response, especially in low-volume fraction composites, tests need to be conducted under closed-loop control with CMOD as the controlled variable. Figure 15.2 represents the flexural test setup. In this test, the CMOD is measured across the face of the notch using an extensometer. The extensometer, which provides the controlled rate of crack opening, measuring CMOD at the notch is shown in Figure 15.2b. In this test method, the monotonic loading procedure measures the mechanical fracture parameters G_f. The rate of loading is controlled by a constant rate of increasing CMOD such that the peak load is reached in approximately 10 min.

CASE A: TENSION DATA ARE UNAVAILABLE

In cases where no tension data are available, one can use the flexural test data to compute equivalent tension tests results. The verification of these material properties in the context of applicability and size effect can only be determined if a sufficient number of samples of various sizes and fiber contents are tested to provide a general understanding of the response. In many of the responses simulated in these samples, little effort is needed to obtain the best fit curve, since those algorithms can be easily implemented using optimization tools and software packages.

Case A1: Inverse Analysis of Load–Deflection Response of Polymeric Fibers

Grace Construction Products conducted flexural tests of FRC materials with two beam sizes at the two volume fractions 0.33% and 0.5% using synthetic polypropylene/polyethylene blend fibers (STRUX 90/40). The mixtures used a conventional concrete mixture produced by a ready mix operation with a 35-MPa specification (A. Reider Klaus, W. R. Grace Company, private communication). The results of these data were used to study the effect of specimen size on the residual strength determination. All specimens were demolded after 24 h and stored in lime-saturated water for at least 28 days. For each volume fraction, six $100 \times 100 \times 350$ mm beams and four $150 \times 150 \times 500$ mm beams were tested at clear spans of 300 and 450 mm. The concrete compressive strengths were 41.3 to 41.9 MPa, and the flexural strengths, measured by four-point bending tests, were in the range of 4.31 to 4.82 MPa.

Because there was no uniaxial tension test results for this investigation, the Young modulus and the cracking strain were estimated by back calculation from the load–deflection response of four-point bending by the elastic relationships:

$$E = \frac{108}{54} \frac{mL^3}{bd^3} \tag{15.1}$$

where m is the slope of elastic load–deflection response, and

$$\varepsilon_{cr} = \frac{P_{cr}L}{bd^2 E} \tag{15.2}$$

where P_{cr} is the load at cracking. The ultimate tensile strain, ε_{tu}, is assumed to be as high as 0.03 in order to allow for the large deflections, observed in the flexural test response, and the ultimate compressive strain, ε_{cu}, is assumed to be 0.004. Because the load–deflection response is not sensitive to yield compressive stress, an assumed value of $0.85f_c'$ was used. The parameters for the normalized residual tensile strength μ were obtained by trial and error until the predicted postpeak response in the flexural test data matched the experimental results. The complete set of parameters used in simulation are discussed by Soranakom and Mobasher [2] and Lim et al. [3], who used a bilinear tension-bilinear compression stress–strain model.

Figures 15.3a–d show the best fit of two tension models for two beam sizes containing a volume fractions of 0.33% and 0.5% that predict the experimental flexural stress–eflection responses in Panels a and b. It is observed that the larger section has lower normalized residual tensile strength for both volume fractions.

According to both the RILEM method and the Japan Concrete Society [4], the load versus beam deflection response can be analyzed according to ASTM C1018-97, JCI-SF4, and TR34 [5]. Estimation of the equivalent flexural responses is possible by normalizing the loads in the postpeak response by the nominal elastic section modulus. For example, according to JSCE [2], the residual flexural strength $f_{e,2}$ or $f_{e,3}$ can be obtained from the experiments. The last two columns in Table 15.1 compare the difference between the residual uniaxial tensile strength $\sigma_r = \mu \varepsilon_{cr} E$, which was used in the tension model, and the residual flexural strength $f_{e,2}$ or $f_{e,3}$, which was obtained from the experiments, according to JSCE [4]. Table 15.1 shows the values of the residual strength computed as 1.02 to 2.19 MPa, which is significantly higher than the tensile residual capacity (defined under $\mu E \varepsilon_{cr}$) of approximately 0.23 to 0.65 MPa, as was observed in the simulation.

Soranakom and Mobasher implemented a step-by-step procedure in an Excel worksheet for ACI Committee 544 [6] to use in the back calculation and design of FRC materials based on the deflection response of bending tests. The worksheet is applicable to the averaged residual strength (ARS) test method by considering only the reloading response of a cracked beam. For each load–deflection response, the normalized postcrack tensile strength, μ, is determined by trial and error, such that the predicted postcrack response best fits the experimental results at deflection levels, allowing one to back calculate the postcrack tensile strength from the load–deflection response.

Alternatively, one can use an approach proposed in Soranakom and Mobasher [7] to calculate the parameters of strain-softening and strain-hardening materials. Because of the extra parameters

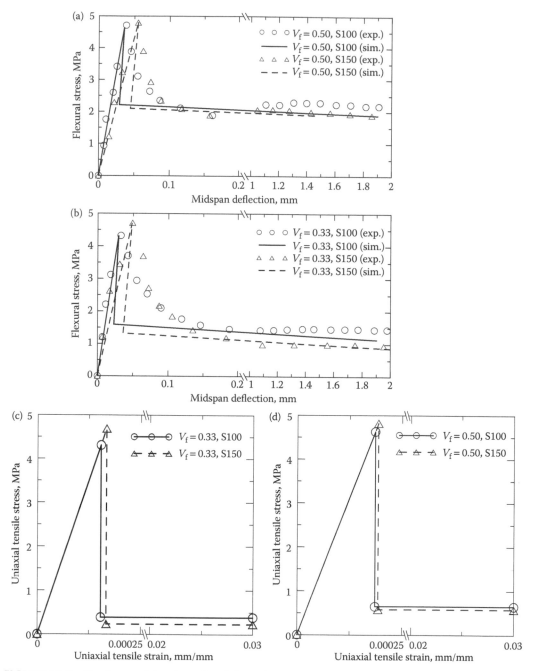

FIGURE 15.3 Tension model and predicted flexural stress deflection response of synthetic FRC for two sizes; (a) fitted uniaxial tensile model for $V_f = 0.33\%$ and $V_f = 0.5\%$. (b) Flexural stress deflection response for $V_f = 0.33\%$ and $V_f = 0.5$. (Data from Soranakom, C., "Multi-Scale modeling of fiber and fabric reinforced cement based materials," PhD dissertation, Arizona State University, Tempe, AZ, 2008.)

involved so that the stiffness of the postcracking phase and the ultimate strain capacity can be gauged an optimization of the model that involves fitting up to four parameters may require programming. However, a simple solution lies in the back calculation of the tensile properties using the bilinear softening model proposed by JSCE [4], which focuses on the main parameters (tensile strain capacity, ε_{tu}, and the residual tensile strength, μ). Note that in order to fit the load deflection

TABLE 15.1
Parameters Used in Simulation of the Four-Point Bending Responses of STRUX FRC

		Parameters Used								**Additional Information**		
Set	V_f (%)	Section	L (mm)	E (GPa)	ε_{cr}	ε_{tu}	μ	ω	ε_{cu}	f'_c (MPa)	σ_{cr} (MPa)	$\mu E \varepsilon_{cr}$ (MPa)	$f_{e,2}$ or $f_{e,3}$ (MPa)
1	0.33	100×100	300	28	153.9	0.03	0.09	8.3	0.004	41.9	4.31	0.39	1.51
2	0.33	150×150	450	28	167.5	0.03	0.05	7.6	0.004	41.9	4.69	0.23	1.02
3	0.50	100×100	300	25	185.2	0.03	0.14	7.6	0.004	41.3	4.63	0.65	2.19
4	0.50	150×150	450	25	192.8	0.03	0.12	7.3	0.004	41.3	4.82	0.58	1.90

Source: Soranakom, C., and Mobasher, B., Closed form solutions for flexural response of fiber reinforced concrete beams. *Journal of Engineering Mechanics*, 133(8), 933–941, 2007.

Note: $f_{e,2}$ or $f_{e,3}$ refers to the JSCE residual strength parameters.

response, up to five parameters = need to be specified, including E and ε_{tu} as the first two parameters that define the linear range. The nonlinearity beyond the first cracking point is addressed by two additional parameters, the stiffness of the cracked material, η, and the range of strain capacity, α, until the ultimate strength is reached. The parameters for the normalized residual tensile strength, μ, which represents the postpeak response, is very much affected by the fiber content.

Using the closed-form equations for moment curvature in addition to deflection computation from curvature values can, however, allow for the easy generation of the load–deflection curve for a variety of materials under various geometry and loading conditions.

Case A2: Inverse Analysis Load–Deflection Response of Macro-PP-FRC (English System)

Two types of polypropylene FRC mixtures were used: a fibrillated fiber, under a trade name Fibrashield manufactured by OCV reinforcements Phoenix, Arizona, and a macrosynthetic fiber manufactured by the Euclid Chemical Company Cleveland, OH. The fibrillated fibers and macrofibers were compared at the rate of $V_f = 5$ lb./yd.[3] The details of the samples and parameters for the simulation are listed in Table 15.2.

Three-point bend flexural tests were performed on $18 \times 6 \times 6$ in. ($457 \times 152 \times 152$ mm) beam specimens with an initial notch of 0.75 in. (19 mm). A test span of 16 in. (406 mm) was used. The deflection of the beam was also measured using a spring-loaded linear variable differential transformer with a 0.1-in (2.54-mm) range. The flexural load deformation responses are shown in Figures 15.4a and 15.4b. Figure 15.4a compares the two fiber types: fibrillated and macrosized.

TABLE 15.2
Postcrack Mechanical Parameters PPFRC Samples after 14 and 28 Days

Fiber Type	Fiber Content (lb./yd.[3])	Age	First Crack Tensile Strength, σ_{cr} (psi)	Normalized Postcrack Tensile Strength, μ	Postcrack Tensile Strength, σ_p (psi)	Average Postpeak Load, P_p (lb.)	ARS (psi)
Macro	5	14	185	0.2	37	346	169
Macro	5	28	245	0.23	56.4	485	238
Macro	5	28	220	0.30	66	615	301
Fibrillated	5	14	205	0.2	41	382	187
Fibrillated	5	14	200	0.23	46	414	203
Fibrillated	5	28	255	0.27	68.9	600	294
Fibrillated	5	28	245	0.29	71	605	296

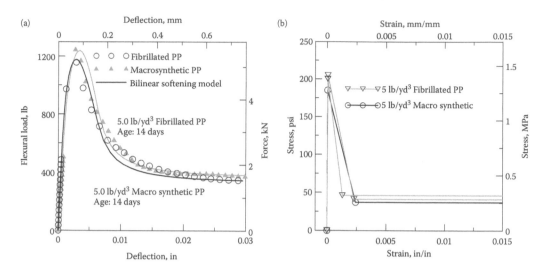

FIGURE 15.4   (a and b) Simulation of flexural response of polypropylene fiber reinforced (PP-FRC) comparison of fibrillated and macrosynthetic fibers; (c) back-calculated tensile response.

As the crack initiates and propagates, the sample exhibits marginal ductility, even though the majority of fiber contribution is in the postcrack range. The reduction in load after the peak is followed by a rather constant postpeak stress load carrying capacity.

After the initial slope of the curve and its termination point are characterized by (E and ε_{cr}), only two additional parameters (μ and ω) are needed to estimate the entire load displacement response. Parameter estimations are presented in Table 15.2.

DATA REDUCTION BY THE ARS METHOD AND RILEM TEST METHOD

The ARS was proposed by Banthia and Dubey [8, 9] and the ASTM C1399 [10]. The postcrack strength can be obtained by a simple open-loop testing machine available in many material testing laboratories. First, a steel plate is placed underneath a concrete beam and the specimen is loaded under a four-point bending machine until the concrete cracks. Then, the steel plate is removed and the cracked specimen is reloaded to obtain postcrack flexural strengths at deflection levels of 0.02, 0.03, 0.04, and 0.05 in. Finally, the equivalent stress results are averaged to represent an ARS value. This parameter has been used as a method to compare different material formulations, but many designers have been using it as a tensile strength measure. It is imperative to note that the ARS value is not an equivalently elastic stress and cannot to be associated with the postcrack tensile strength or the tensile residual strength parameter, σ_{cst}, in Figure 15.1a. The ARS method has been extended to ASTM C1399 [11]. As the postpeak response is averaged, the residual load is used in the elastically equivalent flexural stress with the section modulus of the uncracked beam to calculate a stress measure. In doing so, the load is divided by the equivalent elastic section modulus. The ARS value is an equivalently elastic stress in a specimen that is no longer elastic. The properties are measured in accordance to a flexural neutral axis, which is assumed to be at the centroid of the specimen. Because of cracking, the neutral axis for the specimen has shifted significantly toward the compression zone. Similar to the ARS method, the residual flexural strength, $f_{e,2}$, reported in Table 15.2 and based on the RILEM method, overestimates the residual uniaxial tensile strength, $\mu E \varepsilon_{cr}$, obtained by the present approach.

To show the fundamental problems of using this type of data reduction, the stress distribution during the late stage of composites is presented across the depth of the cross section. The strain distribution across the cross section is shown in Figure 15.5. Note that, according to the ARS method, the stress distribution is linear and the neutral axis remains at the center of the section. However, in

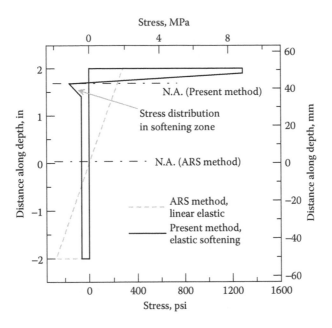

FIGURE 15.5 Comparison of stress distribution in the present approach with the ARS method.

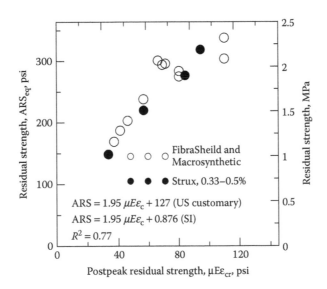

FIGURE 15.6 Correlation of residual strength with postcrack tensile strength measurement.

accordance to the bilinear elastic plastic model, or bilinear softening model, the neutral axis moves toward the compression zone and a uniform tensile stress distribution is distributed over the tensile zone. Figure 15.5 implies that the use of residual flexural strength as a residual tensile strength in design is unconservative, in that it is just an equivalent form of stress and not a material property. It is also shown that the present method captures the location of neutral axis, the linear compressive stress, and the residual tensile stress.

The relationship between the residual flexural strength and the residual uniaxial tensile strength of the four sets of polymeric specimens studied previously is examined next. Direct correlation of residual strength from ASTM C1399 and the present method are shown in Figure 15.6 and indicates that the ARS method overestimates the residual tensile strength by as much as three times. Because the ARS method assumes that the neutral axis is still at the centroid of the specimen, the stress distribution is linear throughout. This leads to very high nominal flexural stress levels in tensile layers, that are far higher than tensile strength, therefore this approach is unreasonable and unrealistic. Extreme caution must be exercised in applying the ARS method to the design and analysis of FRC sections.

Considering the range of specimens studied in the previous two cases, the ratio of uniaxial to flexural tensile residual strength varies from 0.23 to 0.31, with an average value of 0.27. From these measurements, it is clearly seen that the use of residual strength, obtained from the flexural stress deflection diagram as a material property, seriously overestimates the residual tensile strength, σ_r. In fact, the flexural stress is just an equivalent elastic stress of the true nonlinear stress of material. Thus, the parameter $f_{e,2}$ or $f_{e,3}$ can calculated solely by a residual tensile strength index. The misuse of $f_{e,2}$ or $f_{e,3}$ as a material property for residual tensile strength in design will result in the overestimation of actual capacity. This difference is best demonstrated by showing the stress distribution across the section and will be discussed further in Figure 15.5.

CASE B: TENSION DATA ARE AVAILABLE, FORWARD AND BACK CALCULATION

Case B1: Glass FRC

Marikunte et al. [12] and Singla [13] studied the effects of aging on the mechanical response of glass fiber reinforced cement. The proportions of their mix were 100:100:25:44:12 of cement, sand, metakaolin, water, and polymer, respectively. An alkali-resistant (AR) glass fiber content of 5% by weight of composite was used. Compression test data was not included. A set of specimens cured for 28 days (unaged) was selected for the simulation. Uniaxial tension specimens were approximately $25 \times 10 \times 225$ mm ($1 \times 0.4 \times 9$ in.) with two 6.25 mm (0.25 in.) notches. The flexural specimens for the four-point bending test were approximately $50 \times 10 \times 225$ mm ($2 \times 0.4 \times 9$ in.) with a clear span of 175 mm (7 in.).

Figure 15.7a shows the tension test results of glass fiber reinforced cement along with the fitted tension model and the modified strain model [6]. Figure 15.7b predicts the load–deflection

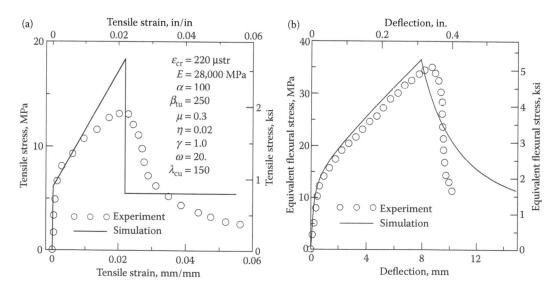

FIGURE 15.7 Glass FRC (a) homogenized material model (only shows tension) and (b) equivalent flexural stress deflection. (Based on ASTM C 1399, *Standard Tests Method for Obtaining Average Residual Strength of Fiber-Reinforced Concrete*, American Society of Testing and Materials, Philadelphia, 1999.)

responses, following the same trends as the ferrocement and fabric reinforced composites. The model completed on the basis of experimental tension data underpredicts the load by as much as 30% when direct tensile data are used. The ratio of 1.27 (experiment/predicted = 35 MPa/27 MPa [5.08 ksi/4.00 ksi]) is obtained. An increased level of stress–strain response by as much as 33% was necessary for the direct prediction of the load–deflection diagrams.

The difference between the tension and flexural data is attributed to the strain distribution profile of the two experiments. In the tension test, the entire volume of the specimen is a potential failure initiation zone. Comparatively, in the flexural test, only a small fraction of the tension region is subjected to an equivalent ultimate tensile strain. From a size effect point of view, this discrepancy is similar to the experimental plain concrete, which exhibits a higher modulus of rupture in bending tests than tensile uniaxial tests [14].

Ferrocement

Paramasivam and Ravindrarajah [15] studied the effect of the reinforcement arrangement on mechanical properties of ferrocement using both tensile and flexural samples. Test data from specimens with six layers of evenly distributed reinforcement were selected to simulate the load–deflection response. Tension test results were obtained from "dog-bone" specimens of $50 \times 25 \times 300$ mm ($2 \times 1 \times 12$ in.) net dimensions with an enlarged width of 100 mm (4 in.) at the ends. The flexural specimens for the four-point bending test were $100 \times 25 \times 380$ mm ($4 \times 1 \times 15.2$ in.) with a clear span of 300 mm (12 in.). The mortar mix ratio was 1.0:1.5:0.45 by weight (cement–sand–water). The mechanical properties of matrix measured at seven days included the compressive strength of f_{mc}' = 50 MPa (7.25 ksi), the modulus of rupture f_{mr} = 4.8 MPa (0.70 ksi), and the modulus of elasticity E_m = 27 GPa (3916 ksi). The reinforcement was steel wire mesh with a grid size of 8.5×8.5 mm (0.33 $\times$ 0.33 in.), diameter 0.87 mm (0.034 in.), yield strength f_{sy} = 245 MPa (35.53 ksi), ultimate tensile strength f_{su} = 371 MPa (53.8 ksi), and modulus of elasticity E_s = 140 GPa (20305 ksi). Six layers of wire mesh were evenly distributed in the specimen with a 3-mm (0.12 in.) cover at the top and bottom resulting in the longitudinal direction: V_L = 1.51% and the total reinforcement V_R = 3.02%. Results of this simulation are discussed in detail by Soranakom [16].

Case B2: Simulation of Steel FRC

Four sets of Steel FRC (SFRC) specimens containing hook-end fibers at the three volume fraction levels of 0.5%, 1.0%, and 1.5% and tested under three and four-point bending have been selected to demonstrate the discuss the back-calculation procedure [3, 17]. A brief discussion of these samples was presented Soranakom and Mobasher [2], but the full summary of the data is presented here. The details of these four series and the parameters used in the simulations are listed in Table 15.3, in which the specimens are designated as H22, H3, H1, and H21.

TABLE 15.3
Summary of the Back-Calculation Results of SFRC Specimens

Set	L (mm)	V_f (%)	Test	L (mm)	L_p (mm)	E (MPa)	ε_{cr} (×10⁻⁶)	ε_{tu}	μ	ω	ε_{cu}
H22	30	1.0	3P	300	N/A	25,400	116.2	0.015	0.44	9.8	0.004
H22*					100		247.5		0.3		
H3	30	1.5	3P	300	N/A	25,500	121.9	0.015	0.73	9.3	0.004
H3*					N/A		121.9		0.73		
H1	30	0.5	4P	750	N/A	25,400	110.6	0.015	0.24	10.3	0.004
H1*					N/A		178.1		0.24		
H21	50	1.0	4P	750	N/A	25,400	116.2	0.015	0.83	9.8	0.004
H21*					N/A		126.7		0.83		

Figures 15.8a and 15.8b show the tension model and the predicted flexural response for specimens H22 with $V_f = 1.0\%$ under three point bending. The solid line in Figure 15.8a is obtained by fitting the tension model to the experimental tension test data, and parameters for the model are listed in the first line of Table 15.3. The solid curve in Figure 15.8b represents the predicted flexural response obtained by the simulation process. It clearly reveals that the direct use of the tension data under-predicts the flexural response. By examining the load–deflection response shown in Figure 15.8b and Table 15.3, the nominal flexural stress at failure is 6.3 MPa, which is computed from the load at cracking $P_{cr} = 14{,}000$ N, clear span $L = 300$ mm, beam width $b = 100$ mm, and beam depth $d = 100$ mm. When compared with the uniaxial tensile strength $\varepsilon_{cr}E$ of 2.95 MPa in Table 15.3, the ratio of flexural stress to uniaxial stress is 2.13. By factoring the tension model by this ratio, the predicted response will match the experimental result. A uniform increase in tension capacity can be achieved by increasing the first cracking strain, ε_{cr}, because other strains and related stress

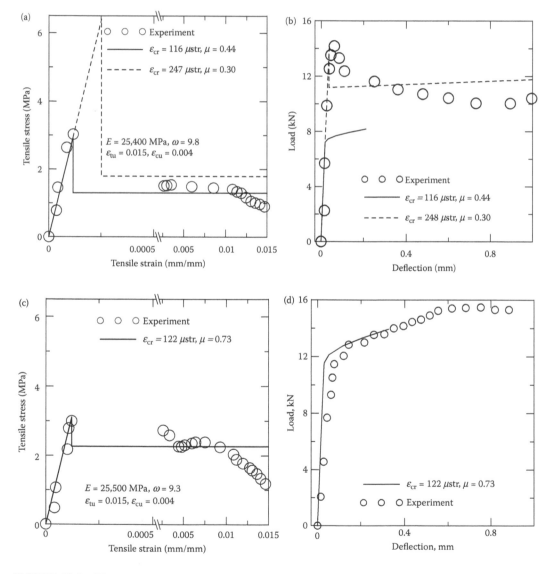

FIGURE 15.8 The tension models and the predicted load deflection response of SFRC under three-point bending: (a) tension model for H22; (b) load deflection response for H22; (c) tension model for H3; (d) load deflection response for H3.

measures in Figure 15.1 will be amplified by the same factor. In this case, however, there is a uniform increase in tension capacity by raising ε_{cr} by factor of 2.13 while maintaining $\mu = 0.30$, which then results in a relatively good fit of the load–deflection results as of shown by the dash line in Figure 15.8b. Figures 15.8c and 15.8d show the use of tension data to predict the flexural response of mixture H3 with $V_f = 1.5\%$. The overall prediction is reasonable, with the simulation exhibiting a slightly stiffer response than the experimental data. Note that no modification to tension model is necessary for this case, because the material is quite ductile; therefore, the brittleness effects are diminished. A detailed discussion of the various mixtures and the applicability of this model to predict the effect of fiber loading is presented by Soranakom and Mobasher [2].

Figure 15.9 shows the direct use of the tensile response to predict the flexural response of a four-point bending test. Figure 15.9a shows the uniaxial tension test result of mix H1 ($V_f = 0.5\%$), the fitted tension model, and the modified tension model, which yields the prediction and the best fit to the flexural test data. Figure 15.9b confirms that the fitted model underestimates the experimental load deflection curve. At cracking load $P_{cr} = 6000$ N, clear span $L = 750$ mm, beam width $b = 100$ mm and beam depth $d = 100$ mm, the nominal flexural stress for the four-point bending is 4.50 MPa, which is 1.61 times the uniaxial tensile strength. For this mix, H1 with a low fiber content $\mu = 0.24 < \mu_{crit} = 0.35$, the modification of the tension model by raising ε_{cr} with a factor of 1.61 results in

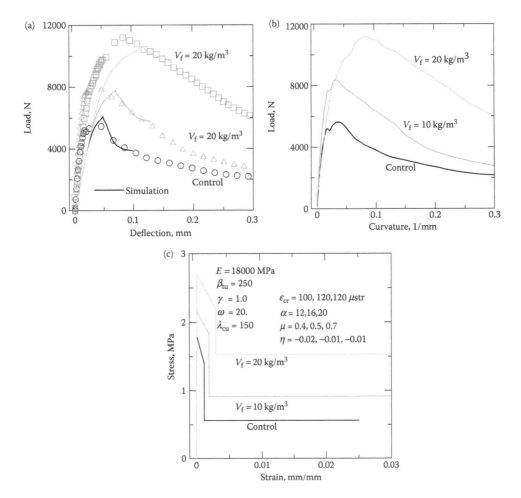

FIGURE 15.9 Simulation of the flexural load–deflection responses of beams with a different volume content of short AR glass fibers; (a) fitting of the load–deflection response, (b) moment–curvature response, (c) back-calculated tensile stress–strain response.

a reasonable prediction, as we can see by the dashed line shown in Figure 15.9b. For mix H21 ($V_f = 1.0\%$), which is shown in Figures 15.9c and 15.9d, the fitted tension model slightly underpredicted the experimental result, and, with a modifying factor of 1.09 to ε_{cr} for increasing tensile capacity, it leads to the best fit with the predicted response.

Three observations should be noted from the simulations in this section. First, there is a discrepancy between the uniaxial tension and nominal flexural strength results of a material such that the uniaxial test yields a lower tensile strength than the flexural test. This is related to the stress gradient in the flexural test both along the span and depth of the beam. Simply stated, the uniform stress in uniaxial test has a higher probability of localizing at a defect in the material and initiate cracking than the triangular stress in the bending test. Secondly, the variation of modified parameters presented in the Table 15.3 indicates that three-point bending varies to a wider degree than the four-point bending test. This verifies the accepted knowledge that the four-point-bending test is a better representative of the tension test than three-point-bending test because of the uniformity in stress distribution. Finally, the geometrical effect seems to decrease in both three and four-point bending when normalized postpeak tensile strength increases. Higher postpeak strength implies there are more fibers in material to suppress cracks initiation at critical defects so that the uniaxial test and flexural test become less sensitive to the probability distribution of defects.

AR GLASS FIBER CONCRETE

High performance HP-type AR glass fibers obtained from VETROTEX, Cem-FIL SAINT-GOBAIN Phoenix, Arizona, were used in reinforcing ordinary concrete mixtures at a wide range of fiber loadings. The concrete mixture had a water-to-cement ratio of 0.4, a characteristic strength of 60 MPa, and a slump of 50 to 55mm. The dosage of HP glass fibers was in the range of 5 to 20 kg/m^3 and was aimed at improving both long-term strength and ductility. Several different lengths of fibers, 6, 12, 24, and 40 mm, were used and have been discussed in detail by Desai et al. [18].

Flexural beam specimens $368 \times 101 \times 101$ mm in size with an initial notch were tested in a three-point bending configuration. A test span of 304 mm and a notch depth of 12.7 mm were also used. The maximum load was normalized with respect to the modified section modulus of the specimen and was referred to as the nominal flexural stress, as reported in Table 15.4. The area under the load deflection curve was calculated by the numerical integration of the load–deflection response. Fracture energy, G_f, was defined as the area under the entire load–deflection curve and normalized with respect to the crack ligament area.

A summary of the results of the flexural tests are shown in Table 15.4. Figure 15.9a shows the flexural response of the concrete with the volume fraction of AR fibers used. As expected, the concrete with 20 kg/m^3 AR glass fibers exhibits a higher flexural strength and ductility after 28 days than the other mixes. All the samples containing 6, 12, and 24 mm fiber lengths behave in a relatively similar manner [10]. Furthermore, the flexural response increases with the addition of

TABLE 15.4
Summary of the Flexural Test Results

Mix ID	Age (Days)	L (mm)	V_f (kg/ m^3)	Max Load (N)	Nominal Flexural Strength (MPa)	Average CMOD ($\times 10^{-2}$) (mm)	Average Deflection ($\times 10^{-2}$) (mm)	Average Toughness (N·mm)
Control	28	NA	NA	5663 (66.5)	2.15	3.06 (0.47)	3.77 (0.58)	1090 (270.5)
HP12_10	28	12	10	7943 (100)	3.10	3.14 (0.16)	4.02	1868
HP12_20	28	12	20	10867 (76.3)	4.12	3.29	8.25	4429

Note: Values in parenthesis reflect the standard deviation of three replicate samples.

fibers, regardless of the length of the fiber used; however, the magnitude of strength is only slightly improved with increasing the fiber lengths.

COMPARISON WITH THE RILEM APPROACH

Full-scale beam tests from the Brite Euram project, BRPR-CT98-0813 "Test and design methods for steel fiber reinforced concrete", were used in the model verification [19]. The first set of the experimental program studied the effects of concrete strength, fiber dosage, and span length on SFRC beams. Two grades of Normal-Strength Concrete (NSC) and High-Strength Concrete (HSC) were used. NSC used the fiber-type RC 65/60 BN at 25 and 50 kg/m³ (42.1 and 84.3 lb./yd.³), whereas HSC used fiber-type RC 80/60 BP at 60 kg/m³ (101.1 lb./yd.³). Two span lengths of 1.0 and 2.0 m (3.33 and 6.67 ft.) were used for the same cross section of 0.20 × 0.20 m (8 × 8 in.). Two replicate samples were tested under four-point bending for each span length, while the spacing between the two point loads was kept constant at 0.2 m (8 in.).

Material properties were characterized according to the RILEM model shown in Figure 15.10 and presented in Table 15.5 [20]. Key strength parameters used in the design were computed as shown in Table 15.5: $\sigma_{cy} = 0.85f_c'$, $\sigma_{cr} = \sigma_1$, $\sigma_p = (\sigma_2 + \sigma_3) / 2$. Using these definitions, μ, ω, M_{cr}, and m_∞ can be calculated [16, 21, 22]. Table 15.6 reports the average test results of two replicates of the six beam (B1–B6) series for three mixtures and two span lengths. To compare the test results with the nominal moment capacity, M_n, the ultimate moment of the section $M_{u,exp}$ was calculated from the

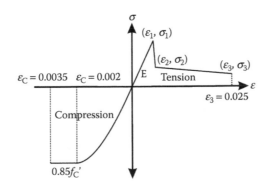

FIGURE 15.10   RILEM material model for SFRC.

TABLE 15.5
SFRC Parameters for RILEM Model

Mix	W_f, kg/m³ (lb./yd.³)	E, Mpa (psi)	f_c', MPa (psi)	σ_1, MPa (psi)	σ_2, MPa (psi)	σ_3, MPa (psi)	ε_1 (%)	ε_2 (%)	ε_3 (%)
NSC	25 (42)	31,854 (4.62E+06)	30.2 (4380)	3.5 (508)	1.1 (160)	0.8 (116)	0.011	0.21	25
NSC	50 (84)	30,564 (4.43E+06)	26.6 (3858)	4.2 (609)	2.0 (290)	1.2 (174)	0.014	0.24	25
HSC	60 (101)	38,411 (5.57E+06)	52.9 (7673)	6.2 (899)	3.1 (450)	3.1 (450)	0.016	0.26	25

Source: Soranakom, C., Yekani-Fard, M., and Mobasher, B., "Development of design guidelines for strain softening fiber reinforced concrete," *7th International Symposium of Fiber Reinforced Concrete: Design and Applications* BEFIB, 2008, ed. R. Gettu, 513–523, September 2008; Soranakom, C., and Mobasher, B., "Flexural design of fiber reinforced concrete," *ACI Materials Journal*, 461–469, September–October 2009. With permission.

Note: Strain at compressive yield stress, $\varepsilon_{cy} = 0.133\%$, and ultimate compressive strain, $\varepsilon_{cu} = 0.35\%$, for all mixes.

TABLE 15.6
Equivalent SFRC Parameters

Mix	W_f, kg/m³ (lb./yd.³)	σ_{cy}, Mpa (psi)	σ_{cr}, MPa (psi)	σ_p, MPa (psi)	ω	μ	m_∞	b, m (in.)	h, m (in.)	M_{cr}, kN·m (kips-ft.)	M_n, kN·m (kips-ft.)
NSC	25 (42)	30.2 (4,381)	3.5 (508)	1.1 (160)	8.63	0.31	0.91	0.2 (8.0)	0.2 (8.0)	4.67 (3.44)	4.25 (3.13)
NSC	50 (84)	26.6 (3,858)	4.2 (609)	2.0 (290)	6.33	0.48	1.33	0.2 (8.0)	0.2 (8.0)	5.60 (4.13)	7.44 (5.49)
HSC	60 (101)	52.9 (7,673)	6.2 (899)	3.1 (450)	8.53	0.50	1.42	0.2 (8.0)	0.2 (8.0)	8.27 (6.10)	11.71 (8.64)

Source: Soranakom, C., Yekani-Fard, M., and Mobasher, B., "Development of design guidelines for strain softening fiber reinforced concrete," *7th International Symposium of Fiber Reinforced Concrete: Design and Applications* BEFIB, 2008, ed. R. Gettu, September 2008, 513–523; Soranakom, C., and Mobasher, B., "Flexural design of fiber reinforced concrete," *ACI Materials Journal*, 461–469, September–October 2009.

Note: Strain at compressive yield stress, $\varepsilon_{cy} = 0.133\%$, and ultimate compressive strain, $\varepsilon_{cu} = 0.35\%$, for all mixes.

maximum experimental load P_{max},

$$M_{u,exp} = \frac{P_{max}\left(L - S_{mid}\right)}{4} \tag{15.3}$$

where L is the clear span and S_{mid} is the spacing between the load points defined earlier. The experimental capacity, $M_{u,exp}$, was compared with the proposed nominal moment capacity, M_n, in Figure 15.8, and they show good agreement with some variation. By using the recommended reduction factor, φ_p of 0.7, the reduced moment capacity $\varphi_p M_n$ is obtained well below the experimentally obtained values.

CONCLUSION

By applying back-calculation procedures to the flexural tests using the present algorithm, the computed values of tensile stress–strain overpredict the experimentally obtained results. This discrepancy can be explained by the nature of the tensile stress distribution in uniaxial and flexural tests. The uniform stress in uniaxial tension testing has a higher probability of localizing at a defect to initiate cracks than the linear stress distribution in bending. The model presents the expressions for the ratio of the measured tensile strength from the bending test to the uniaxial tension test. The higher the tensile residual strength, the less pronounced of size effect. The four-point bending test is more comparable with the uniaxial tension test than three-point bending, because both models have a zone of constant tensile stress. Finally, the residual uniaxial tensile strength is approximately 27% of the residual flexural tensile strength. The use of residual flexural tensile strength as a material property in design leads to un-conservative designs. The residual flexural tensile strength has been represented as an equivalent form of elastic stress, whereas the actual stress distribution is nonlinear and much lower than this assumed level.

According to the strain-softening model, the critical postcrack tensile strength, $\mu_{crit} = 0.35$, characterizes two subclasses of materials: deflection softening ($\mu < 0.35$) and deflection hardening ($\mu > 0.35$). The discrepancy between the present method and the ARS method is due to a difference in stress distribution between the two test methods. Because the flexural design of concrete member is generally governed by the tensile strength of material, the load deformation response can provide us with correct material parameters that agree with experimental observations and can be used for design applications.

REFERENCES

1. Gopalaratnam, V. S. (1995), "On the characterization of flexural toughness in fiber reinforced concretes," *Cement and Concrete Composites*, 17, 239–254.
2. Soranakom, C., and Mobasher, B. (2007), "Closed form solutions for flexural response of fiber reinforced concrete beams," *Journal of Engineering Mechanics*, 133(8), 933–941.
3. Lim, T. Y., Paramasivam, P., and Lee, S. L. (1987), "Analytical model for tensile behavior of steel-fiber concrete," *ACI Materials Journal*, 84(4), 286–298.
4. JSCE, "Methods of tests for flexural strength and flexural toughness of steel fiber reinforced concrete," JSCE-SF4, Japan Society of Civil Engineers, Concrete Library International, No. 3, Part III-2, 1984, 58–61.
5. The Concrete Society (2003), Technical Report 34, Concrete in industrial ground floors: A guide to design and construction, 3rd ed., Camberley.
6. Soranakom, C., and Mobasher, B. (2008), "Moment–curvature response of strain softening and strain hardening cement based composites," *Cement and Concrete Composites*, 30(6), 465–477.
7. Soranakom, C., and Mobasher, B. (2007), "Closed-form moment-curvature expressions for homogenized fiber reinforced concrete," *ACI Materials Journal*, 104(4), July–August, 351–359.
8. Banthia, N., and Dubey, A. (1999), "Measurement of flexural toughness of fiber-reinforced concrete using a novel technique. Part 1: Assessment and calibration," *ACI Materials Journal*, 96(6), 651–656.
9. Banthia, N., and Dubey, A. (2000), "Measurement of flexural toughness of fiber-reinforced concrete using a novel technique. Part 2: Performance of various composites," *ACI Materials Journal*, 97(1), 3–11.
10. ASTM Standard C1399/C1399M-10, "Standard test method for obtaining average residual-strength of fiber-reinforced concrete," ASTM International, West Conshohocken, PA, 2010, DOI: 10.1520/C1399_C1399M-10, http://www.astm.org.
11. ASTM Standard C1609/C1609M-10, "Standard test method for flexural performance of fiber-reinforced concrete (using beam with third-point loading)," ASTM International, West Conshohocken, PA, 2010, DOI: 10.1520/C1609_C1609M-10, http://www.astm.org.
12. Marikunte, S., Aldea, C., and Shah, S. P. (1997), "Durability of glass fiber reinforced cement composites: Effect of silica fume and metakaolin," *Advanced Cement Based Materials*, 5(3–4), 100–108.
13. Singla, N. (2004), "Experimental and theoretical study of fabric cement composites for retrofitting masonry structures," MS thesis, Arizona State University.
14. Wee, T. H., and Lu, H. R., and Swaddiwudhipong, S. (2000), "Tensile strain capacity of concrete under various states of stress," *Magazine of Concrete Research*, 52(3), 185–193.
15. Paramasivam, P., and Ravindrarajah, R. S. (1988), "Effect of arrangements of reinforcements on mechanical properties of ferrocement," *ACI Structural Journal*, 85(1), 3–11.
16. Soranakom, C. (2008), "Multi-scale modeling of fiber and fabric reinforced cement based composites," PhD dissertation, Arizona State University, Tempe, AZ.
17. Lim, T. Y., Paramasivam, P., and Lee, S. L. (1987), "Bending behavior of steel-fiber concrete beams," *ACI Structural Journal*, 84(6), 524–536.
18. Desai, T., Shah, R., Peled, A., and Mobasher, B. "Mechanical properties of concrete reinforced with AR-glass fibers," *Proceedings of the 7th International Symposium on Brittle Matrix Composites (BMC7)* in Warsaw, October 13–15, 2003.
19. Dupont, D. (2003), "Modelling and experimental validation of the constitutive law (σ–ε) and cracking behaviour of steel fiber reinforced concrete," PhD dissertation, Catholic University of Leuven, Belgium.
20. Soranakom, C., Yekani-Fard, M., and Mobasher, B., "Development of design guidelines for strain softening fiber reinforced concrete," *7th international Symposium of Fiber Reinforced Concrete: Design and Applications BEFIB 2008*, ed. R. Gettu, September 2008, 513–523.
21. Soranakom, C., and Mobasher, B. (2008), "Moment–curvature response of strain softening and strain hardening cement based composites," *Cement and Concrete Composites*, 30(6), 465–477.
22. Soranakom, C., and Mobasher, B. (2009), "Flexural design of fiber reinforced concrete," *ACI Materials Journal*, Sep.-Oct. 6(5), 461–469.

16 Modeling of Fiber Reinforced Materials Using Finite Element Method

INTRODUCTION

Fiber reinforced concrete (FRC) has gained popularity in recent years for its direct role in reducing construction time and labor costs. Many structural elements can now be reinforced with fibers as partial or total substitution for conventional reinforcement such as rebars or welded mesh [1]. In addition to the cost savings in terms of construction efficiency, labor, and materials, quality aspects and the serviceability condition of the final products are quite important. FRC allows for crack control and distributed cracking with much smaller crack widths that ultimately lead to durability enhancements.

New construction materials require standards in their mechanical properties and computational tools for structural design. Procedures for design with FRC materials are becoming more commonly available, as various research groups have been working to develop methods and design guides [2–5].

One of the simplest structural test conditions used both for design and also material property determination, is the biaxial bending of a circular panel. Several types of panel tests [6–8] have been developed to simulate the stress state in plate type specimen. These tests have been used to evaluate the flexural capacity of FRC. Test results produce a load–deflection response that can be used to calculate the energy absorption as a material index. As the size, thickness, and fixity of the panel changes, the load, deflection, and energy absorption also change. Therefore, the quality of FRC produced in various sizes, fiber dosage rates, and shapes cannot be equally compared because of the ambiguity of test methods. Unfortunately, there is a general lack of theoretical models to predict the flexural behavior, even in this simple case.

Figure 16.1a shows the most recent ASTM test method, C1550, in which a round panel on three pivoted supports is subjected to a point load at the center. This test has been adopted as a new standard in ASTM [3]. The silent feature of this test is in its consistency of the test results. With three supports, the bending moment is highest along the line between two supports, which leads to predictable three crack patterns, as shown in Figure 16.1b. The consistent load–deflection response, as well as the predictable crack patterns obtained in this test, serve as an index for quality control purposes. Alternatively, one can conduct round panel tests on a continuous support, as shown in Figure 16.2a. Assuming that there is a sufficient amount of fibers present to help in developing a redundant structure, this test results in six to eight lines of cracks (Figure 16.2b). Despite the consistency of the test results, round panel tests do not provide direct engineering material properties, since these properties have to be obtained using a back-calculation technique [9, 10].

The objective of this chapter is to develop a back-calculation procedure that serves to reduce the experimental data necessary in the use of models to formulate the crack patterns observed in various tests. A theoretical model with a constitutive model to correlate the experimental results using different boundary conditions is presented in here. One can expect that, if the number of supports increases from 3 to 4, or to N, the discrete support condition will, in the limit

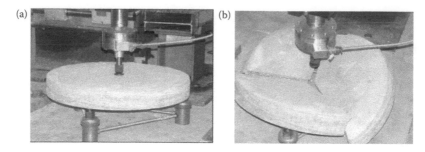

FIGURE 16.1   Round panel test on three pivoted support; (a) test setup, (b) crack patterns.

FIGURE 16.2   Round panel test on simply support; (a) test setup, (b) crack patterns of six samples tested (six to eight cracks).

case, approach the theoretical continuous simple support case. On the other hand, if the number of supports decreases from the infinity to a finite number, the variation in the radial moment distribution increases. This indicates that the bending moment along the radial line midpoint between two supports will be higher than the line from the center to each support. It is, hence, expected that the flexural behavior of an uncracked round panel with as many as eight discrete supports will behave sufficiently close to the limit case of a round panel on continuous simple support. At higher load levels, the bending stress along the line between supports can trigger up to eight lines of cracks. With this expectation, the elastic solution for a round panel on simple support can be used to predict the elastic behavior of a round panel on eight supports, and the yield line theory can be used to explain the flexural behavior of the cracked panel. Finite element software, ABAQUS [11, 12], is further used to create virtual testing of the round panel test. There four main tasks of this kind of testing is as follows:

1. Use general purpose finite element program ABAQUS to conduct an inverse analysis of FRC in order to find the set of material parameters for concrete models such so that it can predict the flexural behavior of the experimental load–deflection response. This process is necessary, because with only round panel test load–deflection results, structural analysis tools are needed to correlate the material properties with the structural tests.
2. Model an FRC round panel on a continuous support and eight discrete supports subjected to a point load at the center. This task helps in better understanding the flexural behavior of the round panel test and yields the load–deflection response and moment–curvature relationship for use in the round panel rigid crack model.
3. Model FRC three-point bending test as an alternative approach to obtain a moment–curvature relationship for the round panel rigid crack model.
4. Develop procedures for rigid crack models to predict the flexural behavior of the finite element result.

MODEL CONCRETE STRUCTURE WITH ABAQUS

This section addresses the use of finite element software, ABAQUS, to create a numerical model for the simulation of a round panel test for Tasks 1, 2, and 3, as mentioned in the previous section. Several numerical aspects are discussed in each subsection.

IMPLICIT OR EXPLICIT ANALYSIS TYPES

The modeling of concrete structures involves a potential displacement localization in the postpeak and softening load–deformation response. Both methods of implicit or explicit analysis differ by convenience or by computational efficiency, but yield comparable results if the models are appropriately calibrated. In some cases, however, only one option is open, since the other choice does not produce the results (because of the uniqueness of solution or loss of positive definiteness) needed from the global structural stiffness matrix.

In many implicit analysis approaches, the global stiffness matrix, K, and force vector, F, are assembled at every time increment, and the Newton–Raphson method is used to iteratively solve the system of equations (starting with Equation 16.1), until the solution D meets the convergence criteria,

$$KD = F \tag{16.1}$$

When the stress state at the integration points of an element moves beyond its maximum level and enters the softening region, the effective material stiffness becomes negative, subsequently leading to a negative global stiffness. Standard matrix inversion tools are used for the solution of systems of equation failures, A negative eigenvalue warning is issued, that implies that the global stiffness matrix is not positive and definite, leading to the nonuniqueness of the solution, and convergence problems, especially when the algorithm encounters a highly nonlinear stage. Concrete in a cracking model presents such a highly nonlinear problem, that it makes implicit approaches an incessant nuance.

Explicit analysis is preferred for modeling problems with both ascending and softening responses, because it does not form a global stiffness matrix, but solves dynamic equilibrium one equation at a time. The total time steps required to complete the analysis can be divided to several smaller units. The solution at each step is solved explicitly on the basis of the previous stress state such that the iterative procedure is not necessary. One major aspect of using this kind of explicit analysis is that there is a choice of a stable time increment Δt_{stable}, which allows for the presence of a stable solution throughout the analysis. Algorithms allow for the automatic calculation of the stable time step on the basis of the minimum characteristic length of the smallest element in the model L^e and wave speed of the material c_d,

$$\Delta t_{stable} = \frac{L^e}{c_d} \tag{16.2}$$

$$c_d = \sqrt{\frac{E}{\rho}} \tag{16.3}$$

where E is the Young's modulus and ρ is the mass density. If the stable time step is very small (10^{-5}–10^{-6}), double precision will avoid an excessive accumulation of error. All problems were analyzed by ABAQUS Explicit using double precision and an automatic stable time step.

ELEMENT

The size and type of elements can significantly affect the analysis results. For the same size of subdomains, second-order elements are more accurate than the first order, because the displacement field is more accurately approximated by the second-order polynomial function. Unfortunately, because of the number of time steps, only first-order elements are available in the present formulation. To obtain an acceptable accuracy, a finer mesh must be used, even though stable time increments will decrease as the mesh size decreases, leading to excessive time in which to run the model. Thus, accuracy may be sacrificed for computational efficiency or *vice versa*. Relatively uniform small elements are preferable in explicit analysis, because the stable time increment is governed by the smallest element in the model, as mentioned in Equations 16.2 and 16.3.

The proper choice of element types can greatly increase computational efficiency while maintaining accuracy. To model a round panel test, one can use either brick elements (tetrahedron C3D4 or cube C3D8) or a shell element S4. A good rule of thumb is to choose the simplest element that describes the physical behavior of the problem and yields the information needed for the particular case in question. In this case, the load–deflection response and moment–curvature relationship are needed, so a shell element is the better choice here. If brick elements are used, the moment–curvature response must be manually calculated from the stress–strain output, which may be inconvenient. The shell element is also more efficient than brick elements for modeling plate problems, since only one layer of the shell can sufficiently represent a flat panel subjected to bending load, while it requires several layers of brick elements to obtain similar results. More efficiency can be gained by using a reduced integration version S4R instead of the full integration S4, because the reduced integration requires that only one point to be used in the evaluation of element stiffness and stress–strain results, while the full integration requires four points.

QUASI-STATIC SIMULATION

The quasi-static experiments on round panels subjected to center point loading are operated a under displacement control at sufficiently low speeds. An experiment could last up to 5 to 10 min; hence, its simulation using real time steps is almost impossible for explicit analysis. An alternative quasi-static (dynamic analysis that minimizes the inertia effect) is used here instead. To model a nonlinear problem, stable time increments cannot be determined from the mesh size and material density alone. Other aspects, such as material constitutive behavior, crack propagation, geometry, and boundary conditions, affect stable time increments at various time periods. Therefore, the simulated speed of imposed deformations that produce negligible inertial force must be determined by numerical simulation.

The option "smooth step" is used to minimize the acceleration at the beginning of the testing and at the end of loading so that the inertial effects are kept at the minimum level. Figure 16.3a

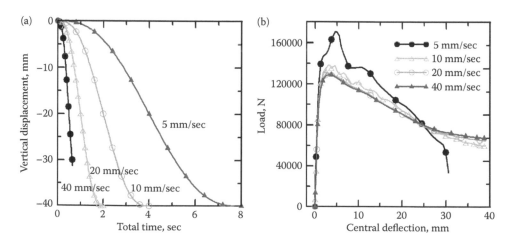

FIGURE 16.3 Quasi-static testing to determine the optimal rate of loading; (a) smooth step option for displacement control loading, (b) load–deflection responses.

shows four different speeds (vertical displacement U3 per 1, 2, 4, and 8 s representing 40, 20, 10, and 5 mm/s loading rates). Figure 16.3b shows the load–deflection response caused by the four speeds of loading. As expected, the fastest speed, 40 mm/s, introduces some inertia effect to the load–deflection response so that the load capacity is the highest and the response fluctuates in the postcrack region. At lower speed levels, 40, 20, 10, and 5 mm/s, the material response becomes smoother and less fluctuation is observed. The nominal loading rate for the simulation of an actual static load test is 10 mm/s.

Mass scaling can also be used to increase the computational efficiency of the model. The artificial increase in mass (density) by a factor of f^2 in Equation 16.3 can increase the stable time step by a factor of f in Equation 16.2, resulting in fewer steps needed to run in the explicit analysis. However, an excessive increase in artificial mass can produce an undesirable inertia force. To use this mass scaling option, a more complicated study of the choosing of the appropriate mass scaling parameters for a given speed was conducted. Because the models used are not rate-dependent materials (viscoelastic or viscoplastic), the rate of loading only affects the inertial force and not the material responses.

Concrete Model in ABAQUS

ABAQUS has three models: concrete with smeared cracking, a cracking model for concrete, and concrete with damaged plasticity. For a simple problem that only has tensile cracking as a predominant mode of failure, any concrete model can be used in the analysis. Concrete damaged plasticity suitable for the modeling of monotonic and cyclic loading is used here.

Because the load–deflection response of FRC is mainly controlled by the tensile capacity, material parameters for other aspects (such as compression response and shear) are less relevant. Only specimen size, boundary conditions, rate of loading, and load–deflection response are used as the primary input to the problem. Using trial and error. Material parameters can be found by an inverse analysis, such that the predicted response closely matches the experimental results while the material parameters are in a reasonable range, according to engineering experience. In this case, the two most critical parameters for concrete, Young's modulus and the tensile stress–crack width model, are the design variables in the inverse analysis; other material properties are assumed to have default values: compression is in elastic, flow potential eccentricity $e = 0.1$, the ratio of equi-biaxial compressive yield stress to initial uniaxial compressive yield stress $\sigma_{co}/\sigma_{co} = 1.16$, and the ratio of the second stress invariant on the tensile meridian, $q_{(TM)}$, to that on the compressive meridian $q_{(CM)}$, $\mathbf{K} = 2/3$.

Crack propagation and the tensile response of concrete structures can be modeled by two approaches: stress–strain or stress–crack width, as shown in Figures 16.4a and 16.4b, respectively. Both approaches work well in reinforced concrete structures when there is a sufficient amount of reinforcement to redistribute stress. However, the stress–strain approach becomes mesh-size-dependant for plain and FRC. Tensile stress tends to localize in a small region as the mesh size decreases, and no unique solution can be found. One remedy to mesh size dependence is to use a stress–crack width approach that uses fracture mechanics to define the amount of energy per unit area G_f in creating the crack surface. This approach is a unique solution and is used in the modeling of a round panel test. In fitting the experimental data, one needs to prescribe and define tensile stress, σ_t, and crack width, u_t^{ck}, for the model depicted in Figure 16.4b. For cyclic loading, tensile damage parameter, d_t, can also be specified to convert cracking displacement value, u_t^{ck}, to plastic displacement value, u_t^{pl}, by:

$$u_t^{pl} = u_t^{ck} - \frac{d_t}{\left(1 - d_t\right)} \frac{\sigma_t l_0}{E_0} \tag{16.4}$$

where the specimen length, l_0, is assumed to be one unit length, $l_0 = 1$. For the monotonic loading of a round panel test, the damaged parameter, d_t, is not as relevant; thus, by setting $d_t = 0$, the unloading path is the same as the loading path.

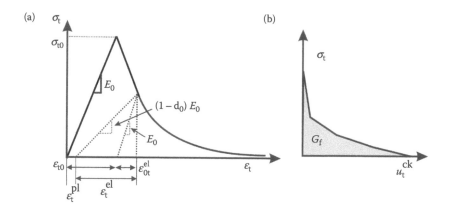

FIGURE 16.4 Concrete model in ABAQUS: (a) tensile stress–strain model, (b) stress–crack width model.

CALCULATION OF MOMENT–CURVATURE RESPONSE

This section briefly explains the procedure to use central difference approach to calculate section forces, moments, strains, and curvatures at the integration point of a thin shell element.

Nodal Calculation

The dynamic equilibrium at the beginning of time increment can be written as:

$$M \ddot{u}_t = P_t - I_t \tag{16.5}$$

where M is the nodal mass matrix, $\ddot{u}_t$ is the acceleration, P_t is the external applied force, and I_t is the internal element forces at time t. By solving the dynamic equilibrium, one would obtain:

$$\ddot{u}_t = M^{-1}\left(P_t - I_t\right) \tag{16.6}$$

Using this value of nodal acceleration and the central difference equations, one can calculate for the magnitude of velocity $\dot{u}_{t+\frac{\Delta t}{2}}$ at intermediate time $t + \Delta t/2$ by:

$$\dot{u}_{t+\frac{\Delta t}{2}} = \dot{u}_{t-\frac{\Delta t}{2}} + \frac{1}{2}\left(\Delta t_{t+\Delta t} + \Delta t_t\right)\ddot{u}_t \tag{16.7}$$

Similarly, the displacement at the end of time step $u_{t+\Delta t}$ is calculated as follows:

$$u_{t+\Delta t} = u_t + \Delta t_{t+\Delta t}\dot{u}_{t+\frac{\Delta t}{2}} \tag{16.8}$$

Element Calculation

At the element level, local velocity field $\dot{d}$ can be approximated by multiplying shape function N to the nodal velocity $\dot{u}$.

$$\dot{d}_{t+\frac{\Delta t}{2}} = N\dot{u}_{t+\frac{\Delta t}{2}} \tag{16.9}$$

The velocity gradient $\dot{E}$ is then obtained by applying the differential matrix, B, to the local velocity field d:

$$\dot{E}_{t+\frac{\Delta t}{2}} = B\, d_{t+\frac{\Delta t}{2}} \tag{16.10}$$

where $\dot{E} = (\partial / \partial t)\{\varepsilon_{11}, \varepsilon_{22}, \gamma_{12}, \kappa_{11}, \kappa_{22}, \kappa_{12}\}$ and $[B] = [\partial][N]$. The velocity gradient increment for each time step is:

$$\Delta E = \Delta t_{t+\Delta t}\, \dot{E}_{t+\frac{\Delta t}{2}} \tag{16.11}$$

Finally, the section force and moment increments ΔS can be found by the relationship of the element stiffness matrix $\partial S / \partial E$ and the velocity gradient increment ΔE as follows:

$$\Delta S = \begin{Bmatrix} \Delta N_1 \\ \Delta N_2 \\ \Delta N_{12} \\ \Delta M_1 \\ \Delta M_2 \\ \Delta M_{12} \end{Bmatrix} = \begin{bmatrix} A_{11} & A_{12} & A_{16} & B_{11} & B_{12} & B_{16} \\ A_{12} & A_{22} & A_{26} & B_{12} & B_{22} & B_{26} \\ A_{16} & A_{26} & A_{66} & B_{16} & B_{26} & B_{66} \\ B_{11} & B_{12} & B_{16} & D_{11} & D_{12} & D_{16} \\ B_{12} & B_{22} & B_{26} & D_{12} & D_{22} & D_{26} \\ B_{16} & B_{26} & B_{66} & D_{16} & D_{26} & D_{66} \end{bmatrix} \begin{Bmatrix} \Delta\varepsilon_{11} \\ \Delta\varepsilon_{22} \\ \Delta\gamma_{12} \\ \Delta\kappa_1 \\ \Delta\kappa_2 \\ \Delta\kappa_{12} \end{Bmatrix} \tag{16.12}$$

where substiffness matrices A, B, and D are the integral of material stiffness of plane stress, Q, from the bottom to the top surface of a shell element, which are defined as:

$$[A] = \int_{-h/2}^{h/2} [Q]\,dz, \quad [B] = \int_{-h/2}^{h/2} [Q]\,z\,dz, \quad [D] = \int_{-h/2}^{h/2} [Q]\,z^2\,dz,$$

$$[Q] = \begin{bmatrix} \dfrac{E_1}{1 - v_{12}v_{21}} & \dfrac{v_{12}E_2}{1 - v_{12}v_{21}} & 0 \\ \dfrac{v_{12}E_2}{1 - v_{12}v_{21}} & \dfrac{E_2}{1 - v_{12}v_{21}} & 0 \\ 0 & 0 & G_{12} \end{bmatrix} \tag{16.13}$$

where h is the total thickness of a shell element. During the analysis, one can use either the Simpson rules or the Gaussian numerical integration to evaluate the coefficient matrices **A**, **B**, and **D**. In this study, the Simpson rules with 21 points are used to calculate section stiffness for a thin shell S4R. At the end of each time step, the section force and moment increments are added to the previous values and result in the section forces and moments at the end of time step $t + \Delta t$,

$$S_{t+\Delta t} = S_t + \Delta S \tag{16.14}$$

The generalized section strain E is updated in the same way:

$$E_{t+\Delta t} = E_t + \Delta E \tag{16.15}$$

IMPLEMENTATION OF THE USER MATERIAL MODEL

Implementation of the user material model in ABAQUS Explicit requires developers to write their source code in subroutine UGENS, which is called-up at every time step increment. Throughout the analysis, ABAQUS provides the section state variables at the start of the increment (section forces and moments, S_t; generalized section strains, E_t; and solution-dependent state variables), the generalized section strain increments ΔE, and the time increment Δt.

The subroutine, UGENS, performs two functions: it updates the section forces and moments at the end of time step $t + \Delta t$ (as shown in Equation 16.7), and it updates the material tangent stiffness $(\partial S/\partial E)_{t+\Delta t}$ (as shown in Equation 16.4) according to the state variables at time $t + \Delta t$ for use in the next increment. To update the material model Q in Equation 16.3, users must first develop their material models and select the state variables, such as total strain and/or confining pressure, to be used in the updating process.

INVERSE ANALYSIS OF FRC

Two FRC mixtures containing steel fibers at loading rates of 80 kg/m³ (Mix80) and 100 kg/m³ (Mix100) were cast into round panel specimens with a thickness of 150 mm, radius of 790 mm, and were simply supported at a radius of 750 mm, as shown in Figure 16.2b. Six samples were tested at the University of Brussels in Belgium [13–15]. Only Mix80 is used in the finite element model here to demonstrate the inverse analysis procedure in obtaining material parameters. A comparison of the results of both Mix80 and Mix100 is discussed elsewhere [1].

Figure 16.5 (a) shows the finite element model for a round panel test on a continuous simply support. The roller boundary conditions (U3 = 0) are placed at r = 750 mm. The local axes 1′, 2′, and 3′ are the radial, circumferential, and vertical directions, respectively. At the center, the vertical displacement U3 is forced downward at the rate of 10 mm/s up to a deflection of 40 mm in order to simulate the movement of an actuator under displacement control. To prevent rigid body rotation about the center (r = 0), eight pin boundary conditions (local axis coordinate U2′ = U3′ = 0) are placed 22° apart and act to preserve symmetry.

Figure 16.6 (a) shows the averaged experimental response of six samples for use in the inverse analysis. Poisson's ratio of FRC was assumed as 0.15. A Young's modulus of $E = 20$ GPa was found by trial and error to fit the initial load–deflection response of the averaged experimental curve. The tensile stress–crack width relationship shown in panel b was found by trial and error in order for the simulation process to capture the peak load and postpeak response as shown in panel a. Inverse analysis of round panel test to obtain material parameters for concrete model in ABAQUS; (a) load–deflection response of mix ($V_f = 80$ kg/m³), (b) back-calculated tensile stress–crack width model from inverse analysis.

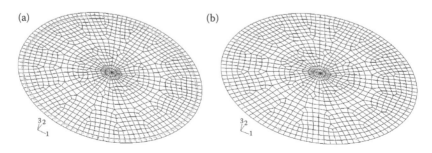

(a)   (b)

FIGURE 16.5 Finite element model for a round panel test; (a) simply support model, (b) eight discrete simply support model.

FINITE ELEMENT SIMULATION OF ROUND PANEL TEST

After an inverse analysis of Mix80 was performed to obtain the material parameter for a concrete model, the same finite element configuration was modified from the continuous simply support shown in Figure 16.5a to the eight discrete simply supports shown in Figure 16.5b. With the only difference being in the support condition, the analysis can be repeated and the results compared (Figure 16.6).

SIMULATION RESULT

Figure 16.7 confirms the anticipated results, in that the elastic response of a round panel test on eight supports can be well approximated with the response of a round panel test on simply support. This simplification allows one to use the existing analytical solution for a round panel test on simply support to predict the response of a round panel test on eight supports. However, the response becomes different when the load exceeds the maximum capacity where the cracks start

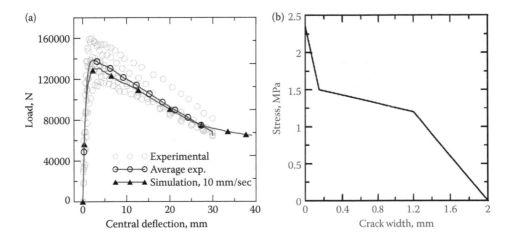

FIGURE 16.6 Inverse analysis of round panel test to obtain material parameters for concrete model in ABAQUS; (a) load deflection response of mix ($V_f = 80$ kg/m³); (b) Backcalcuated tensile stress crack width model from inverse analysis.

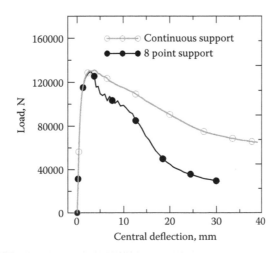

FIGURE 16.7 Comparison of load–deflection responses measured at the center between simply support model and eight-support model.

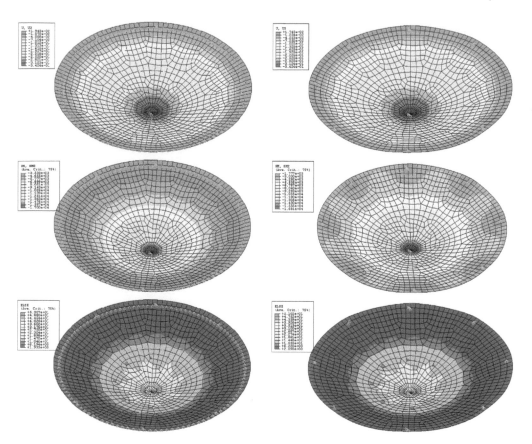

FIGURE 16.8 Comparison of elastic responses (taken at a central point deflection of U3 = 0.34 mm) between simply support model and eight-support model: (a and b) vertical displacement U3, (c and d) circumference moment SM2, and (e and f) element strain energy density. (a–f) The key elastic responses and comparison of the vertical deflection (U3), circumferential moment (SM2), and element strain energy density (ELSE) between the simply support model and the eight-support model. It can be seen that all three responses of the two models are comparable in magnitude. It should be noted that the circumferential moment of the eight-support model in Panel d is more localized along the line between the supports, which leads to the symmetry of eight predictable cracks in the plastic stage.

taking place. As expected, the continuous support model predicts higher load capacities than the eight-support model. This is due to the fact that continuous support has indefinite potential cracks paths, providing tougher crack resistance than the eight-support model, which has eight predicatble lines of cracks. For the same reason, one can expect different results if the number of supports are reduced to 7, 6, 5, and 4. As the number of cracks is reduced, a lower load capacity is obtained, because the energy required for total crack propagation decreases (Figures 16.8 and 16.9).

Figure 16.10 shows moment and curvature responses in two local directions: Radial 1 and Circumferential 2, which are the main sources of strain energy for the flexural type of problems. From Figures 16.10a and 16.10b, it can be seen that the circumference curvature (SK2) dominates a much smaller the radial curvature (SK1). With a comparable magnitude of the radial moment (SM1) and circumferential moment (SM2) shown in Figures 16.11c and 16.11d, it can be concluded that the primary source of strain energy comes from the circumferential moment SM2 and curvature SK2, which can then be used to balance the external energy caused by a central point load and deflection. Figure 16.10b shows that the curvature SK2 is approximately constant for most of the yield line (r = 200–800 mm); therefore, the rotation between two crack segments along the yield line can be approximated as a constant rotation.

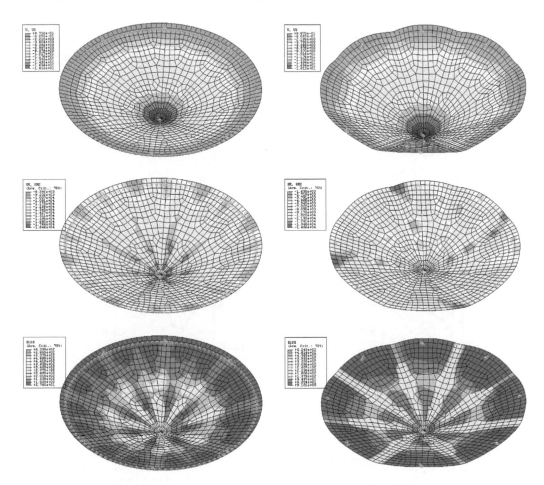

FIGURE 16.9 Comparison of plastic responses (taken at central point deflection of U3 = 18.13 mm) between simply support model and eight-support model: (a) vertical displacement U3, (b) circumference moment SM2, and (c) element strain energy density. (a–f) The key responses in plastic range and comparison of the vertical deflection (U3), circumferential moment (SM2), and strain energy (ELSE) between the simply support model and the eight-support model. Panel b confirms that the deformed shape of a round panel on eight supports can be approximated with rigid crack deformation. (e and f) The strain energy ELSE in eight-support model is more localized within the lines of cracks than the simply support model does; thus, the assumption that all strain energy is concentrated at the yield line is more accurate in the case of the eight-support model.

MOMENT–CURVATURE RELATIONSHIP FOR RIGID CRACK MODEL

To predict the load–deflection response of a three-point bending beam or a round panel test with the rigid crack models, the moment–curvature relationship of a panel and the size of a crack bandwidth must be provided as input. This information can be obtained directly from the yield line of a round panel test on eight supports. Another possible source is to obtain the fundamental moment–curvature response from the three-point bending test, which is a simpler and more economical method. If moment–curvature response and crack bandwidth can be estimated from the three-point bending test and can successfully predict the load–deflection response of a round panel test, there is no need to conduct costly round panel tests at all.

Figure 16.11 shows the simplified finite element model of a three-point bending test, which uses the same configurations as the round panel model: material parameters, rate of loading, mesh size, and element types. The numerical specimen has a thickness of 150 mm, a width of 150 mm, and a

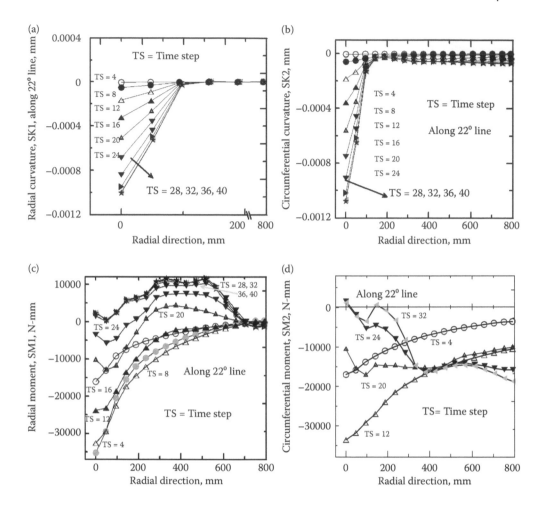

FIGURE 16.10 Comparative distribution of moment and curvature along the yield line (at angle 22°) for the simply support model; (a) radial curvature SK1, (b) radial moment SM1, (c) circumferential curvature SK2, (d) circumference moment SM2.

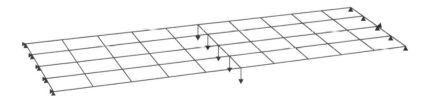

FIGURE 16.11 Finite element model for three-point bending test.

clear span of 450 mm. The boundary conditions at the left support are U1 = U3 = 0; additionally, U2 = 0 at the mid support is imposed to prevent the movement in Direction 2. The boundary condition at the right end is U3 = 0 with an additional U2 = 0 imposed at the mid support. At the center of the beam, U3 is forced to move downward for a distance of 40 mm in 4 s, which is the same rate of loading as a round panel test [16].

Figure 16.12 compares the moment–curvature response from two sources. The response from the eight-support model is obtained by averaging the circumferential moment and curvature at eight locations: $r = 300$ mm and angles $\alpha = 22.5n$ degree (where $n = 1,\ldots, 8$). The response of three-point bending

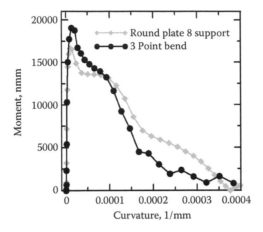

FIGURE 16.12 Moment–curvature response obtained from the round panel test on eight supports and a three-point bending test.

model is obtained by averaging the moment and curvature at several nodes along the center line. Both responses are comparable, implying that the three-point bending test could be used to obtain the moment–curvature diagram, which can then be used in a round panel rigid crack model to predict the flexural behavior of round panel testing. Because the same constitutive law formulation applies to both cases, one can conduct both three-point bending tests and round panel tests and back-calculate the material properties in terms of the moment–curvature relationship. The moment–curvature response can also be used in a simplified plasticity and/or limit analysis-based computations.

MODELING OF ROUND PANEL TEST WITH RIGID CRACK MODEL

An alternative approach to the solution is to assume a limit analysis approach. Initially, the structure is assumed to be elastic so that elasticity equations apply. The load may increase until, at some ultimate point, a collapse mechanism takes place. it is assumed that a full-scale rigid crack model applies and that Figures 16.13a and 16.13b show the top and side views of a round panel test and the expected crack pattern from the point load. Figure 16.13c shows a representative crack segment. Using the principle of virtual work, one can develop an upper bound to the failure mechanism by equating the external energy (because of the point load, P/n, and the deflection, δ) to the internal strain energy caused by the circumferential moment and curvature along the yield lines. This approach results in the derivation of an equation to predict the load–deflection response shown in Figure 16.13d. The equations needed to predict the load–deflection response of a round panel test on eight supports are divided into two stages: elastic analysis and plastic deformation.

ELASTIC RANGE ($M_\alpha < M_{CR}$)

In the elastic range, the circumferential moment M_α of a round panel test on eight supports can be approximated by measuring the elastic solution of a round panel test on simply support. The approximations introduced here rely on the solutions of the round plate with simply support as the limit case. According to Timoshenko and Woinowski-Keieger [17], the radial and tangential moment distribution M_r and M_α along a plate of radius R in dimension that is subjected to a point load at the center, is calculated by:

$$M_\alpha = \frac{P}{4\pi}\left[(1+v)\log\left(\frac{R}{r}\right)+1-v\right] \qquad M_t = \frac{P}{4\pi}(1+v)\log\left(\frac{R}{r}\right) \tag{16.16}$$

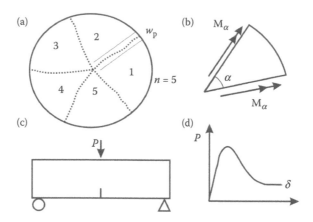

FIGURE 16.13 A round panel test; (a) a round panel test on top view with crack patterns, (b) a round panel test on side view, (c) representative crack segment, and (d) load–deflection response.

The deflection δ along any point can also be predicted by the elastic solution, given as:

$$\delta = \frac{P}{16\pi D}\left[\frac{(3+v)}{(1+v)}\left(R^2 - r^2\right) + 2r^2 \log\left(\frac{r}{R}\right)\right] \qquad (16.17)$$

where r is the variable representing location along the radial location, v is the Poisson ratio, and R is the total radius of a plate. Parameter D represents the stiffness of the plate and is represented as:

$$D = \frac{Eh^3}{12(1-v^2)} \qquad (16.18)$$

where h is the thickness. Equations 16.17 and 16.18 are valid only in the elastic range, in which the circumferential moment M_α at short distance $r = 5$ mm stays below the cracking moment M_{cr}. Note that the radius $r = 5$ mm is an arbitrary location chosen to check the initiation of yield line cracking and switching from elastic solution to plastic solution.

PLASTIC RANGE ($M_\alpha > M_{CR}$)

When the circumferential moment M_α exceeds the cracking moment M_{cr}, it is assumed that complete yield lines appear. One can approach the problem incrementally, but the yield line theory assumptions require the formation of all yield lines as a precondition to imposing a kinematically compatible displacement field. It is also assumed that crack segments deform rigidly so that the rotation along the yield lines between two segments is uniform over the length and constant. From Figure 16.14, one can derive the relationship between the central deflection δ and yield line rotation θ by using the cross product of two rigid segments to get a vector normal to a plane of crack segments. For design applications, however, the load and geometry will be different from the round determinate setup, and the model needs to be modified for each application. This can be done after the rigid crack model is verified. Parameter α is the angle of a crack segment between any two yield lines.

Figure 16.15 shows a schematic drawing of a rigid crack model for any two consecutive crack segments of any total N segments. Hence, the model is general for all round panel test that have N cracks. Vector v_0, v_1, and v_2 define crack lines (yield lines) as the following:

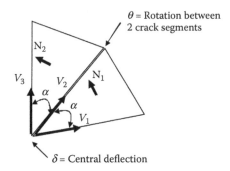

FIGURE 16.14   Rigid rotation of two crack segments.

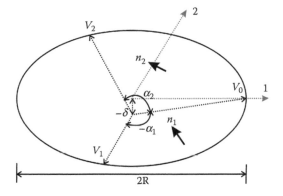

FIGURE 16.15   Multicrack panel with varying angles of rotation.

$$
v_0 = \left\{ \begin{array}{c} R \\ 0 \\ \delta \end{array} \right\} \quad v_1 = \left\{ \begin{array}{c} R\cos\alpha_1 \\ -R\sin\alpha_1 \\ \delta \end{array} \right\} \quad v_2 = \left\{ \begin{array}{c} R\cos\alpha_2 \\ R\sin\alpha_2 \\ \delta \end{array} \right\} \tag{16.19}
$$

Assuming that each crack segment is rigid (no bending), the vector normal to crack segments 1 and 2 can be found by calculating the cross product of the two adjacent vectors:

$$
n_1 = v_1 \times v_0 = \left\{ \begin{array}{c} -R\sin\alpha_1 \\ -\delta R(\cos\alpha_1 - 1) \\ R^2\sin\alpha_1 \end{array} \right\} \quad \text{and} \quad n_2 = v_0 \times v_2 = \left\{ \begin{array}{c} -\delta R\sin\alpha_2 \\ \delta R(\cos\alpha_2 - 1) \\ R^2\sin\alpha_2 \end{array} \right\} \tag{16.20}
$$

Rotation around the crack line (axis θ) is the angle between the two normal vectors and is determined by the dot product:

$$
\theta = \cos^{-1}\left\{ \frac{n_1 \times n_2}{|n_1||n_2|} \right\}
$$

$$
= \cos^{-1}\left\{ \frac{\delta^2\left(\sin\alpha_1\sin\alpha_2 - \cos\alpha_1\cos\alpha_2 + \cos\alpha_1 + \cos\alpha_2 - 1\right) + R^2\sin\alpha_1\sin\alpha_2}{\sqrt{(1-\cos\alpha_1)(R^2\cos\alpha_1 + 2\delta^2 + R^2)}\sqrt{(1-\cos\alpha_2)(R^2\cos(\alpha_2 + 2\delta^2 + R^2)}} \right\} \tag{16.21}
$$

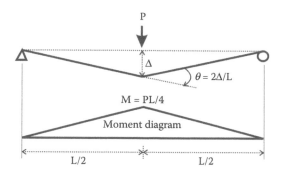

FIGURE 16.16 Collapse mechanism for a three-point bending beam.

If all crack segments have an equal crack angle, that is, $\alpha_1 = \alpha_2 = \alpha$, then the equation is simplified to the standard rigid crack model,

$$\theta = \cos^{-1}\left\{\frac{\cos(\alpha)(2\delta^2 + R^2) + R^2}{\cos(\alpha)R^2 + 2\delta^2 + R^2}\right\} \tag{16.22}$$

From the principle of virtual work as applied to the three-point bending specimen shown in Figure 16.16 or the case of round panel shown in Figure 16.15, the summation of external work done is equal to the internal work at the crack lines as shown in Equation 16–23. The incremental central deflection of the panel δ is used to calculate the rotation of the crack line θ. The moment per unit width m can be calculated from the moment rotation obtained from the three-point bending test. For N number of crack segments having a radius R, we can obtain the relationship between the load and deflection of the panel as follows:

$$W_{ext} = W_{int} \tag{16.23}$$

$$P\delta = R\sum_{i=1}^{N} m_i \theta_i$$

$$P = \frac{R}{\delta}\sum_{i=1}^{N} m_i \theta_i \tag{16.24}$$

The moment rotation per unit width can be obtained directly from a three-point bending test of the beam having the same depth as the panel. If the depths are different, the approach needs to be developed to obtain the $M-\theta$ for different depth,

$$M = \frac{PL}{4}; \quad m = \frac{M}{b}; \quad \theta = \frac{2\Delta}{L} \tag{16.25}$$

Expressing the principal of virtual work in incremental terms, one can measure both the external virtual work and internal virtual work,

$$\delta W_{ext} = \delta W_{int}, \quad Pd\delta = nRM_p d\theta \tag{16.26}$$

where n is number of crack segments that is equal to the number of yield lines, and M_p is the postcracking moment per unit length obtained from a given moment–curvature relationship. By substituting $n = 2\pi/\alpha$ and curvature $\kappa = d\theta/dx = \Delta\theta/w_p$, where w_p is the crack bandwidth (as

TABLE 16.1

Plastic Hinge Lengths as a Function of Geometry and Material Properties of Flexural Beams

Reference	Hinge Length (l_p)	Hinge Length Variables
Baker [18]	$k(z/d)^{1/4}d$	Span, depth
Sawyer [19]	$0.25d + 0.075z$	Span, depth
Corley [20]	$0.5d + 0.2\sqrt{d}(z/d)$	Span, depth
Mattock [21]	$0.5d + 0.005z$	Span, depth
Priestley and Park [22]	$0.08L + 0.022d_b\, f_y$	Span, bar diameter
Panagiotakos and Fardis [23]	$0.18L + 0.021d_b\, f_y$	Span, bar diameter

shown in Figure 16.13a into Equation 16.4), the relationship of the postcracking load–deflection response of a round panel test (noting that the term $d\delta$ correlates with the transverse direction t), can be obtained as:

$$P = nRM_p \frac{d\theta}{dx}\left(\frac{d\delta}{dx}\right)^{-1} \quad \text{and} \quad P = \frac{2\pi R}{\alpha} M_p \kappa \left(\frac{d\delta}{dt}\right)^{-1} \tag{16.27}$$

Assuming that the displacement rate along the radial crack length is constant, one can find the magnitude of the load at each curvature level. The procedure to use the moment–curvature relationship to incrementally obtain the force displacement is as follows:

1. Assume N and α representing the number and the arc angle associated with each cracked segment.
2. Increment the delta δ using a stepwise approach.
3. Compute θ using Equation 16.21 or 16.22.
4. Compute κ, estimated as the average curvature.
5. Compute M from the normalized moment–curvature relationship, as postulated by Soranakom and Mobasher [9, 10] and discussed in Chapter 15.
6. Get P from Equation 16.27.

The parameter that has not been evaluated in detail in this case is the average width of the localized zone, w_p. A review of the available literature indicates that a variety of empirical approaches are available. The determination of hinge length has been addressed by several authors (listed in Table 16.1). Major hinge length variables are functions of the depth of slab d, the diameter of the reinforcing bar d_b, and the yield strength of the reinforcing bar, which most likely controls the rotation of the hinge region. The span (z or L), also contributes to the rotation in the nonhinge region.

PREDICTION OF LOAD–DEFLECTION RESPONSE

A simplified routine was used to calculate the complete load–deflection response from uncracked elastic to cracked plastic. Equations 16.16 and 16.17 are used to calculate the load–deflection response in the elastic range until the circumferential moment M_α at $r = 1$–5 mm (assumed not to exceed the 0.5% of the diameter of the plate) exceeds the cracking moment M_{cr}. After the yield lines appear, it can be assumed that every point along the yield line rotates by the amount calculated by Equation 16.19. To use the moment–curvature relationship, a crack bandwidth w_p must be assumed in order to obtain the curvature by $\kappa = \theta/w_p$. With a known curvature, the postcracking moment per unit length M_p can be drawn from a given moment–curvature relationship, while Equation 16.20

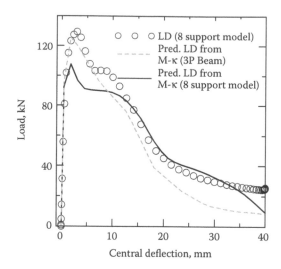

FIGURE 16.17 Comparison of the predicted response using a rigid crack model versus the finite element method (FEM) eight-support model; (a) average moment–curvature response at $r = 300$ mm from eight-support model, (b) the elastic-rigid crack model prediction.

is used to predict the level of applied load for a given central deflection δ. Parameters used for the simulation of the load–deflection of a round panel of Mix80 are as follows:

- Angle, $\alpha = \pi/4$ ($n = 8$ pieces of crack segments)
- Young's modulus, $E = 20000$ MPa
- Poisson ratio, $v = 0.15$
- Radius of a round panel, $R = 750$ mm
- Thickness, $t = 150$ mm
- Plastic crack bandwidth, w_p is assumed to 0.85 of thickness $= 0.85 \times 150 = 127.5$ mm
- Central deflection is measured at $r = 1$ mm, because the solution is singular at $r = 0$ mm
- A cracking criteria M_α ($r = 5$ mm) $> M_{cr} = 12000$ N·mm to switch from elastic to plastic yield line analysis was used.

Figure 16.17 compares three load–deflection responses: one is obtained directly from the eight-support model and two are predicted responses using the moment–curvature diagram from a round panel of the eight-support model and from a three-point bending beam, as shown in Figure 16.13. The two predicted responses show the same elastic part, because both use the same Young's modulus and Poisson ratio, which exactly match the load–deflection response obtained directly from the eight-support finite element model. After cracking has taken place, the predicted response from the moment–curvature relationship (obtained from three-point bending) matches the peak load better than the response using the moment–curvature relationship from an eight-support model.

SUMMARY

A round panel test under various configurations presents a potential new testing mechanism that can be used in evaluating the material quality of FRC. The proposed test method in eight supports results in predictable crack lines with a more consistent load–deflection response. The round panel rigid crack model, which is formed on the basis of the elastic solution and yield line theory, successfully predicts the flexural behavior of a round panel test with an acceptable degree of accuracy. Results of this test correlate very well with the three-point bending test, which could also be used as a potential verification method of back-calculation for material properties. The moment–curvature response obtained from three-point bending and the round panel test can be also used to predict the flexural response of plate specimens. Additional research on the effect of crack bandwidth on the yield line is warranted.

REFERENCES

1. Soranakom, C., Mobasher, B., and Destrée, X., "Numerical simulation of FRC round panel tests and full scale elevated slabs," Deflection and Stiffness Issues in FRC and Thin Structural Elements, ACI SP-248-3, 2008, 31–40.
2. di Prisco, M., Failla, C., Plizzari, G. A., and Toniolo, G., "Italian guidelines on SFRC," in di Prisco, M., and Plizzari, G. A. (eds.), *Fiber Reinforced Concrete: From Theory to Practice*, Proceedings of the International Workshop on Advances in Fibre Reinforced Concretes, Bergamo (Italy), September 24–25, 2004, 39–72.
3. RILEM TC 162-TDF (2000), "Test and design methods for steel fibre reinforced concrete: Recommendations. Bending test," *Materials and Structures*, 33(225), 3–5.
4. RILEM Final Recommendation TC 162-TDF (2003), "Test and design methods for steel fibre reinforced concrete; $\sigma-\varepsilon$ design method," *Materials and Structures*, 36, 560–567.
5. Vandewalle, L., "Test and design methods for steel fibre reinforced concrete proposed by RILEM TC 162-TDF," *Proceeding of the International Workshop on Fiber Reinforced Concrete: From Theory to Practice,* Bergamo, September 24–25, 2004, 3–12.
6. EFNARC, European Specification for Sprayed Concrete, European Federation of National Association of Specialist Repair Contractors and Material Suppliers for the Construction Industry, Aldershot, UK, 1996.
7. Bernard, E. S. (2000), "Behaviour of round steel fibre reinforced concrete panels under point loads," *Materials and Structures*, 33, 181–188.
8. ASTM Standard C1550-10a (2010), "Standard test method for flexural toughness of fiber reinforced concrete (using centrally loaded round panel)," ASTM International, West Conshohocken, PA, 2010, DOI: 10.1520/C1550-10A, www.astm.org.
9. Soranakom, C., and Mobasher, B. (2008), "Moment–curvature response of strain softening and strain hardening cement based composites," *Cement and Concrete Composites,* 30(6), 465–477.
10. Soranakom, C., and Mobasher, B. (2007), "Closed form solutions for flexural response of fiber reinforced concrete beams," *ASCE Journal of Engineering Mechanics*, 133(8), August, 933–941.
11. Hibbitt, B. K., and Sorensen, P. (2004), *ABAQUS Analysis User's Manual Version 6.5*, ABAQUS, Volumes II–V. Inc., Pawtucket, RI.
12. Hibbitt, K., and Sorensen, P. (2004), *Getting Started with ABAQUS*, ABAQUS Version 6.5. Inc., Pawtucket, RI.
13. Destrée, X., "Free suspended elevated slabs of steel fibre reinforced concrete: Full scale test results and design," *7th international Symposium of Fiber Reinforced Concrete: Design and Applications BEFIB 2008,* ed. R. Gettu, September 2008, 941–950.
14. Destrée, X., and Mandl, J. (2008), "Steel fibre only reinforced concrete in free suspended elevated slabs: Case studies, design assisted by testing route, comparison to the latest SFRC standard documents," in Walraven, J., and Stoelhorst, D. (eds.), *Tailor Made Structure*, International FIB 2008 Symposium Amsterdam, Boca Raton, FL: CRC Press, 111.
15. Destrée, X., "Planchers structurels en béton de fibres métalliques: Justification du modèle de calcul," Université Laval, CRIB, Troisième Colloque International Francophone sur les Bétons Renforcés de Fibres Métalliques, Juin 11, 1998.
16. Soranakom, C., and Mobasher, B., "Correlation of tensile and flexural behavior of fiber reinforced cement composite," *Proceeding of the Eighth International Symposium and Workshop on Ferrocement and Thin Reinforced Cement Composites,* Bangkok, Thailand, February 2006, 621–632.
17. Timoshenko, S. P., and Woinowski-Keieger, S. (1970), *Theory of Plates and Shells*, 2nd ed., New York: McGraw-Hill, 580.
18. Baker, A. L. (1956), *Ultimate Load Theory Applied to the Design of Reinforced and Prestressed Concrete Frames*. London: Concrete Publications Ltd., 91.
19. Sawyer, H. A., "Design of concrete frames for two failure states," *Proceedings of the International Symposium on the Flexural Mechanics of Reinforced Concrete*, ASCE-ACI, 1964, 405–431.
20. Corley, G. W. (1966), "Rotation capacity of reinforced concrete beams," *Journal of Structural Engineering*, 92(ST10), 121–146.

21. Mattock, A. H. (1967), "Discussion of rotational capacity of reinforced concrete beams, by W. D. G. Corley," *Journal of the Structural Division*, 93(2), 519–522.
22. Priestley, M. J. N., and Park, R. (1987), "Strength and ductility of concrete bridge columns under seismic loading," *ACI Structural Journal*, 84, 61–76.
23. Panagiotakos, T. B., and Fardis, M. N. (2001), "Deformations of reinforced concrete members at yielding and ultimate," *ACI Structural Journal*, 98(2), 135–148.

17 Flexural Design of Strain-Softening Fiber Reinforced Concrete

INTRODUCTION

Despite the fact that fiber reinforced concrete (FRC) has been used in construction industry for more than four decades, applications are still limited to a few market sectors. This is mainly due to the lack of standard guidelines for design procedures. To facilitate the design process, technical guidelines for FRC have been developed by RILEM committee TC 162-TDF for steel FRC (SFRC) [1–3] during the past 15 years. The committee proposed a three-point bending test of a notched beam specimen for material characterization. The elastically equivalent flexural strength at specific crack–mount–opening–displacement is empirically related to the tensile stress–strain model. The compression response is described by a parabolic-rectangular stress–strain model. The strain compatibility analysis of a layered beam cross section is required to determine the ultimate moment capacity. Similar to the RILEM, the German guidelines for design of flexural members use the strain compatibility analysis to determine the moment capacity [4]. In the United Kingdom [5], the practice of FRC traditionally followed the Japanese Standard JCI-SF4 [6]; however, it has recently shifted toward and the RILEM design methodology. The Italian guideline is also based on load–deflection curves deduced from flexural or direct tension test [7]. The current U.S. design guidelines for flexural members are based on empirical equations of Swamy et al. [8] and Fischer [9]. The particular types of fibers and nature of concrete were not specified in the guidelines. Henager and Doherty [10] proposed a tensile stress block for SFRC that is comparable with the ultimate strength design of ACI 318-05 [11].

This chapter proposes a design methodology for strain-softening FRC and consists of two parts: design for ultimate strength, and, design for serviceability. The design procedures are based on simplified and also extended theoretical derivations of Soranakom and Mobasher [12, 13], in addition to ACI 318-05 [11] and RILEM TC 162-TDF [3]. Topics include nominal moment capacity, minimum reinforcement for flexural cracking, minimum postcrack tensile strength, tensile strain limit, short-term deflection calculations, and a conversion design chart to correlate traditional reinforced concrete and FRC systems. Several design examples are presented to illustrate the application of the proposed methods to typical structural members.

The proposed design guideline provides computational efficiency as compared with the commonly used strain compatibility analysis of a layered beam in determining moment capacity of FRC members. The closed-form equations and guidelines are compatible with the ACI 318 design procedures, while, allowing deflection and serviceability criteria to be calculated based on fundamentals of structural mechanics. The approaches proposed can easily be applied to strain-hardening and hybrid reinforced concrete containing continuous rebars as well. These computations allow engineers to reliably analyze, design, and compare the overall performance of conventional reinforced concrete system and FRC.

Strain-Softening FRC Model

Tensile and compressive response of strain-softening FRC such as SFRC and polymeric FRC can be simplified to idealized stress–strain models as shown in Figures 17.1a and 17.1b. It is assumed that the addition of fibers has no effect on the first cracking response of matrix and there is no precritical crack growth in the ascending portion of stress–strain response. The contribution of fibers is mostly apparent in the postpeak tensile region, where the response is described by a simplified and constant residual stress level. This average constant postcrack tensile strength σ_p for the softening response can be correlated to the fiber length, volume fraction, and bond characteristics.

The following assumptions are made in the development of the material models: (a) Young's modulus E for compression and tension are equal, (b) tension model (Figure 17.1a) consists of a linear stress–strain response up to the cracking tensile strain ε_{cr}, followed by a constant postcrack tensile strength $\sigma_p = \mu E \varepsilon_{cr}$ with parameter μ ($0 < \mu < 1$) representing the postcrack strength as a fraction of the cracking tensile strength $\sigma_{cr} = E \varepsilon_{cr}$, and (c) compression model defined by an elastic perfectly plastic model (Figure 17.1b) using a yield compressive strain $\varepsilon_{cy} = \omega \varepsilon_{cr}$ with a parameter ω ($\omega > 1$) representing the compressive–cracking tensile strain ratio.

To minimize the number of material parameters, the tensile strength and the Young's modulus are assumed to be marginally affected by fiber type and content and conservatively estimated by the relationship governing normal concrete using ACI 318 Section 11.2 and Section 8.5.1, respectively,

$$\sigma_{cr} = E \varepsilon_{cr} = 0.56\sqrt{f_c'} \ (\text{MPa}) \quad \left(\text{or} = 6.7\sqrt{f_c'} \ (\text{psi}) \right) \tag{17.1}$$

$$E = 4733\sqrt{f_c'} \ (\text{MPa}) \quad \left(\text{or} = 57{,}000\sqrt{f_c'} \ (\text{psi}) \right) \tag{17.2}$$

where f_c' is the ultimate uniaxial cylinder compressive strength. First crack tensile strain for FRC can be calculated assuming Hooke's law as

$$\varepsilon_{cr} = \frac{\sigma_{cr}}{E} = \frac{0.56\sqrt{f_c'}}{4733\sqrt{f_c'}} = \frac{6.7\sqrt{f_c'}}{57000\sqrt{f_c'}} = 118 \ \mu\text{str} \tag{17.3}$$

The ultimate moment capacity of FRC is significantly affected by the normalized postcrack tensile strength parameter μ while less sensitive to the compressive–tensile strength ratio ω [14, 15]. According to the RILEM model [1] shown in Figure 17.2, the ultimate compressive strain ε_{cu} is

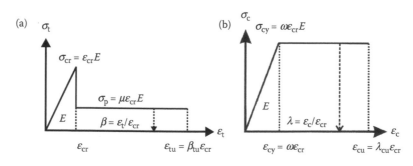

FIGURE 17.1 Idealized material models for strain-softening FRC: (a) tension model and (b) compression model.

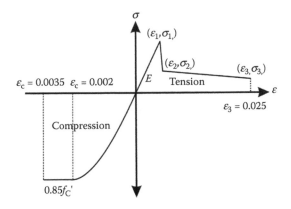

FIGURE 17.2 RILEM material model for SFRC.

0.0035, which is the lower bound value of typical SFRC [16, 17], and the ultimate tensile strain is defined as 0.025. The yield compressive strength for FRC is adopted as

$$\sigma_{cy} 0.85 f_c' \left(\text{psi and MPa} \right) \tag{17.4}$$

The two normalized parameters used in the material models (Figures 17.1a and 17.1b) are summarized as follows:

$$\mu = \frac{\sigma_p}{E\varepsilon_{cr}} = \frac{\sigma_p}{\sigma_{cr}} \tag{17.5}$$

$$\omega = \frac{\varepsilon_{cy}}{\varepsilon_{cr}} = \frac{\sigma_{cy}}{E\varepsilon_{cr}} = \frac{\sigma_{cy}}{\sigma_{cr}} = 1.52\sqrt{f_c'} \ \text{MPa} \ \left(\text{or } 0.127\sqrt{f_c'} \ \text{psi} \right) \tag{17.6}$$

Equation 17.6 implies that the normalized yield compressive strain ω is also a compressive–tensile strength ratio. Thus, these terms can be used interchangeably. For typical f_c' between 20 and 65 MPa, ω varies between 6.8 and 12.8. The tensile and compressive responses terminate at the normalized ultimate tensile strain β_{tu} and compressive strain λ_{cu}, respectively.

$$\beta_{tu} = \frac{\varepsilon_{tu}}{\varepsilon_{cr}} = \frac{0.025}{118 \times 10^{-6}} \approx 212 \tag{17.7}$$

$$\lambda_{cu} = \frac{\varepsilon_{cu}}{\varepsilon_{cr}} = \frac{0.0035}{118 \times 10^{-6}} \approx 30 \tag{17.8}$$

Note that the terms β and λ without subscript refer to normalized tensile strain ($\varepsilon_t/\varepsilon_{cr}$) and compressive strain ($\varepsilon_c/\varepsilon_{cr}$), respectively, are functions of imposed curvature on a section.

MOMENT–CURVATURE RESPONSE

For a generalized rectangular section, the derivations for neutral axis depth ratio k, normalized moment m, and normalized curvature ϕ are described in Chapter 14, and simplified by using parameter "alpha" = 1. In this formulation, the compressive fiber strain is used as the independent parameter. Figure 17.3

shows three ranges of applied top compressive strain $0 < \varepsilon_c < \varepsilon_{cr}$, $\varepsilon_{cr} < \varepsilon_c < \varepsilon_{cy}$ and $\varepsilon_{cy} < \varepsilon_c < \varepsilon_{cu}$ or in dimensionless form $0 < \lambda < 1$, $1 < \lambda < \omega$, and $\omega < \lambda < \lambda_{cu}$. The location of neutral axis parameter k is derived by solving the equilibrium of internal forces. The moment is computed from taking the force about the neutral axis, whereas the curvature is obtained by dividing top compressive strain with the depth of neutral axis. The corresponding closed-form solutions for normalized neutral axis, moment, and curvature (k, m, and ϕ) are presented in Table 17.1. Using these expressions, the moment M and curvature Φ represented in terms of their first cracking values are defined as

$$M = mM_{cr} \quad \text{and} \quad M_{cr} = \frac{\sigma_{cr}bh^2}{6} \tag{17.9}$$

$$\Phi = \phi\,\Phi_{cr} \quad \text{and} \quad \Phi_{cr} = \frac{2\varepsilon_{cr}}{h} \tag{17.10}$$

where M_{cr} and Φ_{cr} are the cracking moment and the curvature and b and h are the width and the height of beam, respectively. The expressions for the moment–curvature and neutral axis depth are obtained as shown in Table 17.1 [14].

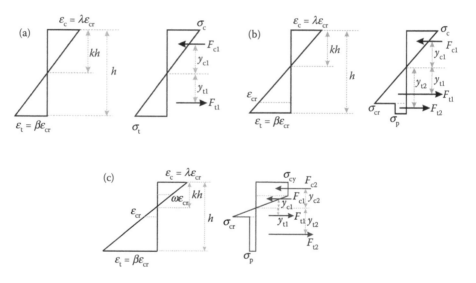

FIGURE 17.3 Stress–strain diagram at three ranges of normalized top compressive strain λ: (a) elastic for compression and tension ($0 < \lambda < 1$), (b) elastic for compression but nonlinear for tension ($1 < \lambda < \omega$), and (c) plastic for compression and nonlinear for tension ($\lambda > \omega$).

TABLE 17.1
Neutral Axis Depth Ratio and Normalized Moment–Curvature Expression for Three Ranges of Applied Normalized Top Compressive Strain λ

Range	k	m	ϕ
$0 \le \lambda \le 1$	$\dfrac{1}{2}$	$\dfrac{\lambda}{2k}$	$\dfrac{\lambda}{2k}$
$1 \le \lambda \le \omega$	$\dfrac{2\mu\lambda}{\lambda^2 + 2\mu(\lambda+1)-1}$	$\dfrac{(2\lambda^3 + 3\mu\lambda^2 - 3\mu + 2)k^2}{\lambda^2} - 3\mu(2k-1)$	
$\omega \le \lambda \le \lambda_{cu}$	$\dfrac{2\mu\lambda}{-\omega^2 + 2\lambda(\omega+\mu)+2\mu-1}$	$\dfrac{(3\omega\lambda^2 - \omega^3 + 3\mu\lambda^2 - 3\mu+2)k^2}{\lambda^2} - 3\mu(2k-1)$	

The moment capacity at ultimate compressive strain ($\lambda = \lambda_{cu}$) is very well approximated by the limit case of ($\lambda = \infty$). Using the expression for k in range 3 of Table 17.1, one obtains the neutral axis parameter at infinity k_∞ [4]:

$$k_\infty = \frac{\mu}{\omega + \mu} \tag{17.11}$$

By substituting $k = k_\infty$ and $\lambda = \infty$ in the expression for m in range 3 of Table 17.1, the ultimate moment capacity, m_∞, is also obtained,

$$m_\infty = \frac{3\omega\mu}{\omega + \mu} \tag{17.12}$$

BILINEAR MOMENT–CURVATURE DIAGRAM

For sufficiently high postpeak tensile capacity, the flexural response of FRC shows no drop in moment capacity after cracking and is referred to as strain softening and deflection hardening. Figure 17.4 shows a typical moment–curvature response for this class of material generated from Table 17.1. The smooth response can be approximated as a bilinear response using an optimization approach in the curve fitting. The termination point of the bilinear model was designated as m_{cu} and assumed to be equal to m_∞ given by Equation 17.12. The intersection point is defined as the bilinear cracking point (ϕ_{bcr}, m_{bcr}) and is higher than the original cracking point (ϕ_{cr}, M_{cr}). With the predetermined bilinear cracking points from material database covering the possible ranges of FRC behavior [18, Chapter 14], a linear regression equation was established as:

$$m_{bcr} = 0.743m_{cu} + 0.174 \quad \text{and} \quad \phi_{bcr} = m_{bcr} \tag{17.13}$$

The curvature at the ultimate compressive strain ϕ_{cu} can be determined by substituting a relatively large λ_{cu} value in the expression for k and ϕ in range 3 presented in Table 17.1. For example, a simplified expression for ϕ_{cu} at $\lambda_{cu} = 30$ is

$$\phi_{cu} = \frac{-\omega^2 + 60\omega + 62\mu - 1}{4\mu} \tag{17.14}$$

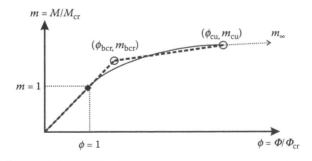

FIGURE 17.4 Normalized moment–curvature response for strain-hardening, deflection-hardening material and its simplified bilinear model.

The bilinear model can be used to obtain the curvature distribution according to a given moment profile. The slope in the elastic part is $\phi_{br}/m_{br} = 1$ while the slope in the postcrack region is

$$\theta_{pcr} = \frac{\phi_{cu} - \phi_{bcr}}{m_{cu} - m_{bcr}} \tag{17.15}$$

Finally, the normalized curvature–moment relationship can be expressed as

$$\phi = \begin{cases} m & \text{for } 0 < m \leq m_{bcr} \\ \phi_{bcr} + \theta_{pcr}(m - m_{bcr}) & \text{for } m > m_{bcr} \end{cases} \tag{17.16}$$

ALLOWABLE TENSILE STRAIN

For sufficiently high-fiber fractions and a well developed interface bond characteristics, the ultimate moment capacity of the strain-softening FRC can be as high as 2.6 times the first cracking moment [4]. There may, however, be a need to design based on a limit to an allowable tensile strain, and/or crack width. Because many deflection-hardening FRC show multiple cracking, the nominal tensile strain, which may be obtained by averaging deformations along several cracks that are spaced apart, is proposed as the serviceability criterion. This section only addresses the effect of lower and upper bounds of the allowable tensile limit and their effect on service moments.

From the linear strain distribution diagrams in the postcrack ranges (Figures 17.3b and 17.3c), the relationship between normalized allowable tensile strain β_a and the corresponding normalized compressive stain λ_a at a balanced condition can be written as

$$\frac{\lambda_a \varepsilon_{cr}}{kh} = \frac{\beta_a \varepsilon_{cr}}{h - kh} \tag{17.17}$$

Equation 17.17 is solved in conjunction with the neutral axis parameter k defined in Table 17.1 for two possible ranges 2 or 3. This results in two possibilities of λ_a:

$$\lambda_a = \begin{cases} \sqrt{2\mu\beta_a - 2\mu + 1} & \text{for } \beta_a \leq \beta_{crit} \\ \dfrac{2\mu\beta_a - 2\mu + \omega^2 + 1}{2\omega} & \text{for } \beta_a > \beta_{crit} \end{cases} \tag{17.18}$$

$$\beta_{crit} = \frac{\omega^2 + 2\mu - 1}{2\mu} \tag{17.19}$$

where β_{crit} is the critical tensile strain. When $\beta_a < \beta_{crit}$, the parameter λ_a will be in between 1 and ω (range 2), and when $\beta_a > \beta_{crit}$, the parameter λ_a will be greater than ω (range 3).

Two levels of normalized allowable tensile strain $\beta_a = 20$ and 60 (corresponding to 2360 and 7080 µstr for the cracking strain of 118 µstr defined in Equation 17.3 and the lower and upper bound compressive–tensile strength ratio $\omega = 6$ and 12) are evaluated. The closed-form solutions for the allowable moments corresponding the combination of these β_a and ω can be derived by first substituting the values in Equation 17.18, then substitute the obtained λ_a and/or ω in the expressions for k and m in ranges 2 and 3 in Table 17.3. The final form of allowable moments in range 2 or 3 (m_{2a} and m_{3a}) are presented in Table 17.2, depending on the value of β_a compared with β_{crit} as shown in Equation 17.19.

Figure 17.5 presents the normalized moment for postcrack tensile strain μ in the range of 0.0 to 1.0. The increase in allowable tensile strain β_a from 20 to 60 for each level of ω slightly increases the

TABLE 17.2
Normalized Allowable Moment

Range		$\omega = 6$	$\omega = 12$
m_{2a} $1 \le \lambda \le \omega$	$\beta_a = 20$	$\dfrac{76\mu\sqrt{38\mu+1}+2\sqrt{38\mu+1}+1197\mu+2}{\left(20+\sqrt{38\mu+1}\right)^2}$	
	$\beta_a = 60$	$\dfrac{236\mu\sqrt{118\mu+1}+2\sqrt{118\mu+1}+10797\mu+2}{\left(60+\sqrt{118\mu+1}\right)^2}$	
m_{3a} $\omega \le \lambda \le \lambda_{cu}$	$\beta_a = 20$	$\dfrac{18\left(1444\mu^2+12388\mu-343\right)}{\left(277+38\mu\right)^2}$	$\dfrac{36\left(1444\mu^2+30172\mu-6591\right)}{\left(625+38\mu\right)^2}$
	$\beta_a = 60$	$\dfrac{18\left(13924\mu^2+95108\mu-343\right)}{\left(757+118\mu\right)^2}$	$\dfrac{36\left(13924\mu^2+206972\mu-6591\right)}{\left(1585+118\mu\right)^2}$

TABLE 17.3
Detailed Calculation of Moment–Curvature Response as a Function of Applied Strain

Stage	λ	κ	ϕ'	M'	ϕ	M
1	0.00	0.500	0.000	0.000	0.00E+00	0
1	1.00	0.500	1.000	1.000	4.21E–05	16,432
2	1.67	0.363	2.304	1.034	9.70E–05	16,992
2	2.68	0.259	5.173	1.130	2.18E–04	18,563
2	4.36	0.176	12.391	1.216	5.22E–04	19,986
2	6.04	0.133	22.652	1.261	9.54E–04	20,715
2	7.72	0.107	35.958	1.287	1.51E–03	21,153
3	8.90	0.096	46.334	1.298	1.95E–03	21,321
3	10.08	0.089	56.711	1.303	2.39E–03	21,406
3	11.25	0.084	67.087	1.306	2.82E–03	21,455
3	12.43	0.080	77.464	1.308	3.26E–03	21,485

allowable moment, with the maximum difference of only 8.8% at $\mu = 1.0$. Thus, use of lower bound value $\beta_a = 20$ as a tensile strain criterion is reasonably safe for preventing excessive cracking while the moment capacity is slightly reduced. Note that at $\beta_a = 20$, the allowable moment is insensitive to changes of ω between 6 and 12, whereas at $\beta_a = 60$, only small differences are observed. On the basis of this simplification, a conservative case of $\beta_a = 20$ and $\omega = 6$ as presented in Table 17.2 is proposed as a tensile strain criterion and summarized as

$$m_a = \begin{cases} \dfrac{76\mu\sqrt{38\mu+1}+2\sqrt{38\mu+1}+1197\mu+2}{\left(20+\sqrt{38\mu+1}\right)^2} & \text{for } \beta_a = 20 \le \dfrac{35+2\mu}{2\mu} \\[4mm] \dfrac{18\left(1444\mu^2+12388\mu-343\right)}{\left(277+38\mu\right)^2} & \text{for } \beta_a = 20 > \dfrac{35+2\mu}{2\mu} \end{cases}$$

(17.20)

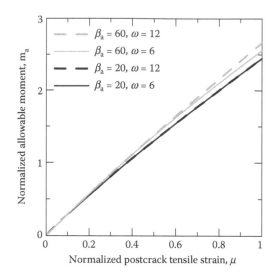

FIGURE 17.5   Normalized allowable moment.

ULTIMATE MOMENT CAPACITY

An LRFD-based ultimate strength design philosophy is based on the reduced nominal moment capacity $\phi_p M_n$ exceeding the ultimate factored moment M_u that is determined by linear elastic analysis and load coefficients in accordance to ACI 318-05 Section 9.2. The reduction factor ϕ_p addresses the uncertainty of using postcrack tensile strength in predicting ultimate moment capacity. On the basis of the statistical analysis of limited test data [19], a value $\phi_p = 0.7$ was used currently, however, as experience is gained, this value may be increased to as high as 0.9. The nominal moment capacity M_n can be obtained by using Equations 17.9 and 17.12 with the reduction factor:

$$\phi_p M_n = \phi_p m_\infty M_{cr} = \frac{3\omega\mu}{\omega+\mu}\phi_p M_{cr} \geq M_u \tag{17.21}$$

Alternatively, the nominal moment capacity can be expressed as a function of postcrack tensile strength μ and compressive strength f_c' by substituting Equation 17.6 in Equation 17.21:

$$\phi_p M_n = \left[\frac{6\mu\sqrt{f_c'}}{\xi\mu+2\sqrt{f_c'}}\right]\phi_p M_{cr} \quad (\xi=1.32 \text{ for } f_c' \text{ in MPa}, \xi=15.8 \text{ for } f_c' \text{ in psi}) \tag{17.22}$$

The postcrack tensile strength necessary to carry the ultimate moment can be obtained from Equation 17.22 as

$$\mu = \frac{2M_u\sqrt{f_c'}}{6\phi_p M_{cr}\sqrt{f_c'}-\xi M_u} \quad (\xi=1.32 \text{ for } f_c' \text{ in MPa}, \xi=15.8 \text{ for } f_c' \text{ in psi}) \tag{17.23}$$

MINIMUM POSTCRACK TENSILE CAPACITY FOR FLEXURE

To prevent a sudden drop of moment capacity after initial cracking, a minimum fiber loading is required. The postcrack tensile strength that maintains a load capacity equivalent to the cracking strength level ($M_n = M_{cr}$) is defined as μ_{crit} and obtained by solving Equation 17.21 with a reduction factor $\phi_p = 1$,

$$\mu_{crit} = \frac{\omega}{3\omega - 1} \tag{17.24}$$

For typical FRC materials with compressive–tensile strain ratio ω ranging from 6 to 12 results in $\mu_{crit} = 0.353$–0.343. A conservative value of $\mu_{min,flex} = 0.35$ therefore ensures postcrack moment capacity higher than the first cracking moment.

HYBRID REINFORCEMENT CONVERSION DESIGN CHART

An equivalent flexural FRC system can be substituted for minimum or even flexural reinforcement in reinforced concrete structures. The procedures for the development of such design charts are based on the use of closed-form solutions for the moment–curvature response of a reinforced concrete section containing fibers. The theoretical basis for this section was derived in Chapter 14, where a hybrid reinforcement of short fibers and continuous reinforcement was used to obtain the moment–curvature response of a rectangular section. Parametric results are shown in Figure 17.6, where the effect of postpeak residual strength on the moment capacity of a singly reinforced concrete is shown. It is noted that an increase in either the reinforcement ratio, or the residual tensile strength tends to increase the moment capacity. The chart can be used to obtain a combination of continuous and short fiber systems that would provide an equivalent bending capacity. This approach can be further used to develop another conversion design chart to help designers replace the reinforced concrete system with an FRC system, based on a with a specified postcrack flexural capacity. The nominal moment capacity of a singly reinforced concrete section can be determined by the compressive stress block concept (ACI Section 10.2.7),

$$M_n = A_s f_y \left(d - \frac{a}{2} \right) \tag{17.25}$$

where $a = A_s f_y / (0.85 f_c' b)$ is the depth of compressive stress block, $A_s = \rho_g bh$ is the area of tensile steel, ρ_g is the reinforcement ratio per gross section, and α is the normalized effective depth (d/h).

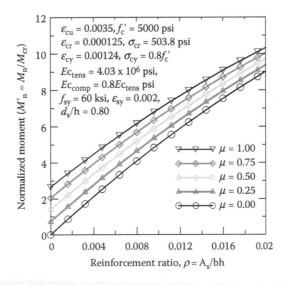

FIGURE 17.6 Design chart for reinforced concrete with the matrix phase containing various levels of postpeak strain-softening response.

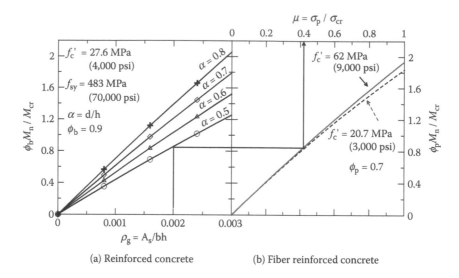

FIGURE 17.7 Conversion chart between reinforced concrete system and FRC system.

The reduction factors $\phi_b = 0.9$ and $\phi_p = 0.70$ are used in the conversion chart to address the reliability of two reinforcing mechanisms. For any grade of steel and concrete strength, a conversion chart can be generated by Equations 17.22 and 17.28 as shown by the left side of the Figure 17.7. The reinforcement ratio ρ_g together with the effective depth determines the moment capacity that is transferred to the FRC chart shown on the right side of the diagram [14]. By drawing a vertical line up to the abscissa, one can determine the equivalent normalized postcrack tensile strength, μ.

DEFLECTION CALCULATION FOR SERVICEABILITY

An important aspect of serviceability-based design is in accurate calculation of deflections under service load. The present approach can be used to compute the deflections by integration of the curvature along the beam length. Geometric relationships between curvature and deflection have been derived by Ghali [20, 21]. The curvature distribution along the length can be arbitrary; however, a parabolic or a linear shape results in accurate results whereas other shapes result in approximate values. The sign convention for curvature is the same as the convention used for moments. Two typical cases of a simple beam (or continuous beam) and cantilever beam are presented in Figure 17.8. The midspan deflection δ of a simple (or continuous beam) can be computed by

$$\delta = \frac{L^2}{96}\left(\Phi_1 + \Phi_2 + \Phi_3\right) \tag{17.26}$$

The tip deflection of the cantilever beam can be computed by

$$\delta = -\frac{L^2}{6}\left(2\Phi_2 + \Phi_3\right) \tag{17.27}$$

where L is the span length and Φ_1, Φ_2, and Φ_3 are the curvature at left end, center, and right end, respectively. For short-term deflection, the curvatures $\Phi_1 - \Phi_3$ due to moment at service loads can be estimated from Equations 17.10 and 17.16.

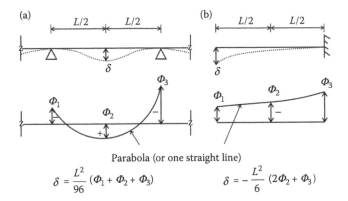

Parabola (or one straight line)

$$\delta = \frac{L^2}{96}(\varPhi_1 + \varPhi_2 + \varPhi_3) \qquad \delta = -\frac{L^2}{6}(2\varPhi_2 + \varPhi_3)$$

FIGURE 17.8 Geometric relationship between curvature and deflection. (From Ghali, A., "Deflection of reinforced concrete members: A critical review," *ACI Structural Journal*, 90(4), 364–373, July–August 1993; and Ghali, A., and Favre, R., *Concrete Structures: Stresses and Deformations*, Chapman and Hall, London, 352, 1986. With permission.)

For a simply supported beam, the maximum deflection is therefore computed as

$$\delta = \frac{L^2 \varPhi_{max}}{96} \qquad (17.28)$$

MINIMUM POSTCRACK TENSILE STRENGTH FOR SHRINKAGE AND TEMPERATURE

Because of the large surface-to-volume ratio of an unreinforced concrete slab, severe cracking because of restrained shrinkage conditions can take place. To control crack width within acceptable range, the minimum shrinkage and the temperature reinforcement must be placed perpendicular to the main flexural reinforcement. According to the ACI Section 7.12.2.1, the minimum ratio of reinforcement to gross section area ρ_{min_ST} is:

$$\rho_{min\_ST} = \begin{cases} 0.0020 & 40 \text{ ksi} < f_y < 50 \text{ ksi, deformed bar} \\ 0.0018 & f_y = 60 \text{ ksi, welded-wire fabric (smooth or deformed),} \\ \dfrac{0.0018 \times 60,000}{f_y} > 0.0014 & f_y > 60 \text{ ksi at } \varepsilon_{sy} = 0.35\% \end{cases}$$

$$(17.29)$$

where f_y is the yield strength of steel. When steel rebar or welded wire mesh is replaced with FRC, the minimum normalized postcrack tensile strength μ_{min_ST} is determined using a statically equivalent approach in terms of yielding force. The tensile performance can be determined by the equivalence of tensile capacity of the two systems, each using their own reliability index,

$$\phi_b \rho_{min\_ST} f_y bh = \phi_p \mu_{min\_ST} \sigma_{cr} bh \qquad (17.30)$$

It is assumed in the reduction factor for structural members that the failure is controlled by reinforcement, which is the same as the factor for bending $\phi_b = 0.90$ (ACI Section 9.3.2.1). On the other hand, the reduction factor $\phi_p = 0.70$ is used for the member in which the tensile failure is controlled by postcrack capacity of FRC. The use of higher and lower reduction factors in Equation 17.21 ensures that the reliability index is maintained if we have to assume that more experience is available with steel reinforcement as compared with FRC. Conservative values of $\rho_{min\_ST} = 0.0018$ and fy = 60,000 psi are substituted in Equation 17.21 to produce highest tension force and solve for μmin_ST.

$$\mu_{min\_ST} = \frac{140}{\sigma_{cr}} \, (\text{psi}) \quad \left(or = \frac{0.97}{\sigma_{cr}} \, (\text{MPa}) \right) \tag{17.31}$$

DESIGN EXAMPLES

The design procedure for strain-softening FRC is best suited for thin structural applications, such as slab systems, in which the size effect is minimal and the internal forces are relatively low compared with its moment capacity. An example of slab on grade is presented to demonstrate the design calculations. Typically, slabs on grade are designed on the basis of minimum shrinkage and temperature steel. The loads on slab are not critical and normally transferred directly to stiff compacted base materials. These slabs are allowed to crack but not to disintegrate. Other types of slabs on grade and pavement that are designed on the basis of applied load, and subgrade modulus are not considered here.

DESIGN EXAMPLE 1: SLAB ON GRADE

Design a concrete slab that is 5 in. thick, reinforced at mid-depth with steel rebar no. 4 at 18 in. (12.7 × 457 mm). The materials used are as follows: concrete compressive strength f_c' of 3000 psi (20.7 MPa) and steel yield strength f_y of 60 ksi (414 MPa). Replace this existing design with an SFRC that has compressive strength f_c' of 4000 psi (27.6 MPa).

The slab is designed on the basis of a 1-ft. strip (254 mm), and the amount of reinforcement A_s is calculated by

$$A_s = \frac{\pi d^2}{4} \frac{12}{spacing} = \frac{\pi 0.5^2}{4} \frac{12}{18} = 0.131 \, \text{in.}^2/\text{ft. (or 277 mm}^2/\text{m)}$$

Calculate the plastic compressive zone according to ACI stress block concept,

$$a = \frac{A_s f_y}{0.85 f_c' b} = \frac{0.131 \times 60}{0.85 \times 3 \times 12} = 0.257$$

The factored ultimate moment M_u is equal to the reduced nominal moment capacity $\phi_b M_n$,

$$M_u = \phi_b M_n = \phi_b A_s f_y \left(d - \frac{a}{2} \right)$$

$$= 0.9 \times 0.131 \times 60 \left(2.5 - \frac{0.257}{2} \right) \frac{1}{12} = 1.40 \, \text{kip} \cdot \text{ft./ft.(or 6.23 kN} \cdot \text{m/m)}$$

Equivalent Moment Capacity with SFRC, f_c' = 4000 psi (27.6 MPa)

Calculate the cracking tensile strength of SFRC according to Equation 17.14, $\sigma_{cr} = 6.7 \sqrt{f_c'} = 6.7\sqrt{4000} = 424$ psi (2.92 MPa). Calculate the cracking moment according to Equation 17.7, $M_{cr} = \sigma_{cr} bh^2/6 = (424 \times 12 \times 5^2/6)(1/12,000) = 1.77$ kip · ft/ft. (or 7.87 kN·m/m). Calculate the compressive–tensile strength ratio by Equation 17.2, $\omega = \sigma_{cy}/\sigma_{cr} = 0.85 \times 4000/424 = 8.02$. Determine the normalized postcrack tensile strength by Equation 17.13,

$$\mu = \frac{M_u \omega}{3 \omega \phi_p M_{cr} - M_u} = \frac{1.40 \times 8.02}{3 \times 8.02 \times 0.7 \times 1.77 - 1.40} = 0.395$$

It can be verified by Equation 17.17 that the reduced nominal moment capacity $\phi_b M_n$ of the SFRC slab is equal to the ultimate moment M_u determined from the reinforced concrete slab,

$$\phi_p M_n = \left[\frac{\mu\sqrt{f_c'}}{15.8\mu + 2\sqrt{f_c'}} \right] \phi_p \sigma_{cr} bh^2$$

$$= \left[\frac{0.395\sqrt{4000}}{15.8(0.395) + 2\sqrt{4000}} \right] 0.7(424)(12)(5)^2 \frac{1}{12000} = 1.40 \text{ kip ft./ft. kip} \cdot \text{ft. or}$$

$$= \left[\frac{0.395\sqrt{27.6}}{1.32(0.395) + 2\sqrt{27.6}} \right] 0.7(2.92)(1000)(127)^2 \left(1\times10^{-6}\right) = 6.23 \text{ kN} \cdot \text{m/m}$$

However, the required μ must be checked against the minimum postcrack tensile strength for flexure defined in Equation 17.19, $\mu_{min\_flex} = 0.40$. Thus, use $\mu = 0.40$.

Equivalent Tensile Capacity

Assume plain concrete in reinforced concrete slab has no residual strength; thus, only the amount of reinforcement will be replaced with SFRC having the same tensile capacity,

$$\phi_b A_s f_y = \phi_p \mu \sigma_{cr} bh$$

The postcrack tensile strength is calculated as

$$\mu = \phi_b A_s f_y / \phi_p \sigma_{cr} bh = 0.9 \times 0.131 \times 60 / 0.7 \times 0.424 \times 12 \times 5 = 0.397$$

The minimum normalized postcrack tensile strength for shrinkage and temperature can be calculated by Equation 17.22, $\mu_{min\ ST} = 140/\sigma_{cr} = 140/424 = 0.33$.

The calculations in this example point out to several potential design approaches. If the goal is to replace the existing reinforced concrete slab with SFRC system and still have the same performance and reliability index as the original design, the postcrack tensile strength must be 0.40 based on the minimum flexural strength. However, if only shrinkage and temperature cracking in service condition is of concern, the required strength can be reduced to 0.33. In this case, specifying $\mu = 0.40$ based on equivalent performance to the original reinforced concrete slab. It should be pointed out that the specified postcrack tensile strength for material testing must be calculated by the same cracking tensile strength, σ_{cr}, used in the design calculations, which may be different from the actual value obtained from test. According to the definition in Equation 17.1, the postcrack tensile strength is $\sigma_p = \mu\sigma_{cr} = 0.40 \times 424 = 170$ psi (or 1.17 MPa).

DESIGN EXAMPLE 2: EQUIVALENT REINFORCED SLAB WITH VARIOUS STEEL YIELD STRENGTHS

Develop an equivalently reinforced concrete slab that replaces rebars with steel fibers. The reinforced concrete slab information is as follows: concrete compressive strength, $f_c' = 3000$ psi; steel yield strength, $f_y = 40$ or 60 ksi; thickness, $h = 5$ in.; depth of reinforcement, $d_s = 2.5$ in.; and reinforcement, rebar no. 4 spacing at 18 in.

Step 1: Calculate Existing Moment Capacity Based on 1-Ft. Strip

$$A_s = \frac{\pi d^2}{4} \frac{12}{S} = \frac{\pi 0.5^2}{4} \frac{12}{18} = 0.131 \text{ in.}^2$$

For $f_{sy} = 40$ ksi, $\quad a = \frac{A_s f_y}{0.85 f_c' b} = \frac{0.131 \times 40}{0.85 \times 3 \times 12} = 0.171$

$$M_n = A_s f_y \left(d - \frac{a}{2}\right) = 0.131 \times 40 \left(2.5 - \frac{0.171}{2}\right) \frac{1}{12} = 1.05 \text{ kip} \cdot \text{ft.}$$

$$\text{For } f_{sy} = 60 \text{ ksi,} \quad a = \frac{A_s f_y}{0.85 f'_c b} = \frac{0.131 \times 60}{0.85 \times 3 \times 12} = 0.257$$

$$M_n = A_s f_y \left(d - \frac{a}{2} \right) = 0.131 \times 60 \left(2.5 - \frac{0.257}{2} \right) \frac{1}{12} = 1.56 \text{ kip} \cdot \text{ft.}$$

Step 2: Calculate Normalized Ultimate Moment

Assume tensile strength of SFRC, $\sigma_{cr} = 6\sqrt{f'_c} = 6\sqrt{3000} = 329$ psi

Cracking moment, $M_{cr} = \dfrac{1}{6} \sigma_{cr} bh^2 = \dfrac{1}{6} 329 \times 12 \times 5^2 \dfrac{1}{12000} = 1.37$ kip·ft.

Ultimate moment for $f_y = 40$ ksi and $f_y = 60$ ksi (reduction factor for bending, $\phi_b = 0.90$),

$$M_{u\_fy40} = \frac{M_n}{\phi_b} = \frac{1.05}{0.9} = 1.17 \text{ kip} \cdot \text{ft.}$$

$$M_{u\_fy60} = \frac{M_n}{\phi_b} = \frac{1.56}{0.9} = 1.73 \text{ kip} \cdot \text{ft.}$$

Normalize ultimate moment with cracking moment,

$$M'_{u\_fy40} = \frac{M_{u\_fy40}}{M_{cr}} = \frac{1.17}{1.37} = 0.854$$

$$M'_{u\_fy60} = \frac{M_{u\_fy60}}{M_{cr}} = \frac{1.73}{1.37} = 1.26$$

Step 3: Determine Postcrack Tensile Strength Using Simplified Equation

Ultimate strength design is defined as $M'_u = \phi_t M'_n = \phi_t (3\omega\mu / \omega + \mu)$ or $\mu_{required} \geq M'_u \omega / (3\phi_t \omega - M'_u)$. Note that ϕ_b is replaced with ϕ_t. Factor ϕ_b is based on steel reinforcement while ϕ_t is based on the present approach and may be used somewhat conservatively. Because of the lack of a code specification for the present design procedure, assume the reduction factor for flexural capacity on the basis of residual tensile strength of concrete $\phi_t = 0.65$.

Plastic compressive–tensile strength ratio, ω:

$$\omega = \frac{\sigma_{cy}}{\sigma_{cr}} \approx \frac{0.85 f'_c}{6\sqrt{f'_c}} \approx 0.142\sqrt{f'_c} = 0.142\sqrt{3000} = 7.78$$

Required postcrack tensile strength ratio, μ:

$$\mu_{\_fy40} = \frac{M'_u \omega}{3\phi_t \omega - M'_u} = \frac{0.854 \times 7.78}{3 \times 0.65 \times 7.78 - 0.854} = 0.464$$

$$\mu_{\_fy60} = \frac{M'_u \omega}{3\phi_t \omega - M'_u} = \frac{1.26 \times 7.78}{3 \times 0.65 \times 7.78 - 1.26} = 0.705$$

Check that the reduced nominal moment capacity must greater than or equal to the factored moment,

$$\phi_t M_n = \phi_t M_n' M_{cr} = \phi_t \frac{3\omega\mu}{\omega+\mu} M_{cr} \geq M_u$$

$$\phi_t M_{n\_fy40} = 0.65\frac{3\times7.78\times0.464}{7.78+0.464}1.37$$
$$= 1.17\,\text{kip}\cdot\text{ft.} \geq M_{u\_fy40} = 1.17\ \text{kip}\cdot\text{ft.}$$

$$\phi_t M_{n\_fy60} = 0.65\frac{3\times7.78\times0.705}{7.78+0.705}1.37$$
$$= 1.73\,\text{kip}\cdot\text{ft} \geq M_{u\_fy60} = 1.73\ \text{kip}\cdot\text{ft.}$$

Nondimensional moment–curvature parameters are $M_u' = 1.308$, $\phi_u' = 77.464$, $M_{cr} = 11{,}831$ lb./in, and $\phi_{cr} = 3.51\text{E}{-}05$ lb./in. Note that the simplified equation predicts the section moment capacity of $M = 1.17/0.65 \times 12{,}000 = 21{,}600$ lb./in. But the more accurate equations as presented in Table 17.3 predict 1.308 kip·ft./ft. or 21,485 lb./in. There is very little difference between the simplified equation used in the design procedure and more accurate equations, which can be easily implemented in an Excel spreadsheet.

It is concluded that the final results depend on the grade of steel used in reinforced concrete slab for both $f_y = 40$ and 60 ksi. On the basis of a section of 12×5 in. reinforced with rebar no. 4 at 18 in. spacing, the ultimate moment capacity $M_u = 1.17$ kip·ft./ft. for $f_y = 40$ ksi and 1.73 kip·ft./ft. for $f_y = 60$ ksi. If reinforcement are to be replaced with SFRC, it requires the postcrack tensile strength ratio μ of 0.464 for $f_y = 40$ ksi and 0.705 for $f_y = 60$ ksi, respectively. The postcrack tensile strength $\mu = 0.705$ is relatively high, which may require large amount of fiber content. An alternative solution is to increase the depth from 5 to 6 in. and to reduce μ from 0.705 to 0.49. The calculations of alternative solution are not shown here but can be easily developed in a spreadsheet format. Please note that the design does not provide the reduction factor for concrete under tensile failure mode ϕ_t. While in the present calculation, this factor is assumed to be 0.65, other values between 0.5 and 0.7 are also reasonable on the basis of the quality of materials and importance of structures.

DESIGN EXAMPLE 3: SIMPLY SUPPORTED SLAB WITH SERVICEABILITY CRITERIA

An example is presented to demonstrate the design calculations for a one-way slab with a single span of 3.5 m (11.67 ft.) subjected to a uniformly distributed live load of 2.0 kN/m² (41.8 lb./ft.²) and superimposed dead load of 0.7 kN/m² (14.6 lb./ft.²). A point load of 4.0 kN/m (0.274 kip·ft.) is applied at the center. The design requires use of SFRC with a compressive strength f_c' of 45 MPa (6531 psi) and unit weight of 24 kN/m³ (153 lb./ft.³).

Ultimate Moment Capacity

The one-way slab is designed on the basis of 1.0 m strip (3.33 ft.). The self-weight for an assumed thickness of 0.15 m (6 in.) is $w_{sw} = 0.15\times24 = 3.6\,\text{kN/m}^2$ (75.2 lb./ft.²). The factored loads according to ACI Section 9.2.1 are $w_u = 1.2(3.6+0.7)+1.6(2.0) = 8.36\,\text{kN/m}^2$ (174.6 lb./ft.²) and $P_u = 1.6(4.0) = 6.4$ kN/m (0.439 kip·ft.). The maximum moment at midspan due to the uniform and point loads is

$$M_u = \frac{w_u L^2}{8} + \frac{P_u L}{4} = \frac{8.36\times3.5^2}{8} + \frac{6.4\times3.5}{4} = 18.4\ \text{kN}\cdot\text{m/m (4.14 kip}\cdot\text{ft./ft.)}$$

Tensile strength and cracking moment are estimated by Equations 17.1 and 17.9, respectively: $\sigma_{cr} = 0.56\sqrt{45} = 3.75$ MPa (544 psi) and

$$M_{cr} = \sigma_{cr}bh^2/6 = (3.75\times 10^3)\times 1.0 \times 0.15^2/6 = 14.06 \text{ kN-m/m (3.16 kip-ft./ft.)}$$

The required postcrack tensile strength for the ultimate moment M_u by Equation 17.23 is

$$\mu = \frac{2M_u\sqrt{f_c'}}{6\phi_p M_{cr}\sqrt{f_c'} - \xi M_u} = \frac{2\times 18.4\sqrt{45}}{6\times 0.7\times 14.06\sqrt{45} - 1.32\times 18.4} = 0.66$$

The postcrack tensile strength is determined by Equation 17.1, $\sigma_p = \mu\sigma_{cr} = 0.66 \times 3.75 = 2.48$ MPa (360 psi).

Check Tensile Strain Limit

Because the allowable tensile strain $\beta_a = 20 > (35 + 2 \times 0.66)/(2 \times 0.66) = 27.5$ in Equation 17.20, the allowable moment is calculated as

$$m_a = \frac{18(1444\mu^2 + 12388\mu - 343)}{(277 + 38\mu)^2} = \frac{18(1444\times 0.66^2 + 12388\times 0.66 - 343)}{(277 + 38\times 0.66)^2} = 1.67$$

Unfactored loads are used to calculate the moment at service condition at the midspan, $M_s = (wL^2/8) + (PL/4) = ((3.6 + 0.7 + 2.0) \times 3.5^2/8) + (4.0 \times 3.5/4) = 13.15$ kN · m/m (2.96 kip·ft./ft.). The normalized moment at service load is $m_s = M_s/M_{cr} = 13.15/14.06 = 0.935 < m_a = 1.67 = $ "passed."

Short-Term Deflection

To calculate the deflection, the bilinear curvature–moment relationship is generated. The compressive–tensile strength ratio ω computed by Equation 17.6 is $\omega = 1.52\sqrt{f_c'} = 1.52\sqrt{45} = 10.2$. Two data points (ϕ_{br}, m_{br}) and (ϕ_{cu}, m_{cu}) and the slope θ_{pcr} in the postcrack region can be determined by Equations 17.13 through 17.16,

$$m_u = \frac{3\omega\mu}{\omega + \mu} = \frac{3\times 10.2\times 0.66}{10.2 + 0.66} = 1.86$$

$$\phi_{cu} = \frac{-\omega^2 + 60\omega + 62\mu - 1}{4\mu} = \frac{-10.2^2 + 60\times 10.2 + 62\times 0.66 - 1}{4\times 0.66} = 207.5$$

$$m_{bcr} = 0.743m_u + 0.174 = 0.743\times 1.86 + 0.174 = 1.56$$

$$\phi_{bcr} = m_{bcr} = 1.56,$$

$$\theta_{pcr} = \frac{\phi_u - \phi_{bcr}}{m_u - m_{bcr}} = \frac{207.5 - 1.56}{1.86 - 1.56} = 686.5$$

For a simple beam case, the curvature at both ends (ϕ_1 and ϕ_3) is zero and the curvature at midspan ϕ_2 is determined by Equations 17.10 and 17.16. Because m_s is less than m_{bcr}; thus, $\phi_2 = m_s$,

$$\Phi_2 = \phi_2\Phi_{cr} = m_s\frac{2\varepsilon_{cr}}{h} = 0.935\frac{2\times 118\times 10^{-6}}{150} = 1.471\times 10^{-6}\text{mm}^{-1} \ (3.736\times 10^{-5} \text{ in.}^{-1})$$

Finally, the midspan deflection of the beam is calculated by the geometric relationship between curvature and deflection defined in Equation 29.1, $\delta = (L^2/96)(\Phi_1 + \Phi_2 + \Phi_3) = (3500^2/96)$ $(1.47 \times 10^{-6}) = 0.188$ mm (0.0074 in.). Note that to check the deflection limit for each application, long-term effects such as creep and shrinkage must be taken into account. This aspect is beyond the scope of this section.

Stress Distributions

To demonstrate the differences between the present method and the commonly used elastically equivalent flexural strength $\sigma_{e,flex}$, the stress distribution across the beam section at ultimate compressive strain and at infinite strain is compared with the elastic flexural stress. The neutral axis at ultimate compressive strain ε_{cu} of 0.0035 in addition to the ultimate moment is obtained by substituting $\lambda_{cu} = 30$ in the expressions for k and m in range 3 of Table 17.1:

$$k_{cu} = \frac{2\mu\lambda_{cu}}{-\omega^2 + 2\lambda_{cu}(\omega+\mu) + 2\mu - 1} = \frac{2 \times 0.66 \times 30}{-10.2^2 + 2 \times 30(10.2 + 0.66) + 2 \times 0.66 - 1} = 0.0723$$

$$k_{cu}h = 0.0723 \times 150 = 10.85 \text{ mm}$$

$$m_{cu} = \frac{(3\omega\lambda_{cu}^2 - \omega^3 + 3\mu\lambda_{cu}^2 - 3\mu + 2)k_{cu}^2}{\lambda_{cu}^2} - 3\mu(2k_{cu} - 1)$$

$$= \frac{\left(3 \times 10.2 \times 30^2 - 10.2^3 + 3 \times 0.66 \times 30^2 - 3 \times 0.66 + 2\right)}{30^2} 0.0723^2$$

$$-3 \times 0.66(2 \times 0.0724 - 1) = 1.858$$

$$M_{cu} = m_{cu}M_{cr} = 1.858 \times 14.06 = 26.12 \text{ kN} \cdot \text{m/m} \ (5.87 \text{ kip} \cdot \text{ft./ft.})$$

The neutral axis parameter and the moment at infinite compressive strain are obtained by Equations 17.9, 17.11, and 17.12:

$$k_\infty h = \frac{\mu}{\omega + \mu} h = \frac{0.66}{10.2 + 0.66} \times 150 = 9.12 \text{ mm} \ (0.36 \text{ in.})$$

$$M_\infty = m_\infty M_{cr} = \frac{3\omega\mu}{\omega + \mu} M_{cr} = \frac{3 \times 10.2 \times 0.66}{10.2 + 0.66} \times 14.06 = 26.15 \text{ kN} \cdot \text{m/m} \ (5.88 \text{ kip} \times \text{ft./ft.})$$

The elastically equivalent flexural strength corresponding to the nominal moment of 26.15 kN·m/m is determined by the flexure formula,

$$\sigma_{e,flex} = \frac{M_u c}{I} = \frac{(26.15 \times 10^3) \times 0.15/2}{1.0 \times 0.15^3 / 12} = 6.97 \text{ MPa} \ (1012 \text{ psi})$$

Stress distributions calculated by the three approaches are compared in Figure 17.9. It is observed that the postcrack tensile strength between the idealized material models at ultimate compressive strain ($\varepsilon_{cu} = 0.0035$) and at infinity ($\varepsilon_{cu} = \infty$) is the same at $\sigma_p = 2.48$ MPa (360 psi). This level of postcrack stress is much smaller than the elastically equivalent flexural strength $\sigma_{e,flex}$ of 6.97 MPa (1012 psi). On the other hand, the compressive stress at ultimate strain and at infinity are the same at $\sigma_{cy} = 38.25$ MPa (5552 psi), which is much higher than $\sigma_{e,flex}$ of 6.97 MPa (1012 psi). This example points out the inadequacies of several inverse analysis techniques that have been used to obtain residual tensile capacity such as the average residual stress method (ASTM C 1399-04) [22],

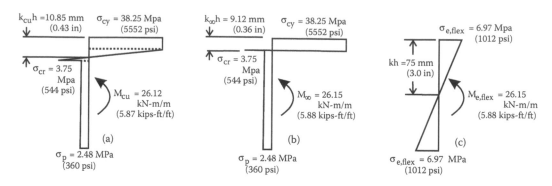

FIGURE 17.9 Stress distribution at ultimate moment: (a) idealized material models at ultimate compressive strain, (b) idealized material models at infinite compressive strain, and (c) elastically equivalent flexural stress. (Adapted from Soranakom, C., and Mobasher, B., "Flexural design of fiber reinforced concrete," *ACI Materials Journal*, 461–469, September–October 2009.)

which report material strengths in terms of equivalent elastic values. Designers should be aware of the shortcomings of these methods and approaches that determine member capacity [23].

The neutral axis of the idealized model at ultimate compressive strain is slightly higher than the neutral axis at infinity. However, the moment capacity is quite close to one another (26.12 vs. 26.15 kN·m/m). This is due to the elastic stress regions near the neutral axis decreasing while the plastic tensile regions increase.

Design Example 4: Four-Span Floor Slab

Design a four-span floor slab resting on steel stringers, subjected to live load of 75 pcf and supper imposed dead load of 15 psf as shown in Figure 17.10. Use SFRC compressive strength f_c' of 4000 psi and ignore the width of steel stringers in calculation of clear span between each support.

The floor slab will be designed as one-way slab based on a 1-ft. strip. Because the slab will be constructed with homogeneous SFRC, the maximum forces and their locations in the four-span slab will govern the design section. Assume that SFRC slab thickness can be estimated in the same way as reinforced concrete beam according to ACI Section 9.5.2.1, Table 9.5a. On the basis of the critical end span, the thickness is determined as $h = l/24 = 8 \times 12/24 = 4$ in. Estimate self-weight of SFRC floor slab by $w_{sw} = 4/12 \times 150 = 50$ lb./ft.^{2}. Calculate factor load according to ACI Section 9.2.1, $w_u = 1.2(50+15)+1.6(75) = 198$ psf.

Moment Capacity

Because the live load is less than three times of the dead load, adjacent spans are not different by more than 20% and other requirement of ACI Section 8.3.3 are met, the ACI moment coefficient is applicable to calculate the ultimate moment at the critical section, which is the first interior support, $M_u = C_m w_u l_n^2 = (1/10)0.198 \times 8^2 = 1.27$ kip·ft./ft. Calculate cracking tensile strength by Equation 17.3, $\sigma_{cr} = 6.7\sqrt{f_c'} = 6.7\sqrt{4000} = 424$ psi. Calculate the cracking moment by Equation 17.11,

$$M_{cr} = \frac{\sigma_{cr}bh^2}{6} = \frac{424 \times 12 \times 4^2}{6} \frac{1}{12,000} = 1.13 \text{ kip·ft.}$$

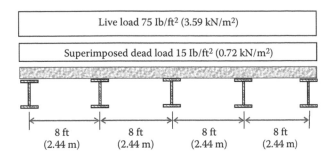

FIGURE 17.10 Four-span floor slab.

Calculate the normalized compressive yield strain by Equation 17.9, $\omega = \sigma_{cy}/\sigma_{cr} = (0.85 \times 4000)/6.7 \times \sqrt{4000} = 8.02$. Determine the postcrack tensile strength for the ultimate moment M_u by Equation 17.10,

$$\mu = \frac{M_u\omega}{3\omega\phi_p M_{cr} - M_u} = \frac{1.27 \times 8.02}{3 \times 8.02 \times 0.7 \times 1.13 - 1.27} = 0.573$$

The minimum requirement for flexure defined in Equation 17.15, μ_{min_flex}, is 0.40, and the minimum for shrinkage and temperature defined in Equation 17.19, μ_{min_ST}, is $140/424 = 0.33$. Thus, use $\mu = 0.573$ to determine the specified postcrack tensile strength by Equation 17.1, $\sigma_p = \mu\sigma_{cr} = 0.573 \times 424 = 243$ psi.

Shear Capacity

Critical shear is at the first interior support, $V_u = C_v(w_u l_n/2) = 1.15 \times 0.198 \times 8/2 = 0.91$ kip·ft. Check the shear capacity of SFRC by Equation 17.20, $\phi_v V_{FRC} = \phi_v 2\sqrt{f'_c}bh = 0.75((2\sqrt{4000} \times 12 \times 4)/1000) = 4.55$ kip·ft. $> V_u$ of 0.91 kip·ft. Therefore, the slab has adequate shear capacity. Specify postcrack tensile strength $\sigma_p = 243$ psi.

DESIGN EXAMPLE 5: RETAINING WALL

The objective of this exercise is to redesign a reinforced concrete retaining wall as shown in Figure 17.11 with FRC materials. The problem is solved using customary English units. The total height is 10 ft., and there is an imposed 150 pcf surcharge in the front side. The soil has density of 120 lb./ft.[3] and frictional angle of 30°. A drainage system will be installed to reduce water pressure such that water pressure can be ignored in the design. Assume that the proportion of the footing has been checked for stability, only the vertical part is left for the redesign with SFRC having compressive strength f'_c of 4000 psi.

Calculate the active pressure coefficient, $C_a = \tan^2(45 - (\phi/2)) = \tan^2(45 - (30/2)) = 0.333$. Calculate lateral pressure due to the surcharge, $p_s = C_a S = 0.333 \times 150 = 50$ psf. Calculate lateral earth pressure at 9 ft. below the ground level, $p_e = C_a \gamma_s H = 0.333 \times 120 \times 9 = 360$ psf. Conservatively estimate the critical shear section at 9 ft. below the ground level; thus, the ultimate shear force is

$$V_u = 1.6\left(p_s H + \frac{1}{2}p_e H\right) = 1.6\left(0.05 \times 9 + \frac{1}{2}0.36 \times 9\right) = 3.31 \text{ kip·ft.}$$

The ultimate moment at the critical section 9 ft. below the ground level is

$$M_u = 1.6\left(p_s H \frac{H}{2} + \frac{1}{2}p_e H \frac{H}{3}\right) = 1.6\left(0.05 \times 9 \times \frac{9}{2} + \frac{1}{2}0.36\frac{9}{3}\right) = 4.1 \text{ kip·ft./ft.}$$

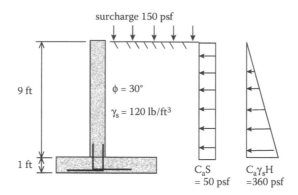

FIGURE 17.11   Design specifications for a retaining wall.

Typically, the thickness of the retaining wall is approximately 1/10 of the height. For relatively low height, deflection and overturning due to later pressure are less critical. Thus, try a thickness of 10 in. For compressive strength, $f'_c = 4000$ psi, the cracking tensile strength and the normalized compressive yield strain are the same as those used in Example 2, Thus, $\sigma_{cr} = 424$ psi and $\omega = 8.02$.

Calculate the cracking moment according to Equation 17.11,

$$M_{cr} = \frac{\sigma_{cr}bh^2}{6} = \frac{424 \times 12 \times 10^2}{6} \frac{1}{12000} = 7.07 \text{ kip·ft./ft.}$$

Determine the normalized postcrack tensile strength by Equation 17.10,

$$\mu = \frac{M_u \omega}{3\omega \phi_p M_{cr} - M_u} = \frac{4.1 \times 8.02}{3 \times 8.02 \times 0.7 \times 7.07 - 4.1} = 0.286$$

The minimum postcrack tensile strength for flexure by Equation 17.15, $\mu_{\text{min\_flex}}$, is 0.40 (governed). Thus, use $\mu = 0.40$.

Check shear capacity of the FRC section by using Equation 17.19,

$$\phi_v V_{FRC} = \phi_v 2\sqrt{f'_c}\, bh = 0.75 \times \frac{2\sqrt{4000} \times 12 \times 10}{1000} = 11.4 \text{ kip·ft.} > V_u = 3.31 \text{ kip·ft.}$$

The Dowel bar that connects footing to the vertical wall must be designed such that it can transfer the ultimate moment of 4.1 kip·ft./ft. Because SFRC can reduce crack width significantly, it assumes the concrete cover 2 in. is sufficient. If bar no. 6 is used, the effective depth is calculated as $d = 10 - 2 - 0.5 \times 3/4 = 7.625$ in. For typical slab, estimate moment arm $jd \sim 0.9d$, $A_s = M_u / \phi_b f_{sy} jd = 4.1 \times 12/0.9/0.9 \times 60 \times 0.9 \times 7.625 = 0.133$ in.^{2}/ft. Because the computed A_s is very low, check the minimum flexural reinforcement required by ACI Section 10.5.1,

$$A_{s,min} = \text{max of } \left(\frac{3\sqrt{f'_c}}{f_y}b_w d, \frac{200}{f_y} \right) = \left(\frac{3\sqrt{4000} \times 12 \times 7.625}{60000}, \frac{200 \times 12 \times 7.625}{60000} \right) = 0.3 \text{ in.}^2/\text{ft.}$$

Thus, the minimum reinforcement controls and use of bar no. 6 at 15 in. ($A_s = 0.353$ in.^{2}/ft.) to connect footing and the vertical component as shown in the Figure 17.11. Because the parameter μ used is $0.40 > \mu_{\text{min\_ST}} = 140/424 = 0.33$, reinforcement is not required to control shrinkage and temperature. Any amount of transverse reinforcement can be used to hold the vertical steel in position.

Because the thickness of the wall is greater than 5 in., the size-dependent safety factor must be calculated by Equation 17.6, $k_h = 1.0 - 0.03\,(10^{-5}) = 0.85$. The specified strength $\sigma_p = 0.40 \times 424 = 170$ psi must be requested with size-dependent safety factor $k_h = 0.85$.

DESIGN EXAMPLE 6: DESIGN WITH MACROPOLYMERIC FIBERS

Problem Formulation

Develop an alternative design methodology for reinforced concrete slabs that are in the range of 7 to 8 in. The present reinforced concrete design methodology requires concrete compressive strength f_c' of 4000 psi and steel reinforcement with a yield strength f_y of 60 ksi and placed at bar no. 4 at 12 in. at mid-depth. Develop an equivalent design with macro-polypropylene (FibraShield) FRC that has compressive strength f_c' of 4000 psi.

Proposed Approach

The procedure is to calculate the capacity moment of the reinforced concrete section with conventional reinforcement, and then based on the moment capacity obtained, compute the required residual tensile strength. Using an experimental program, one can calculate and compare the amount of residual strength obtained from back calculation of flexural tests performed. Using a data base of flexural tests, the appropriate volume fraction of fibers to provide the same level of capacity is calculated.

Moment Capacity of a 7-In.-Thick Reinforced Concrete Slab

The slab is designed based on a 1-ft. strip, and the amount of reinforcement A_s is calculated by $A_s = (\pi d^2 / 4)(12 / \text{spacing}) = (\pi 0.5^2 / 4)(12 / 18) = 0.131$ in.^{2}/ft. Calculate the plastic compressive zone according to ACI stress block concept $a = A_s f_y / 0.85 f_c' b = 0.131 \times 60 / 0.85 \times 4 \times 12 = 0.193$ in. The factored ultimate moment M_u is equal to the reduced nominal moment capacity $\phi_b M_n$,

$$M_u = \phi_b M_n = \phi_b A_s f_y \left(d - \frac{a}{2} \right) = 0.9 \times 0.131 \times 60 \left(3.5 - \frac{0.193}{2} \right) \frac{1}{12} = 2.07 \text{ kip} \cdot \text{ft./ft.}$$

Replace the Moment Capacity with Macropolymeric Fiber, $f_c' = 4000$ psi

Assume a cracking tensile strength of macropolymeric fiber, $\sigma_{cr} = 6.7\sqrt{f_c'} = 6.7\sqrt{4000} = 424$ psi. Calculate the cracking moment, $M_{cr} = \sigma_{cr} b h^2 / 6 = (424 \times 12 \times 7^2 / 6)(1/12,000) = 3.462 \text{ kip} \cdot \text{ft./ft.}$ Calculate the normalized compressive yield strain, $\omega = \sigma_{cy} / \sigma_{cr} = (0.85 \times 4000)/6.7\sqrt{4000} = 8.02$. Determine the normalized postcrack tensile strength. Assume a safety factor of $\phi_p = 0.85$,

$$\mu_{req} = \frac{M_u \omega}{3\omega\phi_p M_{cr} - M_u} = \frac{2.07 \times 8.02}{3 \times 8.02 \times 0.7 \times 3.462 - 2.07} = 0.24 \quad \text{and}$$

$$\sigma_{p,req} = \mu_{req} \times \sigma_{cr} = 0.24 \times 424 = 101 \text{ psi}$$

Replace Tensile Capacity

Assume the plain concrete in reinforced concrete slab (with conventional rebars) has no residual strength; thus, only the amount of reinforcement will be replaced with macropolymeric fibers FibraShield having the same tensile capacity, $\phi_b A_s f_y = \phi_p \mu \sigma_{cr} b h$. The postcrack tensile strength is calculated,

$$\mu_{req} = \frac{\phi_b A_s f_y}{\phi_p \sigma_{cr} b h} = \frac{0.9 \times 0.131 \times 60}{0.85 \times 0.424 \times 12 \times 7} = 0.23$$

Moment Capacity of an 8-In.-Thick Reinforced Concrete Slab

The same procedure is followed, so A_s is calculated by $A_s = (\pi d^2/4)(12/\text{spacing}) = (\pi 0.5^2/4)(12/18) = 0.131 \text{ in.}^2/\text{ft}$. Calculate the plastic compressive zone according to the ACI stress block concept, $a = (A_s f_y)/(0.85 f'_c b) = (0.131 \times 60)/(0.85 \times 4 \times 12) = 0.193$. The factored ultimate moment M_u will be

$$M_u = \phi_b M_n = \phi_b A_s f_y \left(d - \frac{a}{2} \right) = 0.9 \times 0.131 \times 60 \times \left(4 - \frac{0.193}{2} \right) \times \frac{1}{12} = 2.3 \text{ kip} \cdot \text{ft./ft.}$$

Replace the Moment Capacity with Macrofibers, $f'_c = 4000$ psi

Calculate the cracking tensile strength of macrofibers, $\sigma_{cr} = 6.7\sqrt{f'_c} = 6.7\sqrt{4000} = 424$ psi. Now, calculate the cracking moment, $M_{cr} = (\sigma_{cr} bh^2)/6 = (424 \times 12 \times 8^2/6) \times (1/12,000) = 4.522 \text{ kip} \cdot \text{ft./ft.}$ Calculate the normalized compressive yield strain, $\omega = \sigma_{cy}/\sigma_{cr} = 0.85 \times 4000/6.7\sqrt{4000} = 8.02$. Determine the normalized postcrack tensile strength,

$$\mu_{req} = \frac{M_u \omega}{3\omega\phi_p M_{cr} - M_u} = \frac{2.3 \times 8.02}{3 \times 8.02 \times 0.85 \times 4.522 - 2.30} = 0.205$$

and

$$\sigma_{p,req} = \mu_{req} \times \sigma_{cr} = 0.205 \times 424 = 87 \text{ psi}$$

Replace Tensile Capacity

On the basis of the same concept mentioned in the example 1, the postcrack tensile strength is calculated, $\mu_{req} = \phi_b A_s f_y / \phi_p \sigma_{cr} bh = (0.9 \times 0.131 \times 60)/(0.85 \times 0.424 \times 12 \times 8) = 0.2$. If the goal is to replace the existing reinforced concrete slab with polymeric fibers while maintaining the same performance and reliability index, the postcrack tensile strength for 7 and 8 in. must be 101 and 87 psi, respectively. At this stage one has to refer to a data base of flexural test data that correlate for a specific fiber type and content with postcrack tensile capacity. Using a back-calculation technique proposed in Chapter 15, correlation between fiber type, content, and postcrack residual strength can be obtained and used in the form of the design chart as shown in Figure 17.12. (See Figure 15.4,

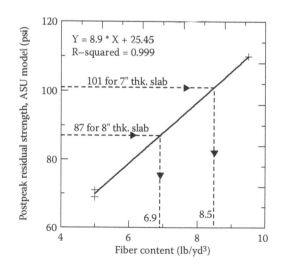

FIGURE 17.12 Development of fiber content based on required postpeak residual strength capacity.

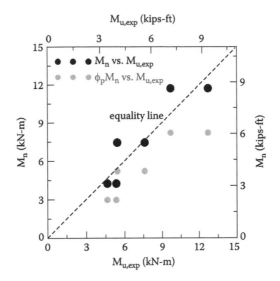

FIGURE 17.13 Predicted nominal moment capacity versus experimental ultimate moment.

Case A2 Chapter 15). Assuming to have the same first crack tensile strength for both 5 and 9.5 lb./yd.³ (conservatively), with increasing the nondimensional postpeak residual strength factor (μ) from approximately 0.28 to 0.44, the postpeak residual strength versus fiber content curve is obtained and shown in Figure 17.12. With a knowledge of required moment capacity, it seems that for the 7- and 8-in.-thick slabs, 6.9 and 8.5 lb./yd.³ fiber content are required. This procedure is shown in Figure 17.12. The safety factor of 0.85 for calculation of μ_{req} is considered.

Figure 17.13 shows the plot of experimentally obtained bending capacity versus the theoretically calculated results of the material parameters used in the model. To make sure that all design bases underestimate the experimental results, a material reduction factor parameter ϕ_p can be used to reduce the analytical expressions, so that there is a one-to-one relationship of theory versus experiments, and the design tools do not overestimate the experiment. In the present case, a value of $\phi_p = 0.7$ is used as the basis for reduction of materials strength so that there is a match between the estimated values and the experimentally obtained ultimate bending capacity values.

CONCLUSIONS

A design guideline for strain-softening FRC is presented using closed-form analytical equations that relate geometrical and material properties to moment and curvature capacity. Conservative reduction factors are introduced for using postcrack tensile strength in design, and a conversion design chart is proposed for developing FRC systems equivalent to traditional reinforced concretes.

The moment–curvature response for a strain-softening deflection-hardening FRC can be approximated by a bilinear model, whereas a geometric relationship between curvature and deflection can be used for serviceability deflection checks.

REFERENCES

1. Vandewalle, L., et al. (2000a), "Test and design methods for steel fibre reinforced concrete: Recommendations for bending test," *Materials and Structures*, 33(225), 3–5.
2. Vandewalle, L., et al. (2000b), "Test and design methods for steel fibre reinforced concrete: Recommendations for σ–ε design method," *Materials and Structures*, 33(226), 75–81.
3. Vandewalle, L., et al. (2003), "Test and design methods for steel fibre reinforced concrete-σ-ε design method—final recommendation," *Materials and Structures*, 36(262), 560–567.

4. Teutsch, M. "German guidelines on steel fiber concrete," *Proceedings of the North American/European Workshop on Advances in Fiber Reinforced Concrete, BEFIB 2004*, Bergamo, Italy, September 2004, 23–28.

5. Barr B., and Lee, M.K. "FRC guidelines in the UK, with emphasis on SFRC in floor slabs," *Proceedings of the North American/European Workshop on Advances in Fiber Reinforced Concrete, BEFIB 2004*, Bergamo, Italy, September 2004, 29–38.

6. Japan Society of Civil Engineers JSCE-SF4, "Methods of tests for flexural strength and flexural toughness of steel fiber reinforced concrete," Concrete Library International, Part III-2, Vol. 3, 1984, 58–61.

7. di Prisco, M., Toniolo, G., Plizzari, G. A., Cangiano, S., and Failla, C., "Italian guidelines on SFRC," Proceedings of the North American/European Workshop on Advances in Fiber Reinforced Concrete, BEFIB 2004, Bergamo, Italy, September 2004, 39–72.

8. Swamy, R.N., Mangat, P.S., and Rao, C. V. S. K. "The mechanics of fiber reinforced cement matrices," Fiber Reinforced Concrete ACI SP-44, 1975, 1–28.

9. Fischer, G., "Current U.S. guidelines on fiber reinforced concrete and implementation in structural design," *Proceedings of the North American/European Workshop on Advances in Fiber Reinforced Concrete, BEFIB 2004*, Bergamo, Italy, September 2004, 13–22.

10. Henager, C. H., and Doherty, T. J. (1976), "Analysis of reinforced fibrous concrete beams," *Journal of the Structural Division*, 102, 177–188.

11. ACI Committee 318 (2005), *Building Code Requirements for Structural Concrete, ACI Manual of Concrete Practice*. Detroit: American Concrete Institute.

12. Soranakom, C., and Mobasher, B. (2007), "Closed-form solutions for flexural response of fiber-reinforced concrete beams," *Journal of Engineering Mechanics*, 133(8), 933–941.

13. Soranakom, C., and Mobasher, B., "Development of design guidelines for strain softening fiber reinforced concrete," *7th RILEM International Symposium on Fibre Reinforced Concrete, BEFIB 2008*, Chennai (Madras), India, September 2008, 513–523.

14. Soranakom, C. (2008), "Multi-scale modeling of fiber and fabric reinforced cement based composites," PhD dissertation, Arizona State University, Tempe, AZ.

15. Soranakom, C., and Mobasher, B., "Development of design guidelines for strain softening fiber reinforced concrete," *7th RILEM International Symposium on Fibre Reinforced Concrete, BEFIB 2008*, Chennai (Madras), India, September 2008, 513–523.

16. Swamy, R. N., and Al-Ta'an, S. A. (1981), "Deformation and ultimate strength in flexural of reinforced concrete beams made with steel fiber concrete," *ACI Structural Journal*, 78(5), 395–405.

17. Hassoun, M. N., and Sahebjam, K., "Plastic hinge in two-span reinforced concrete beams containing steel fibers," *Proceedings of the Canadian Society for Civil Engineering*, 1985, 119–139.

18. Soranakom, C. (2008), "Multi-scale modeling of fiber and fabric reinforced cement based composites," PhD dissertation, Arizona State University, Tempe, AZ.

19. Soronakom, C., and Mobasher B., "Design flexural analysis and design of textile reinforced concrete textile reinforced structures," *Proceedings of the 4th Colloquium on Textile Reinforced Structures (CTRS4) und zur 1.* Anwendertagung, SFB 528, Technische Universität Dresden, Eigenverlag, 2009, ISBN 978-3-86780-122-5, 273–288.

20. Ghali, A. (1993), "Deflection of reinforced concrete members: A critical review," *ACI Structural Journal*, 90(4), 364–373.

21. Ghali, A., and Favre, R. (1986), *Concrete Structures: Stresses and Deformations*, London: Chapman and Hall, 352.

22. ASTM C 1399-04 (2004), "Standard tests method for obtaining average residual-strength of fiber-reinforced concrete," ASTM International, PA, 6 pp.

23 Soranakom, C., and Mobasher, B. (2009), "Flexural design of fiber reinforced concrete," *ACI Materials Journal*, 461–469.

18 Fiber Reinforced Aerated Concrete

INTRODUCTION

Cellular or aerated concrete (AC) is a lightweight, noncombustible cement-based material manufactured from a mixture of portland cement, fly ash, or other sources of silica, quick lime, gypsum, water, and aluminum powder or paste. This material can be autoclaved for accelerated strength gain, in which case it is referred to as autoclaved AC (AAC). The air pores in AC are usually 0.1 to 1 mm in diameter and are formed by a few different methods. The most common technique is the addition of aluminum powder at approximately 0.2% to 0.5% by weight of cement. Approximately 80% of the volume of the hardened material is made up of pores, including 50% being air pores and 30% being micropores [1].

The special cellular structure with its associated physical and mechanical properties enables some outstanding applications for construction industry. A novel variation on AAC is aerated fiber reinforced concrete (AFRC), which is reinforced with short polymeric fibers such as polypropylene. The improved properties need to be characterized to expand the potential areas of appli cations for this highly ductile material system. The approach for studying and using AFRC is based on similarities of this material to AAC and fiber reinforced concrete (FRC). By characterizing and correlating the basic properties such as bulk density, compressive strength, tensile strength, and ductility, new areas of opportunity can be identified. This chapter reviews the properties of AAC from literature and introduces the characteristics of AFRC.

The chemical reaction caused by addition of aluminum powder or paste to a hydrating portland cement microstructure generates hydrogen gas as shown by Equation 18.1. Formation of hydrogen gas makes the mixture expand to about twice its volume, resulting in a highly porous structure [2].

$$2Al + 3Ca(OH)_2 + 6H_2O \rightarrow 3CaO \cdot Al_2O_3 \cdot 6H_2O + 3H_2 \uparrow \qquad (18.1)$$

The special properties of AC include its low density and strength compared with normal weight concrete. The classification of this material is based on the dry density and compressive strength, which are regularly 400 to 800 kg/m³ (25–50 lb./ft.³) and 2–6 MPa (290–870 psi), respectively [3]. The pore structure of AAC controls its light density and is responsible for its special properties compared with conventional construction materials. The high porosity of AAC makes it a good thermal insulator with extremely low thermal conductivity coefficients. For example, for AAC with dry density of 400 kg/m³, thermal conductivity is reported to be 0.07 to 0.11 W/m °C, which is approximately 10 times less than normal weight concrete [2]. The low density is of course associated with a relatively low compressive strength, which limits its applicability to many structural applications; however, it can easily be used in construction of 1 to 2 storey buildings. The pore structure of AAC also affects the acoustic quality resulting in superior sound insulation [4]. The pore structure of AC consists of a variety of sizes from micropores, macropores, and air pores [5] as shown in Figure 18.1, in which r is the pore diameter and $f_v(r)$ is the pore volume distribution.

AAC is an ideal material for sustainable and green construction, especially when produced with high-volume fly ash substitutions. Life cycle analysis of construction materials generally includes

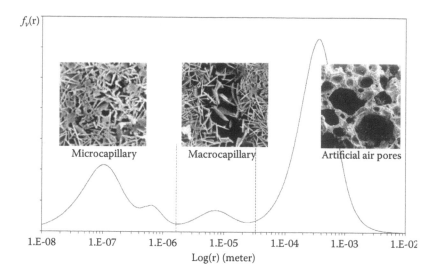

FIGURE 18.1   Pore volume distribution of AAC and SEM image of various pore sizes. (Adapted from Roels, S., Sermijn, J., and Carmeliet, J., "Modelling unsaturated moisture transport in autoclaved aerated concrete: A microstructural approach," *Building Physics 2002—6th Nordic Symposium*, Trondheim, Norway, 2002, 167–174.)

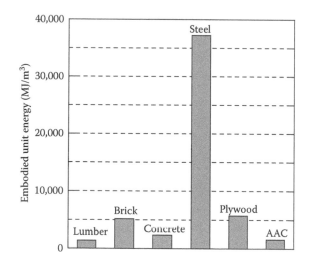

FIGURE 18.2   Embodied energy of typical construction materials by volume. (From Nunez, E., Nunez, S. A., and Fouad, F. H., "Sustainability in autoclaved aerated concrete (AAC) construction," in Limbachiya and Roberts (eds.), *Autoclaved Aerated Concrete*, Taylor & Francis Group, London, 2005. With permission.)

considerations such as energy consumed in production, transportation, construction, and life span and/ or recycling after use. On the basis of the comparative evaluation of the embodied energy data in the literature as shown in Figure 18.2, it is apparent that AAC is a viable alternative to similar construction materials such as blocks, bricks, or lightweight panels [6]. The use of high-volume fly ash in the production of AC contributes to additional benefits in reuse and recycle of coal combustion by-products. Aerated concrete with fly ash can use as much as 400,000 tons of fly ash only in block production [7].

The chemical composition, pore structure, and mechanical properties of AC strongly depend on the raw materials, method of aerating, and curing techniques. The characteristics observed in AAC with and without fly ash are different because the capillary porosity is much finer in fly ash blended

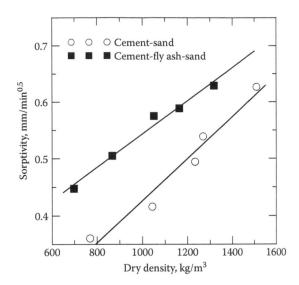

FIGURE 18.3 Variation of sorptivity with density in AC. (Adapted from Nambiar, E. K. K., and Ramamurthy, K., "Sorption characteristics of foam concrete," *Cement and Concrete Research*, 37, 1341–1347, 2007.)

AAC [3]. It is also observed that autoclaving provides higher strength because of the better crystallinity of the reaction products when compared with moist curing.

Transport properties such as absorption, diffusion, and sorptivity are of importance as they are primarily related to porosity and affect the durability and long-term properties of concrete. In the case of aerated and foamed concrete, the moisture movement behavior becomes more complex as such concretes contain much larger volumes of air voids [8]. Figure 18.3 presents the sorptivity–density relationship for AC with and without fly ash substitution. Aerated concrete had less sorptivity values at lower densities, whereas fly ash blended system has a relatively higher sorptivity value.

Because AAC and AFRC are solids with cellular structure, their compressive behavior can be described as hyperelastic, meaning that they carry a considerable amount of residual strength after the peak strength is reached [9]. These materials respond elastically up to very large strains and account both for nonlinear material behavior and also for nonlinear kinematics. The main areas of application for AAC products are similar to polymeric foams. An example of the closed-loop compressive stress–strain response of AAC materials is presented in Figure 18.4 [10]. The response is quite linear up until the compressive strength; beyond this level, there is a sudden drop in the magnitude of the stress. The residual strength after the peak strength in longitudinal and circumferential direction is approximately 30% of the peak strength. Note that the constant level in the postpeak response extends for a large range of up to 0.5% in axial; however, the diametrical dilatation is significant and can reach as high as 1% lateral expansion due to pore crushing. The total amount of energy consumed in this process is as much as 7.63 N·m in axial and 26.78 N·m in circumferential direction, which for a 150 × 300-mm cylinder corresponds to an equivalent volumetric level of 1440 N/m² in axial and 5052 N/m² in volumetric terms.

The mechanical properties of AC (compressive strength, flexural strength, and fracture energy) are related to the material's porosity in addition to pore size distribution. Relationships between AAC density and porosity were studied as well as the density's relationship with mechanical properties, including compressive strength, flexural strength, and fracture energy. The experiment results are shown in Figure 18.5 [12]. One can see that the total porosity for AAC materials decreases linearly from 80% to 50% as the density varies from 600 to 1400 kg/m³. The mechanical properties including compressive strength, flexural strength, and fracture energy increase as the material becomes denser.

Fracture parameters of AAC have been studied as a function of direction of expansion in fresh AAC [13]. The specific fracture energy G_f is higher, and the tensile strength f_t is lower when

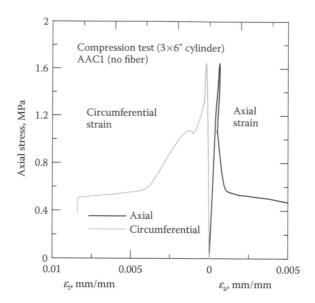

FIGURE 18.4 Compressive stress strain response of AAC showing the axial and circumferential strain. (Adapted from Bonakdar A., and Mobasher, B., *Mechanical properties of aerated concrete and fiber reinforced aerated concrete*, manuscript in preparation.)

measured parallel to the expanding direction. These values are in the range of 5.6 to 6.6 N·m and 0.27 to 0.37 MPa. Blast protection panels have been developed with AAC masonry walls covered with textiles and exposed to high-intensity blast. These ideas proved their capacity to withstand lateral loading. The resistance of the exterior AAC masonry walls was enhanced by applying different reinforcing fabrics [14].

AFRC PRODUCTION

AFRC is manufactured through a process similar to AAC, except that the autoclave process is replaced by normal curing. This trade-off introduces variability in the test data as the curing is nonuniform. The raw materials including portland cement, fly ash, water, aluminum paste, polypropylene fibers, and other chemical admixtures are mixed. The fresh slurry is then poured into large 8 × 1.2 m steel mold with a depth of 0.6 m as shown in Figure 18.6. Within a few minutes, chemical reactions of hydrogen gas generation cause the fresh mixture to rise and fill the mold. The mixing water has a temperature of approximately 55°C (100°F) to accelerate the reaction, which in turn generates more heat through exothermic reactions. The heat generated from the reaction allows the sample temperature to reach high as 100°C, which decays after approximately 24 h. This change of temperature, however, occurs at various rates at different locations throughout the depth of the mold and may induce cracking if not controlled. Once AFRC cures and gains minimum required strength (5–7 days from casting), the entire slab is demolded and loaves are cut using diamond blades into blocks or panels of various sizes and shapes.

DENSITY AND COMPRESSIVE STRENGTH RELATIONSHIP

According to ASTM C1386 [3] and similar to AAC, the main properties of AFRC used for classification and engineering applications are its dry density and compressive strength. The density and compressive strength of AFRC obtained from both laboratory and field tested samples are discussed along with basic statistical analyses of these data for quality control purposes. Several

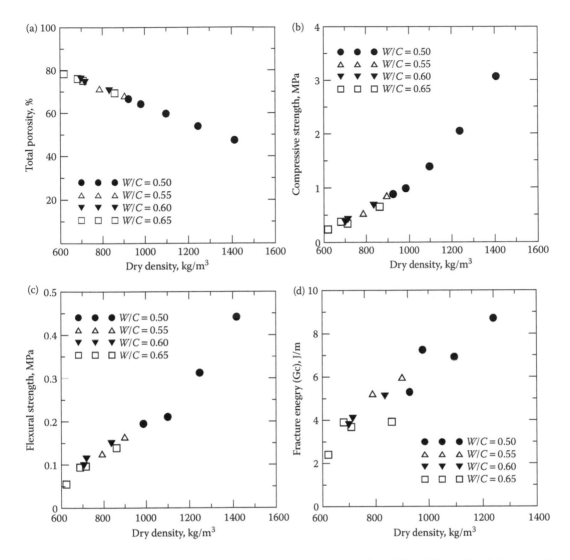

FIGURE 18.5 Relationship between density and mechanical properties. (Adapted from Bonakdar A., and Mobasher, B., *Mechanical properties of aerated concrete and fiber reinforced aerated concrete*, manuscript in preparation.)

randomly selected AFRC blocks 250 × 250 × 600 mm (10 × 10 × 24 in.) in dimension were cut into 100 × 100 × 100 mm (4 × 4 × 4 in.) cubes. Samples were selected from the central core sections. The porosity of the top portion is relatively higher (lower density) than the bottom portion of the mold; therefore, on the basis of their location from the bottom of the mold, the five cubes are named as top, top–middle, middle, middle–bottom, and bottom and tested. Results for dry density of the top–middle and middle–bottom cubes with an average of 540 kg/m³ (33.4 lb./ft.³) and a standard deviation of 12 kg/m³ (0.8 lb./ft.³). The data are also plotted in Figure 18.7a. Plots of compressive strength has a function of location within the loaf are shown in Figure 18.7b. Note that as the material rises in the loaf, the density and the strength both decrease.

Figure 18.8 shows the compressive test results of three AFRC cubes. Note the compressive strength of approximately 4.5 MPa and a residual strength of approximately 3 MPa, which is as much as 60% of the peak strength. This level of postfailure strength may be attributed to the ability of the fibers to maintain the tensile cracks in the sample allowing for more load carrying capacity

FIGURE 18.6 Fresh slurry AFRC rises inside the mold because of the hydrogen generation.

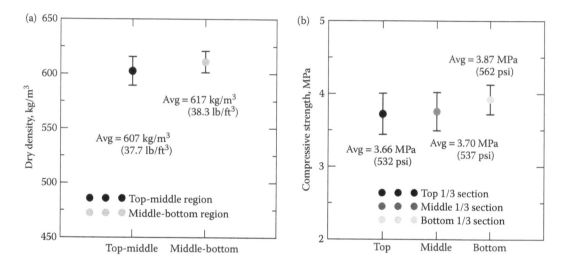

FIGURE 18.7 Average plots of (a) density and (b) compressive strength and for AFRC blocks as a function of location within the loaf.

than the standard AAC. Once again, the crushing capability of the material is dominated and explained by the crushable foam analogy.

The data obtained from field testing were also used for statistical analysis of density and strength of a large sample of AFRC. For a general observation of the properties of manufactured blocks with different mix designs and curing regimes, the results of more than 800 cubes were used. Figure 18.9 shows the normal distribution of density (nondry density) and compressive strength of AFRC cubes. The average value of density for tested samples is 657 kg/m³ (40.8 lb./ft.³), with a standard deviation of 26 kg/m³ (1.6 lb./ft.³). The average value of strength for these samples is 3.2 MPa (454 psi), with a standard deviation of 0.73 MPa (104 psi).

FLEXURAL RESPONSE

The main difference between AAC and AFRC is the flexural response of these materials. This difference is attributed to the short fibers bridging the cracks under tensile loads and adding to the AFRC's energy absorption capacity. AAC and AFRC can have similar values of modulus

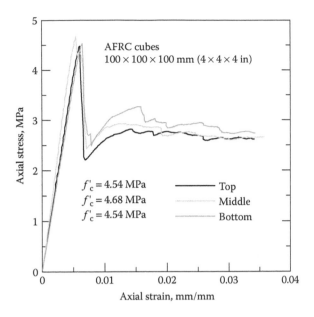

FIGURE 18.8 Typical compression test results for AFRC.

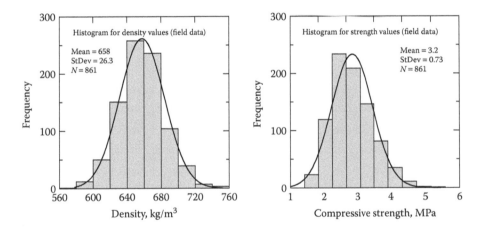

FIGURE 18.9 Histogram of field data (density and strength) for AFRC cubes.

of rupture with significantly different toughness values. For measuring the flexural properties, 152 × 152 × 460 mm (6 × 6 × 18 in.) beams were prepared with a 25-mm notch at the mid-span for a three-point bending test procedure. LVDT and CMOD gages were used to equip the test as shown in Figure 18.10. The results for four replicate AAC samples are presented in Figure 18.11 and compared with the AFRC. In the case of AAC, the material behaves in a brittle manner, and failure occurs after the peak strength is reached. On the other hand, the comparison between the flexural responses of AAC and AFRC indicates that the peak strength for AFRC is reached in the nonlinear portion of the curve where strain hardening is taking place. The large area under the curve (the toughness of AFRC) is obtained by the bridging action of fibers with large deformations as shown. Experimental parameters of the flexural test are summarized in Table 18.1 for AFRC and AAC beams. Although the flexural strength of AAC is approximately 80% that of AFRC, the flexural toughness at 100-mm deflection is more than 40 times higher in AFRC beams because of the effect

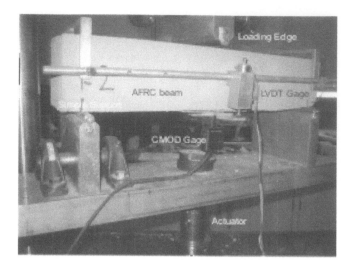

FIGURE 18.10   Three-point bending test setup (equipped with LVDT and CMOD).

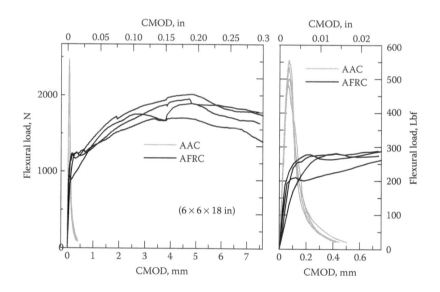

FIGURE 18.11   Typical AFRC and AAC flexural test results (comparative).

of fibers in absorbing energy. This property can be used for any application where energy absorption and high deformation capacity is required in the structural system.

PORE STRUCTURE

Pore structure of AFRC can be studied at different scales. A potential approach to approximate the size and distribution of air pores as well as the total porosity of AC is by means of the principles of stereology. Figure 18.12 shows a 50 × 50-mm (2 × 2-in.) image of a cut surface of AFRC with pores in the size range of 0.5 to 1.5 mm. At a larger scale, the pore structure was studied using a high-resolution digital scanner and converting the grayscale image to a binary image. A MATLAB® code was developed next to measure the size of the circular holes through the chord lengths. The measured pore lengths are in fact the chord size of cut spheres in two-dimensional surface and should

TABLE 18.1
Flexural Parameters for AFC Flexural Tests

Experimental Properties	AFRC Average	SD	AAC Average	SD	AFRC/ AAC Ratio
Load at crack point (N)	900	135	2211	170	0.41
Deflection at crack point (mm)	0.068	0.019	0.066	0.005	1.03
Max load (N)	1906	114	2308	145	0.83
Deflection at max load (mm)	2.59	0.52	0.075	0.006	34
Elastic flexural stiffness (N·mm)	13,727	4265	33,868	2116	0.40
Max deflection (mm)	4.15	1.10	0.23	0.02	18
Max flexural strength (MPa)	0.53	0.03	0.64	0.04	0.83
Toughness at 1 mm deflection (N·mm)	765	20	168	15	4.55
Toughness at 10 mm deflection (N·mm)	6661	1672	168	15	41

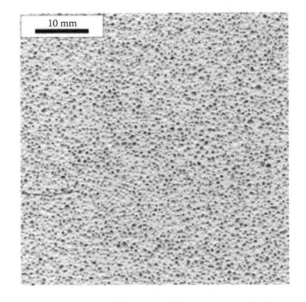

FIGURE 18.12 Grayscale image of a 2 × 2-in. AFRC sample used for image analysis.

be transformed into three-dimensional spheres. On the basis of a stereological methods [15], having the diameter of two-dimensional pores and their distribution in a known unit area, one can estimate the diameter of three-dimensional pores as described in Equation 18.2:

$$\overline{D} = \frac{\pi}{2} H(d)$$

$$N_V = \frac{N_A}{\overline{D}}$$

(18.2)

where d is the diameter of individual two-dimensional pores, $\overline{D}$ is the mean diameter of three-dimensional pores, N_A is the number of two-dimensional pores in unit area, N_V is the number of three-dimensional pores in unit volume, and $H(d)$ is the harmonic mean of two-dimensional diameters. As an example, the diameters of the 20 circles were measured in a 5×5-mm ($\frac{1}{4} \times \frac{1}{4}$-in.) AFRC sample with an average value of 0.67 mm. For this sample, the estimated diameter is calculated to be approximately 0.9 mm, which results in an approximate total porosity of 80%.

It is noted that total porosity can also be calculated by having the bulk density (ρ_b) and pure density (ρ_p) of AFRC. In this case, the approximate values are 600 and 2600 kg/m^3, respectively, the total porosity is calculated following Equation 18.2 as 77% [16],

$$P = \left(1 - \frac{\rho_b}{\rho_p}\right) \times 100\% \qquad (18.3)$$

REFERENCES

1. Neithalath, N., and Ramamurthy, K. (2000), "Structure and properties of aerated concrete: A review," *Cement and Concrete Composites*, 20, 321–329.
2. Holt, E., and Raivio, P. (2005), "Use of gasification residues in aerated autoclaved concrete," *Cement and Concrete Research*, 35, 796–802.
3. ASTM C1386-07 (2007), "Standard specification for precast autoclaved aerated concrete (AAC) wall construction units," ASTM International, PA.
4. Laukaitis, A., and Fiks, B. (2006), "Acoustical properties of aerated autoclaved concrete," *Applied Acoustics*, 67, 284–296.
5. Roels, S., Sermijn, J., and Carmeliet, J., "Modelling unsaturated moisture transport in autoclaved aerated concrete: A microstructural approach," *Building Physics 2002—6th Nordic Symposium*, Trondheim, Norway, 2002, 167–174.
6. Nunez, E., Nunez, S. A., and Fouad, F. H. (2005), "Sustainability in autoclaved aerated concrete (AAC) construction," in M. C. Limbachiya and J. J. Roberts (eds.), *Autoclaved Aerated Concrete*, London: Taylor & Francis Group.
7. Sutton, M. E., "Autoclaved cellular concrete, the future of fly ash," *Proceedings of International Ash Utilization Symposium*, Paper No. 73, University of Kentucky, 1999.
8. Nambiar, E. K. K., and Ramamurthy, K. (2007), "Sorption characteristics of foam concrete," *Cement and Concrete Research*, 37, 1341–1347.
9. Gibson, L. J., and Ashby, M. F. (1997), *Cellular Solids, Structure and Properties*, Cambridge: Cambridge University Press.
10. Marzahn, G. A. (2002), "Extended investigation of mechanical properties of masonry units," LACER No. 7, Leipzig, Germany.
11 Bonakdar A., and Mobasher, B., "Mechanical properties of aerated concrete and fiber reinforced aerated concrete." Manuscript in review. *Materials Journal of American Concrete Institute*.
12. Eden, N. B., Manthorpe, A. R., Miell, S A., Szymanek, P. H., and Watsont, K. L. (1980), "Autoclaved aerated concrete from slate waste—Part 1: Some property/density relationships," *International Journal of Cement Composites and Lightweight Concrete*, 2, 95–100.
13. Trunk, B., Schober, G., Helbling, A. K., and Wittmann, F. H. (1999), "Fracture mechanics parameters of autoclaved aerated concrete," *Cement and Concrete Research*, 29, 855–859.
14. Yankelevsky, D. Z., and Avnon, I. (1998), "Autoclaved aerated concrete behavior under explosive action," *Construction and Building Materials*, 12, 359–364.
15. Hilliard, J. E., and Lawson, L. R. (2003), *Stereology and Stochastic Geometry*, Boston: Kluwer Academic Publishers.
16. Kadashevich, I., Schneider, H. J., and Stoyan, D. (2005), "Statistical modeling of the geometrical structure of the system of artificial air pores in autoclaved aerated concrete," *Cement and Concrete Research*, 35, 1495–1502.

19 Sisal Fiber Reinforced Composites

INTRODUCTION

Natural fiber reinforced cement (FRC) composites provide an opportunity for development of construction materials using an agriculture based economy in arid environments. The economic incentives particularly in developing countries are enormous because the availability and production of natural fiber reinforcement requires a small capital investment and a low degree of industrialization. Furthermore, in comparison with the most common synthetic reinforcing fibers, natural fibers require less energy to produce and are the ultimate green products.

Natural fibers have been traditionally used in the form of chopped, short, or pulp for the production of thin roofing and cladding elements [1–4]. An increased use of these materials for applications such as internal and external partitioning walls is possible and leads to the development of low cost-sustainable materials [5–8]. Natural fiber cement composites have been used for many years, yet their application is still limited because of two main challenges, lack of durability and ductility.

A vast range of fibers of sizes from particulates to long strands trace their beginning from a natural feedstock such as pulp (cellulose), agave (sisal), coconut, hemp, flax, jute, or kenaf. In North America, cellulose-based FRC have found increasing applications in residential housing and non-structural components such as siding and roofing, flat panel applications such as underlayment, and tile backer board. Additionally, lumber substitutes such as trim, fascia, and corner boards are possible. Cellulose-based FRC products have had limited exterior applications because of degradation from ambient wetting and drying.

Vegetable fiber cement composites produced with ordinary portland cement matrices undergo an aging process in humid environments during which they may suffer a reduction in postcracking strength and toughness [3, 4, 9–16]. The aging process is due to fiber mineralization and results in reducing the tensile strength and decreasing the fiber pullout ligament after fracture. This mineralization process is a result of migration of hydration products (mainly $Ca(OH)_2$) to the fiber structure. Efforts to develop durable and structural cement composite laminates reinforced with long sisal fibers have shown promise recently [17, 18]. Matrices that lower calcium hydroxide production (only 50% portland cement as compared with conventional systems) increase the long-term durability of natural fiber, reduce CO_2 emissions, and present an economical and sustainable approach. Recently, developed modified matrices have shown no strength and toughness reduction in accelerated aging tests [18].

The sisal plant leaf is a functionally graded composite structure, which is reinforced by three types of fibers: structural, arch, and xylem fibers. The first occurs in the periphery of the leaf providing resistance to tensile loads. The others present secondary reinforcement, occurring in the middle of the leaf, as well as a path for nutrients (see Figure 19.1). The fibers have an irregular cross section with mean area ranging from 0.04 to 0.05 mm^2 and a mean density, elastic modulus, and tensile strength of 0.9 g/cm^3, 19 GPa, and 400 MPa, respectively [19].

Sisal fibers contain numerous elongated fiber cells approximately 6 to 30 µm in diameter [20]. The microstructure of individual fiber cells, as shown in Figure 19.2, are linked together by means of the middle lamella, which consist of hemicellulose and lignin. The lumen varies in size but is usually

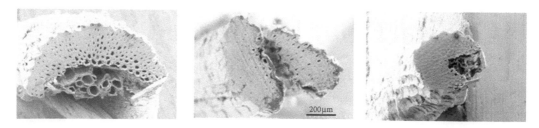

FIGURE 19.1 Sisal fiber morphology. The diverse geometry may result in different fiber–matrix bond adhesion.

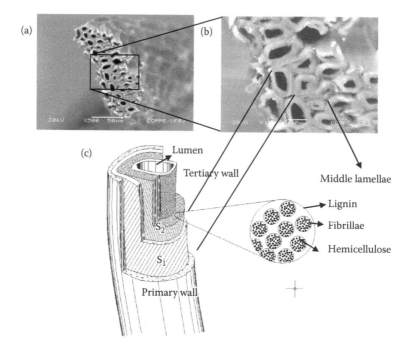

FIGURE 19.2 Fiber–cell microstructure: (a) cross-sectional view showing the fiber cells, lumens, and middle lamellae; (b) magnification of the cross section; and (c) schematic drawing showing the different layers of an individual fiber cell. (After Toledo Filho R. D., Silva, F. A., Fairbairn, E. M. R., Melo Filho, J. A., "Durability of compression molded sisal fiber reinforced mortar laminates," *Construction Building Materials*, 23, 2409–2420, 2009.)

well defined (see Figure 19.2b). Each individual fiber cell is made up of four main parts, namely, the primary wall, the thick secondary wall, the tertiary wall, and the lumen (see Figure 19.2c).

Continuous sisal FRC-based composite system with superior tensile strength and ductility has been developed [21]. The enhanced strength and ductility is primarily governed by the composite action that exists such that the fibers bridge the matrix cracks and transfer the loads, allowing a distributed microcrack system to develop. These materials are strong enough to be used as load bearing structural members, in applications such as structural panels, impact and blast resistance, repair and retrofit, earthquake remediation, strengthening of unreinforced masonry walls, and beam–column connections [22, 23].

SISAL FIBER COMPOSITES

To increase the durability of the sisal fibers in the alkaline environment, a cementitious matrix consisting of 50% portland cement, 30% metakaolin, and 20% calcined waste crushed clay brick was used [24]. The matrix was produced using the portland cement CPII F-32 with a 28-day compressive strength of 32 MPa, metakaolin, and calcined waste crushed clay brick (calcined at 850°C). River sand with a maximum diameter of 1.18 mm and a density of 2.67g/cm³ and a naphthalene superplasticizer Fosroc Reax Conplast SP 430 with content of solids of 44% were also used. The mortar matrix used a mix design 1:1:0.4 (cementitious material–sand–water by weight). Wollastonite fiber (JG class), obtained from Energyarc, was used as a microreinforcement in the composite production ($V_f = 5\%$).

The sisal fibers were characterized as having an irregular cross section with mean area ranging from 0.04 to 0.05 mm² and a mean density, elastic modulus, and tensile strength of 0.9 g/cm³, 19 GPa, and 400 MPa, respectively [19]. These fibers were extracted from the sisal plant in a farm located in the city of Valente, state of Bahia, Brazil. More information on the sisal fibers mechanical properties and its morphology can be obtained elsewhere [25]. For the production of the laminates, the mortar mix was placed in a steel mold, one layer at a time, followed by single layers of long unidirectional aligned fibers (up to five layers). The samples were consolidated using a vibrating table operated at a frequency of 65 Hz, which resulted in a sisal fiber volume fraction of 10%. After casting, the composites were compressed at 3 MPa for 5 min. The specimens were covered in their molds for 24 h before moist curing for 28 days in a curing chamber with 100% RH and 23°C ± 1°C.

The following sections describe the mechanical behavior under static and dynamic loading of the continuous sisal FRC composite system developed [21]. The interface bonding mechanisms and fatigue response under tensile load are also addressed.

STRESS–STRAIN BEHAVIOR AND CRACKING MECHANISMS

Typical tensile stress–strain response of the sisal fiber reinforced composite is shown in Figure 19.3. From a macroscopic perspective, the bend over point (BOP) corresponds to the formation of matrix cracking. Five distinct zones are identified using roman numerals with two zones before and three zones after the BOP. Figure 19.4a shows the relationship between the strain gage reading and the

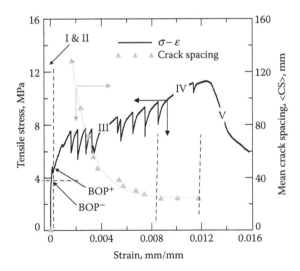

FIGURE 19.3 Tensile response of the sisal fiber reinforced composite system: tensile stress and crack spacing versus strain. (After Silva, F. A., Mobasher, B., and Filho, R. D. T., "Cracking mechanisms in durable sisal fiber reinforced cement composites," *Journal of Cement and Concrete Composites*, 31, 721–730, 2009, 10.1016/j.cemconcomp.2009.07.004.)

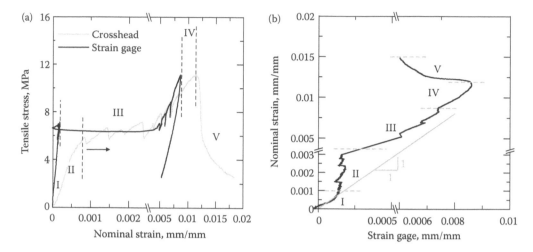

FIGURE 19.4 (a) Comparison of tensile stress versus strain from strain gage measurements and cross-head displacement. (b) Relationship between the strain gage and the strain measured by the stroke. (After Silva, F. A., Mobasher, B., and Filho, R. D. T., "Cracking mechanisms in durable sisal fiber reinforced cement composites," *Journal of Cement and Concrete Composites*, 31, 721–730, 2009, 10.1016/j.cemconcomp.2009.07.004.)

strain measured by the stroke. Both the initial response and also the overall response of the stress–strain curve are shown using a multiscale axis representation. Zone I corresponds to the elastic-linear range where both the matrix and the fiber behave linearly. Because of the low volume fraction of fibers (≤10%), the stiffness of the composite is dominated by matrix properties, and this zone is limited to strain measures of up to 150 to 175 µstr. The initial stress–strain response is marked by a limited range of linear elastic portion as the two strain measures are almost the same, and the specimen exhibits the highest stiffness. The deviation from linearity occurs at around 150 µstr because of initiation and propagation of first cracks. The sensitivity of the stroke displacement in this range is within the instrumentation error; hence, the strain gage response is far more reliable than the stroke in this range. The linear zone is terminated by initial crack formation in the matrix phase (reported as of $\sigma_{BOP}-$ from experiments) as shown in Figure 19.3. After the initiation of cracks in the matrix, its load-carrying capacity does not vanish as the cracks are bridged by the longitudinal fibers.

Immediately after the initiation of the first matrix crack, other cracks also initiate throughout the specimen at approximately regular intervals and begin to propagate across the width [26]. The strain recorded by the resistance gage remains relatively constant, indicating a steady state condition of several cracks that initiate and propagate across the width as shown in Figures 19.4a and 19.4b. The strain range within zone II is associated with formation of matrix cracks; however, no single crack has traversed the entire width. The term defined as BOP+ corresponds to the stress level at which the first matrix crack completely propagates across the width. The linear behavior terminates at the $\sigma_{BOP}- = 3.63$–4.80 MPa. The BOP ranges from the beginning of nonlinearity at 4.80 MPa to a point where the slope drastically decreases ($\sigma_{BOP}+ = 4.80$–5.59 MPa). Zone II is therefore defined as the stable cracking range between the two stress levels of $\sigma_{BOP}-$ and $\sigma_{BOP}+$.

The post-BOP stage is characterized by formation of distributed cracking in zone III. In this homogenization phase, as the applied strain increases, more cracks form and the spacing decreases in an exponential manner as presented empirically by Equation 19.1. The strain measured by the strain gage remains constant while several cracks form throughout the section as shown in Figure 19.4b. The decrease in crack spacing can be empirically represented as a function of three parameters, and its initiation is represented by parameters S_0, S_1, α, and ε_{mu} (Equation 19.1) [27].

$$S(\varepsilon_i) = S_1 + S_0 e^{-\alpha(\varepsilon_i - \varepsilon_{mu})}, \varepsilon_i > \varepsilon_{mu} \tag{19.1}$$

where $S(\varepsilon_i)$ is the crack spacing as a function of strain, ε_{mu} is the average strain at the BOP level or where the first set of measurements were obtained, ε_i is the independent parameter representing strain in the specimen, S_0 and α are the constants representing the initial length of the specimen and rate of crack formation as a function of strain, and S_1 is the saturation crack spacing. The stiffness of the sisal FRC composite system is sufficiently high and keeps the newly formed cracks from widening, thus promoting multiple cracking behavior as shown in Figure 19.1. This stiffness affects the rate of reduction of crack spacing or α parameter. Individual, mean, and standard deviation values of the S_1, S_0, ε_{mu}, and α for the composites studied under tension and bending loads are presented in several papers for different systems [24, 28, 29]. Significant variations in the value of α parameter were observed in the tensile tests. Less variability was obtained for S_1, S_0, and α from the bending tests.

The crack spacing measurements show a general reduction in spacing during loading until a steady state condition is reached. This zone covers a large range at the end of zones II and III and remains constant throughout zone IV. This constant level of crack spacing is defined as saturation crack spacing. Beyond this point, reduction in crack spacing is not observed because new cracks do not form, whereas additional imposed strain results in widening of the existing cracks.

Zone IV corresponds to the completion of cracking phase and initiation of debonding. The strain gage recording fails to increase as the same rate of the overall strain measure and no additional cracks are formed. As the cracking saturates in the specimen, the progressive damage state is extended to a crack widening stage, ultimately leading to failure by fiber pullout. This zone is asymptotically terminated at the saturation crack spacing represented by parameter S_1. The dominant mechanism of failure during stage IV is crack widening, which is associated with fiber debonding and pullout. The postpeak response occurs in zone V where a residual strength of approximately 2 MPa is observed. A considerable difference between the initial ($E_{initial}$) and postcrack (E_{pc}) moduli from cross-head displacement data and strain gages are observed. Results for $E_{initial}$ and E_{pc} from strain gages are, respectively, 3.6 and 2.60 times greater than that of the cross head. This discrepancy is attributed to the spurious deformation, slipping, and localized damage at the grips, indicating that initial stiffness computation from the cross-head displacement is significantly erroneous.

The average ultimate tensile strain of the composite is 1.53% (measured from cross-head displacement), which shows the capacity of the sisal fibers to cause crack distribution. Strain values ranging from 1.15% to 2.2% was obtained for individual tests. The average ultimate tensile strength of 12 MPa and an initial modulus of 34.7 GPa (computed from strain gage measurements) are indicative that sisal FRC composite presents a mechanical performance applicable to structural level applications. Nevertheless, variability was observed when addressing individual tensile tests with ultimate strength ranging from 10.6 to 14.7 MPa.

FLEXURAL RESPONSE

Figure 19.5a shows the typical bending response of the sisal fiber reinforced composite and its crack spacing measurements. Using the same methodology as for the direct tension, the bending curve was divided into five regions identified by roman numerals. Zone I corresponds to the elastic-linear range where both matrix and the fiber behave linearly. The lower and upper bounds of the limit of proportionality (LOP) delimit the zone II. The mean values of 8.89 and 9.60 MPa for σ_{LOP}^- and σ_{LOP}^+, respectively, were obtained. The post-LOP range (zone III) is characterized by a multiple cracking formation that can be represented by an exponential decay function. Figure 19.5a shows that the crack spacing initially drops abruptly from 180 to 60 mm up to the deflection of 10 mm. The crack spacing saturates at the end of zone IV (40 mm), which happens at a deflection of approximately 23 mm. Zone V is characterized by the strain-softening response due to the localization and widening of a major crack. No new cracks appear at this stage and crack spacing remains constant. The initial elastic modulus of 29.55 GPa shows that the use of calcined clays resulted in a matrix of sufficiently high stiffness.

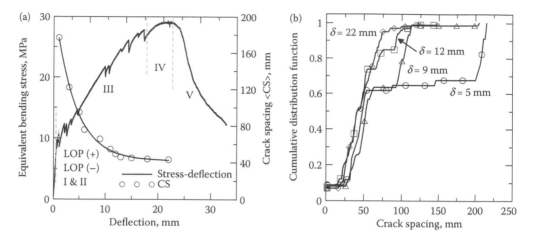

FIGURE 19.5 Bending response of the sisal fiber reinforced composite system: (a and b) bending stress and crack spacing versus displacement and (c) cumulative distribution function for crack spacing. (After Silva, F. A., Mobasher, B., and Filho, R. D. T., "Cracking mechanisms in durable sisal fiber reinforced cement composites," *Journal of Cement and Concrete Composites*, 31, 721–730, 2009, 10.1016/j.cemconcomp.2009.07.004.)

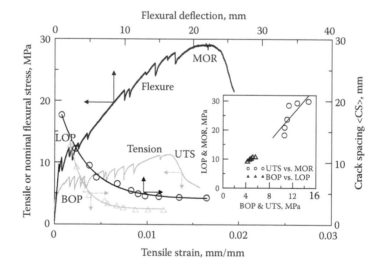

FIGURE 19.6 Comparison of flexural versus tensile response for sisal fiber composites. (After Silva, F. A., Mobasher, B., and Filho, R. D. T., "Cracking mechanisms in durable sisal fiber reinforced cement composites," *Journal of Cement and Concrete Composites*, 31, 721–730, 2009, 10.1016/j.cemconcomp.2009.07.004.)

The bending and tensile responses with their respective crack distribution are compared in Figure 19.6. The inset plot shows the relationship between LOP versus BOP and modulus of rupture versus ultimate tensile stress (UTS). It can be seen that under bending, loads associated with the formation of the first crack occur at stress levels twice as those observed for the direct tension tests. The correlation of tensile and bending test results has been documented through theoretical modeling in Chapter 14 [30]. Flexural results are affected by tension stiffening effects, and normalization of the flexural load with an elastic section modulus may result in apparent tensile strength, which is as high as 2.8 times the tensile strength. The values reported for modulus of rupture is approximately two times greater than that of the UTS. A procedure to theoretically validate this ratio has been shown for a variety of strain hardening cement composite systems is discussed in Chapter 15. Under flexural loads, the saturation crack spacing is twice as large as that of tensile loads.

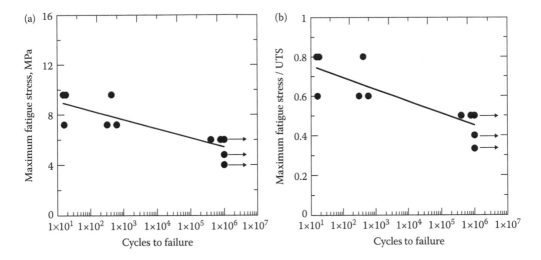

FIGURE 19.7 Stress versus cycles fatigue curve for composites subjected to maximum stress levels ranging from 4 to 9.8 MPa at constant R ratio of 0.2. Fatigue runout was taken at 10^6 cycles. The maximum stress was normalized by the UTS of the composites in panel b. (After Silva, F., Mobasher, B., Toledo Filho, R., "Fatigue behavior of sisal fiber reinforced cement composites," *Materials Science and Engineering A. Materials Science and Engineering*: A, 527, 21–22, 5507–5513, August 20, 2010.)

The analysis of tensile and compressive stress and the strain distribution are important because they provide an insight into the role that cracking plays in redistributing the forces applied onto the member. Flexural loads are carried by means of tensile cracking and redistribution of stresses. Simplifying assumptions that are based on uncracked section modulus leads to an equivalent elastic stress that is not conservative and significantly overestimates the materials' strength.

FATIGUE

Specimens were tested at maximum fatigue stress (MFS) level of 6 MPa up to 45,000, 200,000, and 10^6 cycles. These specimens were then tested under monotonic tensile load up to complete failure. Figure 19.7 shows the stress versus cycles behavior of the sisal reinforced cement composite tested at various maximum stresses (4–9.8 MPa). The composites survive 10^6 cycles up to 6 MPa, which represents 50% of the UTS. The stress level of 6 MPa can be considered a threshold limit where composites may present fatigue failure at cycles close to 10^6. Beyond 6 MPa, all the composites failed below reaching 10^3 cycles. It was observed that for high stress levels (i.e., >6 MPa), all the cracks are formed at the first cycles; however, the number of cracks are the same ($N = 12$) as the ones observed in monotonic tensile tests. After the crack saturation phase, one or more cracks start to widen. Fatigue cycles at these high stress levels degraded the fiber–matrix interface, which increased the rate of cracking opening leading to complete failure even at low cycles (i.e., <10^3).

Composites that survived 10^6 cycles were tested under monotonic tensile load, and the results are presented in Figure 19.8. When comparing the UTS from monotonic to postfatigue tensile tests, a slight decrease was observed. Nevertheless, this decrease lies in the standard variation range of the monotonic tests, and no significant variation among the postfatigue UTS for different stress levels was observed. Stiffness degradation is observed when calculating the modulus for the postfatigue tensile tests. Figure 19.8b shows that samples subjected to an MFS level of 4 MPa had a higher modulus than the one calculated for the monotonic tensile tests. Above 4 MPa, the modulus decreases from approximately 11.5 to 2 GPa (calculated from cross-head displacement data). The first crack strength was also computed for postfatigue tensile tests. The composites showed an increased first crack strength for higher MFS levels (see Figure 19.8b), indicating that the matrix is becoming stiffer after 10^6 cycles when increasing the fatigue stress levels but the interfacial transition zone is degrading.

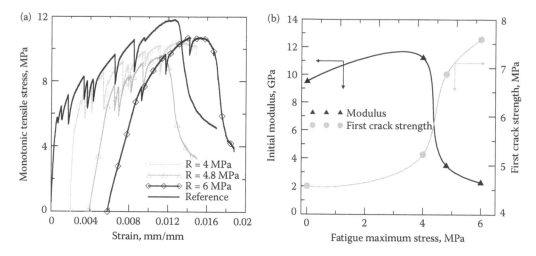

FIGURE 19.8 Monotonic tensile behavior of composites that have survived 10^6 cycles: (a) stress–strain curves of composites subjected to MFS of 4, 4.8, and 6 MPa, and (b) effect of the cycles on modulus of elasticity and first crack strength. (After Silva, F., Mobasher, B., Toledo Filho, R., "Fatigue behavior of sisal fiber reinforced cement composites," *Materials Science and Engineering A. Materials Science and Engineering*: A, 527, 21–22, 5507–5513, August 20, 2010.)

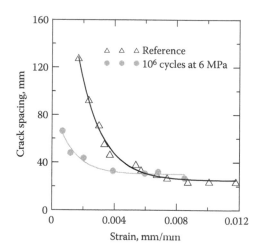

FIGURE 19.9 Effect of the cycles on the crack spacing of composites tested to monotonic tensile load after being subjected to MFS of 6 MPa for 10^6 cycles. (After Silva, F., Mobasher, B., Toledo Filho, R., "Fatigue behavior of sisal fiber reinforced cement composites," *Materials Science and Engineering A. Materials Science and Engineering*: A, 527, 21–22, 5507–5513, August 20, 2010.)

The crack spacing measured during a postfatigue monotonic tensile test on a sample that survived 10^6 cycles at a stress level of 6 MPa is shown in Figure 19.9, indicating a significant difference in the crack spacing behavior of a monotonic when compared with a postfatigue test. The decrease in crack spacing can be empirically represented as a function of three parameters S_0, S_1, and α and a threshold stress σ_{mu} (Equation 19.1), where S(ε_i) is the crack spacing as a function of strain, ε_{mu} is the average strain at the first cracking level σ_{mu} where the first set of measurements were obtained, ε_i is the independent parameter, S_0 and α are the constants representing specimen length and the rate of crack formation, and S_1 is the saturation crack spacing.

A decrease in S_0 and an increase in α parameter (see Table 19.1) is explained by the fact that fatigued specimens had all the cracks already initiated. Therefore, when the specimen was tested

TABLE 19.1
Crack Spacing versus Strain Equation for Monotonic Tension (Reference) and Postfatigue Monotonic Tension

	$S_1 + S_0 e^{-\alpha(\varepsilon_l - \varepsilon_{mu})}$			
Tests	S_1	S_0	α	ε_{mu} (mm/mm)
Reference	25.2	321.8	673.5	0.00155
10^6 cycles at 6 MPa	30.02	60.76	854.7	0.00066

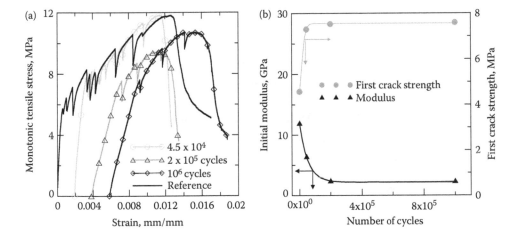

FIGURE 19.10 (a) Postfatigue monotonic tensile behavior of composites tested after a stress level of 6 MPa. The fatigue tests were stopped at 4.5×10^4, 2×10^5, and 10^6 cycles. (b) Effect of fatigue cycles on the modulus of elasticity and first crack strength. (After Silva, F., Mobasher, B., Toledo Filho, R., "Fatigue behavior of sisal fiber reinforced cement composites," *Materials Science and Engineering A. Materials Science and Engineering*: A, 527, 21–22, 5507–5513, August 20, 2010.)

under monotonic load after the fatigue, the majority of cracks appeared at low strain levels although the composite asymptotically approached the same saturation crack spacing of 20 mm.

Test results shown in Figure 19.10 indicate a decrease in the UTS for the postfatigue specimens. This decrease is in the error range of the nonfatigue UTS. A decrease in the modulus from 12 to 2 GPa is observed (see Figure 19.10b). Although a slight increase in the first crack strength is observed when comparing samples fatigued to 45000 and 10^6 cycles, significantly higher values are observed if compared with monotonic tensile tests.

Stress–strain hysteresis measurements were conducted at various stress levels as shown in Figure 19.11 using composites that survived 10^6 cycles at maximum stress levels of 4, 4.8, and 6 MPa. For the maximum stress level of 4 MPa, the individual hysteresis loops are shown and indicate a measure of inelastic damage or energy during a given cycle. No stiffness degradation was observed for this stress level. When increasing the MFS to 4.8 and 8 MP, an increase in the bounded area of the hysteresis loops, which is due to the widening of several cracks in the late stages is observed (Table 19.2).

Stiffness degradation and increment in strain are presented in Figure 19.12. Higher degradation was observed for the composites cycled at a stress of 6 MPa such that at 10^6 cycles, the maximum strain was 0.8% and the Young's modulus was 4.2 GPa. At the 4.8-MPa stress level, a maximum strain of 0.23% and a modulus of 2 GPa at the 10^6 cycles was observed. Four stages in the fatigue cycles can be observed and classified:

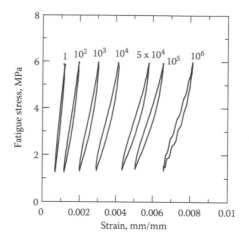

FIGURE 19.11 Hysteresis behavior of composites up to 10^6 cycles at stress level of 4 MPa.

TABLE 19.2
Summary of Fatigue and Postfatigue Monotonic Tensile Tests

Max Fatigue Stress (MPa)	Postfatigue Monotonic Tensile Test			Dynamic Modulus (GPa)					
	UTS (MPa)	First Crack Strength (MPa)	Young's Modulus (Gpa)	Cycle 1	Cycle 10^2	Cycle 10^3	Cycle 10^4	Cycle 10^5	Cycle 10^6
4	10.39	5.22	11.22	16.81 (46.61)	–	–	–	–	16.69 (44.37)
4.8	9.51	6.87	3.47	16.30	15.52	10.2	9.92	6.31	5.13
6	10.72	7.59	2.26	8.70	5.60	3.21	2.55	2.12	1.96
7.2	–	–	–	3.11	2.08	–	–	–	–
9.8	–	–	–	2.62	–	–	–	–	–

Note: Results in parentheses were computed from strain gage measurements.

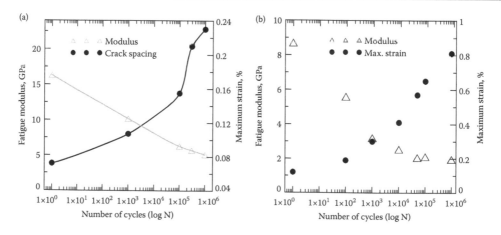

FIGURE 19.12 Effect of the cycles on the fatigue modulus and maximum strain of composites subjected to maximum stresses of (a) 4.8 MPa and (b) 6 MPa. (After Silva, F., Mobasher, B., Toledo Filho, R., "Fatigue behavior of sisal fiber reinforced cement composites," *Materials Science and Engineering A. Materials Science and Engineering*: A, 527, 21–22, 5507–5513, August 20, 2010.)

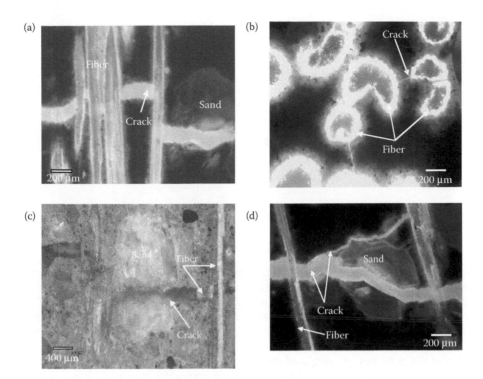

FIGURE 19.13 Cracks in composites subjected to 10^6 cycles at maximum fatigue level of 6 MPa observed using fluorescent and conventional optical microscopy. The higher contrast in the fluorescent optical microscopy allows the visualization of small cracks (<20μm). (After Silva, F., Mobasher, B., Toledo Filho, R., "Fatigue behavior of sisal fiber reinforced cement composites," *Materials Science and Engineering A. Materials Science and Engineering*: A, 527, 21–22, 5507–5513, August 20, 2010.)

  (i) MFS ≤4 MPa—no fatigue and no damage to the material
 (ii) 4 MPa < MFS < 6 MPa—no fatigue with moderate damage to the material
(iii) MFS = 6 MPa—fatigue at high cycles (10^6) or no fatigue with high damage to the material
(iv) MFS > 6 MPa—fatigue at low cycles (i.e., <1000 cycles)

Samples that survived 10^6 cycles at a stress level of 6 MPa were investigated using fluorescent optical microscopy. Figure 19.13 shows the capacity of the fibers to arrest and bridge the cracks formed during fatigue cycles in lateral (Figures 19.13a, 19.13c, and 19.13d) and front (Figure 19.13b) cross-sectional views. This behavior attests the high efficiency in the fiber matrix bond adhesion of the composite system even when subjected to 10^6 cycles at a maximum stress of 6 MPa. Two ranges of crack widths were observed in the micrographs: (i) from 1 to 20 μm and (ii) from 150 to 200 μm at a deformation of 0.4%. The use of fluorescent dye opens possibilities for the visualization of microcracks with width less than 20 μm, which was not observed in conventional optical microscopy. The enhancement in contrast was also achieved with the fluorescent microscopy. Only the cracks, voids, and fibers are green, whereas the matrix is black.

FIBER MATRIX PULLOUT BEHAVIOR

A typical pullout force–slip curve is shown in Figure 19.14a. Four distinct regions are indentified by roman numerals. Region I corresponds to the elastic-linear range with a rapid increasing of the load. As the load increases beyond the linear region, a certain degree of nonlinearity is observed. Region II starts at this nonlinearity and defines the initial point of fiber debonding. The peak response is reached at region III where the pullout force reaches a maximum value (P_{au}). The slip

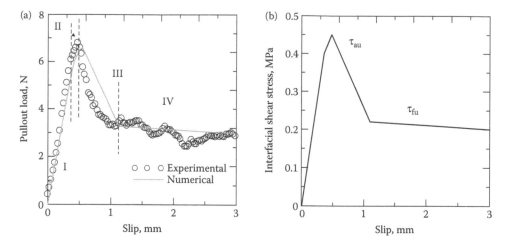

FIGURE 19.14 Fiber pullout test results of a sample tested at 3 days of curing with an embedded length of 20 mm. (a) Comparison of the experimental and numerical result from the finite difference model. (b) Interface constitutive relation obtained from the model. (After Flávio de Andrade, S., Mobasher, B., Soranakom, C., and Toledo Filho, R., "Effect of fiber shape and morphology on the interface mechanical characteristics in sisal fiber cement based composites," in press, *Journal of Cement and Concrete Composites*, 2011, 10.1016/j.cemconcomp.2011.05.003.)

of the fiber at the peak can be correlated with the critical length of the debonded fiber. The peak pullout force depends on the embedded length, the diameter of the fiber, and the curing period. The shear strength computed at P_{au} is defined as the adhesional strength (τ_{au}). In region IV, the load drops to a fixed value after which it remains quite constant. The immediate postpeak region is governed by the shear strength of the fiber and continues till the fiber is completely debonded from the interface. The shear strength at this region is defined as frictional resistant strength (τ_{fu}). The adhesional and frictional components were quantified in terms of stress τ_{au} and τ_{fu}, respectively, and can be calculated using either the model described in Chapter 8 on pullout modeling or a simplified shear strength approach as shown in Equation 19.2:

$$\tau = \frac{P}{2\pi r l} \tag{19.2}$$

where τ is the interface bond strength, P is the pullout force, r is the fiber radius, and l is the embedded length.

Alternatively, one can use Soranakom and Mobasher's [31] finite difference model. In this model, the interfacial bond constitutive law and the tensile stress–strain response of the fiber are described as piecewise linear functions. For the straight fiber, the restraint at the cross junction of disparities along the fiber length that would serve as anchor points was not considered.

The model equilibrium equations are derived from free body diagrams of the nodes as defined in Figure 19.15. The equations are expressed as coefficient and the unknown variable slip s, defined as the relative difference between the elongation of the longitudinal yarn and matrix,

$$s = \int_{x_i}^{x_{i+1}} \left(\varepsilon_y - \varepsilon_m \right) dx \tag{19.3}$$

where ε_y and ε_m are fiber and matrix strains, respectively, and dx is a finite length between two consecutive nodes i and $i + 1$ along the longitudinal x-axis. For typical low fiber volume fraction, the axial stiffness of the yarn $A_y E_y$ is considerably lower than the matrix term $A_m E_m$, and the contribution of matrix elongation to slip is ignored. Thus, the slip s and the fiber strain σ_y are simplified to

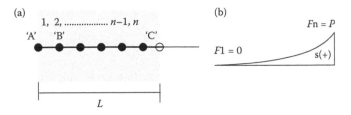

FIGURE 19.15 Schematics of the finite difference pullout model.

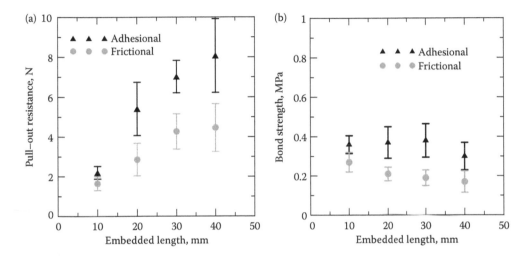

FIGURE 19.16 Influence of embedded length on the fiber–matrix interfacial bond strength: (a) pullout resistance increase with embedded length and (b) bond strength versus embedded length. (After Flávio de Andrade, S., Mobasher, B., Soranakom, C., and Toledo Filho, R., "Effect of fiber shape and morphology on the interface mechanical characteristics in sisal fiber cement based composites," in press, *Journal of Cement and Concrete Composites*, 2011, 10.1016/j.cemconcomp.2011.05.003.)

$$s = \int_{x_i}^{x_{i+1}} \varepsilon_y \, dx \quad \text{and} \quad \varepsilon_y = s' = \frac{ds}{dx} \tag{19.4}$$

The embedded length L is discretized into n nodes with equal spacing of h. The bond stress is assumed constant over the small spacing h for each node within each linear domain. At the left end, the force in the fiber is imposed to be zero, simulating stress free condition, implying that the fiber strain or derivative of slip vanishes. At the right end, the nodal slip is prescribed incrementally, simulating displacement control. As the loading progresses, the part of the fiber that slips out of the matrix has no frictional bond resistance; thus, fiber elongation is the only term in that section. The extruding part can be easily implemented by checking the amount of slip versus the embedded length of each node [31]. The model has shown good correlation with experimental sisal fiber pullout behavior as can be seen in Figure 19.14a. The interfacial constitutive relation was determined from the model and presented in Figure 19.14b. The piecewise straight values of the adhesional and frictional bond strength are shown.

Pullout tests were carried out on embedded lengths of 10, 20, 30, and 40 mm after 3 days of curing and simulated using the model as shown in Figures 19.16a and 19.16b. Increasing the embedded length increases the pullout force from approximately 2 to 8 N. At an embedded length of 40 mm, no significant increase was observed in the pullout force. The standard deviation of

TABLE 19.3
Summary of the Results for Pullout Samples Tested with Different Embedded Lengths

Embedded Length (mm)	Type of Fiber	P_{au} (N)	τ_{au} (MPa)	P_{au} (N)	τ_{fu} (MPa)	P_{au} Average (N)	τ_{au} Average (MPa)	P_{fu} Average (N)	τ_{fu} Average (MPa)
10	(i)	2.23	0.37	1.38	0.23	2.2 (0.64)	0.36 (0.09)	1.65 (0.69)	0.27 (0.10)
	(ii)	2.23	0.35	1.88	0.30				
	(iii)	–	–	–	–				
20	(i)	4.41	0.31	3.07	0.22	5.41 (2.67)	0.37 (0.16)	2.86 (1.64)	0.21 (0.07)
	(ii)	5.29	0.36	2.21	0.17				
	(iii)	7.06	0.45	3.21	0.23				
30	(i)	6.00	0.40	2.99	0.14	7.02 (1.63)	0.38 (0.17)	4.27 (1.77)	0.19 (0.08)
	(ii)	7.97	0.36	6.32	0.29				
	(iii)	7.14	0.38	4.03	0.15				
40	(i)	6.5	0.25	3.39	0.13	8.07 (3.7)	0.30 (0.14)	4.46 (2.4)	0.17 (0.11)
	(ii)	7.53	0.30	3.23	0.12				
	(iii)	9.49	0.34	6.00	0.23				

Note: Values in parentheses are standard deviation.

the fibers tested at 30 mm was in the range as of those tested with 40 mm. The adhesional bond strength computed on the basis of the pullout force shows a constant value for the embedded lengths used (see Table 19.3). This value can thus be used as a material property because its magnitude is independent of geometry. A slight decrease was observed for frictional bond strengths by increasing the embedded length as found in the literature for polypropylene and steel fibers. Singh et al. [32] studied the pullout behavior of polypropylene fibers from cementitious matrix for embedded lengths of 19, 25, and 38 mm and reported that while the pullout force increases with embedded length, the adhesional bond strength was almost constant around 0.5 MPa. Shannag et al. [33] investigated steel fibers at embedded lengths of 6, 12, and 18 mm and reported a significant increase in the pullout adhesional load. The frictional bond strengths for densified small particle (4.4 MPa) and conventional mortar matrices (1.4 MPa) showed no improvement with increasing embedded length.

The effects of curing time for ages from 3 to 28 days are presented in Figure 19.17. The bond strength reaches its maximum capacity at 14 days and further curing of the matrix shows no effect on ages of 21 and 28 days. Average adhesional bond strength after 15 days ranged from 0.59 to 0.67 MPa. The largest variability is observed at 14 days from 0.35 to 1.29 MPa. Frictional bond strength has less scatter with 0.37 to 0.44 MPa after 14 days. The fiber–matrix bond strength results are in the range of some synthetic fibers such as carbon fibers, with mean adhesional bond strength for diameters of 10 and 46 μm ranging from 0.52 to 1.29 MPa and from 0.39 to 3.02 MPa, respectively [34]. Polypropylene fibers have an adhesional bond strength of 0.5 MPa [33].

Pullout test results can be separated in terms of the fiber morphology. Sisal fibers are naturally twisting and are not circular in shape, thus enabling different mechanical components of bond. They can be divided into three types of bond mechanical components: (i) horseshoe shape—representing the majority of the fibers in the sisal plant leaf with a small areas; (ii) arch shape—have larger areas and less abundant than the horseshoe shape; and (iii) twisted arch shape—resulting from fiber extraction process. These different morphologies make a difference in the bond strength, with fiber type iii having the highest values [35] as shown in Table 19.4. Fiber type ii shows, especially in the curing age investigation, a tendency of presenting higher values than type i. The influence of bond strength on the sisal fiber critical length was computed using Equation 19.5:

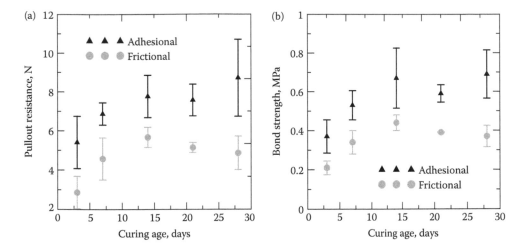

FIGURE 19.17 Influence of curing age on the fiber–matrix interfacial bond strength. After 14 days, there is no further increase in (a) pullout resistance and (b) bond strength. (After Flávio de Andrade, S., Mobasher, B., Soranakom, C., and Toledo Filho, R., "Effect of fiber shape and morphology on the interface mechanical characteristics in sisal fiber cement based composites," in press, *Journal of Cement and Concrete Composites*, 2011, 10.1016/j.cemconcomp.2011.05.003.)

TABLE 19.4
Summary of the Results for Pullout Samples Tested at Different Curing Time

Age (days)	Type of Fiber	P_{au} (N)	τ_{au} (MPa)	P_{fu} (N)	τ_{fu} (MPa)	P_{au} Average (MPa)	τ_{au} Average (MPa)	P_{fu} Average (MPa)	τ_{fu} Average (MPa)
3	(i)	4.41	0.31	3.07	0.22	5.41 (2.67)	0.37 (0.16)	2.86 (1.64)	0.21 (0.07)
	(ii)	5.29	0.36	2.21	0.17				
	(iii)	7.06	0.45	3.21	0.23				
7	(i)	5.46	0.47	3.38	0.29	6.87 (1.14)	0.52 (0.06)	4.57 (2.14)	0.34 (0.12)
	(ii)	8.34	0.59	7.04	0.48				
	(iii)	6.82	0.51	3.28	0.24				
14	(i)	6.64	0.52	5.06	0.35	7.77 (2.18)	0.62 (0.12)	5.67 (1.03)	0.44 (0.08)
	(ii)	6.39	0.58	5.10	0.46				
	(iii)	10.28	0.75	6.86	0.50				
21	(i)	–	–	–	–	7.58 (1.64)	0.59 (0.09)	5.14 (0.51)	0.39 (0.005)
	(ii)	7.06	0.51	5.51	0.40				
	(iii)	7.72	0.61	4.78	0.39				
28	(i)	5.54	0.48	2.31	0.2	8.72 (3.97)	0.67 (0.25)	4.86 (1.71)	0.37 (0.11)
	(ii)	7.01	0.56	5.57	0.46				
	(iii)	12.76	0.92	5.79	0.42				

Note: Values in parentheses are standard deviation.

$$L_c = \frac{r\sigma_f}{\tau_{fu}} \tag{19.5}$$

where L_c is the fiber critical length, r is the fiber radius, σ_f is the fiber ultimate tensile strength, and τ_{fu} is the fiber frictional bond strength. Figure 19.18 shows the influence of bond strength on the critical fiber length using a fiber radius of 0.1 mm and various fiber UTS of 100 to 400 MPa. For a

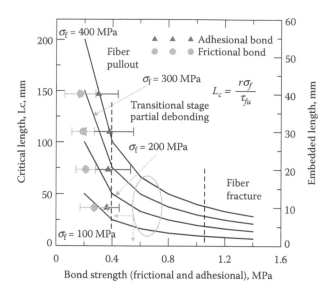

FIGURE 19.18 Influence of bond strength on the fiber critical length. (After Flávio de Andrade, S., Mobasher, B., Soranakom, C., and Toledo Filho, R., "Effect of fiber shape and morphology on the interface mechanical characteristics in sisal fiber cement based composites," in press, *Journal of Cement and Concrete Composites*, 2011, 10.1016/j.cemconcomp.2011.05.003.)

frictional average bond strength selected from curing ages more than 14 days (around 0.4 MPa), a critical length L_c of approximately 100 mm is obtained.

TENSION STIFFENING MODEL

The finite difference tension stiffening model developed by Soranakom and Mobasher [36, 37] simulates the crack spacing and stress–strain response under static and dynamic loads. In this model, a cracked tension specimen is idealized as a series of one-dimensional segments consisting of fiber, matrix, and interface elements. The matrix is treated as brittle with no strain-softening response. As the load on the composite increases such that the cracking stress of the matrix is reached, cracked planes form sequentially while the load at the cracked planes is solely carried by the longitudinal fibers by means of interface elements. The individual pullout segments continue to carry load at cracked locations. In nonlinear analysis, an iterative solution algorithm is used to enforce load deformations to follow the material constitutive laws. Once the slip distributions are solved and corresponding stress and strain responses are identified, results are added to represent the overall tensile response. This approach has been used to simulate the response of high speed tension tests discussed in Chapter 17 and also static tension tests discussed in Chapters 7, 10, and 11 using different fiber–matrix interface models as shown in Figure 19.19a. The static model has a high bond strength and a low slip range, whereas the dynamic model presents elastic-plastic frictional shear with a longer slip range. The material parameters are held constant for all the simulations and are described as follows: fiber modulus = 19 GPa, fiber UTS = 400 MPa, matrix modulus = 35 GPa, and Young's modulus efficiency factor of 0.6.

The simulation is done under load control, and the procedure can be summarized as follows:

- The length of specimen is discretized into N nodes with equal spacing. Nodes are divided into two groups, those in the end grip zone which are not allowed to crack and those within the clear length.
- Both deterministic and stochastic crack patterns can be used. Either uniform matrix strengths with predetermined sequential cracking locations can be specified. Alternatively,

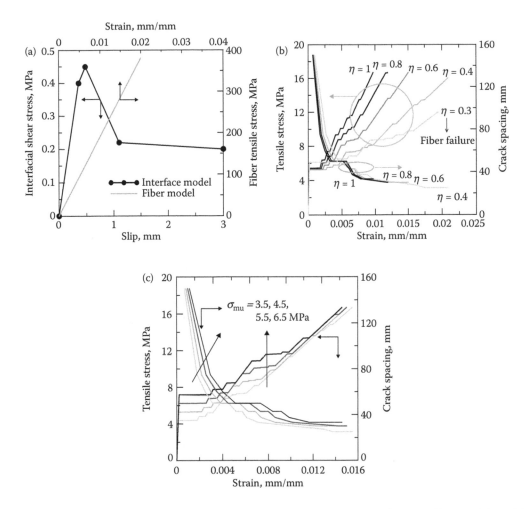

FIGURE 19.19 Prediction of tensile response and crack spacing: (a) interface and fiber model used in the finite difference simulation, (b) effect of efficiency factor of fiber modulus and strength, and (c) effect of matrix first crack strength. (After Flávio de Andrade, S., Mobasher, B., Soranakom, C., and Toledo Filho, R., "Effect of fiber shape and morphology on the interface mechanical characteristics in sisal fiber cement based composites," in press, *Journal of Cement and Concrete Composites*, 2011, 10.1016/j.cemconcomp.2011.05.003.)

random matrix strengths at nodes along the length of specimen are generated and used as cracking criterion. The cracking criterion in the present research was random matrix strengths. A fiber efficiency factor is introduced to take into account the imperfect nature of the bond and inability of the fiber stiffness to be used within the test.

- Although the applied load is less than the first cracking limit, tensile response is calculated by the rule of mixtures and strain compatibility.
- When the first crack appears, the section is divided into two parts and each segment is modeled as a pullout problem and solved independently. As the load increases, additional cracks form at locations where the strength criterion is satisfied. The cracked specimen is represented by a number of pullout segments; each is solved independently, and the solutions are combined to represent the entire specimen.
- The analysis is terminated if the stress in yarn reaches its ultimate tensile strength or a solution is not found due to slip instability (very large slip value).

It is observed from Figure 19.19b that by decreasing the values of efficiency of the interfacial bond η, the total strain of the composite increases up to a point that the fiber starts to fracture. Fiber failure occurs when η equals 0.3. An efficiency factor of 0.6 corresponded with the ultimate strain of approximately 1.5% (close to experimental results). The crack spacing continues to extend to smaller values because of significant debonding and slip when the efficiency factor η decreases.

The influence of the matrix first crack strength on the tensile and crack spacing is shown in Figure 19.19c. An increase in ultimate strain and an increase in crack spacing when increasing the matrix first crack strength ranging from 3.5 to 6.5 MPa are observed. No effect on the UTS is observed. The influence of the fiber–matrix interfacial model was studied. A model that is presented in Figures 19.35a and 19.35b obtained from a pullout test performed at 21 days was used. This specific test resulted in higher adhesional and frictional bond strengths than the one previously used. It was noticed that there was a slight decrease in ultimate strain and no effect on crack spacing and UTS when using an interfacial model with higher bond strengths.

The predicted crack spacing response obtained from the tension stiffening model shows a good correlation with the experimental calculation, as the model accurately predicts the saturated crack spacing. The predicted stress–strain response also shows a good fit with experimental values therefore validating the used model.

Figure 19.20 shows the comparison between experimental and predicted tension and cracking spacing behavior. It used an efficiency factor $\eta = 0.6$, a matrix first crack strength of 7 MPa, and the interfacial bond model presented in Figure 19.19a. An accurate correlation with the strain gage response is seen in the linear elastic region up to the first crack formation (see Figure 19.20b). In the beginning of the multiple cracking zone, the model correlates well with a lower-bound experimental curve up to a strain of 0.4%. After that point, it fits the upper bound experimental curve overestimating the UTS. The numerical crack spacing shows an accurate prediction up to a strain value of 0.005%. At the region of crack saturation, a crack spacing of approximately 30 mm is obtained using the model and 23 mm for experiment. The model has satisfactory predicted the crack spacing, and tensile behavior and has been further extended to computation of tension stiffening effect [37].

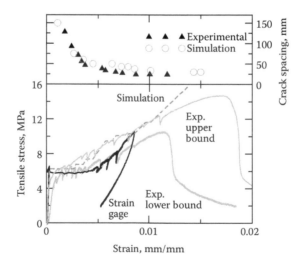

FIGURE 19.20 Comparison of experimental and numerical results of the composite tensile response and crack spacing. (a) Two experimental tensile curves are shown: a lower and upper bound. (b) Numerical simulation shows a perfect fit with the strain gage response in the linear elastic range up to the first crack formation. (After Flávio de Andrade, S., Mobasher, B., Soranakom, C., and Toledo Filho, R., "Effect of fiber shape and morphology on the interface mechanical characteristics in sisal fiber cement based composites," in press, *Journal of Cement and Concrete Composites*, 2011, 10.1016/j.cemconcomp.2011.05.003.)

REFERENCES

1. Swift, D. F., and Smith, R. B. L. (1979), "The flexural strength of cement-based composites using low modulus (sisal) fibers," *Composites*, 6(3), 145–148.
2. Coutts, R. S. P., and Warden, P. G. (1992), "Sisal pulp reinforced cement mortar," *Cement and Concrete Composites*, 14(1):17–21.
3. Toledo Filho, R. D., Ghavami, K., England, G. L., and Scrivener, K. (2003), "Development of vegetable fiber-mortar composites of improved durability," *Cement and Concrete Composites*, 25(2), 185–196.
4. Kelly, A. (1966), *Strong Solids*, Oxford: Clarendon Press.
5. Toledo Filho, R. D., Scrivener, K., England, G. L., and Ghavami, K. (2000), "Durability of alkali-sensitive sisal and coconut fibers in cement mortar composites," *Cement and Concrete Composites*, 22(2), 127–143.
6. Silva, F. A., Ghavami, K., and d'Almeida, J. R. M., "Toughness of cementitious composites reinforced by randomly sisal pulps," *Eleventh International Conference on Composites Engineering*, Hilton Head Island, SC, 2004.
7. Silva, F. A., Ghavami, K., and d'Almeida, J. R. M., "Bamboo–Wollastonite hybrid cementitious composites: Toughness evaluation," *Joint ASME/ASCE/SES Conference on Mechanics and Materials*, Baton Rouge, 2005.
8. Silva, F. A., Ghavami, K., and d'Almeida, J. R. M., "Behavior of CRBP-AL composites subjected to impact load," *17th ASCE Engineering Mechanics Conference*, Delaware, 2004.
9. Mohr, B. J., Nanko, H., and Kurtis, K. E. (2005), "Durability of thermomechanical pulp fiber–cement composites to wet/dry cycling," *Cement and Concrete Research*, 35, 1646–1649.
10. Mohr, B. J., Nanko, H., and Kurtis, K. E. (2005), "Durability of kraft pulp fiber-cement composites to wet/dry cycling," *Cement and Concrete Composites*, 27, 435–448.
11. Bentur, A., and Akers, S. A. S. (1989), "The microstructure and ageing of cellulose fibre reinforced cement composites cured in a normal environment," *International Journal of Cement Composites and Lightweight Concrete*, 11(2), 99–109.
12. Gram, H.-E. (1988), "Durability of natural fibres in concrete," in R. N. Swamy (ed.), *Natural Fibre Reinforced Cement and Concrete*, Blackie and Sons Ltd. Glasgow, 143–172.
13. Gram, H.-E. (1983), *Durability of Natural Fibres in Concrete*, Stockholm: Swedish Cement and Concrete Research Institute, 1–255.
14. Bergstrom, S. G., and Gram, H.-E. (1984), "Durability of alkali-sensitive fibres in concrete," *International Journal of Cement Composites and Lightweight concrete*, 6(2), 75–80.
15. John, V. M., Cincotto, M. A., Sjostrom, C., Agopyan, V., and Oliveira, C. T. A. (2005), "Durability of slag mortar reinforced with coconut fibre," *Cement and Concrete Composites*, 27, 565–574.
16. Canovas, M. F., Selva, N. H., and Kawiche, G. M. (1992), "New economical solutions for improvement of durability of portland cement mortars reinforced with sisal fibers," *Materials and Structures*, 25, 417–422.
17. Silva, F. A., Melo Filho, J. A., Toledo Filho, R. D., and Fairbairn, E. M. R., "Mechanical behavior and durability of compression moulded sisal fiber cement mortar laminates (SFCML)," *Proceedings of the 1st International RILEM Conference on Textile Reinforced Concrete (ICTRC)*, Aachen, 2006, 171–180.
18. Toledo Filho, R. D., Silva, F. A., Fairbairn, E. M. R., and Melo Filho, J. A. (2009), "Durability of compression molded sisal fiber reinforced mortar laminates," *Construction and Building Materials*, 23, 2409–2420.
19. Silva, F. A., Chawla, N., and Toledo Filho, R. D. (2008), "Tensile behavior of high performance natural (sisal) fibers," *Composites Science and Technology*, 68, 3438–3443.
20. Mukherjee, K. G., and Satyanarayana, K. G. (1984), "Structure and properties of some vegetable fibres. Part 1: Sisal fibre," *Journal of Materials Science*, 19, 3925–3934.
21. Silva, F. A., Mobasher, B., and Filho, R. D. T. (2009), "Cracking mechanisms in durable sisal fiber reinforced cement composites," *Journal of Cement and Concrete Composites*, 31, 721–730.
22. Silva, F. A., Zhu, D., Mobasher, B., Soranakom, C., and Toledo Filho, R. D. (2010), "High speed tensile behavior of sisal fiber cement composites," *Materials Science and Engineering. A,* 527, 544–552.
23. Silva, F. A., Zhu, D., Mobasher, B., and Toledo Filho, R. D. (2011), "Impact behavior of sisal fiber cement composites under flexural load," *ACI Materials Journal*, 108(2), 168–177.
24. Toledo Filho, R. D., Silva, F. A., Fairbairn, E. M. R., and Melo Filho, J. A. (2010), "Physical and mechanical properties of durable sisal fiber–cement composites," *Construction and Building Materials*, 24(5), 777–785.

25. Silva, F. A., Chawla, N., and Toledo Filho, R. D. (2009), "Physical and mechanical properties of durable sisal fiber–cement composites," *Materials and Science Engineering. A*, 516, 90–95.
26. Mobasher, B., Castro-Montero, A., and Shah, S. P. (1990), "A study of fracture in fiber reinforced cement-based composites using laser holographic interferometry," *Experimental Mechanics*, 30, 286–294.
27. Mobasher, B., Peled, A., and Pahilajani, J. (2006), "Distributed cracking and stiffness degradation in fabric-cement composites," *Materials and Structures*, 39, 317–331.
28. Mobasher, B., Peled, A., and Pahilajani, J. "Pultrusion of fabric reinforced high flyash blended cement composites," *Proceedings of the RILEM Technical Meeting*, BEFIB, 2004, 1473–1482.
29. Mobasher, B., Pahilajani, J., and Peled, A. (2006), "Analytical simulation of tensile response of fabric reinforced cement based composites," *Cement and Concrete Composites*, 28(1), 77–89.
30. Soranakom, C., and Mobasher, B. (2008), "Correlation of tensile and flexural responses of strain softening and strain hardening cement composites," *Cement and Concrete Composites*, 30, 465–477.
31. Soranakom, C., and Mobasher, B. (2008), "Geometrical and mechanical aspects of fabric bonding and pullout in cement composites," *Materials and Structures*, 42, 765–777, DOI 10.1617/s11527-008-9422-6.
32. Singh, S., Shukla, A., and Brown, R. (2004), "Pullout behavior of polypropylene fibers from cementitious matrix," *Cement and Concrete Research*, 34, 1919–1925.
33. Shannag, M. J., Brincker, R., and Hansen, W. (1997), "Pullout behavior of steel fibers from cement-based composites," *Cement and Concrete Research*, 27, 925–936.
34. Katz, A., Li, V. C., and Kazmer, A. (1995), "Bond properties of carbon fibers in cementitious matrix," *Journal of Materials in Civil Engineering*, 7, 125–128.
35. Soranakom, C., and Mobasher, B. (2009), "Geometrical and mechanical aspects of fabric bonding and pullout in cement composites," *Materials and Structures*, 42, 765–777, DOI 10.1617/s11527-008-9422-6.
36. Soranakom, C., and Mobasher, B. (2010), "Modeling of tension stiffening in reinforced cement composites: Part I: Theoretical modeling, *Materials and Structures*, 43, 1217–1230, DOI 10.1617/s11527-010-9594-8, 2010.
37. Soranakom, C. (2008), "Multi-scale modeling of fiber and fabric reinforced cement based composites," PhD dissertation, Arizona State University, Tempe, AZ.

20 Restrained Shrinkage Cracking

INTRODUCTION

This chapter deals with the role of fibers in extending the cracking resistance of concrete subjected to drying shrinkage. In hot and low-humidity environments, concrete shrinks due to loss of moisture from capillary and gel pore microstructure. When concrete is restrained from free shrinkage, tensile stresses develop and may result in cracking if the stress exceeds materials' low tensile strength. This is more dominant when at an early age the tensile strength is quite low and rate of moisture evaporation is high. The main objective of this chapter is to study effect of fibers on responses of a well-known restrained shrinkage test.

The methodologies of restrained drying shrinkage tests are described. Materials and mixture properties of different concrete samples including plain and fiber reinforced concrete (FRC) samples are investigated experimentally. The experiments are performed according to the ring-type restrained shrinkage test method and strain history in the steel ring is monitored. A systematic methodology based on image analysis approach is used to measure crack width growth in concrete ring specimen. An analytical approach that relates key influential parameters of modeling including diffusion, shrinkage, creep, aging material properties, and restraining effect is presented.

REVIEW OF DRYING SHRINKAGE TESTING METHODS

Understanding the mechanism of shrinkage cracking is essential to design of durable structures. Plastic shrinkage occurs during the early age period when the strength of the paste is quite low, and drying volume changes occur because of depletion of water due to evaporation and chemical reactions in addition to chemical shrinkage. Although the fibers may not affect the evaporation rate, their addition increases the strength and strain capacity sufficiently during the early ages so that the potential for tensile cracking is minimized. Fiber addition to concrete therefore reduces cracking potential due to restrained shrinkage. As the concrete hardens, high stiffness fibers such as steel and macrosynthetic fibers serve to increase the strength, crack growth resistance, and strain carrying capacity, providing a mechanism for additional restraint to distribute volumetric shrinkage and maintain a low crack width.

Different test methods can be used to evaluate the plastic and restrained shrinkage cracking behavior of mortar and concrete. Bentur and Kovler [1] present four categories of experimental tests: ring, plate, longitudinal, and substrate restraint. Free drying shrinkage test conducted in accordance to ASTM C 157 [2] recommends a prismatic specimen of 25, 75, or 100 mm^2 cross section and 285 mm in length. Assuming that the length of the specimen is much larger than the cross-sectional dimensions, shrinkage measurement is attributed to length change only. In a free shrinkage test, no tensile stress and consequently no cracks develop in the specimen. Therefore, this test method does not differentiate the contribution of fibers in increasing the resistance of concrete to cracking.

Two tests have been standardized by ASTM to evaluate early age cracking. The first test method (ASTM C-1579) developed on the basis of the work of Berke and Dalliare [3] focuses on assessing the plastic shrinkage cracking [4] using a stress riser to simulate the effect of evaporation, autogenous shrinkage, and settlement. The stress riser focuses the crack in a small region in the center of the sample and simulates the conditions that occur above a reinforcing bar in practice [5–7]. The second test method (ASTM C 1581) [8] uses the restrained ring geometry to describe restrained drying shrinkage cracking behavior of hardened concrete. This approach has been used for nearly a century [9] and also adopted using a provisional AASHTO method [10].

Restraint provided along the length direction generates tensile stresses due to compatibility conditions. Interpretation of the results of linear specimens is rather straightforward; however, it is difficult to provide sufficient restraint when cross-sectional dimensions are large [11]. Additionally, processes such as localization of deformation and rotation of the uniaxial specimen due to unsymmetric crack growth make it difficult to maintain a uniaxial tensile conditions [2]. Paillere et al. [12] used long specimens (1.5 m) with small cross-sectional dimensions (70 × 100 mm) and flared restrained ends. Other researchers have used similar methods on linear specimens to assess the shrinkage cracking [13–16]. Because of difficulties associated with providing sufficient end restraint, these test methods however are generally not used for quality control procedures [6, 17].

Plate-type specimens have been used to simulate cracking due to restrained shrinkage by Kraai [18], Shaeles and Hover [19], Opsahl and Kvam [20], and Padron and Zollo [21]. When restraint is provided in two directions, a biaxial state of stress is produced. Restrained shrinkage test using a steel ring was done as early as 1942 by Carlson and Reading [22]. As a result of drying, a concrete ring shrinks, but the steel ring prevents the radial contraction and generates tensile stresses in concrete. More recently, instrumented rings have been used to measure the magnitude of tensile stresses [2, 23–27]. Because of its simplicity and economy, the ring test has been developed into both AASHTO [28] and ASTM [29] standards. The main difference between these standards is the relative ratio of the concrete to steel ring thickness, which influences the degree of restraint provided to the concrete.

PLASTIC SHRINKAGE CRACKING

The effects of fibers on plastic shrinkage cracking behavior is controlled by the fiber dispersion throughout the matrix [4, 6]. Fine microfibers with a high-fiber surface area are particularly effective in reducing plastic shrinkage cracking. Najm and Balaguru [30] studied the effect of polymeric fibers on the reduction of crack width caused by plastic and drying shrinkage. Image analysis has been used to characterize the size of the plastic shrinkage cracks [4, 5].

RESTRAINED SHRINKAGE CRACKING

Uniaxial restrained shrinkage tests and tensile tests conducted on large-scale steel FRC specimens with fiber contents ranging from 0 to 100 kg/m³ indicate that multiple cracking affects the overall response of the steel FRC in the hardened state [31]. However, the ring test is unable to capture the effect of multiple cracking simply because as a single crack forms the boundary conditions and the wall change. Several approaches to evaluate the influence of ring geometry and drying direction on the behavior of the restrain shrinkage test using the steel ring specimen have been proposed [32–34]. These studies demonstrate the use of the steel ring in measuring the residual stress development as well as the stress relaxation once cracking occurs. Kraai [18] developed an experimental and analytical simulation algorithm to study the restrained shrinkage cracking in plain and FRC. A constant humidity chamber holding the restrained shrinkage specimens was used with a fan providing constant flow of air around the specimens. The strain in the restraining steel and the crack width in the concrete samples were monitored continuously. The results are correlated with the specimen geometry, humidity and temperature conditions, stiffness of the steel ring, and concrete stiffness, ductility, shrinkage, and creep characteristics.

Shah and Weiss [35] demonstrated that before cracking, the stresses that develop in a plain and an FRC are very similar. They also developed an analytical procedure for stress development in the steel ring, the stress transfer across the crack, and the crack size [15]. Acoustic emission measurements indicated similar energy release in plain and fiber reinforced specimens; however, fibers appear to delay the development of a localized, visible crack [36]. To eliminate the influence of test conditions, an analytical approach is needed to incorporate influential parameters of shrinkage, creep, aging, and microcracking in the stress analysis of a restrained concrete section. Using the theoretical models, it is possible to calibrate and interpret the experimental test results.

RESTRAINED DRYING SHRINKAGE TEST METHODOLOGY

An experimental study was conducted to evaluate effects of alkali-resistant (AR) glass fibers on shrinkage cracking [37]. An instrumented ring specimen similar to AASHTO PP34 99 [28] was used to quantify the restrained shrinkage and tensile creep behavior of concrete. The specimen was a 66.6-mm-thick annulus of concrete cast around a rigid steel ring 11.2 mm in thickness with an outer diameter of 290 mm and a height of 133.3 mm. A schematic configuration and geometry of shrinkage ring specimen is shown in Figure 20.1. FRC containing AR glass fibers were studied using four concrete mixtures with a water–cement ratio of 0.55 and a lump of 65 to 90 mm. The cement and fine aggregates contents for the three mixtures GRC1.5, GRC3, GRC4, and including a control are designed with 1.5, 3, and 4.5 kg/m³ (4.5 kg/m³ = 0.17%) of glass fibers (24 mm long), respectively. The mixture proportions of all mixes are shown in Table 20.1.

After demolding, the rings are placed in a constant humidity chamber (10% RH) and 40°C. The strain gages attached to the inner surface of the steel ring record the relaxation in steel due to cracking in concrete as shown in Figure 20.2. Three stages are considered. During the

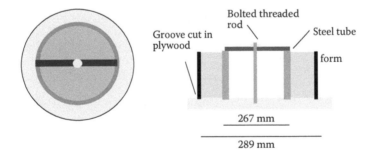

FIGURE 20.1 The configuration and geometry of shrinkage ring specimen.

TABLE 20.1
Mixture Proportions of the Control and GRC samples (kg/m³)

Mix ID	Control	GRC1.5	GRC3	GRC4.5
Portland cement	680	680	680	680
Fine aggregates	1360	1360	1360	1360
Water	374	374	374	374
AR glass fibers	0	1.5	3	4.5
Water–cement ratio	0.55	0.55	0.55	0.55
Sand/cement	2.0	2.0	2.0	2.0

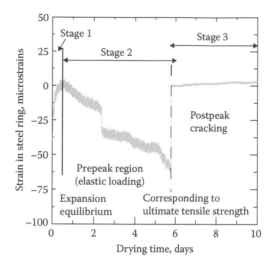

FIGURE 20.2   Typical result of a strain gage attached to steel ring for a plain concrete sample.

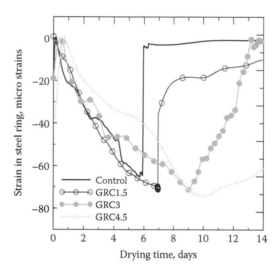

FIGURE 20.3   Effect of AR glass fibers on the results of strain gages attached to steel rings.

first stage, the sample reaches equilibrium within the chamber environment. During the second stage, the response of concrete cast around the steel ring is linear elastic. When shrinkage strain in concrete exceeds ultimate tensile strength, concrete cracks and its effect on steel ring in plain concrete are a drop in the strain values. This drop in strain gage results shows the time of cracking due to very low residual postcrack tensile strength in plain concrete. In this stage (stage 3), strain in steel drops approximately to zero in plain concrete samples, and visible crack width increases by the time. Creep mechanisms are operative. Strain gage versus time response can be used to detect time of cracking in the specimen (e.g., Figure 20.2) 6 days after drying in the chamber.

The response of two strain gages differs after cracking because of their random proximity to crack location. The results of strain gages of different mixtures are shown in Figure 20.3, indicating that using AR glass fibers delayed cracking in concrete samples for nearly 1 or 2 days. Only the control sample and the sample with the lowest amount of fibers show abrupt drop in the strain gage data because of cracking, which occurs between 6 and 8 days for all different samples. Postcrack response of GRC samples with higher fiber dosages are significantly more than the control sample.

The difference between crack widths of the two samples is obvious in Figure 20.4. Using image analysis, it is shown in Figure 20.5 that crack width in GRC3 sample is almost three times smaller than in control. The average crack width and standard deviation obtained for four replicates of control, GRC1.5, GRC3 and GRC4.5 mixes at 14, 21, and 28 days are shown in Table 20.2.

Results of all shrinkage samples show that crack widths increase by increasing drying time. Effect of fiber addition is so significant on reducing crack width dimension that by adding 1.5, 3, and 4.5 kg/m³ AR glass fiber to the plain concrete crack width dimension at 14 days reduces by 51%, 72%, and 82%, respectively. Similar trends are observed after 21 and 28 days of drying in shrinkage chamber as shown in Figure 20.6. The average crack width significantly reduces, even at the 1.5-kg/m³ samples. Addition of 3 kg/m³ glass fibers or more results in much lower deviation in the results and reduces of crack width dimension by as much as 70%.

Modeling Restrained Shrinkage Cracking

Various models have been developed for predicting transverse cracking of concrete ring specimens due to drying shrinkage [38]. Shah et al. [38] used a model based on nonlinear fracture mechanics

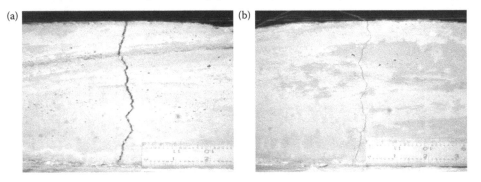

FIGURE 20.4 Transverse cracks due to restrained drying shrinkage: (a) control sample and (b) GRC3 sample after 14 days of drying in the shrinkage chamber.

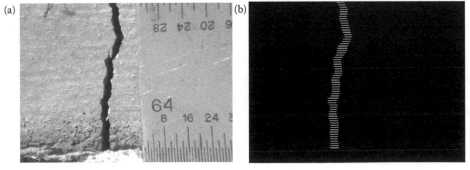

Input grayscale image   Crack width measurement

FIGURE 20.5 Image analysis of a single image taken from the control sample.

TABLE 20.2
Back-Calculated Parameters of the Control and GRC3 Samples at Different Ages

	Age (days)	ε_{t1} (μstr)	ε_{t2} (μstr)	w_{t1} (mm)	w_{t2} (mm)	f_t (MPa)	s_{t3} (MPa)	Gf (MPa/mm)	E_c (GPa)
Control	0					0	0	0	0
	1	120	200	0.02794	0.1524	1.758	0.246	0.041	11.26
	3	120	190	0.0254	0.1524	2.034	0.270	0.043	13.14
	7	110	170	0.02286	0.1143	2.068	0.310	0.046	15.02
	28	110	170	0.02286	0.10922	2.103	0.315	0.048	15.7
GRC3	0					0	0	0	0
	1	170	260	0.04064	1.0668	1.758	0.475	0.255	8.45
	3	120	190	0.03048	0.889	2.034	0.529	0.269	13.14
	7	110	170	0.0254	0.8128	2.068	0.538	0.275	15.02
	28	110	170	0.0254	0.762	2.103	0.547	0.289	15.8

Source: Soranakom, C., Bakhshi M., and Mobasher, B., "Role of alkali resistant glass fibers in suppression of restrained shrinkage cracking of concrete materials," *Proceedings of 15th International Glass Fibre Reinforced Concrete Association Congress*, GRC 2008, Prague, April 20–23, 2008.

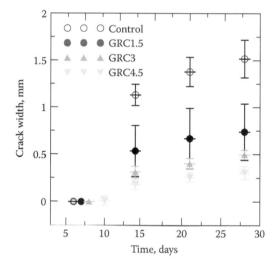

FIGURE 20.6 Mean and standard deviations of crack widths for all mixtures at different times during drying in the shrinkage chamber.

and the fracture resistance curve (R-curve) of the ring specimen. Mobasher et al. [39] used a stress-based approach and defined the compatibility condition that equates the difference between the free shrinkage and estimated creep with the maximum allowable tensile strain. The age at transverse cracking of the specimen subjected to restrained drying shrinkage can be predicted.

Young and Chern [40] proposed a model for shrinkage of steel FRC from the composition of concrete mix, strength, age and environment at drying, size and shape of structures, fiber volume ratio, and aspect ratio of the fiber. The model is based on the well-developed Bazant–Panula model for the shrinkage of plain concrete [41, 42]. All important features of the Bazant–Panula prediction model, such as the diffusion-type size dependence of humidity effects and the square root hyperbolic law for shrinkage, are adopted for steel FRC. Zhang and Li [43] modeled the drying shrinkage of a fiber-reinforced cementitious composite using a free shrinkage expression for the influence of the matrix and fiber properties as well as fiber orientation characteristics.

Lattice Models

Bolander [44] and Bolander and Berton [45] developed irregular lattice models of FRC composites in which the primary material phases (i.e., fiber, matrix, and interfacial zone) are represented as separate entities (see Figures 20.7a and 20.7b). The lattice topology serves to show the distinct values of both the vector field of generalized displacements and the scalar fields (e.g., pore humidity, temperature) that are integral to simulating the durability mechanics of FRC composites. The modeling of moisture diffusion, the convective boundary conditions at exposed surfaces, and the coupling of the stress and diffusion analyses are based on the works of Bazant and Najjar [46] and Martinola and Wittmann [47]. Short fibers are explicitly modeled within the three-dimensional domain, directly accounting for length and orientation efficiency of each fiber as well as wall effects that can bias fiber orientations near the domain boundaries. Fiber distributions can be numerically generated or obtained from actual tests (e.g., via computed tomography).

The lattice model has been used to study drying shrinkage cracking in a fresh cement composite overlay on a mature concrete substrate and the role of fibers in restraining such cracking. Figure 20.7 shows a lattice discretization of the composite system. The lattice topology is based on the Delaunay triangulation of an irregular set of nodal points, whereas the dual Voronoi diagram serves to define the element properties. The overlay, the substrate, and an interfacial layer are defined by their elasticity, hygral, and fracture properties. The upper face of the lattice model is exposed, via a convective boundary condition, to an environment with relative humidity $H_a = 0.5$, whereas the overlay and a portion of the substrate are assigned relative humidity $H = 1.0$. With exposure to the drying environment, the moisture diffusion analysis produces humidity gradients that lead to stress development in the overlay system. Without fiber reinforcement, fracture zones develop rather uniformly along the top of the overlay before 20 days exposure to drying. With shrinkage due to additional drying, localization occurs as characterized by several of the cracks continuing to open while neighboring cracks unload. A maximum crack opening of 0.29 mm occurs after 110 days drying (Figure 20.8a). The pattern of cracks in the overlay as well as the tendency for vertical overlay cracks to turn and run laterally is in good general agreement with the results presented by Martinola and Wittmann [47] (after which these simulations were patterned). When polypropylene fibers (diameter = 0.12 mm, length = 39 mm, E_f = 42.8 GPa, V_f = 2%) are introduced into the overlay material, diffuse microcracking still occurs along the top of the overlay, but only one smaller crack runs through the thickness of the overlay. The fibers restrict the opening of this crack to 0.12 mm after 110 days of drying, and the undesirable lateral branching of cracks is also arrested (Figure 20.8).

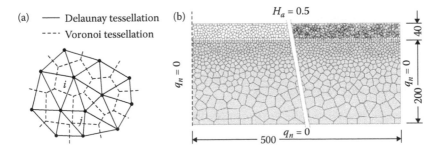

FIGURE 20.7 (a) Element ij within irregular lattice. (b) Computational grid for simulating drying of an overlay system (fiber composite overlay shown on right side; dimensions are in millimeters).

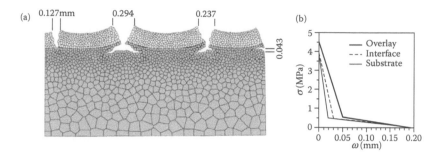

FIGURE 20.8   Unreinforced overlay: (a) damage pattern after 110 days of exposure to drying environment; and (b) stress versus crack opening relations for system components.

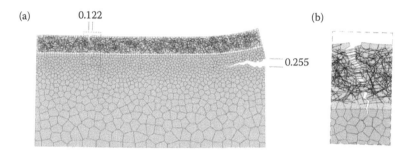

FIGURE 20.9   Fiber-reinforced overlay: (a) damage pattern after 110 days exposure to drying environment; and (b) fibers bridging a crack in the overlay (dimensions in millimeter).

However, because of the lack of stress relief in the overlay, horizontal cracking can develop in the substrate, starting from the edge of the overlay/substrate system (Figure 20.9).

Lamina Model

Simulation of crack width opening is of great importance for durability and serviceability. Various theoretical approaches for modeling the restrained shrinkage cracking of concrete have been developed [27, 48, 49]. These models address the interaction between material properties and shrinkage characteristics of concrete.

Modeling shrinkage requires an analytical approach that relates key influential parameters including diffusion, shrinkage, creep, aging material properties, and restraining effect [50]. The theoretical model is then used to calibrate and interpret the experimental test results. For the model to show effects of fibers on restrained shrinkage cracking, one needs to incorporate the postpeak response of matrix with fibers.

MOISTURE DIFFUSION AND FREE SHRINKAGE

As concrete loses its moisture to the environment, free shrinkage takes place, and the humidity profile $h(z)$ through the thickness of the concrete section is simplified to follow the Fick's law of diffusion as shown in Figure 20.10:

$$h(z) = h_s - (h_s - h_i)erf(z) \qquad (20.1)$$

where z is the distance measured from the outside surface, h_s and h_i represent the humidity at the outside surface and interior section, and erf(z) represents the error function [51].

A cubic function is used to relate the free shrinkage strain as a function of the humidity profile throughout the thickness.

$$\varepsilon_{sh}(z,t) = \varepsilon_{sh}(t)(1 - h(z)^3) \tag{20.2}$$

where $\varepsilon_{sh}(t)$ is the free drying shrinkage strain that can be obtained by experiments or from the empirical relationships proposed by Bazant–Panula [52] for drying shrinkage at infinite time ($\varepsilon_{sh\infty}$) and time-dependent shrinkage formulation suggested by ACI 209 [53] as

$$\varepsilon_{sh}(t) = \frac{t^{\xi}}{f + t^{\xi}} \varepsilon_{sh}(t = \infty) \tag{20.3}$$

where f and ξ are constants and t is time in days. ACI 209R-92 recommends an average value for f of 35 for 7 days of moist curing, whereas an average value of 1.0 is suggested for ξ. In this study, free drying shrinkage is obtained by experiments performed according to ASTM standard C 157-04. Meanwhile, the experimental data of free shrinkage are fitted by the ACI equation as shown in Figure 20.11, and the proper values of f and ξ are selected as 28 and 1.5 days, respectively. As shown

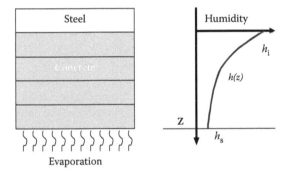

FIGURE 20.10 Fick's law of diffusion to simulate humidity profile through the thickness

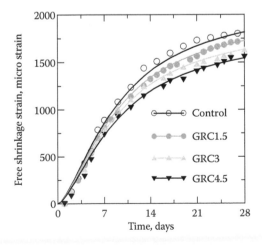

FIGURE 20.11 Experimental free shrinkage data fitted with modified ACI 209R-92 model.

in the figure, the best-fitted values for drying shrinkage at infinite time ($\varepsilon_{sh\infty}$) for control, GRC1.5, GRC3 and GRC 4.5 samples are 2160, 2052, 1944, and 1836 µstr, respectively.

EFFECT OF CREEP IN RESTRAINED SHRINKAGE CRACKING

As concrete is subjected to tensile stress, it produces creep strain with time and offset the strain due to free shrinkage, resulting in relaxation of the elastic stresses in the material. Because of the lack of early age tensile creep model, it is assumed that the creep coefficient of concrete in compression $\nu(t)$ used in ACI 209 report [27] is applicable to the present tensile mode of loading,

$$\nu(t) = \frac{t^{\psi}}{d + t^{\psi}} \nu_u, \qquad (20.4)$$

where ν_u is the ultimate creep coefficient, d and ψ are the constants, and t is time in days and assumed as $\nu_u = 12$ for all samples. ACI 209R-92 recommends an average value of 10 and 0.6 for d and ψ, respectively. However, because the recommendation is for the plain concrete under compression tests, the values of d and ψ are modified to 6 and 1, respectively, for the tensile creep behavior of the control mixture. For the case of GRC1.5, 3, and 4.5, the values of d are chosen as 4, 3, and 1.5, and the values of ψ are chosen as 1.1, 1.3, and 0.5, respectively (Figure 20.12).

AGE-DEPENDENT CONCRETE STRENGTH

To obtain tensile strength properties of the mixtures along the time, monotonic three-point bending tests are performed on the control and GRC3 samples at 1, 3, 7, and 28 days. Figure 20.13 shows the flexural responses of these two mixtures at 1 and 7 days.

To calculate tensile stress–strain parameters from three-point bending tests, a piecewise linear model shown in Figure 20.14, for both plain and GRC samples, was used.

The parameters of the stress–strain back-calculation model of the control and ARG5 samples at 1, 3, 7, and 28 days are summarized in Table 20.2 and shown in Figure 20.15.

Similar to ACI time-dependent compressive strength development function, back-calculated tensile strength, f_t, residual strength, s_{t3}, fracture energy, G_f, and Young's modulus, E_c, are assumed to follow these functions:

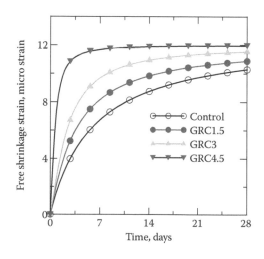

FIGURE 20.12   Assumed creep model for the control and GRC mixes.

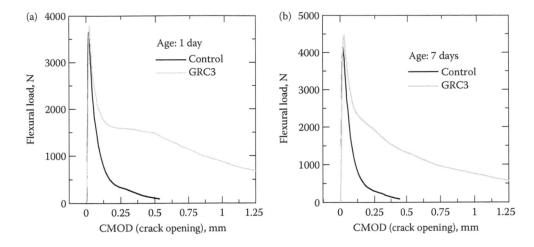

FIGURE 20.13   Flexural response of concrete samples under three-point bending tests

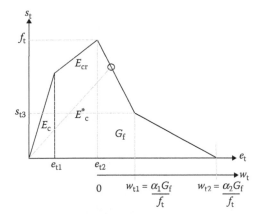

FIGURE 20.14   Tensile stress–strain and crack width model.

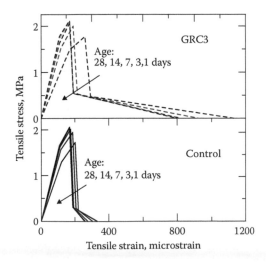

FIGURE 20.15   Back-calculated tensile stress–strain model of two mixtures at different ages.

$$f_t(t) = f_{t28}\left(\frac{t}{a+bt}\right),\ s_{t3}(t) = s_{t3-28}\left(\frac{t}{a+bt}\right),\ G_f(t) = G_{f28}\left(\frac{t}{a+bt}\right),\ E_c(t) = E_{c28}\left(\frac{t}{a+bt}\right) \quad (20.5)$$

where f_{t28}, s_{t3-28}, G_{f28}, and E_{c28} are tensile strength, residual strength, fracture energy, and Young's modulus of concrete at 28 days, respectively. t is the time in days, a and b are the material parameters. ε_{t1} and ε_{t2}, tensile strains at limit of proportionality and peak stress respectively, and w_{t1} and w_{t2}, transition and ultimate crack widths, are assumed to follow these functions:

$$\varepsilon_{t1}(t) = a \times e^{bt} + c,\ \varepsilon_{t2}(t) = a \times e^{bt} + c$$
$$w_{t1}(t) = a \times e^{bt} + c,\ w_{t2}(t) = a \times e^{bt} + c \quad (20.6)$$

In the abovementioned functions, t is the time in days and a, b, and c are the constant material parameters. The best values for the coefficients a, b, and c after fitting Equation 20.5 and 20.6 with the values of Table 20.2 are used in this modeling approach.

EQUILIBRIUM AND COMPATIBILITY CONDITIONS

The tensile strain is defined as positive, whereas compressive and shrinkage strains are negative. The symbol Δ represents the incremental change of quantities between previous time step t_{j-1} and the current time step t_j, and index i is used for the layer number at particular location z_i. The restraining effect of steel ring prohibits concrete shrinkage and is determined by equilibrium of force between the tension force in concrete ΔF_c and the compression force in steel ΔF_{st}:

$$\Delta F_c + \Delta F_{st} = 0 \quad (20.7)$$

In the equilibrium, only the elastic tensile strain component of the concrete $\Delta \varepsilon_{el}$ produces stress, and it is balanced with the compressive stress in steel, which has elastic compressive strain $\Delta \varepsilon_{st}$. Using modulus of materials, the equilibrium of force in Equation 20.7 can be written as

$$\Delta \varepsilon_{el} \overline{E_c^*}(t_{j-1})A_c + \Delta \varepsilon_{st} E_s A_s = 0 \quad (20.8)$$

where $\overline{E_c^*}(t_{j-1})$ is the secant modulus of the concrete at a time step averaged from all layers. This is only an approximation because the modulus at the current time step is not yet known. E_s is the Young's modulus of steel, and A_c and A_s are area of concrete and steel, respectively. By rearranging the terms in Equation 20.8, incremental compressive steel strain $\Delta \varepsilon_{st}$ is obtained:

$$\Delta \varepsilon_{st} = -\frac{\overline{E_c^*}(t_{j-1})A_c}{E_s A_s} \Delta \varepsilon_{el} \quad (20.9)$$

STRESS–STRAIN DEVELOPMENT

Figure 20.16a shows a concrete ring specimen subjected to moisture loss up to cracking. A section of the ring as shown in a box can be approximated as a one-dimensional problem and represented in Figure 20.16b. The free shrinkage strain $\Delta \varepsilon_{sh}$, restrained by steel ring, creates compressive strain

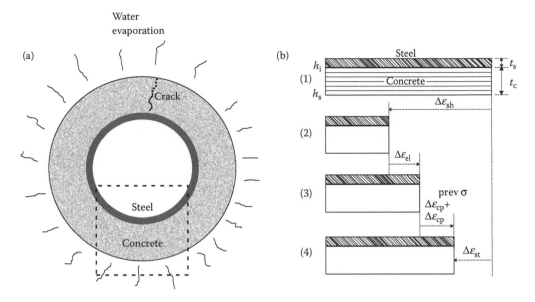

FIGURE 20.16　Schematic drawing for drying shrinkage: (a) ring specimen and (b) strain components.

in the steel $\Delta\varepsilon_{st}$ such that the steel compressive force must be balanced with the tension force in concrete, generating elastic strain in concrete $\Delta\varepsilon_{el}$ (positive). For time interval Δt, the concrete tensile stress will also generate creep strain $\Delta\varepsilon_{cp}$ (positive). In addition, the previous stress history before time t_{j-1} also contributes to the creep strain increment between time t_{j-1} and t_j, $\Delta\varepsilon_{cp}^{(prev\sigma)}$, which can be calculated by using Boltzmann's superposition principle:

$$\Delta\varepsilon_{cp}^{(prev\sigma)} = \left\{ \sum_{k=2}^{j-1} \left| \frac{1}{2} \frac{\sigma_{el_k} - \sigma_{el_{k-1}}}{E_{c28}} \right| v(t_j - t_{k-1}) \right\} - \varepsilon_{cp}(t_{j-1}) \tag{20.10}$$

The equilibrium of incremental strains developed between time step t_{j-1} and t_j can be written as

$$-\Delta\varepsilon_{sh} = \Delta\varepsilon_{el} + \Delta\varepsilon_{cp} + \Delta\varepsilon_{cp}^{(prev\sigma)} - \Delta\varepsilon_{st} \tag{20.11}$$

Substituting the restraining steel strain $\Delta\varepsilon_{st}$ from Equations 20.9 through 20.11 yields

$$-\Delta\varepsilon_{sh} = \frac{\overline{E_c^*}(t_{j-1})A_c}{E_s A_s} \Delta\varepsilon_{el} + \Delta\varepsilon_{el} + \Delta\varepsilon_{cp} + \Delta\varepsilon_{cp}^{(prev\sigma)} \tag{20.12}$$

Using the relationship between tensile stress and creep coefficient $\Delta\varepsilon_{cp} = \Delta\sigma_{el}C(dt)/E_{c28}$ to determine creep strain and substituting the result in Equation 20.12 results in

$$-\Delta\varepsilon_{sh} = \frac{\overline{E_c^*}(t_{j-1})A_c + E_s A_s}{E_s A_s} \Delta\varepsilon_{el} + \Delta\sigma_{el}\frac{v(dt)}{E_{c28}} + \Delta\varepsilon_{cp}^{(prev\sigma)} \tag{20.13}$$

To obtain stress and strain distribution, the thickness of concrete is discretized into N_c sublayers, whereas the steel has only one layer. The incremental elastic stress in layer i located at z_i can be expressed as

$$\Delta\sigma_{el}(z_i) = [\varepsilon_{el}(z_i,t_j) - \varepsilon_{el}(z_i,t_{j-1})]E_c^*(z_i,t_{j-1}) = \Delta\varepsilon_{el}(z_i)E_c^*(z_i,t_{j-1}) \tag{20.14}$$

By substituting incremental stress $\Delta\sigma_{el}(z_i)$ defined in Equations 20.14 through 20.13 and by rearranging the terms, the incremental concrete elastic strain at each sublayer can be expressed as

$$\Delta\varepsilon_{el}(z_i) = \frac{-\Delta\varepsilon_{sh}(z_i) - \Delta\varepsilon_{cp}^{(prev\sigma)}(z_i)}{Q + \dfrac{E_c^*(z_i,t_{j-1})}{E_{c28}}v(dt)}; \quad Q = \frac{\overline{E_c^*}(t_{j-1})A_c + E_s A_s}{E_s A_s} \tag{20.15}$$

In this formulation, the whole thickness with an average concrete secant modulus $\overline{E_c^*}(t_{j-1})$ is considered in accounting for the restraining effect of steel, whereas individual layers with secant modulus of $E_c^*(z_i,t_{j-1})$ at each layer for the effect of free shrinkage and creep are used. The algorithm for strain history in steel ring and crack width at concrete surface is as follows:

1. Calculate free shrinkage strain at each layer i from Equations 20.2 and 20.3 and its increment by

$$\Delta\varepsilon_{sh}(z_i) = \varepsilon_{sh}(z_i,t_j) - \varepsilon_{sh}(z_i,t_{j-1}) \tag{20.16}$$

2. Calculate incremental elastic tensile strain at each layer due to free shrinkage, restraining effect and creep from Equation 20.15.
3. Update total elastic strain, stress, and secant modulus using concrete model,

$$\varepsilon_t(z_i,t_j) = \varepsilon_t(z_i,t_{j-1}) + \Delta\varepsilon_{el}(z_i) \tag{20.17}$$

$$\sigma_t(z_i,t_j) = f[\varepsilon_t(z_i,t_j)] \tag{20.18}$$

$$E_c^*(z_i,t_j) = \frac{\sigma_t(z_i,t_j)}{\varepsilon_t(z_i,t_j)} \tag{20.19}$$

4. If the updated strain $\varepsilon_t(z_i,t_j)$ exceeds the strain at peak stress $\varepsilon_{t2}(t_j)$, the crack width at each layer of concrete is calculated by

$$w_t(z_i,t_j) = \left[\varepsilon_t(z_i,t_j) - \varepsilon_{t\_unload}(z_i,t_j)\right]\pi D(z_i) \tag{20.20}$$

where $\varepsilon_{t\_unload}(z_i,t_j)$ is the unloading strain in the prepeak stress–strain curve corresponding to the same stress level as the strain in postpeak response $\varepsilon_t(z_i,t_j)$ and $D(z_i)$ is the diameter of the concrete ring at location z_i.

5. Sum concrete forces at each concrete layer A_{ci} to obtain the total tensile force, which is equal to compressive the force in steel.

$$F_c(t_j) = \sum_{i=1}^{N_c} A_{ci}\sigma_t(z_i,t_j) \quad \text{and} \quad F_s(t_j) = -F_c(t_j) \tag{20.21}$$

6. Calculate nominal stress and strain in steel, defined by

$$\sigma_{ns}(t_j) = \frac{F_s(t_j)}{A_s} \text{ and } \varepsilon_{ns}(t_j) = \frac{\sigma_{ns}(t_j)}{E_s} \tag{20.22}$$

7. A complete strain history at steel ring and crack width at concrete surface can be obtained by repeating steps 1 to 6 until the time t_j reaches the specified age.

PARAMETRIC STUDY

Figures 20.17a and 20.17b represent the theoretical curve of strain in steel as a parametric study of the effect of fibers. Note that the compressive stress builds up in steel gradually, and as the concrete cracks due to tensile failure mode, there is a gradual relaxation and recovery of strain in steel. Note that the strain in the steel does not decay quite as fast as cracking takes place in the specimen. The first crack formation in the theoretical model is after 8 days, which is quite similar to the experimental results. Beyond the cracking of concrete, its creep properties in the postpeak region contribute to the increase in the strain, causing further reduction of steel strain.

COMPARISON OF EXPERIMENTAL DATA AND SIMULATIONS

The results of simulation by this model and the comparison with the experimental data are shown in Figure 20.18, representing a good correlation between the experimental data including strain in steel ring and crack width opening histories and simulation data. Although the best correlation of the simulation with experimental data is for the control sample, the model overestimates the crack widths of GRC samples after 14 days. However, the model is capable of capturing significant effects of adding glass fibers to control shrinkage cracking in concrete.

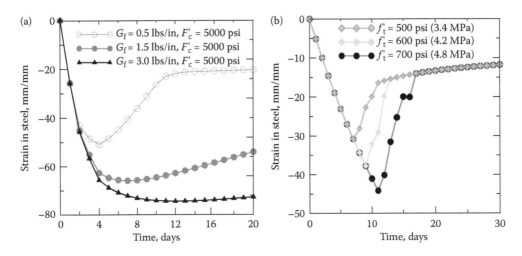

FIGURE 20.17 Comparison of experimental and theoretical curve of shrinkage test. (From Mane, S. A., Desai T. K., Kingsbury, D., and Mobasher, B., "Modeling of restrained shrinkage cracking in concrete materials," *ACI Special Publications*, SP206-14, 219–242, 2002. With permission.) (a) Effect of fracture energy of concrete. (b) Effect of tensile strength of concrete.

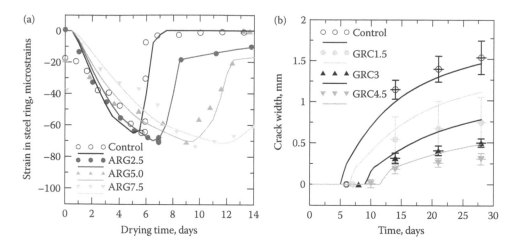

FIGURE 20.18 Comparison of experimental results and model simulation; (a) history of strain in steel ring and (b) crack width dimension history at the outer concrete surface.

CONCLUSIONS

Reducing crack width dimension three times by using low content of AR glass fibers is the most significant effect of adding fibers to the mixture in controlling shrinkage cracking. This effect in addition to delaying crack occurrence by 1 or 2 days represents the low dosage glass fiber concrete as an appropriate construction material to be used in dry and hot regions with the risk of high drying shrinkage. The analytical model for predicting steel strain history and concrete crack width dimension for restrained shrinkage test shows a relatively good correlation with the experimental data. This model is capable of capturing significant effects of fiber addition to control cracking of shrinkage. Collectively speaking, ring-type restrained shrinkage test method is a good preventative of the behavior of concrete materials against restrained drying shrinkage.

REFERENCES

1. Bentur, A., and Kovler, K. (2003), "Evaluation of early age cracking characteristics in cementitious systems," *Materials and Structures*, 36(257), 183–190.
2. ASTM C157-08 (2008), *Standard Test Method for Length Change of Hardened Hydraulic-Cement Mortar and Concrete*, ASTM Book of Standards, Vol. 04.02, Philadelphia: American Society of Testing and Materials.
3. Berke, N. S., and Dalliare, M. P., "The effect of low addition rate of polypropylene fibers on plastic shrinkage cracking and mechanical properties of concrete," Fiber Reinforced Concrete: Development and Innovations, ACI SP-142-2, 1994, 19–41.
4. ASTM C 1579 (2006), *Standard Test Method for Evaluating Plastic Shrinkage 1 Cracking of Restrained Fiber Reinforced Concrete (Using a Steel Form Insert)*, Philadelphia: American Society of Testing and Materials.
5. Qi, C., and Weiss, J. (2003), "Characterization of plastic shrinkage cracking in fiber reinforced concrete using image analysis and a modified Weibull function," *Materials and Structures*, 35(6), 386–395.
6. Qi, C., Weiss, W. J., and Olek, J. (2003), "Characterization of plastic shrinkage cracking in fiber reinforced concrete using semi-automated image analysis," *Concrete Science and Engineering*, 36(260), 386–395.
7. Banthia, N., and Gupta, R. (2006), "Influence of polypropylene fiber geometry on plastic shrinkage cracking in concrete," *Cement and Concrete Research*, 36(7), 1263–1267.
8. ASTM C1581/C1581M-09a (2009), *Standard Test Method for Determining Age at Cracking and Induced Tensile Stress Characteristics of Mortar and Concrete under Restrained Shrinkage*, Philadelphia: American Society of Testing and Materials.

9. Radlinska, A., Moon, J. H., Rajabipour, F., and Weiss, J. (2006), "The ring test: A review of recent developments," *Proceedings of International RILEM Conference on Volume Changes of Hardening Concrete: Testing and Mitigation, RILEM,* Jensen, O., Lura, P., and Kovler, K. eds., Lyngby, Denmark, 205–214.

10. AASHTO PP34-99, *Standard Practice for Estimating the Cracking Tendency of Concrete, AASHTO,* Washington, D.C.

11. Grzybowski, M., and Shah S. P. (1990), "Shrinkage cracking of fiber reinforced concrete," *ACI Materials Journal,* 87(2), 138–148.

12. Paillere, A. M., Buil, M., and Serrano, J. J. (1989), "Effect of fiber addition on the autogenous shrinkage of silica fume Concrete," *ACI Materials Journal,* 86(2), 139–144.

13. Kovler, K. (1994), "Testing system for determining the mechanical behavior of early age concrete under restrained and free uniaxial shrinkage," *Materials and Structures,* 27(6), 324–330.

14. Toma, G., Pigeon, M., Marchand, J., Bissonnette, B., and Bercelo L. "Early-age autogenous restrained shrinkage: Stress build up and relaxation," *Proceedings of the Fourth International Research Seminar on Self-Desiccation and Its Importance in Concrete Technology,* ed. B. Persson and G. Fagerlund, Lund University, 1999, 61–72.

15. Altoubat, A. S., and Lange, D. A. (2002), "Grip–specimen interaction in uniaxial restrained test," in *ACI SP-206 on Concrete: Material Science to Application, A Tribute to Surendra P. Shah,* American Concrete Institute, Farmington Hills, MI, 189–204.

16. Aly, T., Sanjayan, J. G., and Collins F. (2008), "Effect of polypropylene fibers on shrinkage and cracking of concretes," *Materials and Structures,* 41(10), 1741–1753.

17. Shah, S. P., Weiss, W. J., and Yang, W. (1997), "Shrinkage cracking in high performance concrete," *Proceedings of the PCI/FHWA International Symposium on High Performance Concrete,* New Orleans, LA, 217–228.

18. Kraai, P. P. (1985), "A proposed test to determine the cracking potential due to drying shrinkage of concrete," *Concrete Construction,* 30(9), 775–778.

19. Shaeles, C. A., and Hover, K. C. (1988), "Influence of mix proportions and construction operations on plastic shrinkage cracking in thin slabs," *ACI Materials Journal,* 85(6), 495–504.

20. Opsahl, O. A., and Kvam, S. E., "Betong med EE-Stal-Fiber (Concrete with FE-Steel Fibers)," Report No. STF 65 A82036, SINTEF dep, Cement and Concrete Research Institute, Technical University of Norway, Trondheim, June 1982.

21. Padron, I., and Zollo R. F. (1990), "Effect of synthetic fibers on volume stability and cracking of portland cement concrete and mortar," *ACI Materials Journal,* 87(4), 327–332.

22. Carlson, R. W., and Reading, T. J. (1988), "Model study of shrinkage cracking in concrete building walls," *ACI Structural Journal,* 85(4), 395–404.

23. Shah, S. P., Karaguler, M. E., and Sarigaphuti, M. (1992), "Effects of shrinkage reducing admixture on restrained shrinkage cracking of concrete," *ACI Materials Journal,* 89(3), 289–295.

24. Shah, S. P., Marikunte, S., Yang, W., and Aldea, C. (1997), "Control of cracking with shrinkage-reducing admixtures," *Transportation Research Record: Journal of the Transportation Research Board,* 1574, 25–36.

25. Bentur, A. (2003), "Early Age Cracking in Cementitious Systems," Chapter 6.5, *RILEM State of the Art Report TC 181-EAS,* ed. A. Bentur, RILEM Publications, Bagneux, France.

26. Hossain, A. B., and Weiss, J. (2004), "Assessing residual stress development and stress relaxation in restrained concrete ring specimens," *Cement and Concrete Composites,* 26, 531–540.

27. Turcry, P., Loukili, A., Haidar, K., Pijaudier-Cabot G., and Belarbi, A. (2006), "Cracking tendency of self-compacting concrete subjected to restrained shrinkage: Experimental study and modeling," *Journal of Materials in Civil Engineering,"* 18(1), 46–54.

28. AASHTO PP 34–99 (2004), *Standard Practice for Estimating the Crack Tendency of Concrete.*

29. ASTM C1581/C1581M-09a (2009), *Standard Test Method for Determining Age at Cracking and Induced Tensile Stress Characteristics of Mortar and Concrete under Restrained Shrinkage,* Philadelphia.

30. Najm, H., and Balaguru, P. (2002), "Effect of large-diameter polymeric fibers on shrinkage cracking of cement composites," *ACI Materials Journal,* 99(4), 345–351.

31. Bissonnette, B., Therrien, Y., Pleau, R., et al. (2000), "Steel fiber reinforced concrete and multiple cracking," *Canadian Journal of Civil Engineering,* 27(4), 774–784.

32. Swamy, R. N., and Stavarides, H. (1979), "Influence of fiber reinforcement on. restrained shrinkage cracking," *Journal of the American Concrete Institute,* 76(3), 443–460

33. Hossain, A. B., Pease, B., and Weiss, W. J. (2003), "Quantifying restrained shrinkage cracking potential in concrete using the restrained ring test with acoustic emission," *Transportation Research Record, Concrete Materials and Construction,* 1834, 24–33.

34. Hossain, A. B., and Weiss, W. J. (2004), "Assessing residual stress development and stress relaxation in restrained concrete ring specimens," *Cement and Concrete Composites*, 26(5), 531–540.

35. Shah, H. R., and Weiss, W. J. (2006) "Quantifying shrinkage cracking in fiber reinforced concrete using the ring test," *Materials and Structures*, 39(9), 887–899.

36. Kim, B., and Weiss, W. J. (2003), "Using acoustic emission to quantify damage in fiber reinforced cement mortars," *Cement and Concrete Research*, 33(2), 207–214

37. Soranakom, C., Bakhshi, M., and Mobasher, B. (2008), "Role of alkali resistant glass fibers in suppression of restrained shrinkage cracking of concrete materials," *Proceedings of the 15th International Glass Fibre Reinforced Concrete Association Congress,* GRC 2008, Prague, April 20–23, 2008.

38. Shah, S. P., Ouyang, C., Marikunte, S., Yang, W., and Becq-Giraudon, E. (1998), "A method to predict shrinkage cracking of concrete," *ACI Materials Journal*, 95(4), 339–346.

39. Mane, S. A., Desai, T. K., Kingsbury, D., and Mobasher, B., "Modeling of restrained shrinkage cracking in concrete materials," ACI Special Publications, SP206-14, 2002, American Concrete Institute, Farmington Hills, MI, 219–242 .

40. Young, C.,H., and Chern, J. C. (1991), "Practical prediction model for shrinkage of steel fiber reinforced-concrete," *Materials and Structures*, 4(141), 191–201.

41. Bazant, Z. P., Kim, Joong-Koo, and Panula, L. (1991), "Improved prediction model for time-dependent deformations of concrete: Part 1—Shrinkage." *Materials and Structures*, 24(143), 327—345.

42. Bazant, Z. P., and Panula, L. (1978), "Practical prediction of time dependent deformations of concrete: Parts I–VI," *Materials and Structures*, 11, 307–316, 317–328, 425–434.

43. Zhang, J., and Li, V. C. (2001), "Influences of fibers on drying shrinkage of fiber-reinforced cementitious composite," *Journal of Engineering Mechanics,* 127(1), 37–44.

44. Bolander, J. E., "Numerical modeling of fiber reinforced cement composites: Linking material scales," *Proceedings of the Sixth International RILEM Symposium on Fibre-Reinforced Concretes—BEFIB 2004,* eds. M. di Prisco, R. Felicetti, and G. A. Plizzari, RILEM, 2004, 45–60.

45. Bolander, J. E., and Berton, S. (2004), "Simulation of shrinkage induced cracking in cement composite overlays." *Cement and Concrete Composites*, 26, 861–871.

46. Bazant, Z. P., and Najjar, L. J. (1971), "Drying of concrete as a nonlinear diffusion problem," *Cement and Concrete Composites,* 1, 461–473.

47. Martinola, G., and Wittmann, F. H. (1995). "Application of fracture mechanics to optimize repair mortar systems," in *Fracture Mechanics of Concrete Structures*, ed. F. H. Wittmann, Freiburg: Aedificatio Publishers, 1481–1486.

48. Grzybowski, M., and Shah, S. P. (1989), "A model to predict cracking in fiber reinforced concrete due to restrained shrinkage," *Magazine of Concrete Research*, 41(148), 125–135.

49. Shah, S. P., Ouyang, C., Marikunte, S., Yang, W., and Becq-Giraudon, E. (1998), "A method to predict shrinkage cracking of concrete," *ACI Materials Journal*, 95(4), 339–346.

50. Mane, S. A., Desai, T. K., Kingsbury, D., and Mobasher, B. (2002), "Modeling of restrained shrinkage cracking in concrete materials," ACI Special Publications, SP206-14, 219–242.

51. Tuma, J. J. (1998), *Engineering Mathematics Handbook*, 4th ed., New York: McGraw-Hill, 566.

52. Bazant, Z. P., and Panula, L. (1978), "Practical prediction of time-dependent deformation of concrete. Part 1: Shrinkage. Part 2: Creep," *Materials and Structures*, 11(5), 307–328.

53. ACI Committee 209 (2008), *Guide for Modeling and Calculating Shrinkage and Creep in Hardened Concrete (ACI 209.2R-08)*, Michigan: American Concrete Institute.

21 Flexural Impact Test

INTRODUCTION

Structural elements such as wall panels, hydraulic structures, airport pavements, military structures, and industrial floor overlays may be subjected to severe impact loads. Very high stress rates occur during such dynamic loads as a large amount of energy is transmitted to the structure in a short duration. Structural elements respond to earthquake and explosive forces by a variety of interactive mechanical properties. These properties include strength, absorbed energy, deformation capacity, and ductility and must be present to maintain integrity without collapse. Fiber addition to concrete improves its ductility, tensile, impact, and flexural strength.

Proper design of a composite system subjected to high-loading rates requires characterization of strain rate sensitivity, modes of failure, and energy absorption. High ductility of fiber reinforced cement composites makes them ideal for use under blast, impact, and dynamic loads; however, they have rate-dependent mechanical properties such as Young's modulus, ultimate strength, and fracture toughness. Proper knowledge of the constitutive relationship for a wide range of strain rates is required to develop realistic material laws. During the life span of a structure, impact events by debris during a hurricane event, wind and seismic loads, and ballistic projectile are expected to occur. In some cases, impact velocities may be relatively low, but a high projectile mass may cause significant damage [1]. In addition, the potential for use of cement composites as a tool to withstand the high-energy explosives cannot be overlooked [2].

Impact resistance of fiber reinforced concrete (FRC) has been measured by several test methods: Charpy, Izod, drop weight, split Hopkinson bar (SHB), explosive, and ballistic impact [3, 4]. These tests can be either instrumented or heuristically based, and the resistance can be measured by means of fracture energy, damage accumulation, and/or measurement of the number of drops to achieve a determined damage or stress level. Results depend on variables such as size of specimen, machine compliance, strain rate, type of instrumentation, and test setup.

Silva et al. [5] performed Charpy impact tests (strain rate = 14.4 s^{-1}) and static three-point bending tests in sisal pulp fiber reinforced cement composites with a fiber mass fraction of 14%. No significant difference between maximum force obtained from static (0.21 kN) and impact (0.19 kN) were observed. A modified SHB was used by Romano et al. [6] to characterize steel FRC. SHB results showed an increase in both toughness and ultimate strength compared with static compression tests.

Low-velocity impact on fiber reinforced composites has been the subject of many experimental investigations [7–11]. Bindiganavile and Banthia [12, 13] showed that the flexural strength of FRC is higher under impact loading than under quasi-static loading. Energy absorption is also higher under impact for concrete reinforced with polymeric fibers. Lok and Zhao [14] reported that the postpeak ductility of SFRC is lost at strain rates exceeding 50 s^{-1} because fragments can no longer bond onto the steel fibers. Choi and Lim [15] used a linearized contact law approach to address the impact response of composite laminates. Wang et al. [16] identified two different damage mechanisms for FRC drop weight impact. For fiber fractions lower than the critical fiber volume, fiber fracture dominates the failure mechanism, whereas for values higher than the critical fiber volume, the fiber pullout mechanism dominates the response. The critical value for the hooked steel fibers was between 0.5% and 0.75%. Manolis et al. [17] showed that fibrillated polypropylene fibers significantly improved the impact resistance of concrete slabs. Li et al. [18] investigated static and impact behavior of extruded sheets with short fibers by using polyvinyl alcohol and glass fibers. Results indicated that glass fibers were more effective in improving the tensile strength and impact properties, whereas polyvinyl alcohol fibers increased the tensile strain and absorbed energy of specimens. The dynamic compressive strength of SFRC when tested by Lok and Zhao [14] using an SHB increased very slowly from quasi-static to a relatively low strain rate (between 10 and 20 s^{-1}).

A detailed discussion of static properties of textile–cement composites manufactured using the pultrusion process is presented in previous chapters. The behavior of fabric–cement composites exposed to tensile loads are characterized by multiple cracking; therefore, stress–strain curve, toughness, and strength are dependent on the reinforcing fabrics and matrix properties as well as the interface bond and the anchorage of the fabrics [19–21]. Microstructural features such as crack spacing, width, and density allow formulation of the damage evolution as a function of macroscopically applied strain are also dependant on strain rate [22, 23].

An impact test setup on the basis of a free-fall drop of an instrumented hammer on a three-point bending specimen is discussed. The flexural impact behavior of alkali-resistant (AR) glass and polyethylene textile composites made by pultrusion process was investigated to evaluate if the superior static response of these systems can be extended to the dynamic load cases as well. Flexural impact tests of cement-based composites reinforced by AR glass have been performed [24, 25].

Composites with six and eight layers of AR glass, polyethylene, and sisal fabrics were tested in vertical and horizontal positions with respect to the direction of applied impact load. Four drop heights of 50, 100, 200, and 250 mm were used as the initial height of the drop hammer. The impact load is measured by three different methods, that is, acceleration response of specimens, piezoelectric load cell, and conventional strain gage-based load cell. The time histories of impact loads, deflections, accelerations, absorbed energy, and ratio of absorbed energy to input potential energy were discussed in detail. It is observed that for AR glass fabrics flexural strengths as high as 29.9 MPa for plate specimens when subjected to potential energies in the range of 7 to 14 J [24] are obtained. The absorbed energy increased with drop height, whereas the ratio of absorbed energy to potential energy decreased with drop height. Interlaminar shear was the dominant failure mode.

EXPERIMENTAL PROGRAM

MATERIAL PROPERTIES AND MIX DESIGN

AR Glass Composite

The bonded AR glass fabric consisted of a perpendicular set of yarns (warp and weft) glued together at the juntion points and was coated with sizing during fabric production. The fabric had a density of four yarns per centimeter in both warp and weft directions and was manufactured by NEG Glass, as shown in Figure 21.1, using yarns of filament diameter of 13.5 μm, tensile strength of 1270 to 2450 MPa, and modulus of elasticity of 78 GPa. There are 400 filaments of fiber in a yarn resulting in a bundle diameter of 0.27 mm. The mix design of cement paste for the composites is presented in Table 21.1.

The cement-based composites were made with six and eight layers of fabrics, resulting in reinforcement content of approximately 4.0% and 6.0% by volume of AR glass fibers, respectively. These are referred to as ARG6 and ARG8 composites. After curing, the panels were cut to dimensions of 50×150 mm, and 18 specimens were obtained for each type of the composites. The thicknesses of ARG6 and ARG8 panels were approximately 7 to 9 and 11 to 13 mm, respectively.

Sisal Fiber Composites

To increase the durability of the sisal fibers in the alkaline environment, a cementitious matrix consisting of 50% portland cement, 30% metakaolin, and 20% calcined waste crushed clay brick discussed in Chapter 19 was used [26]. The mortar matrix had a mix design of 1:1:0.4 (cementitious material–sand–water by weight). Wollastonite fiber (JG class), obtained from Energyarc, were used as a microreinforcement in the composite production (V_f = 5%). The volume fraction of sisal fibers was 10%.

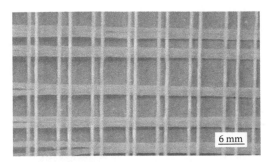

FIGURE 21.1 AR glass fabrics.

TABLE 21.1
Mix Design of Cement Paste

Material	Weight
Cement	8159 g
Water	3263 g
Silica fume	677 g
Superplasticizer	12 mL

Drop Weight Impact Setup

An impact test setup based on a free-fall drop of an instrumented hammer on a three-point bending specimen. The schematic of the system is shown in Figure 21.2. The drop heights range from 1 to 200 cm and can be controlled by means of an electronic hoist and release mechanism. An antirebound system consisting of a pneumatic brake system triggered by a contact type switch was used to stop the hammer after the duration of impact was completed.

The components of experimental setup consist of the following sections. The entire moving part that impacts the specimen includes the free weight, the frictionless bearings along the drop columns, the load cell, the connection plate, and a set of threaded rods. This entire assembly was referred to as the hammer and weighed 134 N. The hammer was released from a predetermined drop height by means of an electronic brake release mechanism. The impact force induced by the free-fall weight was measured by four independent instruments, a load cell with a range of 90 kN mounted on the hammer behind the blunt-shaped impact head, an alternative piezoelectric load cell referred to as a load washer, and also a set of accelerometers mounted on the specimen and hammer. A second strain-gage-based load cell with 90 kN capacity was mounted beneath the support plate and measured the force transmitted to the equipment base. A linear variable differential transformer (LVDT) with a range of +10 mm was connected to the specimen by means of a lever arm. Three accelerometers were used to document the acceleration–time history of the hammer, specimen, and support system. Two accelerometers with a capacity of ±500 g were mounted on the top load cell and tension zone of the specimen. The third accelerometer with the range of ±100 g was placed on bottom base of the equipment beneath the support plate. The data acquisition system consisted of a PC-based National Instruments PCI acquisition card and LABVIEW VIs with trigger function that can record signals from load cell, accelerometers and LVDT simultaneously at sampling rates of up to 100 kHz. The entire duration of the test lasted approximately 20 to 60 ms, depending on the type of test specimen. Several MATLAB programs were developed for data processing to compute the frequency content of the specimen and equipment, to filter and smooth the raw data with a low-pass filter, and to calculate the mechanical properties. A high-speed digital camera (Phantom Version 7) was used to capture pictures of the samples during the impact tests. The damage caused in the samples for the different drop heights was then compared by visual examination.

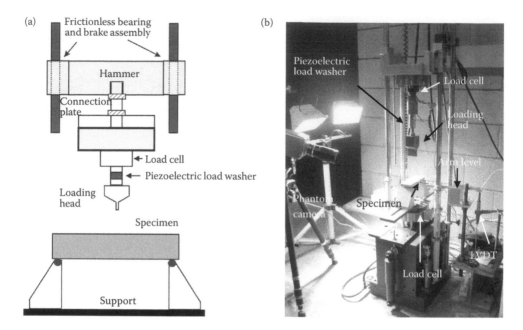

FIGURE 21.2   (a) Schematic diagram and (b) actual picture of the impact test setup.

The ARG6 and ARG8 specimens were tested in both vertical and horizontal positions with respect to the direction of applied impact load (beam and plate, respectively). The fabrics of composites in the vertical orientation (beam type) were parallel to the direction of load, referred to as ARG6B and ARG8B, whereas the fabrics were perpendicular to the direction of load application (plate type) in the horizontal specimens, referred to as ARG6P and ARG8P. The span of composites was 127 mm for three-point flexural impact tests. The sisal composites were tested in horizontal direction.

Dynamic Calibration

The signal acquisition during a high dynamic test is strongly conditioned by the nature of the test procedures [27]. Rapid variation due to acceleration excites vibrations depending on the stiffness and mass of the specimen, support, or the hammer, resulting in signal disturbances. Interpretation of raw signals without prior knowledge of the dynamic characteristics of the system would be questionable. After the system dynamics are identified, it is essential to filter the data to retain the material responses [28].

Modal analysis tests were conducted to identify the predominant frequencies of the hammer and the test specimens. All the tests were recorded with sampling frequency of 20 kHz. Figure 21.3a shows the Fourier amplitude spectra of the hammer acceleration data by using fast Fourier transform. The predominant frequency of the hammer is approximately 5 kHz. Using the same method, one can also get the predominant frequency of specimens which were less than 1 and 2 kHz for ARG6P and ARG6B specimens, respectively, as shown in Figures 21.3b and 21.3c, respectively. Consequently, a low-pass filter with cutoff frequency of 2 kHz was selected to separate the high-frequency response of the instrumentation from the test data. The filtered data were considered to represent the real response of the specimens. The Fourier amplitude spectra of filtered data were plotted in Figure 21.3 as well. Figures 21.4a and 21.4b indicate the difference between the original acceleration and the filtered acceleration of the ARG6B specimen and hammer, respectively. As the high-frequency contents are filtered out, acceleration amplitudes much lower than original values are obtained.

The acceleration–time history of hammer and the specimen were recorded by two individual accelerometers. As shown in Figure 21.5a, the acceleration of hammer at four different drop heights of 50, 100, 200, and 250 mm indicates that the peak acceleration during impact is increasing with

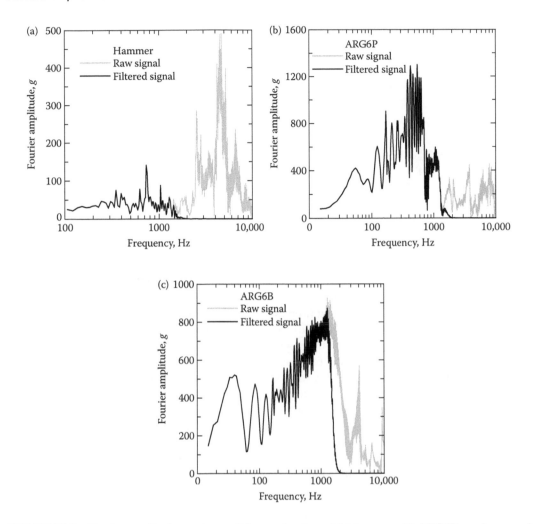

FIGURE 21.3 Fourier amplitude spectrum of the acceleration of (a) hammer, (b) ARG6P specimen, and (c) ARG6B specimen.

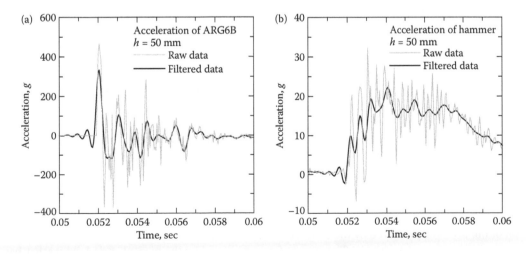

FIGURE 21.4 Comparison of raw and filtered acceleration of (a) ARG6B specimen and (b) hammer.

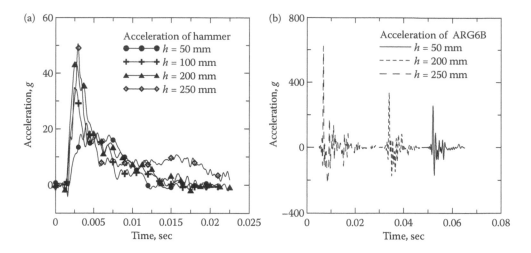

FIGURE 21.5   Acceleration of (a) hammer and (b) ARG6B specimens at different drop heights.

drop height from approximately 20 g at drop height of 50 mm to 50 g at drop height of 250 mm. Figure 21.5b shows the acceleration of ARG6B specimens during impact at three different drop heights of 50, 100, and 200 mm. The transient peak specimen acceleration experienced increases from 270 g at drop height of 50 mm to 610 g at drop height of 250 mm. Because the mass of specimen is around 0.2 kg, according to Newton's second law of motion, an inertial force of 54 N is obtained at drop height of 50 mm. This value is roughly 3% of the impact force measured by the conventional load cell. At higher impact heights such as 250 mm, the inertial force is approximately 120 N, which is still approximately 3% of the impact force measured by the conventional load cell. The inertial force on the specimen during impact was therefore neglected.

The procedure to obtain impact force from hammer acceleration data was used as a verification check. The force history was calculated by applying the Newton's second law to measured acceleration data:

$$F(t) = m \times a(t) \tag{21.1}$$

where $a(t)$ is the acceleration history and m is the mass of the hammer.

Figures 21.6a and 21.6b show the force response measured by the conventional load cell and calculated from acceleration of the hammer according to Equation 21.1, respectively. The acceleration was filtered by a low-pass filter as discussed previously. There is negligible difference between the calculated impact force with that measured by the conventional load cell at drop height of 50 mm. However, at the drop heights of 100 mm and above, the calculated impact force dramatically increases as compared with the conventional load cell measurement. For example, at drop height of 200 mm, the impact force measured by the conventional load cell is 4000 N, whereas the calculated impact force is 6000 N.

The accuracy of force measurement by a conventional load cell is further verified using a piezoelectric load washer (Kistler 9041A) with a capacity of 90 kN and rigidity of 7.5 kN/μm that was installed between the conventional load cell and the blunt-shaped impact head. The response frequency of the load washer is 33 kHz. The load signal was amplified through a Kistler 5010B dual mode charge amplifier. The signals were filtered by a low-pass filter with cutoff frequency of 2 kHz to eliminate high-frequency noise during data processing. Figures 21.7a and 21.7b show the comparison of impact force measured by the conventional load cell and the piezoelectric load washer versus time and deflection, respectively. The impact force measured by the piezoelectric load washer has more oscillations than the conventional load cell during the initial loading range. This behavior is

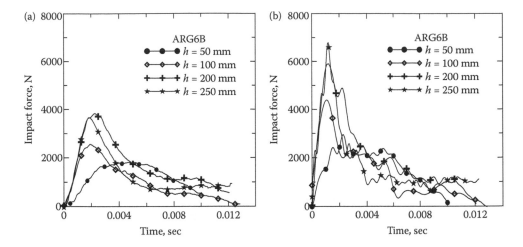

FIGURE 21.6 Impact force variation for ARG6B specimens at different drop heights: (a) measured by conventional load cell and (b) calculated from acceleration of hammer (according to Equation 21.1).

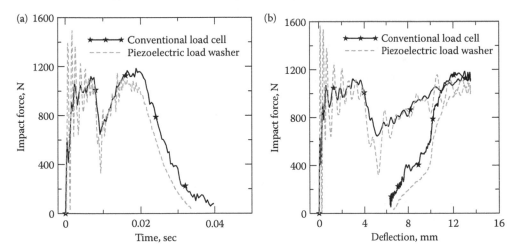

FIGURE 21.7 Comparison of impact forces measured by conventional load cell and piezoelectric load washer: (a) impact force versus time and (b) impact force versus deflection.

due to the specimen and the hammer reacting to the impact by oscillating at their natural frequencies, whereas the piezoelectric load washer has a higher response frequency than the conventional load cell and is able to acquire details of the oscillation. In consideration to the absorbed energy, there is negligible difference between them on the basis of the impact force measured by these two load transducers. Therefore, results listed are based on the impact force measured by the conventional load cell.

The input energy is the potential energy of the hammer and depends on its drop height, mass, and the amount of energy lost during the free-fall drop of the hammer. Some of the input energy is absorbed by the test specimen, whereas the remaining energy is either dissipated by friction, or transferred to the test setup through the supports after the impact event.

The input potential energy of the hammer, U_i, was defined as follows:

$$U_i = mgH = \frac{1}{2}mv_0^2 + U_d = U_k + U_f + U_d \tag{21.2}$$

where m is same as above, g is the acceleration of gravity, H is the drop height of hammer, v_0 represents the hammer velocity before impact, U_d represents the frictional dissipated energy between the time of release of the hammer until just before the impact event, U_k represents the absorbed energy by the specimen, and U_f represents the energy remaining in the system after the failure of the specimen has taken place. This energy may be elastically stored in the sample, and result in the rebound, or transmitted through the specimen to the support.

The absorbed energy (kinetic energy dissipated in the specimen), U_k, was defined as follows:

$$U_k = \int_{t=0}^{t=t^*} P(t)v(t)dt \approx \sum P(t)\Delta d(t) \tag{21.3}$$

where $P(t)$ and $v(t)$ represent the force and velocity history of the impact event, t^* represents the impact event duration, and $\Delta d(t)$ represents the deflection increment history of test specimen.

The maximum flexural stress, σ_f, was measured using the linear elastic small displacement bending equation as follows:

$$\sigma_f = \frac{3}{2}\frac{P_m L}{bh^2} \tag{21.4}$$

where P_m is the maximum load recorded during testing, b and h are the width and thickness of the test specimen, respectively, and L is the specimen span.

The strain rate for three-point bending test was computed in a method based on continuous mechanics proposed by Land [29]. The general equation of the model is:

$$\varepsilon_{max} = \frac{2h(N+2)y}{(L-a)[L+a(N+1)]} \tag{21.5}$$

where h and L are same to the above, N is the creep exponent, and y is the deflection. For a three-point bending configuration and elastically deflected material: $a = 0$ and $N = 1$ and differentiating with respect to time; hence,

$$\dot\varepsilon = \frac{d\varepsilon}{dt} = \frac{6hV}{L^2} \tag{21.6}$$

where t is time and V is the velocity. The impulse was computed using Equation 21.7:

$$I = \int_0^{t_{max}} F(t)dt \tag{21.7}$$

where I is the impulse and $F(t)$ is the force as a function of time.

RESULTS AND DISCUSSIONS

AR Glass Composite

Figure 21.8 represents the time history of impact load, deflection, and acceleration response of a test specimen subjected to low-velocity impact from a drop height of 100 mm. Note that both the acceleration and the deceleration response of the hammer and the specimen indicate that the specimen may accelerate after the initial contact to as high as 330 g and experience loads of up to 2600

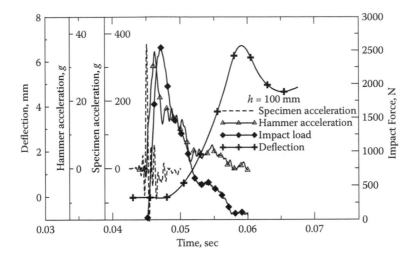

FIGURE 21.8 Time history of load, deflection, and acceleration of AR glass composite specimen subjected to low-velocity impact.

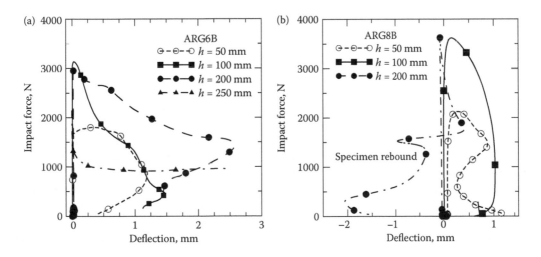

FIGURE 21.9 Impact force versus deflection: (a) ARG6B specimens and (b) ARG8B specimens.

N. There is a phase lag in the deflection signal due to the loading of the specimen as the maximum deflection is achieved while the load drops significantly; the specimen decelerates and comes to rest as the load and deflection signals stabilize after the impact event. Permanent deflection and postfailure oscillations exist after the impact event (Figure 21.9).

EFFECT OF DROP HEIGHT

The ARG6B specimens were tested at hammer drop heights of 50, 100, 200, and 250 mm and the input energy varied from 7 to 35 J. Figure 21.10a shows the impact load–deflection curves of the ARG6B specimens tested for all the drop heights. At drop heights of 50, 100, and 200 mm, the rebound of specimens after impact is evident. In this case, it is shown that the ARG6B specimen respond to the impact loads as a stiff beam. The area under the impact load–deflection curve is the deformation energy that is initially progressively transferred from the hammer to the beam and then given back from the beam to the rebounding hammer, the area included inside the curve refers to

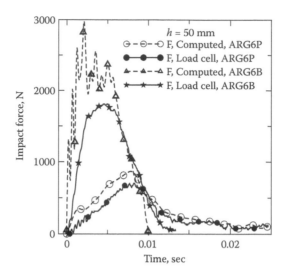

FIGURE 21.10 Effect of specimen orientation, impact forces measured by conventional load cell and calculated from acceleration of hammer according to Equation 21.1.

the energy absorbed by the test specimen during impact [30]. At drop height of 250 mm, the impact load–deflection curve after peak force was flat, indicating no rebound. The area under the curve is the deformation energy that is progressively transferred from the hammer to the specimen, in this case of absence of rebound, is also the energy absorbed during the impact [19].

The ARG8B specimens were tested at hammer drop heights of 50, 100, and 200 mm, and the input energy varied from 7 to 28 J. Figure 21.10b shows the impact force versus deflection curves, indicating that the impact forces increase proportional to the hammer drop height increased of up 100 mm. For the drop height of 200 mm, the ARG8B specimens were slightly stronger than those tested at 100 mm. It also shows that negative deflection (specimen rebound), as much as 1 to 2 mm, was present at midspan of specimen at the drop height of 200 mm. Because the lever part of the deflection measurement system was connected at the midspan, the LVDT measured the deflection as well as the rebound of specimen. The rebound of the ARG8B specimens could not be prevented by pneumatic brake system. Their absorbed energy gradually increased up to the drop height of 100 mm as calculated to the point of rebound. Because of rebound, the maximum deflection and the absorbed energy of test specimen at the drop height of 200 mm were not compared with those at other drop heights.

Table 21.2 summarizes the impact response of all test specimens. For the ARG6B specimens, the impact force increased as the drop height increased initially and remained almost constant at drop heights of 200 and 250 mm. The absorbed energy also increased with the drop height. Although for ARG8B specimens, the impact force increased from 2076 N at the drop height of 50 mm to 3618 N at the drop height of 100 mm; however, there were no significant increases at the drop height of 200 mm. The absorbed energy continuously increased as the hammer drop height increased. It was also observed that the deflections of ARG6B and ARG8B specimens continuously increased as the drop height increased.

The ARG6P and the ARG8P specimens that represented a plate loading condition were tested at drop heights of 50 and 100 mm. Higher drop heights were not attempted because the specimen failed because of interlaminar shear. The impact load, maximum flexural stress, absorbed energy, and maximum deflection of ARG6P specimens increased from 799 to 1062 N, from 17.8 to 29.9 MPa, from 4.03 to 5.04 J, and from 9.23 to 14.67 mm, respectively, when the drop height increased from 50 to 100 mm. The impact load, maximum flexural stress, absorbed energy, and maximum deflection

TABLE 21.2
Summary of Impact Response of AR Glass Composite Specimens

Specimen Type	Drop Height (mm)	Potential Energy (J)	Drop Velocity (m/s)	Max. Impact Force (N)	Max. Stress (MPa)	Max. Deflection (mm)	Absorbed Energy (J)
ARG6P	50	7	1.0	799	17.8	9.23	4.03
	100	14	1.4	1062	29.9	14.67	5.04
ARG6B	50	7	1.0	1802	10.5	1.24	1.68
	100	14	1.4	2501	16.6	1.63	2.29
	200	28	2.0	3771	24.8	2.58	3.93
	250	35	2.2	3569	23.2	3.32	4.90
ARG8P	50	7	1.0	585	12.45	5.58	2.57
	100	14	1.4	934	27.5	10.21	4.89
ARG8B	50	7	1.0	2076	14.3	0.97	2.78
	100	14	1.4	3618	19.7	1.63	2.96
	200	28	2.0	3638	22.0	3.58	3.42

of ARG8P specimens increased from 585 to 934N, from 12.45 to 27.5 MPa, from 2.57 to 4.89 J, and from 5.58 to 10.21 mm, respectively, when the drop height increased from 50 to 100 mm.

Effect of Number of Lamina and Specimen Orientation

It is observed that the maximum flexural stress and absorbed energy of beam specimens increased with the number of fabric layers, whereas in the plate specimens, the same properties decreased with an increase in the number of fabric. The failure in the plate specimens was due to interlaminar shear failure. At the drop height of 50 mm, the maximum impact force of ARG8B specimens was approximately 15% higher than the ARG6B composites as shown in Table 21.2. Because of the higher specimen stiffness in the beam mode compared with the plate mode, beam specimens had lower deflections during impact than plate specimens. At the same drop height, the impact force experienced by beam specimens was much higher than plate specimens regardless of means of force measurement by the conventional load cell or acceleration response of the hammer as shown in Figure 21.11. For ARG6P specimens at the drop height of 50 mm, the maximum impact force measured by the conventional load cell was slightly less than the calculated force. However, ARG6B specimen showed that the maximum impact force was approximately 1800 N measured by load cell and was approximately 2300 N for averaged computed value of multiple peaks. Figure 21.11 also shows that the test duration of ARG6B specimen is only half of that of ARG6P specimen. In beam specimens, complete fracture does not take place as cracks form and close due to rebound, whereas in plate specimens, interlaminar shear is the dominant failure mode. The beam specimens use the fabrics as shear reinforcement as well as flexural reinforcement, whereas in the plate samples the role of fabric is primarily in the flexural reinforcement and there is no resistance offered to shear delamination.

Energy Absorption

The input potential energy was determined as a function of the hammer drop height and weight, assuming that no frictional losses took place during the free fall. Equation 21.2 can be simplified as follows:

$$U_i = mgH = \frac{1}{2}mv_0^2 = U_k + U_f \quad (21.8)$$

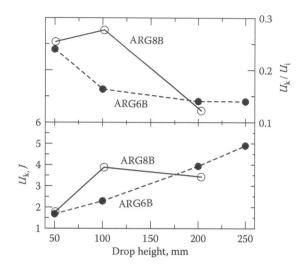

FIGURE 21.11   Effect of fabric layer numbers and drop heights on the absorbed energy and ratio of absorbed energy to the input potential energy for beam specimens.

The ratio of the absorbed energy to the input potential energy was determined for each drop height, defined as follows:

$$\beta = \frac{U_k}{U_i} \tag{21.9}$$

The absorbed energy and its ratio to the input energy were determined for each drop height for ARG6B and ARG8B specimens, as shown in Figure 21.11. The absorbed energy increased with drop height, whereas the ratio of absorbed energy to potential energy decreased with drop height for ARG6B specimens. The absorbed energy and the ratio of absorbed energy to potential energy for ARG8B specimens have similar trends. They increased first and then decreased when the drop height increased from 50 to 100 mm and then to 200 mm. Although ARG8B specimens carried higher impact loads than ARG6B specimens, their absorbed energy levels are comparatively lower for the same drop height levels. This behavior can be attributed to ARG8B specimens displaying less deflection than ARG6B specimens because of stiffer behavior, allowing less damage to take place.

Sisal Fiber Composites

A typical flexural stress versus deflection behavior is presented in Figure 21.12 for drop height of 100 mm. Five distinct zones are represented by roman numerals. Zone 1 is the linear elastic range that ends at the formation of the first crack (limit of proportionality [LOP] region). The LOP range is similar to those obtained by the static tests. Zone 2 starts after the LOP and is characterized by multiple cracking behavior with strain hardening. An average of four parallel cracks was observed at this zone. Stiffness degradation takes place during zone 2, and maximum flexural stress values are obtained at the end of this region. Strain softening behavior characterizes zones 3 and 4. Damage that takes place in the matrix and fiber–matrix interface is a function of drop height and deflection level. Depending on the nature of available energy, if there is sufficient ductility in the sample to absorb the applied energy, some of the stored energy release causes a rebound that is characterized by a reduction in deflection of the sample as the load is decreased. Zone 5 characterizes the rebound of the specimen.

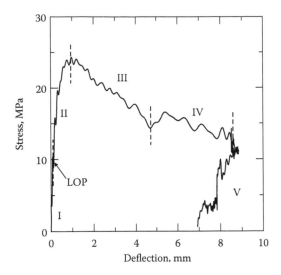

FIGURE 21.12 Different damage zones in a typical stress versus deflection impact curve.

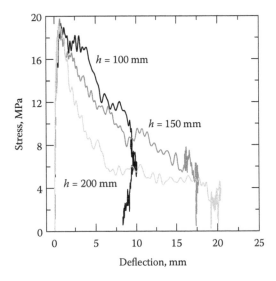

FIGURE 21.13 Effect of drop height on the stress versus deflection response of the sisal fiber reinforced composite.

Figure 21.13 compares the flexural stress versus deflection response for hammer drop height ranging from 100 to 200 mm, representing an input potential energy from 10.55 to 21.1 J. Note that the strain rates calculated range from 3.82 to 5.83 s^{-1}. These tests can be characterized as low-velocity impact. The duration of the impact is between 4 and 6.5 ms, depending on the drop height and the maximum load capacity is around 850 N. No significant effect in the ultimate strength was noticed with the increase in drop height. The average strength decreased from 23 ($h = 100$ mm) to 18 MPa ($h = 200$ mm) with standard deviations ranging from 3.3 to 4.6 MPa, respectively, as shown in Table 21.3. These values are in the same range of the modulus of rupture (MOR) for static bending tests. It is also observed that the maximum deflection increases with increasing drop height. The postpeak stiffness seems to decrease with increasing input potential energy, which may be possibly attributed to a reduction in bond characteristics of fiber–matrix interface. The frictional nature of the bond has been shown to be affected by the loading rate [31, 32].

TABLE 21.3
Summary of Impact Test Results

Height (mm)	Potential Energy (J)	Strain Rate (s^{-1})	Specimen ID	Max. Flexural Stress (MPa)	Deflection at Max. Stress (mm)	Max. Deflection (mm)	Initial Stiffness (N·mm)	Velocity (mm/s)	Peak Absorbed Energy (kJ/m^2)	Total Absorbed Energy (kJ/m^2)
50	10.55	3.82	1	28.01	2.51	7.26	9174	725	3.40	9.2
			2	21.56	1.11	10.13	3248	988	1.18	10.4
			3	25.16	1.96	7.62	1716	870	–	7.5
			4	24.30	0.97	8.84	1969	694	1.08	8.2
			5	19.43	0.49	9.56	5541	996	0.93	7.9
			6	19.35	0.80	10.04	5344	865	0.78	8.5
			Average	22.97 (3.4)	1.31 (0.76)	8.91 (1.2)	4498 (2806)	856 (127)	1.48 (1.1)	8.6 (1.0)
100	15.83	5.46	1	27.57	2.40	13.99	2038	953	1.53	17.9
			2	18.78	0.40	18.19	2846	1503	0.29	15.9
			3	23.40	1.08	17.03	1943	1275	1.28	15.4
			4	19.76	0.68	17.69	2206	972	1.11	11.2
			5	19.22	0.38	20.14	3043	1330	0.95	10.5
			6	14.08	2.23	19.77	2703	1313	1.34	12.4
			Average	20.47 (4.6)	1.20 (0.9)	17.81 (2.2)	2463 (460)	1224 (217)	1.09 (0.4)	13.9 (2.9)
200	21.1	5.83	1	16.26	0.44	20.07	3983	1369	0.98	10.7
			2	21.86	0.79	20.37	1840	1213	0.97	13.1
			3	21.53	1.80	20.40	1189	1184	–	9.2
			4	13.74	1.07	20.46	1784	1462	0.70	8.4
			5	15.78	0.48	20.11	2012	1291	0.60	10.4
			6	19.18	0.55	20.40	2409	1329	1.19	8.5
			Average	18.06 (3.3)	0.85 (0.5)	20.30 (0.2)	2203 (975)	1309 (102)	0.89 (0.2)	10.0 (1.1)

Note: Values in parentheses are standard deviation.

The initial stiffness decreases from 4.5 to 2.2 kN·mm with increasing drop height (see Figure 21.14). Stiffness degradation also takes place after the first crack formation. No significant difference is noticed for stiffness degradation as a function of drop height. The total absorbed energy increases from static test to impact achieving its maximum value of 13.9 kJ/m² at drop height of 150 mm (see Figure 21.15a). By normalizing the total absorbed energy to the input energy, it is observed a constant ratio of 0.5 from a height of 100 to 150 mm with a posterior drop to 0.3. This indicates that the drop height of 150 mm (15.83 J) is a threshold level. The impulse follows similar behavior of absorbed energy. There is an increase from 18 to 20 N·s by increasing the drop height from 101.6 to 152.4 mm with a posterior decrease to 14 N·s (see Figure 21.15b).

Figure 21.16 shows the damage in the composite caused at different input energies. The damage is associated with matrix cracking and delamination and is initiated by the formation of tensile flexural cracks, which may be deflected to shear cracks because of bridging fibers. Tensile cracks

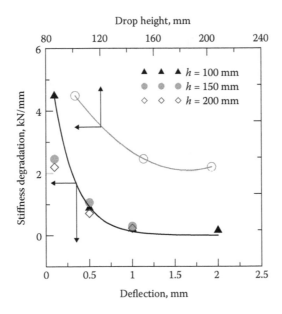

FIGURE 21.14   Effect of drop height on the initial stiffness and degraded stiffness.

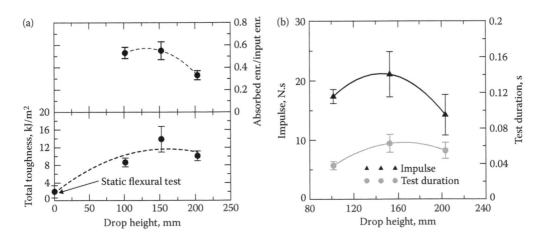

FIGURE 21.15   Effect of drop height on (a) total absorbed energy and (b) impulse. Total Absorbed energy was normalized in respect to the input energy in panel a.

are introduced when in-plane normal stresses exceed the transverse tensile strength of the matrix layer (see Figures 21.16a and 21.b) at the bottom surface of the composite but may not lead to material failure. Shear cracks start as flexural matrix cracks away from the midcenter and convert to delamination cracks as they intersect the continuous fibers. This indicates that transverse shear stresses play a role in their formation. Shear cracks were only observed when the drop height was 150 mm or higher (see Figure 21.16b). Delamination, which is the debonding between adjacent layers, is of most concern because it can reduce the strength of the composite. Damage caused by delamination was only observed at the drop height of 200 mm and resulted in a complete material failure. It can be seen from Figure 21.17 that shear cracks (initiated after 5000 μs) started the delamination process. Nevertheless, even with the formation of shear cracks, the composites presented ductility and strength similar to the lower drop heights reaching MOR of 18.06 MPa at deflection of 0.85 mm.

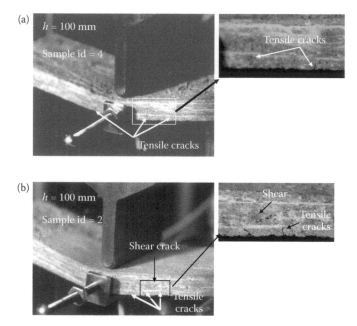

FIGURE 21.16 Damage mechanisms in composites subjected to impact at drop height = (a) 101.6 mm and (b) 152.4 mm. Tensile cracks were observed in the bottom face and shear cracks in the midplane.

FIGURE 21.17 Damage mechanisms in sample subjected to impact at drop height of 203.2 mm. Shear crack (a) initiated the delamination process in panel b.

DISCUSSIONS

The flexural impact behavior of AR glass fabric and sisal fiber composites subjected to low-velocity impact loading that was conducted used an experimental setup that used the impact force, accelerations of hammer and test specimen, and deflection at midspan of composite specimens. The acceleration data of hammer and the specimen provide alternative methods to calculate the impact force experienced by specimen and study the inertial effect during impact loading. The system dynamics need to be identified and a low-pass filter with cutoff frequency of 2 kHz is effective at the strain rates studied to reduce the noise due to oscillation of equipment and specimen at their natural frequencies. Test results to determine the flexural impact behavior in terms of maximum flexural stress, maximum deflection, absorbed energy, and crack patterns of the specimens. The results can be summarized as the follows:

(1) Beam type specimens have a larger load carrying capacity with a lower deflection than plate specimen due to coupled effect of the orientation of cross-section and the direction of fabric.

(2) The maximum flexural stress and absorbed energy of beam specimens increased with the number of fabric layers. The maximum flexural stress and absorbed energy of plate specimens decreased with the number of fabric layers because of interlaminar shear failure. Maximum flexural stress as high as 29.9 MPa for plate specimens was observed when subjected to a range of potential energy in the range of 7 to 14 J and as high as 24.8 MPa for beam specimens when the potential engergy was in the range of 7 to 35 J.

(3) The total absorbed energy increased with drop height, whereas the ratio of absorbed energy to potential energy decreased with drop height for ARG6B specimens. In the ARG8B specimens, the absorbed energy and the ratio of absorbed energy to potential energy have similar trends. They increased first and then decreased when the drop height increased from 50 to 100 mm and then to 200 mm.

(4) The acceleration and the deceleration response of the hammer and the sisal fiber composites indicate that the specimen may accelerate after the initial contact to as high as 100 m/s^2 and may experience loads of up to 850 N for an input energy of 10.55 J. For higher input energies (15.83 and 21.1 J), the sample acceleration increased to 170 m/s^2, but no significant difference was noticed for the load. The maximum flexural stress ranged from 22.97 MPa (input energy of 10.55 J) to 18.06 MPa (21.1 J), with standard deviation ranging from 3.3 to 4.6 MPa. The average static MOR was approximately 24 MPa. The absorbed energy increased from static test to impact achieving its maximum value of 13.89 kJ/m^2 at input energy of 15.83 J.

(5) Initial stiffness decreases from 4.4 to 2.2 kN·mm when increasing the input energy from 10.55 to 21.1 J. Stiffness degradation takes place after the first crack formation, and no significant difference was noticed between stiffness degradation as a function of impact energy.

(6) The damage morphology consisted of matrix cracking and delamination. The damage process is initiated by the formation and distribution of flexural cracks. Tensile and shear cracks were observed. Delamination resulting in total failure was only observed when the impact energy was above 21.1 J.

REFERENCES

1. Abrate, S. (1998), *Impact on Composite Structures*, Cambridge, UK: Cambridge University Press.
2. Romano, G. Q., Silva, F. A., Toledo Filho, R. D., Fairbairn, E. M. R., and Battista, R, "On the removal of steel fiber reinforced refractory concrete from thin walled steel structures using blast loading," in Unified International Technical Conference on Refractories (Unitecr 07), Dresden, 2007, 509–512.
3. Balaguru, P. N., and Shah, S. P. (1992), *Fiber-Reinforced Cement Composites*. Singapore: McGraw-Hill.

4. Silva, F. A. (2004), "Toughness of non conventional materials," MSc thesis, Department of Civil Engineering, PUC-Rio.

5. Silva, F. A., Ghavami, K., and d'Almeida, J. R. M., "Toughness of cementitious composites reinforced by randomly sisal pulps," Eleventh International Conference on Composites Engineering—ICCE-11, Hilton Head Island, SC, 2004.

6. Romano, G. Q., Silva, F. A., Toledo Filho, R. D., Tavares, L. M., Battista, R., and Fairbairn, E. M. R., "Impact loading behavior of steel fiber reinforced refractory concrete exposed to high temperatures," Fifth International Conference on Concrete under Severe Conditions: Environment and Loading, Tours, France, Vol. 2, 2007, 1259–1266.

7. Liu, D., Raju, B. B., and Dang, X. (1998), "Size effects on impact response of composite laminates," *International Journal of Impact Engineering*, 21(10), 837–854.

8. Tang, T. T., and Saadatmanesh, H. P. E. (2003), "Behavior of concrete beams strengthened with fiber-reinforced polymer laminates under impact loading," *Journal of Composites for Construction*, 7(3), 209–218.

9. Bindiganavile, V., Banthia, N., and Aarup, B. (2002), "Impact response of ultra-high strength fiber reinforced cement composite," *ACI Materials Journal*, 99(6), 543–548.

10. Alhozaimy, A. M., Soroushian P., and Mirza F. (1996), "Mechanical properties of polypropylene fiber reinforced concrete and the effects of pozzolanic materials," *Cement and Concrete Composites*, 18, 85–92.

11. Banthia N., Yan C., and Sakai K. (1998), "Impact resistance of fiber reinforced concrete at subnormal temperatures," *Cement and Concrete Composites*, 20, 393–404.

12. Bindiganavile, V., and Banthia, N. (2001), "Polymer and steel fiber-reinforced cementitious composites under impact loading. Part 1: Bond–slip response," *ACI Materials Journal*, 98(1), 10–16.

13. Bindiganavile, V., and Banthia, N. (2001), "Polymer and steel fiber-reinforced cementitious composites under impact loading. Part 2: Flexural toughness," *ACI Materials Journal*, 98(1), 17–24.

14. Lok, T. S., and Zhao, P. J. (2004), "Impact response of steel fiber-reinforced concrete using a split hopkinson pressure bar," *Journal of Materials in Civil Engineering*, 16(1), 54–59.

15. Choi, I. H., and Lim, C. H. (2004), "Low-velocity impact analysis of composite laminates using linearized contact law," *Composite Structures*, 66, 125–132.

16. Wang, N. Z., Sidney, M., and Keith, K. (1996), "Fiber reinforced concrete beams under impact loading," *Cement and Concrete Research*, 26(3), 363–376.

17. Manolis, G. D., Gareis, P. J., Tsono, A. D., and Neal, J. A. (1997), "Dynamic properties of polypropylene fiber-reinforced concrete slabs," *Cement and Concrete Composites*, 19, 341–349.

18. Li, Z., Mu, B., and Chui, S. N. C. (2001), "Static and dynamic behavior of extruded sheets with short fibers," *Journal of Materials in Civil Engineering*, 13(4), 248–254.

19. Mobasher, B., Pivacek, A., and Haupt, G. J. (1997), "Cement based cross-ply laminates," *Journal of Advanced Cement Based Materials*, 6, 144–152.

20. Peled, A., and Mobasher, B. (2004), "Pultruded fabric–cement composites," *ACI Materials Journal*, 102(1), 15–23.

21. Peled, A., Sueki, S., and Mobasher, B. (2006), "Bonding in fabric–cement systems: Effects of fabrication methods," *Cement and Concrete Research*, 36(9), 1661–1671.

22. Mobasher, B., Peled, A., and Pahilajani, J. (2006), "Distributed cracking and stiffness degradation in fabric–cement composites," *Materials and Structures*, 39(3), 17–331.

23. Peled, A., and Mobasher, B. (2006), "Properties of fabric–cement composites made by pultrusion," *Materials and Structures*, 39(8), 787–797.

24. Zhu, D., Gencoglu, M., and Mobasher, B. (2009), "Low velocity flexural impact behavior of Ar glass fabric reinforced cement composites," *Cement and Concrete Composites*, 31(6), 379–387.

25. Gencoglu, M., and Mobasher, B. (2007), "Static and impact behavior of fabric reinforced cement composites in flexure," in *High Performance Fiber Reinforced Cement Composites (HPFRCC5)*, Mainz, Germany: RILEM Publications S.A.R.L., PRO 53, 463–470.

26. Toledo Filho, R. D., Silva, F. A., Fairbairn, E. M. R., and Melo Filho, J. A. (2010), "Physical and mechanical properties of durable sisal fiber–cement composites," *Construction and Building Materials*, 24(5), 777–785.

27. Cheresh, M. C., and McMichael, S. (1987), "Instrumented impact test data interpretation," in Kessler, S. L., et. al. (eds.), *Instrumented Impact Testing of Plastic and Composite Materials*, Philadelphia: STP: ASTM, 963.

28. Cain, P. J. (1987), "Digital filtering of impact data," in Kessler, S.L. et al. (eds.), *Instrumented Impact Testing of Plastic and Composite Materials*, Philadelphia: STP: ASTM, 963.

29. Land, P. L. (1979), "Strain rates for three and four point flexure tests," *Journal of Materials Science*, 14, 2760–2761.
30. Belingardi, G., and Vadori, R. (2002), "Low velocity impact tests of laminate glass–fiber–epoxy matrix composite material plates," *International Journal of Impact Engineering*, 27, 213–229.
31. Sueki, S., Soranakom, C., Peled, A., and Mobasher, B. (2007), "Pullout–slip response of fabrics embedded in a cement paste matrix," *Journal of Materials Engineering*, 19, 718–727.
32. Pacios, A., Ouyang, C., and Shah, S. P. (1995), "Rate effect on interfacial response between fibres and matrix," *Materials and Structures*, 28, 1359–5997.

22 Textile Composites for Repair and Retrofit

INTRODUCTION

New elements are continuously developed for repair and strengthening of unreinforced masonry walls (UMW), beams, columns, and other structural elements [1]. To improve the tensile and flexural performance, reinforcements can be fiber reinforced polymer (FRP) or cement-based textile reinforcements. The main advantage of textiles cement-based composites as reinforcements is in their compatibility with the base material, long-term durability, fire resistance, and ultraviolet degradation resistance. Various production methods allow great flexibility in textile design [2] to control fabric geometry, yarn type, and orientation [3].

A large number of buildings constructed with unreinforced masonry were designed with little to no consideration of the forces due to seismic events. Earthquake introduces lateral in-plane and out-of-plane forces and is detrimental to UMW with typical damage ranging from minor cracking to catastrophic collapse. The low tensile strength of masonry is a limiting factor; hence, reinforcement can overcome this limitation in seismic areas [4]. Conventional rehabilitation and strengthening methods like injection grouting, insertion of reinforcing steel, prestressing, jacketing, and various surface treatments are commonly used [5, 6]. These methods involve use of skilled labor and equipment and are disruptive the normal function of the building.

Seismic risk surveys carried out in the early 2000s identified a high number of government owned masonry structures including as many as 7000 masonry buildings belonging to the U.S. Army in moderate and high seismic zones that are at risk of structural damage in the event of a major earthquake. The U.S. Army Engineer Research and Development Center, Construction Engineering Research Laboratory (ERDC/CERL) in partnership with the FRP Composite Industry, launched a program to investigate the ability of FRP composite systems to strengthen both lightly reinforced and UMW to better withstand damage during earthquakes [7]. The program followed FEMA 273 guidelines for new materials [8] for seismic rehabilitation of buildings and considered strength and deformation capacity when estimating seismic capacity.

The primary objective for seismic rehabilitation is to preserve life safety for building occupants in the event of the earthquake. Traditional strengthening techniques used for existing masonry buildings include diaphragm stiffening, tying walls to floor diaphragms, or adding steel tiles. In such cases, load-bearing masonry walls can exhibit direct shear failure mechanisms [9]. In case of seismic action mode II, shear sliding failure occurs in masonry panels along the unit/mortar interface, particularly for low levels of vertical loads and/or low friction coefficient [10]. Common rehabilitation procedures include ferrocement surface coating, casting shotcrete over a grid of reinforcing bars or infilling door or window openings with masonry. These techniques increase the lateral strength of masonry walls but sacrifice the deformation capacity. New rehabilitation procedures include adhering FRP strips, including surface-mounted FRP rods or adhering cement-based reinforced composites.

The basic material property data in tension and flexure for thin-sheet fabric-reinforced cement composites used for retrofit projects of UMW are discussed first, and the subject of long-term durability is addressed. Test methods are developed and verified using typical specimens. These test procedures support the protocol development for fabric–cement composite systems for reinforced and unreinforced masonry and are listed by the International Code Council (ICC AC 218), and recommendations for system installation in terms of matrix layer thickness, strand alignment, and orientation are provided [11].

A composite consisting of a sequence of layers of cement-based matrix and alkali-resistant (AR) glass-coated reinforcing grid were used. The experimental program included materials and structural tests. Because of the requirements for long-term performance, tensile and flexural tests were carried out on unaged and aged composites. Structural tests included in-plane shear concrete masonry unit (CMU) walls. Results of composite configurations were compared with FRP systems. Retention of tensile properties over time was approximately 75% to 80% after 50 years of service life. Structural tests demonstrated the ability of the cement-based system to strengthen the walls and showed superior performance of textile reinforced cement composite system compared with FRP alternatives. X-cracking failures were observed as there was no delamination of the system from the CMU walls. Because of its advantages and unique properties, this system is a potential alternative to traditional and new FRP masonry rehabilitation and strengthening techniques.

COMPARISON OF FRP SYSTEMS WITH TEXTILE REINFORCED CONCRETE

FRP composite systems have been widely used to retrofit concrete and masonry structures in high seismic zones. However, there are no specifications or design guidance available to support the use of FRP systems for masonry rehabilitation and retrofit in the public sector. By determining the failure modes, mechanisms, and strengths of cement composite laminates, relations to develop design parameters for strengthening walls can be developed. More than a hundred masonry walls were tested by ERDC/CERL in shear tension, in-plane shear, and in-plane rocking using commercially available FRP systems, E-glass and carbon fabrics, near surface-mounted FRP rods, and a polymer-modified AR glass grid reinforced cement-based systems.

FRP composites are commonly used for strengthening and rehabilitation of existing buildings because of high strength-to-weight ratio, ease of installation, high productivity, corrosion resistance, and minimal change in geometry [12]. However, the following disadvantages are associated with polymeric-based repair methods: low fire resistance, high cost of resins, incompatibility of resins with substrate, poor bond to the substrate unless substrate is primed, poor behavior of resins above glass transition temperature, application on wet surfaces or at low temperatures, no recyclability and reversibility of strengthening method, which is of particular relevance for application on historical buildings, ultraviolet degradation, lack of vapor permeability, moisture accumulation at the interface, chemical imbalance at the surface due to concentration gradients, health hazards due to special handling equipment, and training for installation are required.

As an alternate system, cement-based systems present a set of benefits that include the following advantages: fire resistance, compatibility and bond with the substrate masonry or concrete, breathability of the system that allows vapor and moisture transport through the matrix, reversibility, and ease of installation. Because of these key advantages and unique properties, cement-based retrofit systems are alternative to traditional and new FRP rehabilitation and strengthening techniques.

The objectives of the development program were to determine the mechanical properties of textile cement composites and to assess its effectiveness for UMW seismic performance from a load bearing capacity and deflection limits point of view.

EXPERIMENTAL PROGRAM

The experimental program included material tests and full-scale tests. The textile cement composite used consists of a cement-based matrix and two layers of AR-glass-coated reinforcing grid. The grid was developed by Saint-Gobain Technical Fabrics as a coated bidirectional weft insertion knit open grid made of AR glass fiber roving, consisting of machine direction (MD) and cross-direction (CD) strands connected perpendicularly at approximately 1 in. (24.5 mm) typical spacing

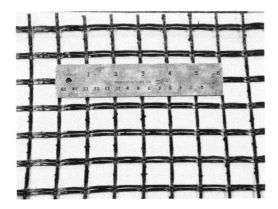

FIGURE 22.1 Coated AR glass grid.

with polyethylene stitch yarns (Figure 22.1). The coating used is a proprietary elastomeric polymer. Tensile strength, as measured by the manufacturer according to ASTM D6637-01 [13], is 255 lb./in. (45 kN·m). The cement-based matrix was a polymer-modified portland-cement-based mortar using 1% by volume of 12.7 mm long chopped AR glass fibers. The mechanical properties of the matrix, as measured by the manufacturer according to ASTM C887-05 [14] after 28 days, are compressive strength 3500 psi (24.1 MPa) and flexural strength 800 psi (5.5 MPa).

MATERIALS TESTS

Cement-based composites consisting of the textile cement composite system using two layers of grid were manufactured. The coupon tension and the flexure tests were carried out on composites in unaged and aged conditions using accelerated aging procedures. Sample-making procedures were explored to assess the effect of quality of installation, for example, strand alignment and orientation, on the mechanical performance of the composite. The effect of grid arrangement was studied. Full details of the materials tests are presented by Mobasher et al. [15] and Singla [16]. Tensile and flexural samples in MD and in CD were cut from panels after 28 days normal curing to the dimensions of 2.5 × 8 × 3/8 in. (60 × 200 × 10 mm). Accelerated aging was simulated by soaking the samples in water saturated with calcium hydroxide at 80°C, according to ASTM C1560-03 [17–19]. The acceleration factor used by the Proctor model [20, 21] states that 1 day of accelerated aging is equivalent to approximately 4.6 years aging in continental climate. A total of 30 samples were tested: 5 samples per combination test (tension), direction (MD, CD), and age: 0 days soaked (unaged) and 14 and 28 days soaked (aged).

STRUCTURAL TESTS

In-plane shear concrete masonry full-scale wall tests were carried out at ERDC/CERL, Champaign, Illinois, to simulate seismic action [22]. Lightly reinforced single-wythe masonry walls were used to simulate typical piers between windows of a building (Figure 22.2). The standard CMU 8 × 8 × 16 in. (20 × 20 × 40 cm) and the masonry mortar type *N*, with a compressive strength of 1885 psi (13 MPa), were used to build the walls.

During the test, a constant 54 kip (240.19 kN) total axial load was applied vertically to the specimen using the vertical actuators, maintaining the horizontal plane of the top of the specimen. The horizontal actuator provided a cyclic load into the wall specimen and forced the wall into a shearing load (Figure 22.3a). Two linear variable differential transformers (LVDTs) and eight linear deflection gages were used to monitor wall specimen movements (Figure 22.3b).

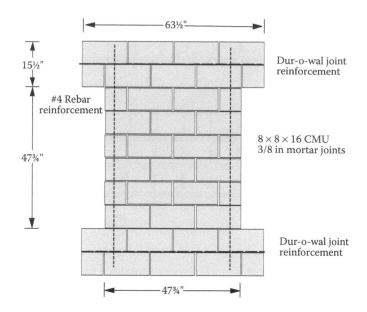

FIGURE 22.2   Substrate UMW.

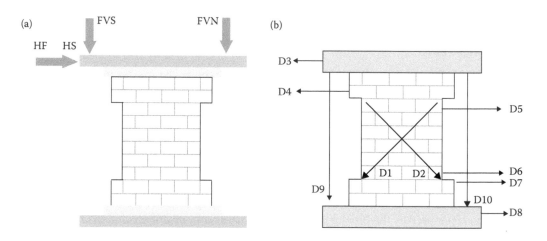

FIGURE 22.3   (a) Hydraulic actuator loading and (b) instrumentation layout.

A number of commercially available FRP systems using E-glass fabrics applied in various reinforcement configurations (Figure 22.4) and the textile cement composites were included in the experimental program. Textile cement composites application was a full coverage, on one side of the wall only. Three walls were tested varying the reinforcement layout to see the effect of the number of plies and orientation on the overall performance: Wall 1—two plies, 0°–90°; the first layer of grid was installed vertically, and the second layer of grid was installed in the horizontal direction, with the corresponding MD and CD strands at −90° with respect to the first ply, thus providing balanced bidirectional reinforcement. Wall 2—two plies, 0°–90°, ±45°; the first layer of grid was installed vertically, and the second layer of grid was installed with the corresponding MD and CD strands at ±45° with respect to the first ply, thus providing a multiaxial reinforcement. Wall 3—three plies 0°–90°, 2 × ±45°; the first layer of grid was installed vertically, the second layer of grid was installed with the corresponding MD and CD strands at +45° with respect to the first ply, and the third layer of grid was installed with the corresponding MD and CD strands at −45° with respect to the first

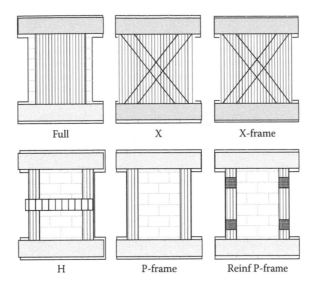

Full   X   X-frame

H   P-frame   Reinf P-frame

FIGURE 22.4 FRP reinforcement configuration.

ply, thus providing a multiaxial reinforcement and a balanced ±45° reinforcement. The walls were tested in cycles, increasing horizontal displacement according to FEMA 273 [9].

TENSILE PROPERTIES

Figures 22.5a and 22.5b represent the tensile response of as is samples in MD and XMD directions, respectively. The dashed regions are indicative of the range of properties that are observed from five replicate samples. Note that because of the variations in placement, a range of properties is observed in the response. However, one can use an average response to indicate the overall performance. Figure 22.6a compares the average flexural response of MD and XMD samples.

Mechanical response of specimens for three different aging cycles of 0, 14, and 28 days were studied in tension and flexure. The stress–strain plot as affected by accelerated aging has a significant difference in both the MD and the XMD strength values especially in the first 14 days of accelerated aging. This may be attributed to the machine direction being more susceptible to aging possibly because of the curvature in the fabric construction. Similar trend is observed in specimens aged for 28 days. The effect of aging in flexural response can be observed as a reduction in maximum deflection capacity followed by a significant reduction in postpeak load carrying capacity and toughness. Figure 22.6b is a comparative representation of typical tensile stress–strain curves for unaged and aged specimens for three different aging cycles of 0, 14, and 28 days. The trends are significantly affected by accelerated aging especially in the first 14 days of accelerated aging. Figure 22.7 presents the effect of aging on tensile properties. The average tensile strength retentions for 14 and 28 days aged specimens compared with unaged specimens are 80.6% and 79.9%, respectively, whereas the average ultimate strain retentions are substantially lower than 75.6% and 60.5%, respectively. Reductions in average ultimate strain are significantly higher than those in average tensile strength. Most of the reductions occur in the first 14 days of accelerated aging. By the 28 days of aging, the decrease in average tensile strength is comparable with that after 14 days, whereas the reduction in average tensile strain continues. This suggests that although decrease in tensile strength after 14 days accelerated aging is limited and slower, the composite undergoes significant embrittlement.

For all the samples studied, the flexural results are as much as three to four times higher than the tensile strength results, indicating that there is significant overprediction of tensile strength

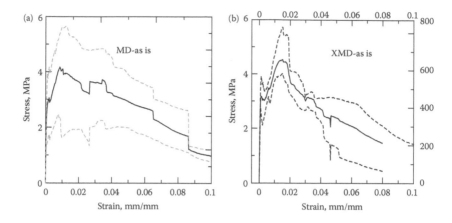

FIGURE 22.5 Typical unaged sample tensile stress–strain curves for (a) machine direction and (b) cross-machine direction.

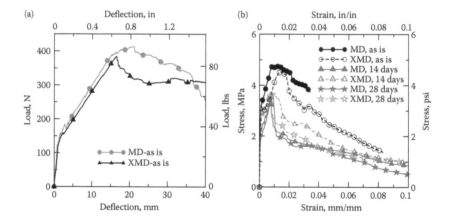

FIGURE 22.6 (a) Comparison of flexural response in machine and cross-machine direction. (b) Comparative effect of aging on tensile properties.

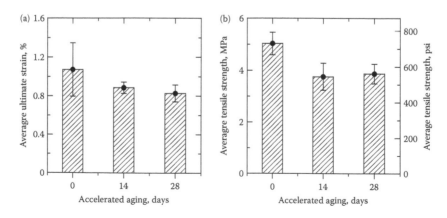

FIGURE 22.7 Effect of aging on tensile properties (a) tensile strength and (b) ultimate tensile strain.

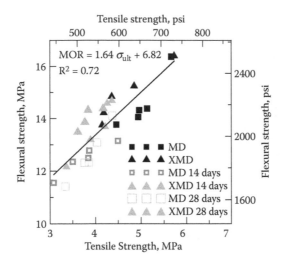

FIGURE 22.8 Correlation of tensile and flexural strength of textile cement composites.

if the flexural data are used in the analysis and design of members. This correlation is shown in Figure 22.8 when the tensile strength is plotted against flexural strength of samples. This subject was discussed in detail in Chapters 14 and 15, showing the means of converting the properties from one test method to the other. The ultimate strain capacity is another main indication of aging, which is a sensitive measure of durability. An average ultimate strain capacity more than 1.20% was observed for unaged specimens in machine direction. However, retention of tensile properties over time was very good, ~75% to 80% after equivalent 50 years service life according to Peled and Mobasher [2] and Mobasher [3].

STRUCTURAL TESTS OF MASONRY WALLS

There are four types of failure mechanisms of unreinforced masonry in shear: bed joint sliding and rocking are typical for walls with lower gravity loads, whereas X-cracking and toe crushing are typical of higher gravity loads. Bed joint sliding occurs when the masonry wall section slides back and forth along a horizontal bed joint. The next most desirable failure mechanism is rocking, when the masonry pier rocks rather than slides along the bed joint. In X-cracking, failure cracks propagate in an X-pattern through the mortar joints or the masonry units themselves. The least favorable mechanism is toe crushing, in which the masonry units are crushed by high gravity loads or large overturning forces.

Figure 22.9 shows a strengthened wall after the in-plane shear test. X-cracking was the failure mode observed for all three walls strengthened with textile cement composites. The bond between the strengthening system and the substrate plays a critical role in providing adequate load carrying capacity to the structural element strengthened. Multiple cracking of textile cement composites surface was observed during the tests, which suggests stress distribution and energy absorption provided by the reinforcing grid. Crack growth was marked with a marker after each cycle. All failures were due to shear between the front and the rear faces of the blocks, and there was no textile composite delamination from the CMU wall. The difference between the front or reinforced face and the unreinforced back at failure was as follows: the reinforced face of the wall held together at failure, whereas the material spalled away from the back. In all tests, the structural integrity of the walls at failure was maintained by textile composites, indicating that its use may prevent the collapse of unstrengthened walls.

Backbone curves were plotted based on load–deformation curves for each wall tested. Backbone curves were drawn through the intersection of the first cycle of each displacement increment curve

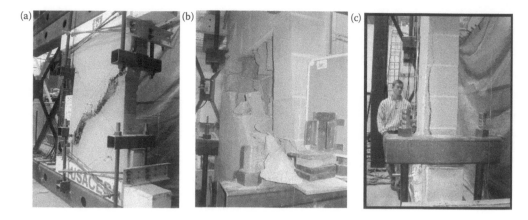

FIGURE 22.9 Typical failure, walls with textile cement composites after in-plane shear tests: (a) front strengthened, (b) back, and (c) side view.

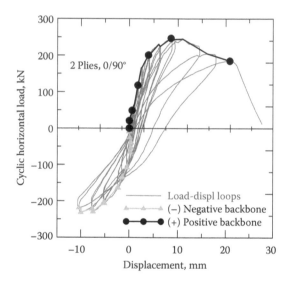

FIGURE 22.10 Typical hysteresis and backbone curve.

and the last cycle of the previous displacement increment. The peak load indicated along the backbone curve is defined as the engineering load [2]. Figure 22.10 shows a typical hysteresis and backbone curve. Figure 22.11 presents typical backbone curves for control and strengthened wall using textile cement composites. Addition of strengthening system significantly improves the performance of the wall. In all cases, the engineered load improved for the strengthened walls tested compared with control: wall 1 performed best, exhibiting a 57% strength increase over control, followed by wall 3, with a 42% increase and wall 2, with a 38% increase [14].

Figure 22.12 compares the engineering load increases as well as the displacement improvement for walls 1, 2, and 3 to the control specimen, with the other CMU wall configurations loaded in-plane shear. Textile cement composites added 47 to 53 kips (324–366 MPa) of horizontal resistance, or 38% to 57% to the engineering strength of the wall specimens in in-plane shear [14]. Among the system configurations, two plies of reinforcing grid aligned at 0 and 90° to each other, and the wall provided the best performance among the CMG configurations as well as other FRP overlay configurations.

Figure 22.12 shows that in addition to providing the maximum strength increase, wall 1 also exhibited the best displacement improvement before failure compared with walls 2 and 3 and also

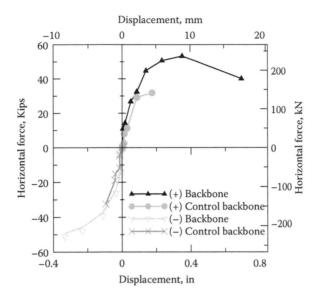

FIGURE 22.11 Typical backbone curves for strengthened (CMG, two plies 0°/90°) versus control.

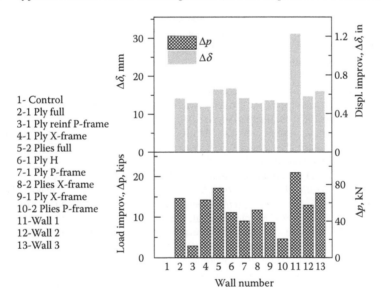

1- Control
2-1 Ply full
3-1 Ply reinf P-frame
4-1 Ply X-frame
5-2 Plies full
6-1 Ply H
7-1 Ply P-frame
8-2 Plies X-frame
9-1 Ply X-frame
10-2 Plies P-frame
11-Wall 1
12-Wall 2
13-Wall 3

FIGURE 22.12 CMG engineering load and displacement improvement in in-plane shear compared with FRP systems.

with the other FRP configurations tested. Overall, textile cement composites increased the horizontal displacement by as much as 0.58 to 1.22 in. (15–31 mm) or 29% to 44% when tested in in-plane shear [14]. The ranking of the three textile cement composites configurations is similar for displacements and engineering load.

CONCLUSIONS

The ability of the textile cement composites to strengthen the walls exceeded the FRP alternatives. The cement-based system not only provided a performance within the range of the FRP systems, but one fabric configuration provided the best overall performance among all systems tested. Failures of all walls reinforced using textile cement composites were due to shear between the front and rear faces of the blocks. There were no delaminations of the system from the CMU walls, and the Textile cement composites held the masonry pier together at failure.

REFERENCES

1. Mayrhofer, C. (2001), "Reinforced masonry walls under blast loading," *International Journal of Mechanical Science*, 44, 1067–1080.
2. Peled, A., and Mobasher, B. (2003), "The pultrusion technology for the production of fabric–cement composites," *Brittle Matrix Composites 7—Proceedings of the 7th International Symposium*, ed. Brandt, A. M., V. C. Li, and I. H. Marshall, Cambridge: Woodhead Publ. Ltd.; Warsaw: ZTUREK Research-Scientific Institute, 2003, 505–514.
3. Mobasher, B., "Micromechanical modeling of angle ply cement based composites," *Proceedings of the 6th International Symposium on Brittle Matrix Comp.* (BMC6), Warsaw, Poland, 2000.
4. Jai, J., Springer, G. S., Kollar, L. P., and Krawinkler, H. (2000), "Reinforcing masonry walls with composite materials—Test results," *Journal of Composite Materials*, 34(16), 1369–1381.
5. Albert, M. L., Elwi, A. E., and Cheng, J. J. R. (2001), "Strengthening of unreinforced masonry walls using FRPs," *Journal of Composites for Construction*, 5(2), 76–84.
6. Galano, L., and Gusella, V. (1998), "Reinforcement of masonry walls subjected to seismic loading using steel X-bracing," *Journal of Structural Engineering*, 124(8), 886–895.
7. Marshall O., and Sweeney S.C. "In-plane shear performance of masonry walls strengthened with FRP," *International SAMPE Symposium and Exhibition Proceedings*, Long Beach, v47, II, 2002, 929–940.
8. NEHRP Guidelines for the Seismic Rehabilitation of Buildings (FEMA Publication 273), Federal Emergency Management Agency, Washington, DC, 1997, 2-44–2-46.
9. Drysdale, R. G., Hamid, A. A., and Baker L. R. (1994), *Masonry Structures, Behaviour and Design*. Englewood Cliffs, NJ: Prentice Hall.
10. Magenes, G., and Calvi, G. M. (1997), "In-plane seismic response of brick masonry walls," *Earthquake Engineering and Structural Dynamics*, 26, 1091–1112.
11. ICC AC218, "Acceptance criteria for cement-based matrix fabric composite systems for reinforced and unreinforced masonry," ICC Evaluation Service, Inc., July 1, 2003.
12. Prota, A., and Marcari, G., Fabbrocino, G., Manfredi, G., and Aldea C. (2006), "Experimental in-plane behaviour of tuff masonry strengthened with cementitious matrix-grid composites," *ASCE Journal of Composites for Construction*, 10(3), 223–233.
13. ASTM D6637-01 (2001), *Standard Test Method for Determining Tensile Properties of Geogrids by the Single or Multi-Rib Tensile Method, Vol. 04.13*, ASTM International, West Conshohocken, PA.
14. ASTM C887-05 (2005), *Standard Specification for Packaged, Dry, Combined Materials for Surface Bonding Mortar, Vol. 04.05*, ASTM International, West Conshohocken, PA.
15. Mobasher, B., Jain, N., Aldea, C. M, and Saranakom, C., "Development of fabric reinforced cement composites for repair and retrofit applications," SP ACI Fall 2005 Convention Kansas City, in progress.
16. Singla, N. (2004), "Experimental and theoretical study of fabric cement composites for retrofitting masonry structures," MS thesis, Arizona State University.
17. ASTM C1560-03 (2009), *Standard Test Method for Hot Water Accelerated Aging of Glass-Fiber Reinforced Cement-Based Composites*, ASTM International, West Conshohocken, PA.
18. Beddows, J., and Purnell, P., "Durability of New Matrix Glassfibre Reinforced Concrete," *GRC 2003 Proceedings of 12th Congress of the GRCA*, ed. J. N. Clarke and R. Ferry, Barcelona, Spain, October 2003, paper 16.
19. Purnell, P., "A positive look at GRC durability," *GRC 2003 Proceedings of the 12th Congress of the GRCA*, ed. J. N. Clarke and R. Ferry, Barcelona, Spain, October 2003, paper 22.
20. Litherland, K. L., Oakley, D. R., and Proctor, B. A. (1981), "The use of accelerated aging procedures to predict the long term strength of GRC composites," *Cement and Concrete Research*, 11, 455–466.
21. Proctor, B. A., Oakley, D. R., and Litherland, K. L. (1982), "Development in the assessment and performance of GRC over 10 years," *Composites*, 13, 173–179.
22. Marshall, O. (2002), *Internal Report*, Champaign, IL: U.S. Army Construction Engineering Research Laboratory (CERL).

23 Retrofit of Reinforced Concrete Beam–Column Joints Using Textile Cement Composites

INTRODUCTION

Beam–column joints are crucial regions of reinforced concrete (RC) structures because reaction forces due to strong ground motions concentrate at the joint. Beam–column joints that are not detailed in accordance with seismic codes pose a serious hazard to the overall ductility of the structure subjected to earthquake shocks. Moderately severe earthquakes such as in 1999 at Kocaeli (Turkey), in 2003 at Bam (Iran), and in 2005 at Kashmir (Pakistan) resulted in major loss of life because of the collapse and heavy damaged of RC structures, which was attributed to deficient transverse reinforcements and poor anchorage of main reinforcements in the joint, ultimately resulting in weak column/strong beam design.

Structures designed and built before the development of or without complying with current seismic codes pose a hazard for life safety and property and need rehabilitation. Development of economical and technologically feasible methods to prevent a brittle shear failure of joints allows shifting the failure plane to a beam–flexural hinge yield mechanism. Several rehabilitation techniques were proposed for strengthening joint shear resistance. These techniques vary from simple to complex and involve use of concrete, metals, or composite jackets, bolted steel plates, or corrugated steel sheets [1–5]. Problems that are encountered in the rehabilitation of beam–column joints are due to the difficulty in providing effective confinement to the joint, compatibility, and adaptability of the material used in strengthening, cost, and availability of materials.

One of the commonly practiced methods in strengthening RC column and beam–column joints is jacketing. Beres et al. [3] and Ghobarah et al. [6] studied the shear resistance of strengthened rectangular RC column and joints using flat or corrugated steel sheets and proposed using mechanic anchors to prevent the bulging problems. Estrada [7] and Biddah [8] proposed the use of steel plates attached to the beam bottom face or sides and anchored by using either steel rods or adhesive anchors. This method however did not improve the joint behavior because of the failure of the anchors and slip of the steel plates.

Use of fiber reinforced polymers has several advantages such as quick application, short construction time, corrosion resistance, and small dimensional perturbations. Gergely et al. [9], Ghobarah and Said [10,11], Antonopoulos and Triantafillou [12], Prota et al. [13,14], Mukherjee and Joshi [15], Ghobarah and El-Amoury [16], and Balsamo et al. [17] investigated the effects of carbon fiber reinforced plastics (CFRP) on the flexural and shear strength and ductility of beam–column joints. Results indicate that with proper detailing, CFRP composites can be successfully applied to strengthening of beam–column joints.

CFRP composites, however, pose several problems when applied to RC structures. The primary mode of failure in several cases is the delamination and debonding attributed to the stiffness mismatch with the substrate. This is coupled by inefficiencies in fire resistance, UV degradation, moisture trapping at the interface leading to osmotic pressures, changes in ionic concentration, and formation of expansive compounds at the interface. An alternative approach to develop economic seismic rehabilitation is by using more compatible textile cement composites in using fibers with less stiff than carbon, that is, alkali-resistant glass fibers.

Two systems of carbon and AR glass fibers are evaluated. The carbon textile is orthotropic carbon–fiber polymer sheets with epoxy resin, the AR glass is used in two different fine (discussed in Chapters 11 and 13 and defined as fine bonded), and the coarse geometries (discussed in Chapter 22, denoted as coarse bonded) were compared. The effects of the rehabilitation techniques in increasing load-carrying capacity and in improving the behavior of RC beam–column joints from view point of ductility, total, dissipated, and recovery energy and stiffness are presented.

EXPERIMENTAL PROGRAM

MATERIAL PROPERTIES

Six half-scaled exterior RC beam–column joints were constructed. One of the RC exterior joints was designed according to ACI 318-02 and was labeled as RCACI318. Other five specimens were designed to represent RC structures built before 1970 code provisions from the view point of transverse reinforcement details and referred to as control (RCNH1). The cross-sectional dimensions of column and beam were 125×200 and 125×300 mm, respectively. Concrete had a maximum aggregate size of 10 mm and compressive strength of 27 MPa. Steel for longitudinal reinforcements was grade 60 with average yield stress of 545 MPa. Grade 40 stirrups with average yield stress of 290 MPa were used.

The first of five specimens labeled as control (RCNH1) was tested under the reversed cyclic loads as four remaining specimens were strengthened as two using CFRP epoxy resin and two as AR glass textile–cement paste. The test specimens were placed to the loading frame such that the columns were horizontal and the beams were in vertical position as shown in Figure 23.1. All main reinforcements of beams were anchored 200 mm inside the joint zone. The reinforcement details and geometry of the beam–column joint specimens are presented in Figure 23.2. The bending

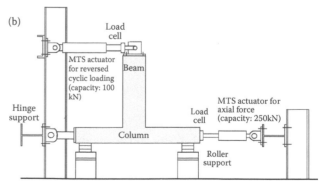

FIGURE 23.1 (a) Biaxial closed-loop servohydraulic test set up. (b) Schematics of the loading setup using a biaxial loading test system.

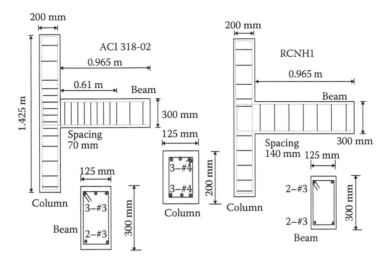

FIGURE 23.2   Geometry of the beam–column joint specimens.

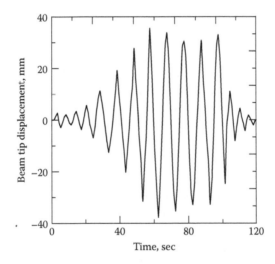

FIGURE 23.3   Displacement loading history of reverse cyclic loading.

moment capacity of columns in the joint was two times higher than beams for all specimens. The height of the column and the length of the beam represent the distance to the points of contraflexure in frame.

The displacement history was based on the reversed cyclic loads applied at the beam tip. The loading history illustrated in Figure 23.3 was slowly applied to eliminate any dynamic effects by MTS actuator with capacity of 250 kN. A constant axial load of 90 kN ($0.13\,A_g f_c'$) was applied to the columns of each specimen.

Two types of AR glass textiles with cement paste as well as CFRP sheets with epoxy resin were compared. A summary of the dimensional aspects of the wrapping parameters is presented elsewhere [18]. The strengthening technique consisted of three different segments. The first segment was an L-shaped connecting the beams and column to compensate the deficiency of tensile reinforcements in column and beam. The U-shaped segment was attached to the web of beam at the joint. The last segment functioned as a stirrup and wrapped the L- and U-shaped segments. Specimens with one and two sheets of L-shaped carbon textile segments were designated as CFRP1

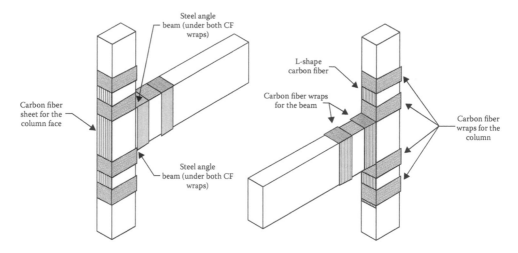

FIGURE 23.4   Schematics of the repair plan using carbon–epoxy system.

and CFRP2, respectively, as shown in Figure 23.4. To eliminate potential for delamination in carbon textile composites, two steel angle beam pieces (125 × 75 × 75 mm) were attached with anchorage bolts at the corners of beam and column strengthened using CFRP. These steel angles prevented the bulging of the CFRP segment because of compressive forces during cyclic loads. In the cement textile composites, U wrapping segment had a single layer, and the shaped segment had six layers (of fine weave) and four layers of coarse weave AR glass textiles, respectively. These two textile RC composites are referred to as ARG6 and SGFA4, respectively (see Figure 23.4).

EXPERIMENTAL RESULTS

BEHAVIOR OF THE SPECIMENS

The displacement-cyclic load responses of all joint specimens are presented in Figures 23.5a through 23.5f. The hysteretic loops of beam–column joint specimens that dominate shear damages in the joint are pinching toward origin. Control specimen failed because of the widening the flexural crack at the column face of beam. Diagonal shear cracks were observed in RCNH1 (control specimen) and ARG6. These cracks also occurred in the joints of CFRP1 and SGFA4 during the final three cycles resulting in pinching toward origin (see in Figures 23.5c and 23.5e). L-shaped fabric–cement paste composites showed no delamination throughout test. No diagonal shear cracks were observed in the joints of the control and the CFRP2 specimen. All of the strengthened specimens have higher displacement-cyclic load responses than both the Control and RCNH1.

The envelopes of displacement-cyclic load hysteretic for all specimens were compared with the displacement-cyclic load responses of the strengthened exterior beam–column joint specimens. The test results of ARG6 and RCNH1 are shown in Figure 23.6a. Figure 23.6b shows the comparison of different envelops of textile cement composites compared with the ACI 318 and the RCNH1 samples. The displacement ductility significantly improves with the use of textile cement composites.

ABSORBED ENERGY

Total, dissipated, and recovery energy capacities of joint specimens were defined and calculated to determine the strengthening effects on the joint behavior. These energies absorbed by beam–column joints directly depend on strength and ductility of joints under reversed cyclic loads.

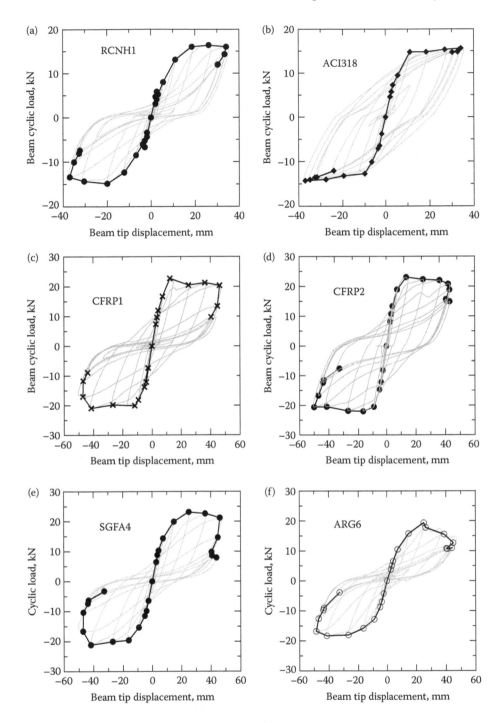

FIGURE 23.5   Cyclic load displacement of beam column sections with retrofit systems.

Total Energy

The total energy is defined as the cumulative area under the force-displacement hysteretic loop up to peak loads in both push and pull cases. Total energy is equal to the summation of dissipated and recovery energies defined in the following sections. The variation of cumulative total energy versus displacements corresponding to peak loads of hysteretic loops and the total energy are shown in

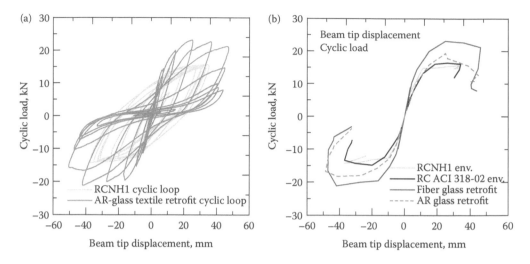

FIGURE 23.6 (a) Comparison of cyclic behavior of ARG6 retrofit sample with the RCNH1 sample. (b) Comparison of beam tip displacement versus load envelop for the cement composites compared with both ACI 318 and RCNH1 samples.

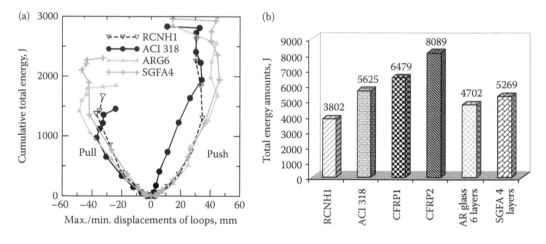

FIGURE 23.7 (a) Cumulative total energy as a function of each cycle of loading for the push and pull cases. (b) The total energy measured for the repair system compared with standard and control specimens.

Figures 23.7a and 23.7b, respectively. Although it is clear that the strengthening technique using CFRP sheets improves the behavior of beam–column joints compared with joints built according to ACI 318-02 from view point of strength and ductility, the specimens using AR glass fabric–cement paste and SGFA fabric–cement paste composite displayed a remarkable improvement over the behavior of RCNH1 as well. These improvements may be sufficient to make beam–column joints detailed according to seismic codes.

Dissipated Energy

The energy dissipated in a beam–column joint through a plastic deformation is the sum of the areas in the beam tip force–displacement hysteretic loops. Ductility of a beam–column joint is defined in terms of the dissipated energy during its plastic deformation. The cumulative and the total amounts of the dissipated energy for all specimens are given in Figures 23.8a and 23.8b, which indicate the beam–column joints strengthened using CFRP dissipate more energy than

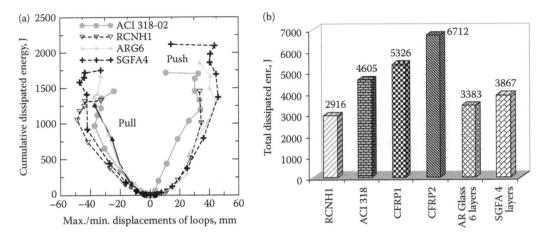

FIGURE 23.8   (a) Cumulative dissipated energy as a function of each cycle of loading for the push and pull cases. (b) The total energy measured for the repair system compared with standard and control specimens.

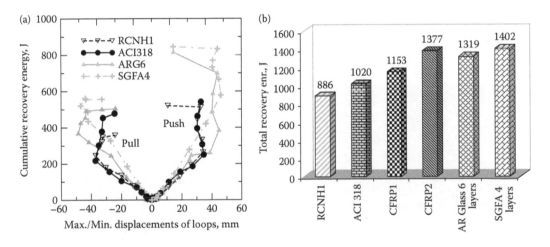

FIGURE 23.9   (a) Cumulative recovery energy as a function of each cycle of loading for the push and pull cases. (b) The recovery energy of various systems compared with standard and control specimens.

other specimens. Figures 23.8a and 23.8b also indicate that the fabric–cement paste composites were also effective in increasing the load capacity and ductility of joints but were not nearly enough to raise up to the load-carrying capacity and ductility levels of beam columns designed in accordance to ACI 318.

Recovery Energy

A portion of the energy stored through the deformation of the specimens during loading is given back to the system in the course of unloading. This is defined as the recovered energy. The recovery energy capacity could be used as an indicator of the unloading capability of a structural element subjected to cyclic loads. The ductility and the load-carrying capacity of beam–column joint are proportional to the residual unloading capability of structural element. The amount of elastic recovery energy for each specimen was calculated by subtracting the amount of total dissipated energy from the amount of total energy. Except for RCNH1 joint specimen, all of the strengthened exterior beam–column joint and RCACI318 joint specimen have almost similar recovery energy. SFGA4 and CFRP2 gave back higher absorbed energy in the course of unloading (Figures 23.9).

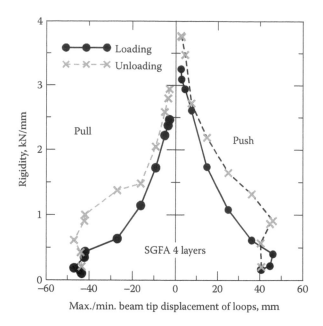

FIGURE 23.10 Stiffness degradation of SGFA4 beam column as a function of applied tip displacement.

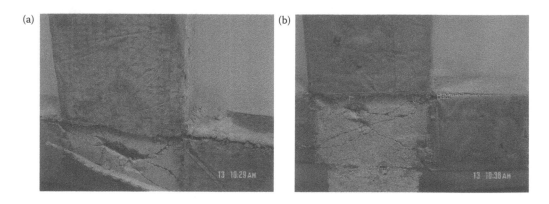

FIGURE 23.11 Mode of failure of (a) SGFA4 sample and (b) ARG6 sample.

STIFFNESS DEGRADATION

Tangent stiffness for beam of exterior beam–column joint specimen during each cycle of the test was determined by calculating the slopes of loading and unloading parts of each cycle for both pulling and pushing positions of cyclic loads. For both loading and unloading cases of cyclic loads, the stiffness degradations of beam of joint specimen with respect to beam tip displacements are presented in Figure 23.10. Moreover, the beam tangent stiffness of the joint specimens and the ratios of beam stiffness of the joint specimens to one of the RCNH1 specimen are given in Table 23.1 for main displacement levels of cycles. From Figure 23.10 and Table 23.1, the effects of strengthening process on the stiffness of joint can be thoroughly realized after the first structural crack in specimen occurred. Hence, the stiffness of strengthened joint specimens (CFRP1, CFRP2, and SGFA4) is almost three and two times higher than the one of the control joint specimen (RCNH1) for the beam tip displacement level of 10 and 34 mm, respectively.

Figure 23.11 indicates the mode of failure after the test and removal of the textile reinforced layer. Figure 23.11a represents the SGF4 samples, and Figure 23.11.b indicates the ARG6 samples.

TABLE 23.1
Summary Results of Various Retrofit Systems Used for Beam–Column Connection

Specimen Name	Max. Cyclic Load (kN)	Max. Beam Tip Displacement (mm)	Initial Stiffness, $\delta = 0$		Initial Stiffness, $\delta = 10$ mm		Initial Stiffness, $\delta = 34$ mm	
			Stiffness (kN·mm)	Ratio	Stiffness (kN·mm)	Ratio	Stiffness (kN·mm)	Ratio
RCNH1	15.62	34.2	2.7313	1	0.7068	1	0.3634	1
ACI 318	16.39	34.2	2.847	1.042	1.6142	2.284	0.4936	1.358
CFRP1	22.69	40.8	3.52	1.29	2.2845	3.232	0.6482	1.784
CFRP2	23.03	42.8	3.5193	1.29	2.6847	3.798	0.9127	2.512
ARG6	19.36	44.8	1.7397	0.64	1.2657	1.791	0.4228	1.163
SGFA4	23.13	46.1	3.252	1.19	2.2078	3.124	0.7065	1.944

In both of these cases, the textile reinforcement that resulted in the formation of a yield hinge that was contained in the RC section of the shear cracking in the form of x in the sample is further supported in which the specimen is able to contain the internal damage at the node and resist the cyclic loads applied throughout the testing. Significant more work in this area is required for use of textile RC in this discipline.

CONCLUSIONS

RC beam–column joints were strengthened by textile RC. Test results of joints under cyclic loads are evaluated in terms of load-carrying capacity, total energy dissipation, stiffness retention, and ductility. After retrofit, sections that violated the transverse requirements at ACI 318-02 are comparable with sections built in accordance to the requirements.

1. CFRP sheets attached to the concrete surfaces of beam–columns by using epoxy resins increase the load-carrying capacity, the ultimate beam tip displacements, and the total and dissipated energy amounts in excess of ACI 318.
2. The cement-based composites with SGFA can improve the behavior of the RC beam–column joints as much as the CFRP sheets.
3. Two layers of the CFRP sheets used at the tension surfaces of beam and column can prevent the shear cracks in the joints that have no transverse reinforcement and thus move the failure section from the column face of beam to the midspan if these CFRP sheets are conveniently attached and wrapped to concrete surfaces.
4. The cement-based composites made from AR glass fabric can effectively improve the behavior of RC beam–column joints.

REFERENCES

1. Migliacci, A., Antonucci, R., Maio, N. A., Napoli, P., Ferretti, A. S., and Via, G., "Repair techniques of reinforced concrete beam–column joints," *Final Report, IABSE Symposium on Strengthening of Building Structures-Diagnosis and Therapy*, Venice, Italy, 1983, 355–362.
2. Corazao, M., and Durrani, A. (1989), "Repair and strengthening of beam-to-column connections subjected to earthquake loading," NCEER Report No. NCEER-89-13.
3. Beres, A., El-Borgi, S., White, R. N., and Gergely, P. (1992), "Experimental results of repaired and retrofitted beam–column joint tests in lightly reinforced concrete frame buildings," NCEER Technical Report No. 92-25, State University of New York, Buffalo, NY.
4. Alcocer, S. M., and Jirsa, J. O. (1993), "Strength of reinforced concrete frame connections rehabilitated by jacketing," *ACI Structural Journal*, 90(3), 249–261.

5. Ghobarah, A., Aziz, T. S., and Biddah, A. (1997), "Rehabilitation of reinforced concrete frame connections using corrugated steel jacketing," *ACI Structural Journal*, 94(3), 283–94.
6. Ghobarah, A., Aziz, T. S., and Biddah, A. (1996), "Seismic rehabilitation of reinforced concrete beam–column connections," *Earthquake Spectra*, 12(4), 761–780.
7. Estrada, J. I. (1990), "Use of steel elements to strengthen a reinforced concrete building," MS thesis, University of Texas at Austin, TX. Fyfe Company LLC (2000), Tyfo® BC composite, http://www.fyfeco. com (accessed July 2000).
8. Biddah, A. M. S. (1997), "Seismic behavior of existing and rehabilitated reinforced concrete frame connections," PhD thesis, McMaster University, Hamilton, Ontario, Canada.
9. Gergely, J., Pantelides, C. P., and Reaveley, L. D. (2000), "Shear strengthening of RC T-joints using CFRP composites," *Journal of Composites for Construction*, 4(2), 56–64.
10. Ghobarah, A., and Said, A. (2001), "Seismic rehabilitation of beam column joints using FRP laminates," *Journal of Earthquake Engineering*, 5(1), 113–129.
11. Ghobarah, A., and Said, A. (2002), "Shear strengthening of beam–column joints," *Engineering Structures*, 24(7), 881–888.
12. Antonopoulos, C. P., and Triantafillou, T. C. (2003), "Experimental investigation of FRP-strengthened RC beam–column joints," *Journal of Composites for Construction*, 7(1), 39–49.
13. Prota, A., Nanni, A., Manfredi, G., and Cosenza, E. (2003), "Capacity assessment of RC subassemblages upgraded with CFRP," *Journal of Reinforced Plastics and Composites*, 22(14), 1287–1304.
14. Prota, A., Nanni, A., Manfredi, G., and Cosenza, E. (2004), "Selective upgrade of underdesigned reinforced concrete beam–column joints using carbon fiber reinforced polymers," *ACI Structural Journal*, 101(5), 699–707.
15. Mukherjee, A., and Joshi, M. (2005), "FRPC reinforced concrete beam–column joints under cyclic excitation.," *Composite Structures*, 70, 185–199.
16. Ghobarah, A., and El-Amoury, T. (2005), "Seismic rehabilitation of deficient exterior concrete frame joints," *Journal of Composites for Construction*, 9(5), 408–416.
17. Balsamo, A., Colombo, A., Manfredi, G., Negro, P., Prota, A. (2005), "Seismic behavior of a full-scale RC frame repaired using CFRP laminates," *Engineering Structures*, 27, 769–780
18. Gencoglu, M., and Mobasher, B. (2007), "The strengthening of the deficient RC exterior beam–clumn joints using CFRP for seismic excitation," *Proceedings of the Third International Conference on Structural Engineering, Mechanics and Computation (SEMC 2007)*, ed. A. Zingoni, 1993–1998.

24 Dynamic Tensile Characteristics of Textile Cement Composites

INTRODUCTION

Fiber reinforcement is undoubtedly one of the most effective means of enhancing the resistance, strength, and energy absorption of cement-based materials subjected to dynamic loading. High-strength woven fabrics are ideal materials for use in systems where large deformations and high-energy absorption are required. High strain rate applications are quite varied and include structural, military, roadway impact, high wind and earthquake, and blast mitigation applications. Fabric materials with their high strength-to-weight ratio and the ability to resist high-speed impacts can be even more efficient than metals. They can be directly applied in load bearing structural members in applications such as structural panels, impact and blast resistance, repair and retrofit, strengthening of unreinforced masonry walls, and beam–column connections. Thus, the response to impulse loading for applications in extreme loading conditions is of great importance.

Although quasi-static tensile strength data for the single fibers are available, results cannot be extrapolated and scaled up for yarns consisting of many fibers, woven, knitted, or bonded fabrics with a two- or three-dimensional microstructure. Furthermore, the strain rates observed in static experiments are orders of magnitude smaller than the strain rates observed in ballistic applications [1]. This chapter discusses the development and the test results of a variety of fabric sand composite systems subjected to high-speed tensile testing.

DYNAMIC TENSILE TESTING

As methods for dynamic tensile tests are still under development, an understanding of the dynamic phenomena, parameter interaction, and their influence on the overall measurements are far from complete. There is a lack of general agreement about the standards and methodology used to conduct dynamic tensile tests [2]. A number of experimental techniques addressing high strain rate material properties include the following: split Hopkinson pressure bar, falling weight devices, flywheel facilities, hydraulic machine, and so forth [3–6]. The use of servohydraulic machines in medium strain rate tensile testing was reported for steel [7, 8], plastics [2, 9], and composite materials [10]. The Society of Automotive Engineers has developed industrial standards and practice guidelines for dynamic tensile testing at medium strain rates [11, 12]. A consortium formed by the International Iron and Steel Institute has developed high strain rate tensile test standards for sheet steel [13], whereas European researchers worked on an ISO standard [14]. Both the International Iron and Steel Institute [13] and the Society of Automotive Engineers [14] projects have recognized the importance of specimen geometry and size in dynamic material testing and provided detailed discussions on the relationship between specimen size, wave propagation, inertia effect, strain measurement technique, loading devices, load measurement, system ringing, gripping devices, and clamping mechanism. The split Hopkinson pressure bar technique, developed over a half century ago [15], continues to be modified to generate reliable experimental data for metals [16], polymers [17, 18], composites [19], ceramics [20], rocks [21], and fibers [22, 23].

Measurement of deformation plays an important role in establishing the dynamic behavior of materials. Traditional strain measuring techniques such as extensometers [24] and strain gages [25] have limitations of range and frequency response. Noncontacting strain measuring techniques include laser Doppler velocimetry and laser extensometer [26, 27], digital laser speckle technique [28], rotating drum high-speed camera [29], diffraction-grating technique [30], and image correlation technique [31, 32]. The woven nature of many textiles results in large displacements and shape changes during tests and make it impractical to use either a noncontact laser extensometer or the digital image correlation (DIC) technique, which requires specimens with even and reflective surfaces.

DYNAMIC TESTING OF CEMENT COMPOSITES

Because of the inherent brittleness and low tensile strength of most cement-based elements, dynamic loading can cause severe damage and cracking [33, 34]. Dynamic properties of cement composites have been studied mainly for short fiber composites under impact condition. Testing dynamic properties is a challenge because the results are highly dependent on the loading rate, the method of testing, and the geometry of the tested element [35–38]. Available literature on plain concrete indicates an increase in tensile strength for increasing strain rates [39–45]. Xiao et al. [39] reported that compared with the quasi-static tensile strength of concrete (strain rate of 10^{-5} s^{-1}), the dynamic tensile strengths of concrete at strain rates of 10^{-4}, 10^{-3}, and 10^{-2} s^{-1} increase by 6%, 10%, and 18%, respectively. Birkimer and Lindemann [40] reported that strength at a strain rate of 20 s^{-1} was between 17.2 and 22.1 MPa, whereas the static tensile strength was 3.4 MPa at the quasi-static strain rate of 0.57×10^{-6} s^{-1}.

Dynamic tensile data on fiber reinforced concrete are limited. Kim et al. [46] investigated the strain rate effect on the tensile behavior of high-performance fiber reinforced cement composites using two deformed high-strength steel fibers, namely, hooked fibers and twisted (Torex) fibers. The strain rate ranged from pseudostatic (strain rate of 0.0001 s^{-1}) to seismic (strain rate of 0.1 s^{-1}). Results showed that the tensile behavior of high-performance fiber reinforced cement composites with twisted fibers is sensitive to the strain rate, whereas hooked fiber reinforced specimens show no rate sensitivity. It was also observed that lower fiber volume fraction ($V_f = 1\%$) reinforced specimens show higher sensitivity than higher volume fractions ($V_f = 2\%$). Maalej et al. [47] performed dynamic tensile tests on engineered cement composites containing 0.5% steel and 1.5% polyethylene (PE) fibers (in volume). The applied strain rate ranged from 2×10^{-6} to 2×10^{-1} s^{-1}. Results indicate a substantial increase in the ultimate tensile strength from 3.1 to 6 MPa with increasing strain rate. The strain capacity does not appear to be affected by the strain rate.

Fabric–cement composites clearly demonstrate a significant improvement in the energy absorption capacity under static loading as compared with plain concrete materials and other fiber cement composites [48–52]. The discussion in this chapter in addition to Chapter 21 on impact shows the potential of TRC materials under high-speed loading [53–55].

EXPERIMENTAL METHODOLOGY

Fabric–Cement Composites

The laminated fabric–cement composites were prepared using the pultrusion process with cement paste of 0.4 water–cement ratio [49]. Boards were made with four layers of fabrics in a laminated sheet with 250×300 mm and thickness of approximately 7 to 12 mm (depending on fabric type). The reinforcing yarns in the composite of each fabric were placed along the pultrusion direction.

All boards were cured in water at room temperature for 28 days and then cut to the specimens of 25 mm in width and 150 mm in length. The average thickness of carbon, alkali-resistant (AR) glass, and PE composites were 11, 6.6, and 12 mm, respectively. Five replicate fabric–cement specimens were used for each fabric category. Aluminum plates of dimensions of $25 \times 50 \times 1$ mm were glued onto the gripping edges of the specimen to minimize localized damage and to provide better load transfer from the grips to the specimen during the high-speed tensile test. The gauge length of each specimen was 50 mm.

Sisal fiber composites were manufactured by a hand layup technique. The matrix in this system contained a blended cementitious matrix consisting of 50% portland cement, 30% metakaolin (MK), and 20% calcined waste crushed clay brick [56]. This composite system is comparable with other continuous textiles reinforced concrete systems as it presents a strain hardening behavior with enhanced strength and ductility governed by the composite action. The fibers bridge the matrix cracks and transfer the loads, allowing a distributed micro-crack system to develop [57].

DYNAMIC LOADING DEVICES AND TECHNIQUE

The dynamic tensile test methodology was developed using an MTS high-rate servohydraulic testing machine [58]. The speed of the stroke is controlled by the rate of flow of hydraulic fluid, resulting in different stroke speeds. The stroke can reach a maximum speed of 14 m/s with a load capacity of 25 kN. Figure 24.1 shows the schematic of the high strain rate testing system. In addition to the loading frame, the MTS Flex SE control panels, a high-speed data acquisition card, a noncontact laser extensometer, and a Phantom Version 7 high-speed digital camera were also used.

Figure 24.1b presents the setup for a dynamic tensile test. The load train consists of a piezoelectric load washer, upper and lower grips, test specimen, and slack adaptor. Proper characterization of the load cell response in a high-rate test is a concern, and piezoelectric load washers are recommended for dynamic tests [2, 11–14]. The slack adaptor eliminates the inertia effect of the lower grip and stroke during the acceleration phase, and a constant velocity is imposed on the specimen. The sudden engagement with the upper portion of the setup generates a high amplitude stress wave, causing oscillations at the system's natural frequency, that is, system ringing [2]. To reduce the inertia effect, lightweight grips are recommended in dynamic tensile tests [2, 11–14].

DATA PROCESSING METHOD FOR DYNAMIC TENSILE TESTING

Figure 24.2 shows a typical displacement and its corresponding velocity history at a nominal velocity of 1 m/s. The test duration is 1 ms and falls well within a nearly constant velocity.

Dynamic Characterization

Dynamic testing at high rates requires proper characterization of the system's range of natural frequencies. System ringing is an artifact of specimen–machine interaction and has been discussed in both quantitative [2] and qualitative [9] terms.

RESULTS AND DISCUSSIONS

UNIDIRECTIONAL SISAL FIBER REINFORCED COMPOSITE

Response of the sisal fiber reinforced composite performed at a strain rate of 24.6 s^{-1} is presented in Figure 24.3. The development of the test procedures is presented elsewhere [59]. The dynamic tensile curve can be divided into five zones. Zone 1 represents the elastic region characterized by an average dynamic modulus of 1 GPa, which is still much lower than the values measured using strain gages. In the elastic region, the stress–strain curve is linear with Young's modulus of the composite defined as the slope of the stress–strain curve in this region.

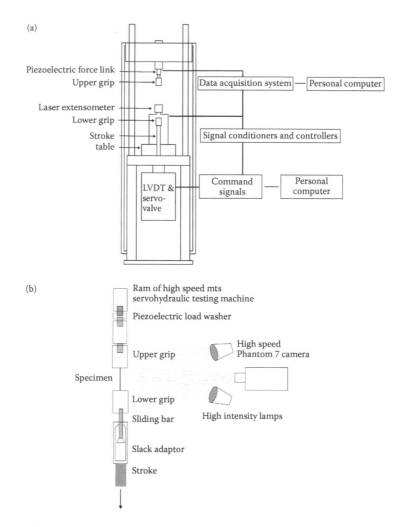

FIGURE 24.1   (a) Schematic of high strain rate testing system. (b) Setup for dynamic tensile test.

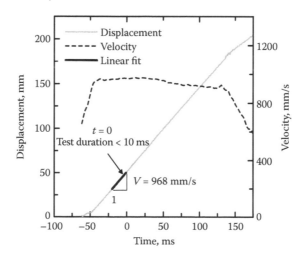

FIGURE 24.2   A typical displacement history curve and corresponding velocity history curve at nominal velocity of 0.968 m/s, generated by the servohydraulic high-rate testing setup.

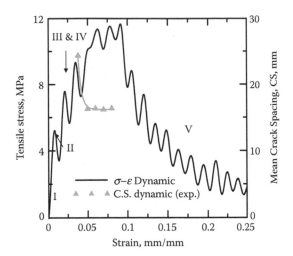

FIGURE 24.3 Experimental high-speed tensile response of sisal fiber cement composite.

Zone 2 is characterized by a bend over point (BOP) ranging from 5.1 to 10.6 MPa. A higher scatter for BOP values was obtained when compared with static results. The differences between the bounds of the BOP values could not be differentiated in the high-speed tests as it was done for the static cases. As in the static tests, zones 3 and 4 are marked by the formation of three cracks, followed by their widening with a strain hardening behavior. Such a distinction between zones 3 and 4 cannot be drawn from dynamic results; however, the widening of cracks which happens in zone 4 was observed. As the composite specimen starts to fail and distributed cracking forms, this leads to a quasi-strain-hardening behavior that is extended until the peak stress, defined as tensile strength. The postpeak range is labeled zone 5.

Strong strain rate dependence was noticed for samples with an average ultimate strain value of 10%. These results are different than the work of Yang and Li [60], who observed that in engineered cement composites, the tensile strain capacity decreased from 3% to 0.5% when the loading rate increased from quasi-static (10^{-5} s^{-1}) to seismic strain rate (10^{-1} s^{-1}). In contrast, Maleej et al. [46] reported that the tensile strain capacity in their tests was insensitive to strain rate.

FABRIC REINFORCED COMPOSITES

The tensile response under high-rate conditions (strain rate higher than 10 s^{-1}) of the different composites with the carbon, AR glass, and PE fabrics are presented in Figure 24.4. A fairly uniform tensile behavior is clearly observed of all composite systems within a narrow range of applied strain rates. Test results under high-rate loading are listed in Table 24.1 along with their standard deviations. Figure 24.5 compares typical stress–strain behavior of the composites with the different fabrics. The highest load carrying capacity response is demonstrated by the carbon composite followed by the AR glass composite and then PE composite. These trends correlate well with the tensile properties of the fabrics (Table 24.1).

There is a linear behavior almost up to the peak in PE and AR glass composites, followed by a reduction in composite strength. On the other hand, the carbon composites show a stiff linear behavior up to the BOP, at stress value of approximately 6 MPa and strain of 0.005 mm/mm. Beyond this point, a pronounced change in the slope of the stress–strain response occurs and the nominal stress increases with a reduced stiffness to as high as 24 MPa. This nominal stress level indicates distributed cracking in carbon composites.

The force versus strain behavior of plain fabrics and their composites are shown in Figures 24.6a and 24.6b. The force response is scaled to represent the equal number of yarns in both fabric and

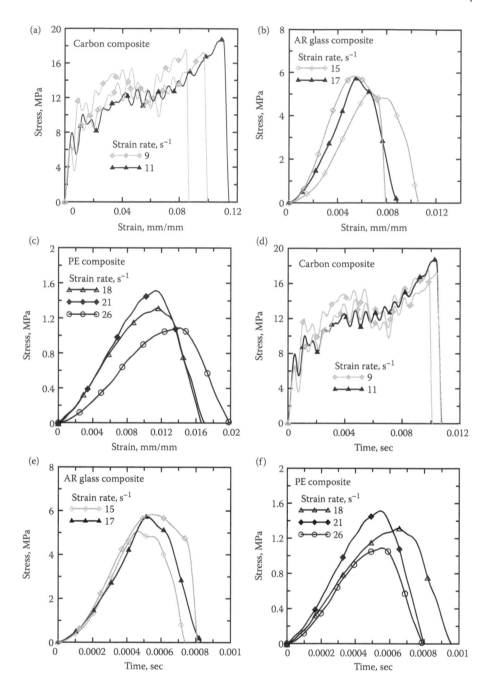

FIGURE 24.4 Tensile responses of composites under high strain rate testing with the various fabric types: (a and d) carbon, (b and e) AR glass, and (c and f) PE.

composite cases. In all cases, the plain fabric can carry more load than when it is part of the composite. The highest difference in load carrying capacity is demonstrated by the carbon followed by the AR glass and then by the PE, indicating that the reinforcing efficiency of the fabric is not fully used in all cases, particularly for the carbon and glass composites. Note that the carbon and glass were made of multifilament yarns where the PE was made of monofilament which can at least to some extent explain these trends. The large strain of the carbon composite as opposed to the more brittle behavior of the individual carbon fabric (Figure 24.6b) may suggest that the obtained loads

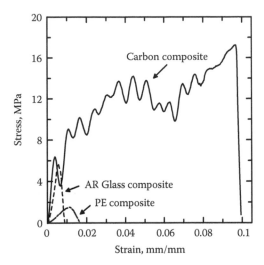

FIGURE 24.5 Comparison of typical stress–strain curves of cement-based composites reinforced with carbon, AR glass, and PE fabrics under high strain rate loading.

TABLE 24.1
Composites Properties under High-Speed and Quasi-Static Loading

Composite	Strain Rate (1/s)	Young's Modulus (MPa)	Strength (MPa)	Toughness (MPa)	Max. Strain (mm/mm)	Strength (MPa)	Toughness (MPa)	Max. Strain (mm/mm)
		High-Speed Loading (1000 mm/s)				Quasi-Static Loading (0.004 mm/s)[a]		
PE [60]	23 (4)	140 (24)	1.31 (0.17)	0.016 (0.007)	0.021 (0.007)	1.36 (0.12)	2.53 (0.23)	1.98 (0.97)
AR glass [60]	18 (3)	1176 (323)	5.56 (0.51)	0.032 (0.009)	0.01 (0.002)	5.11 (0.25)	0.96 (0.05)	1.03 (0.07)
Carbon [60]	10 (1)	2247 (463)	17.86 (0.82)	1.21 (0.14)	0.10 (0.014)	26.63 (2.87)	0.35 (0.05)	0.03 (0.01)
Sisal [58]	24.6	1040 (300)	13.37 (1.2)	0.0455 (0.012)	0.67 (0.09)			

Note: The values in parentheses are standard deviation.
[a] Strain rate: $2.2e^{-5}$/s.

are mainly due to fabric slippage against the matrix and much less due to loads carried by the fabric itself.

Cracking and Failure Behavior

The cracking patterns recorded using a high-speed digital camera indicate the progression of damage in each composite system. In the case of the AR glass composites, three stages are presented as shown in Figure 24.7, with Figure 24.7a representing the intact specimen and Figure 24.7b representing the end of the multiple cracking process. The cracking behavior indicates the stress transfer mechanism at the fabric–cement interface. After this stage, the strain increased and cracks open uniformly until a point of bifurcation where a single dominant crack continues to open while the load begins to decrease. Only the dominant crack can be observed, and all other cracks, although present, unload, close, and are no longer visible by the present magnification (Figures 24.7c through 24.7e]. The main crack continues to widen under load for the duration of the test as the fibers pullout. At this point, the forces were mainly carried by the fabric bridging the main widening crack until all

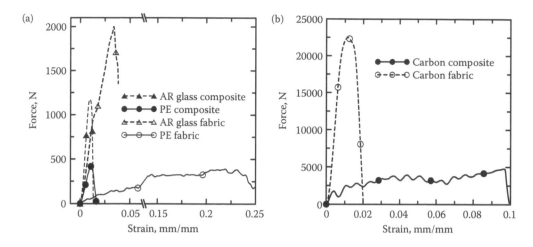

FIGURE 24.6 Comparison of typical tensile force versus strain curves of composites and their respective fabrics (a) PE and AR glass and (b) carbon.

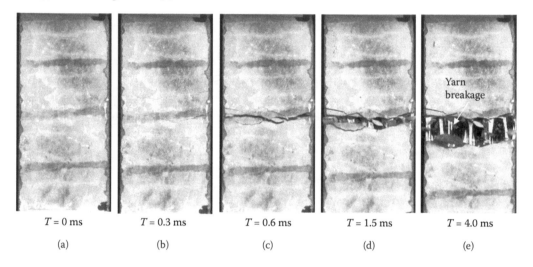

| $T = 0$ ms | $T = 0.3$ ms | $T = 0.6$ ms | $T = 1.5$ ms | $T = 4.0$ ms |
| (a) | (b) | (c) | (d) | (e) |

FIGURE 24.7 AR glass composites: (a) intact, (b) multiple micro cracking, (c and d) main crack widening, (e) failure of yarns.

its yarns were completely broken or pulled out, leading to a complete failure of the tested element. Yarn breakage was clearly demonstrated in Figure 24.7e.

Similar multiple cracking behavior is also observed for carbon composites as shown in Figure 24.8. Figure 24.8a represents the intact specimen, and Figures 24.8b through 24.8e represent the development of multiple cracking. The distributed cracks were more visible than those of the glass system because of a larger crack width for the carbon composites, perhaps due to less efficient unloading mechanism. No brakeage of fibers could be observed after 6 ms into the test (Figure 24.8e).

Figure 24.9 presents the cracking behavior of PE composites. No multiple cracking was observed with this fabric composite as a single major crack was detected. A major crack opened and widened until the fabric failed at that junction. Comparison of the different composites at the end of testing is shown in Figure 24.10, which demonstrates different failure mechanisms within the tested fabrics: yarn breakage for the AR glass and PE systems in Figures 24.9a and 24.9b as opposed to mainly the pullout of the carbon fibers from the matrix in the grip, without significant breakage (Figure 24.10c). A large fraction of the large strains (approximately 0.1) obtained in the carbon

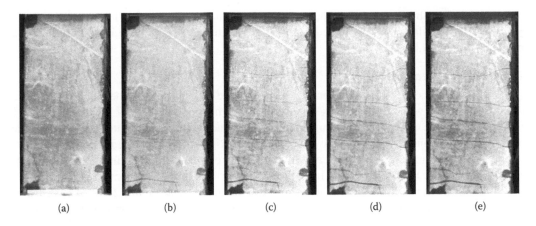

FIGURE 24.8 Carbon composites: (a) intact, (b) multiple cracking, and (c–e) multiple crack widening.

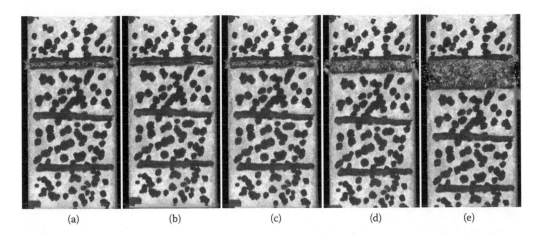

FIGURE 24.9 PE composites: (a) intact, (b) single cracking, (c and d) crack widening, and (e) total failure. Crack presents by white arrow.

FIGURE 24.10 Yarns failure at the end of testing of the different composites: (a) AR glass, (b) carbon, and (c) PE.

composite was attributed to yarn pullout mechanism, which contributed to load carrying capacity. Only after all filaments were pulled out from the matrix, a drastic drop in tensile stress was observed (Figure 24.6b).

Microstructural Features

Figure 24.11 exhibits the PE knitted fabric embedded in the cement matrix. The monofilament yarn is fully surrounded with cement matrix (Figure 24.11a). The loop structure of this fabric shown in Figure 24.11b is a contributor to the toughening as the penetration of paste into this loop provides strong mechanical anchorage of the individual yarns. The tightening and confinement of the matrix by the loops may cause drastic splitting and fragmentation of matrix during failure. Weak mechanical properties of the PE fabric resulted in a low overall performance of the composite.

The microstructure of AR glass fabric composite is observed in Figure 24.12. The AR glass fabric is made from multifilament yarns coated with sizing. Paste penetration in between the filaments of the bundle leads to the bundle acting as a single rigid unit. Because during the high-rate loading the entire bundle is in uniform tension, relative slip and pullout of individual filaments was not expected. The fracture of the bundle was mainly observed at failure. Formation of hydration products at the surface of the coated bundle and penetration of cement matrix in between the filaments mechanically anchors the bundle.

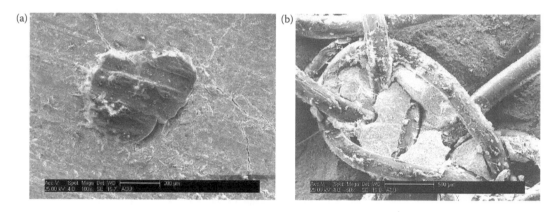

FIGURE 24.11 SEM micrographs of PE yarn in cement matrix and (a) cross section of reinforcing yarn and (b) top view of loop.

FIGURE 24.12 SEM micrographs of AR glass fabric embedded in cement matrix, side view.

FIGURE 24.13 SEM micrographs of carbon fiber bundle embedded in cement matrix: (a) view of cross section and (b) side view.

Two different observations of the carbon fabric composites were seen in Figure 24.13, with a cross-sectional view of the reinforcing bundle embedded in the matrix (Figure 24.13a) and a side view observing the bundle (Figure 24.13b). Poor penetration of the matrix in between the filaments of carbon bundle was obvious in Figure 24.13a. Only few of the filaments at the perimeter were well bonded as they were in contact with the cement matrix. Others had no contact with the matrix. This phenomenon was also observed when looking at the side of the bundle along its filaments. Only the external filaments were in contact with the cement, where the rest of the filaments were free from the hydration products and the empty spaces between the filaments can be clearly seen (Figure 24.13b). This may be attributed to the relatively large size of the cement particles compared with the interstitial spaces of filaments bundle, thus making it difficult to penetrate in between the filaments. Formation of hydration products with aging of the composites, however, may result in filling up of these interstitial spaces, at least to some extent.

These observations explain the differences in the reinforcing efficiency of fabrics, low in all cases but significantly low for the carbon followed by the glass as discussed earlier. Because of the monofilament nature of the PE, each yarn is surrounded with cement matrix at its perimeter when it is part of the composite (Figure 24.11a), and therefore all yarns are carrying the loads during testing. However, with the multifilament yarns, this is not the case because of the lack of penetrability in between the individual filaments; therefore, the loads are carried mainly by the external filaments and much less if any by the core filaments, decreasing the reinforcing efficiency of the whole bundle.

CONCLUSIONS

High-speed tensile tests of various types of fabric–cement composites can provide tools to design composite elements that are subjected to extreme loading conditions during their service life. A good correlation was found between the properties of the composites and fabrics. The fabric with the highest performance provides the best composite behavior in the high-rate tensile test; in this study, it related to carbon fabric. Differences in tensile behavior of the various composites indicate differences in the role of each fabric as reinforcements under high-speed loading. However, in all cases, the individual textiles carry more loads than when it is part of the composite, that is, the reinforcing efficiency of the fabric is not fully used when it is in the composite. This was most significant for the carbon following by the glass, which might be explained based on the multifilament nature of the yarns made these fabrics and the poor cement penetrability in between their filaments. Improving the bond can change this trend.

REFERENCES

1. Farsi, D. B., Nemes, J. A., and Bolduc, M. (2006), "Study of parameters affecting the strength of yarns," *Journal of Physics*, 134, 1183–1188.
2. Xiao, X. R. (2008), "Dynamic tensile testing of plastic materials," *Polymer Testing*, 27, 164–178.
3. Meyers, M. A. (1994), *Dynamic Behavior of Materials*, New York: John Wiley & Sons.
4. Nicholas, T. (1981), "Tensile testing of material at high rates of strain," *Experimental Mechanics*, 21, 177–185.
5. Kenneth, G. H. (1966), "Influence of strain rate on mechanical properties of 6061.t6 aluminum under uniaxial and biaxial states of stress," *Experimental Mechanics*, 6(4), 204–211.
6. Zabotkin, K., O'Toole, B., and Trabia, M., "Identification of the dynamic properties of materials under moderate strain rates," *16th ASCE Engineering Mechanics Conference,* Seattle, WA, 2003.
7. Bastias, P. C., Kulkarni, S. M., Kim, K. Y., and Gargas, J. (1996), "Noncontacting strain measurements during tensile tests," *Experimental Mechanics*, 78, 78–83.
8. Bruce, D. M., Matlock, D. K., Speer, J. G., and De, A. K. (2004), "Assessment of the strain-rate dependent tensile properties of automotive sheet steels," SAE Technical Paper Series, Society of Automotive Engineers, Inc., Troy, MI, 2004-1-0507.
9. Hill, S., and Sjöblom, P. (1998), "Practical considerations in determining high strain rate material properties," SAE Technical Paper 981136, doi:10.4271/981136.
10. Fitoussi, J., Meraghni, F., Jendli, Z., Hug, G., and Baptiste, D. (2005), "Experimental methodology for high strain rates tensile behavior analysis of polymer matrix composites," *Composite Science Technology*, 65, 2174–2188.
11. Hill, S. I. (2004), "Standardization of high strain rate test techniques for automotive plastics project," UDRI: Structural Test Group, University of Dayton Research Institute, Dayton, OH, UDR-TR-2004-00016.
12. SAE (2006), *High Strain Rate Testing of Polymers*, J2749 (draft).
13. Borsutzki, M., Cornette, D., Kuriyama, Y., Uenishi, A., Yan, B., and Opbroek, E. (2004), *Recommendations for Dynamic Tensile Testing of Sheet Steels,* High Strain Rate Experts Group, International Iron and Steel Institute, Brussels, Belgium.
14. ISO (2003), *Plastics—Determination of Tensile Properties at High Strain Rates* (draft of ISO/CD 18872).
15. Kolsky, H. (1949), "An investigation of the mechanical properties of materials at very high rates of loading," *Proceedings of the Physical Society. Section B*, 62, 676–700.
16. Gray, G. T. (2000), *Classic Split-Hopkinson Pressure Bar Testing, Mechanical Testing and Evaluation Handbook, Vol. 8*, Materials Park, OH: American Society for Metals, 488–496.
17. Chen, W., Lu, F., and Cheng, M. (2002), "Tension and compression tests of two polymers under quasi-static and dynamic loading," *Polymer Testing*, 21, 113–121.
18. Chen, W., Zhang, B., and Forestal, M. J. (1999), "A split Hopkinson pressure bar technique for low-impedance materials," *Experimental Mechanics*, 39, 81–85.
19. Chocron Benlooulo, I. S., Rodriguez, J., Martinez, M. A., and Sanchez Galvez, V. (1997), "Dynamic tensile testing of aramid and polyethylene fiber composites," *International Journal of Impact Engineering*, 19(2), 135–146.
20. Johnstrone, C., and Ruiz, C. (1995), "Dynamic testing of ceramics under tensile stress," *International Journal of Solids and Structures*, 32(17), 2647–2656.
21. Frew, D. J., Forrestal, M. J., and Chen, W. (2001), "A split Hopkinson bar technique to determine compressive stress–strain data for rock materials," *Experimental Mechanics*, 41, 40–46.
22. Xia, Y., and Wang, Y. (1999), "The effects of strain rate on the mechanical behaviour of Kevlar fibre bundles: An experimental and theoretical study," *Composites Part A*, 29A, 1411–1415
23. Cheng, M., Chen, W., and Weerasooriya, T. (2005), "Mechanical properties of Kevar KM2 single fiber," *Journal of Engineering Materials and Technology*, 127, 197–204
24. MTS System Corporation (1993), *Extensometers and Clip Gage Catalog*, Minneapolis, MN: MTS System Corporation.
25. Measurements Group (1993), *Strain Gage Technology*, Raleigh, NC: Measurements Group.
26. Bastias, P. C., Kulkarni, S. M., Kim, K. Y., and Gargas, J. (1996), "Non-contacting strain measurements during tensile tests," *Experimental Mechanics*, 36(1), 78–83.
27. Hercher, M., Wyntjes, G., and DeWeerd, H. (1987), "Non-contact laser extensometer," *Industrial Laser Interferometry*, SPIE-The International Society for Optical Engineering, 746, 185–191.
28. Anwander, M., Zagar, B. G., Weiss, B., and Weiss, H. (2000), "Non-contacting strain measurements at high temperatures by the digital laser speckle technique," *Experimental Mechanics*, 40(1), 98–105.

29. Verleysen, P., and Degrieck, J. (2004), "Optical measurement of the specimen deformation at high strain rate," *Experimental mechanics*, 44(3), 247–252.

30. James, F. B. (1967), "On the direct measurement of very large strain at high strain rate," *Experimental Mechanics*, 7(1), 8–14.

31. Choi, D., Thorpe, J. L., and Hanna R. B. (1991), "Image analysis to measure strain in wood and paper," *Wood Science and Technology*, 25, 251–262.

32. Ahmet, H. A., Murat, G., and Tuncer, B. E. (2004), "Use of image analysis in determination of strain distribution during geosynthetic tensile testing," *Journal of Computing in Civil Engineering*, 18(1), 65–74.

33. Gupta, P., and Banthia, N. (2000), "Fiber reinforced wet-mix shotcrete under impact," *Journal of Materials in Civil Engineering*, 12, 81–90.

34. Banthia, N., Bindiganavile, V., and Mindess, S., "Impact blast protection with fiber reinforced concrete," *Proceedings of RILEM Conference on Fiber Reinforced Concrete*, BEFIB, 2004, 31–44.

35. Xu, H., Mindess, S., and Duca, I. J. (2004), "Performance of plain and fiber reinforced concrete panels subjected to low velocity impact loading," *6th RILEM Symposium on Fiber-Reinforced Concretes (FRC)*, BEFIB, Varenna, Italy, 2004, 1257–1268.

36. Zhang, J., Maalej, M., Quek, S. T., and Teo, Y. Y., "Drop weight impact on hybrid-fiber ECC blast/shelter panels," *Proceedings of Third International Conference on Construction Materials: Performance, Innovation and Structural Applications,* Vancouver, Canada, 2005.

37. Bharatkumar, B. H., and Shah, S. P. (2004), "Impact resistance of hybrid fiber reinforced mortar," *International RILEM Symposium on Concrete Science and Engineering: A Tribute to Arnon Bentur,* e-ISBN: 2912143926, RILEM Publication SARL, 2004.

38. Zhu, D., Peled, A., and Mobasher, B. (2011), "Dynamic tensile testing of fabric–cement composites," *Construction and Building Materials,* 25, 385–395.

39. Xiao, S., Li, H., and Lin, G. (2008), "Dynamic behaviour and constitutive model of concrete at different strain rates," *Magazine of Concrete Research,* 60, 271–278.

40. Birkimer, D. L., and Lindemann, R. (1971), "Dynamic tensile strength of concrete materials," *Journal of the American Concrete Institute,* 68, 47–49.

41. Oh, B. H. (1987), "Behavior of concrete under dynamic tensile loads," *ACI Materials Journal*, 84, 8–13.

42. Rossi, P., Van Mier, J. G. M., Toutlemonde, F., Le Maou, F., and Boulay, C. (1994), "Effect of loading rate on the strength of concrete subjected to uniaxial tension," *Materials Structure*, 27, 260–264.

43. Rossi, P., and Toutlemonde, F. (1996), "Effect of loading rate on the tensile behaviour of concrete: Description of the physical mechanisms," *Materials and Structure*, 29, 116–118.

44. Cadoni, E., Labibes, K., Albertini, C., Berra., M., and Giangrasso, M. (2001), "Strain-rate effect on the tensile behaviour of concrete at different relative humidity levels," *Materials and Structure*, 34, 21–26.

45. Malvar, L. J., and Ross, C. A. (1998), "Review of static and dynamic properties of concrete in tension," *ACI Materials Journal*, 95, 735–739.

46. Kim, D. J., El-Tawil, S., and Naaman, A. E. (2009), "Rate-dependent tensile behavior of high performance fiber reinforced cementitious composites," *Materials and Structure*, 42(2), 399–414.

47. Maalej, M., Quek, S. T., Zhang, J. (2005), "Behaviour of hybrid-fibre engineered cemetitious composites subjected to dynamic tensile loading and projectile impact," *Journal of Materials in Civil Engineering*, 17, 143–152.

48. Häußler-Combe, U., Jesse, F., and Curbach, M., "Textile reinforced composites—overview, experimental and theoretical investigations," *Proceedings of the 5th International Conference on Fracture Mechanics of Concrete and Concrete Structures,* Ia-FraMCos 204, Vail, Colorado, 2004, 749–756.

49. Peled, A., and Mobasher, B. (2007), "Tensile behavior of fabric cement-based composites: Pultruded and cast," *Journal of Materials in Civil Engineering*, 19(4), 340–348.

50. Mobasher, B., Peled, A., and Pahilajani, J. (2006), "Distributed cracking and stiffness degradation in fabric–cement composites," *Materials and Structure*, 39(3), 317–331.

51. Peled, A., and Bentur, A. (2003), "Fabric structure and its reinforcing efficiency in textile reinforced cement composites," *Composites Part A*, 34, 107–118.

52. Kruger, M., Ozbolt, J., and Reinhardt, H. W., "A new 3D discrete bond model to study the influence of bond on structural performance of thin reinforced and prestressed concrete plates," *Proceedings of the High Performance Fiber Reinforced Cement Composites (HPFRCC4)*, RILEM, Ann Arbor, MI, 2003, 49–63.

53. Peled, A., "Textiles as reinforcements for cement composites under impact loading," *Workshop on High Performance Fiber Reinforced Cement Composites (RILEM) HPFRCC-5,* ed. H. W. Reinhardt and A. E. Naaman, Mainz, Germany, July 10–13, 2007, 455–462.

54. Butnariu, E., Peled, A., and Mobasher, B., "Impact behavior of fabric–cement based composites," *Proceedings of the 8th International Symposium on Brittle Matrix Composites (BMC8) in Warsaw,* October 23–25, 2006, 293–302.

55. Zhu, D., Gencoglu, M., and Mobasher, B. (2009), "Low velocity impact behavior of AR glass fabric reinforced cement composites in flexure," *Cement and Concrete Composites*, 31(6), 379–387.

56. Toledo Filho, R. D., Silva, F. A., Fairbairn, E. M. R., and Melo Filho, J. A. (2009), *Construction and Building Materials*, 68, 3438–3443.

57. Silva, F. A., Mobahser, B., and Toledo Filho, R. D. (2009), "Cracking mechanisms in durable sisal fiber reinforced cement composites," *Cement and Concrete Composites*, in press.

58. Silva, F., Zhu, D., Soranakom, C., Mobasher, B., and Toledo Filho, R. (2009), "High speed tensile behavior of sisal fiber cement composites," *Materials Science and Engineering. A.*, in press.

59. Naik, D., Sankaran, S., Mobasher, B., Rajan, S. D., and Pereira, J. M. (2009), "Development of reliable modeling methodologies for fan blade-out containment analysis, Part I: Experimental studies," *International Journal of Impact Engineering*, 36 (1), 1–11.

60. Yang, E., and Li, V. C., "Rate dependence in engineered cementitious composites," *Proceedings of the HPFRCC-2005 International Workshop,* Honolulu, HI, 2005.

Index

Note: Page numbers followed by "*f*" and "*t*" denote figures and tables, respectively.